VOLUME SEVENTY

ADVANCES IN INSECT PHYSIOLOGY

Evolutionary Adaptations and Mosquito-Borne Viruses

VOLUME SEVENTY

ADVANCES IN INSECT PHYSIOLOGY

Evolutionary Adaptations and Mosquito-Borne Viruses

Edited by

SASSAN ASGARI
University of Queensland,
Brisbane, QLD, Australia

Series Editor

KAYVAN ETEBARI
Senior Lecturer in Horticulture and Crop Entomology,
The University of Queensland,
Brisbane QLD 4072, Australia

ELSEVIER

ACADEMIC PRESS
An imprint of Elsevier

Academic Press is an imprint of Elsevier
125 London Wall, London, EC2Y 5AS, United Kingdom
50 Hampshire Street, 5th Floor, Cambridge, MA 02139, United States

First edition 2026

ISBN: 978-0-443-47034-9
ISSN: 0065-2806

For information on all Academic Press publications
visit our website at https://www.elsevier.com/books-and-journals

Publisher: Zoe Kruze
Acquisition Editor: Mariana L. Kuhl
Editorial Project Manager: Rehan Ismail
Production Project Manager: Abdulla Sait
Cover Designer: Muthukumaran Thangaraj

Typeset by MPS Limited, India

Contents

Contributors

Omar S. Akbari
Department of Cell and Developmental Biology, School of Biological Sciences, University of California, San Diego, La Jolla, CA, United States

Sassan Asgari
School of the Environment and School of Chemistry and Molecular Biosciences, The University of Queensland, Brisbane, QLD, Australia

Cecília Artico Banho
Laboratório de Pesquisas em Virologia, Faculdade de Medicina de São José do Rio Preto, São José do Rio Preto, São Paulo, Brazil

Joshua B. Benoit
Department of Biological Sciences, University of Cincinnati, Cincinnati, OH, United States

Victoria Bernardi
Laboratório de Pesquisas em Virologia, Faculdade de Medicina de São José do Rio Preto, São José do Rio Preto, São Paulo, Brazil

Laura E. Brettell
School of Science, Engineering and Environment, University of Salford, Manchester, United Kingdom

Rohit Chebbi
Laboratory of Malaria and Vector Research, National Institute of Allergy and Infectious Diseases, National Institutes of Health, Rockville, MD, United States

Igor da Silva Teixeira
Laboratório de Pesquisas em Virologia, Faculdade de Medicina de São José do Rio Preto, São José do Rio Preto, São Paulo, Brazil

Beatriz de Carvalho Marques
Laboratório de Pesquisas em Virologia, Faculdade de Medicina de São José do Rio Preto, São José do Rio Preto, São Paulo, Brazil

Lucy de Guilhem de Lataillade
Departments of Vector Biology and Tropical Disease Biology, Centre for Neglected Tropical Diseases, Liverpool School of Tropical Medicine, Liverpool, United Kingdom

Matthew DeGennaro
Department of Biological Sciences; Biomolecular Sciences Institute, Florida International University, Miami, FL, United States

George Dimopoulos
W. Harry Feinstone Department of Molecular Microbiology and Immunology, Bloomberg School of Public Health, Johns Hopkins University, Baltimore, United States

Shengzhang Dong
W. Harry Feinstone Department of Molecular Microbiology and Immunology, Bloomberg School of Public Health, Johns Hopkins University, Baltimore, United States

Gregory D. Ebel
Department of Microbiology, Immunology and Pathology, College of Veterinary Medicine and Biomedical Sciences, Center for Vector-borne Infectious Diseases, Colorado State University, Fort Collins, CO, United States

Emily N. Gallichotte
Department of Microbiology, Immunology and Pathology, College of Veterinary Medicine and Biomedical Sciences, Center for Vector-borne Infectious Diseases, Colorado State University, Fort Collins, CO, United States

Gladys Gutierrez-Bugallo
W. Harry Feinstone Department of Molecular Microbiology and Immunology, Bloomberg School of Public Health, Johns Hopkins University, Baltimore, United States

Eddie Hill
Department of Cell and Developmental Biology, School of Biological Sciences, University of California, San Diego, La Jolla, CA, United States

Grant L. Hughes
Departments of Vector Biology and Tropical Disease Biology, Centre for Neglected Tropical Diseases, Liverpool School of Tropical Medicine, Liverpool, United Kingdom

Renuka E. Joseph
Laboratory of Malaria and Vector Research, National Institute of Allergy and Infectious Diseases, National Institutes of Health, Rockville, MD, United States

Chloé Lahondère
Department of Biochemistry, Virginia Polytechnic Institute and State University, Blacksburg, VA, United States

Nannan Liu
Department of Entomology and Plant Pathology, Auburn University, Auburn, AL, United States

Takeshi Morita
Department of Biological Sciences; Biomolecular Sciences Institute, Florida International University, Miami, FL, United States

Mauricio Lacerda Nogueira
Laboratório de Pesquisas em Virologia, Faculdade de Medicina de São José do Rio Preto, São José do Rio Preto, São Paulo, Brazil; Department of Pathology, University of Texas Medical Branch, Galveston, TX, United States

Hugo D. Perdomo
Departments of Vector Biology and Tropical Disease Biology, Centre for Neglected Tropical Diseases, Liverpool School of Tropical Medicine, Liverpool, United Kingdom

Robyn Raban
Department of Cell and Developmental Biology, School of Biological Sciences, University of California, San Diego, La Jolla, CA, United States

Nikos Vasilakis
Department of Pathology; Center for Vector-Borne and Zoonotic Diseases; Institute for Human Infection and Immunity, University of Texas Medical Branch, Galveston, TX, United States

Joel Vega-Rodriguez
Laboratory of Malaria and Vector Research, National Institute of Allergy and Infectious Diseases, National Institutes of Health, Rockville, MD, United States

CHAPTER ONE

Insecticide resistance in mosquitoes: Molecular mechanisms and evolutionary dynamics

Nannan Liu*
Department of Entomology and Plant Pathology, Auburn University, Auburn, AL, United States
*Corresponding author. e-mail address: liunann@auburn.edu

Contents

Abstract

Mosquito-borne diseases continue to expand globally, sustaining major health and economic burdens and intensifying reliance on chemical vector control. However, pervasive insecticide resistance threatens the durability of current interventions. Resistance represents one of the most rapid and recurrent examples of contemporary adaptive evolution under strong anthropogenic selection in natural mosquito populations. This review synthesizes advances in the molecular, physiological, and evolutionary foundations of resistance in major vectors (*Anopheles, Aedes,* and *Culex*), framing resistance as an integrated, multilayered adaptive phenotype shaped by persistent insecticide pressure. We examine the dominant mechanisms, enhanced metabolic detoxification and target-site insensitivity, highlighting the modular, polygenic architecture of detoxification systems (cytochrome P450s, carboxyl/cholinesterases,

Advances in Insect Physiology, Volume 70
ISSN 0065-2806, https://doi.org/10.1016/bs.aiip.2026.05.002

glutathione S-transferases, and UDP-glycosyltransferases), as well as constrained yet functionally important mutational landscapes of neuronal targets (VGSC, AChE, and GABA receptors). Beyond mechanism cataloguing, we emphasize regulatory integration, exploring how promoter evolution, transcriptional and post-transcriptional control, and stress-responsive signaling pathways coordinate expression and amplify resistance intensity. Population-level processes, including standing genetic variation, gene flow, selection intensity, and fitness trade-offs, are considered in shaping the emergence, spread, and persistence of resistance alleles and in driving recurrent, convergent evolutionary solutions across taxa. Finally, we outline future directions integrating genome editing, population genomics, epigenomics, and systems-level modeling to resolve causal regulatory architectures and enable predictive, evolution-informed resistance management.

1. Introduction

Mosquitoes (Diptera: Culicidae) are among the most important disease vectors affecting human health, owing to their extraordinary capacity to transmit a wide range of pathogens and impose profound medical, social, and economic burdens worldwide (Gubler, 1998; WHO, 2024). The impact of mosquito-borne diseases is particularly severe in tropical and subtropical regions and is intensified by rapid urbanization, climate suitability for vector proliferation, expanding human mobility, and persistent gaps in infrastructure and public health capacity (Bhatt et al., 2015; WHO, 2024). Within the Culicidae, species of the genera *Anopheles*, *Aedes*, and *Culex*, notably *Anopheles gambiae*, *Aedes aegypti*, *Aedes albopictus*, and *Culex quinquefasciatus*, serve as primary vectors of globally important pathogens, including *Plasmodium* parasites (malaria), dengue virus, chikungunya virus, Zika virus, West Nile virus, and *Wuchereria bancrofti* (Cano et al., 2014; Onen et al., 2023; WHO, 2017, 2024). In recent years, major mosquito-borne disease outbreaks have increased in both frequency and geographic range, repeatedly straining healthcare systems and exposing vulnerabilities in existing control programs (Gao et al., 2025; Venkatesan, 2024). Dengue, in particular, has shown dramatic resurgence, with more than 14.4 million global cases reported in 2024, the highest annual total recorded to date (WHO, 2025). In the Americas alone, over 13 million cases were reported in 2024, representing a more than threefold increase relative to 2023 and marking the most severe epidemic on record in the region (WHO, 2025). These escalating trends underscore the urgent need for effective, sustainable, and locally adapted vector control strategies grounded in ecological and operational realities.

Controlling mosquito populations and interrupting pathogen transmission remain cornerstones of disease prevention. Contemporary control programs employ diverse approaches, including environmental management, biological control, genetic and biotechnological interventions, personal protective measures, and chemical insecticides (Achee et al., 2019; WHO, 2024). Among these, chemical control remains the most widely deployed and immediately effective intervention, particularly during epidemic emergencies. Large-scale insecticide-based strategies, including indoor residual spraying, space spraying, and insecticide-treated materials, have contributed substantially to historic reductions in malaria and other mosquito-borne diseases (Bhatt et al., 2015). However, prolonged and often intensive insecticide use has imposed strong selective pressures on mosquito populations, accelerating the evolution and global dissemination of insecticide resistance. Insecticide resistance therefore represents one of the most rapid and recurrent examples of contemporary adaptive evolution in natural mosquito populations. Resistance has now been documented against all major classes of public health insecticides, including organochlorines, organophosphates, carbamates, and pyrethroids, with emerging evidence for reduced susceptibility even to newer chemistries such as neonicotinoids (Fouet et al., 2024; Liu, 2015; Moyes et al., 2017). The expansion of resistance compromises intervention efficacy, increases programmatic costs, shortens the functional lifespan of control tools, and undermines disease control outcomes (Ranson and Lissenden, 2016; WHO, 2021). Understanding the molecular and evolutionary bases of insecticide resistance has therefore become a central priority in insect physiology, vector biology, and public health.

Insecticide resistance in mosquitoes is best understood as a complex, multifactorial adaptive phenotype arising from interactions among genetic variation, physiological regulation, and ecological selection operating across multiple biological scales (Hemingway et al., 2004; Li and Liu, 2010; Ranson and Lissenden, 2016). Four principal mechanisms have been described: metabolic detoxification, target-site insensitivity, behavioral resistance, and cuticular resistance. Among these, enhanced metabolic detoxification, mediated primarily by cytochrome P450 monooxygenases, carboxyl/cholinesterases, glutathione S-transferases, and UDP-glucuronosyltransferases, and structural alterations in key neuronal targets, including the voltage-gated sodium channel, acetylcholinesterase, nicotinic acetylcholine receptors, and γ-aminobutyric acid receptors, constitute the dominant molecular foundations of resistance in major mosquito vectors

(Fouet et al., 2024; Hancock et al., 2024; Hemingway et al., 2004; Liu, 2015). Importantly, these mechanisms rarely operate in isolation. Rather, these mechanisms are embedded within broader regulatory architectures involving signaling cascades (Li and Liu, 2017, 2019), transcriptional control networks (Ingham et al., 2017; Misra et al., 2011), redox-responsive pathways (Lu et al., 2020), and epitranscriptomic regulation (Yang et al., 2021a) that collectively govern resistance gene expression and amplify resistance phenotypes under sustained selection pressure. Resistance evolution is further shaped by standing genetic variation, de novo mutation, gene flow, and fitness trade-offs, leading to recurrent and often convergent adaptive solutions across mosquito taxa and geographic regions (Moyes et al., 2017; Weetman et al., 2018).

This chapter synthesizes current knowledge on the molecular mechanisms and evolutionary dynamics of insecticide resistance in mosquitoes, with particular emphasis on metabolic detoxification, target-site insensitivity, and the regulatory networks that integrate these processes. By integrating advances in genomics, functional genetics, physiology, and evolutionary theory, this review advances a systems-level evolutionary framework for understanding resistance as an emergent, multilayered adaptive phenotype and outlines future directions for developing more durable, predictive, and evolution-informed vector control strategies.

2. Molecular mechanisms of insecticide resistance in mosquitoes

At the molecular level, insecticide resistance in mosquitoes arises through a limited number of conserved mechanisms that alter either the fate of insecticidal compounds within the organism or their interaction with neuronal target sites. These mechanisms include enhanced metabolic detoxification, which increases the breakdown and elimination of insecticides; structural or functional modifications of insecticide target proteins that reduce binding affinity or disrupt toxic action; changes in cuticular properties that limit or slow insecticide penetration; and physiological or behavioral adaptations that reduce insecticide exposure (Fig. 1). Although multiple mechanisms may coexist within individual mosquitoes or populations, metabolic detoxification and target-site insensitivity represent the most prevalent and extensively characterized molecular bases of resistance across major vector species worldwide (Hemingway et al., 2004; Liu, 2015;

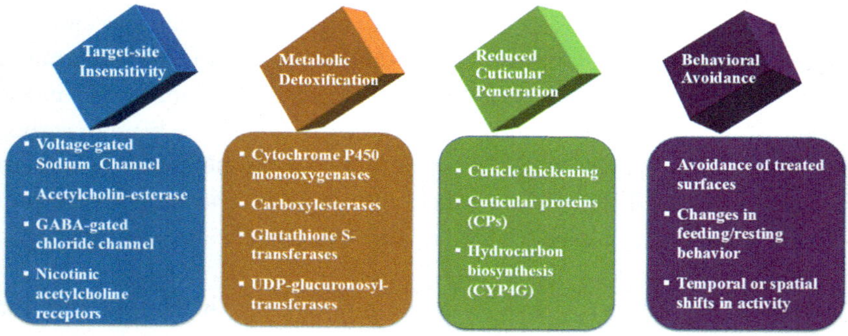

Fig. 1 Schematic overview of the principal molecular and physiological mechanisms underlying insecticide resistance in mosquitoes. Resistance arises through multiple conserved processes, including metabolic detoxification, target-site insensitivity, reduced penetration, and behavioral modification, that may act in parallel and frequently co-occur within individuals or populations.

Liu and Brown 2026; Ranson and Lissenden, 2016; Scott, 2019). As summarized in Fig. 1, insecticide resistance in mosquitoes arises through several conserved molecular and physiological mechanisms. Across insect species, insecticide resistance is rarely driven by a single mechanism; rather, it frequently results from the co-occurrence of multiple mechanisms, most notably target-site insensitivity and enhanced metabolic detoxification, that together contribute to elevating resistance intensity (Hemingway et al., 2004; Ranson and Lissenden, 2016; Scott, 2017). The sections that follow examine these mechanisms in detail, beginning with metabolic detoxification and followed by target-site insensitivity, while highlighting their physiological integration and evolutionary significance.

2.1 Metabolic detoxification as a central mechanism of resistance

At the molecular level, metabolic detoxification represents one of the most widespread and versatile mechanisms underlying insecticide resistance in mosquitoes. This form of resistance arises through enhanced enzymatic transformation, sequestration, or clearance of insecticidal compounds, thereby reducing the effective dose reaching neuronal target sites. Growing body of evidence for the importance of metabolic detoxification emerged from synergist assays (Brown et al., 2025; Huang et al., 2025; Li et al., 2023; Pelloquin et al., 2025; Wang et al., 2023; Xu et al., 2005) and complementary biochemical and enzyme activity analyses (Kasai et al., 1998; Liu, 2015; Vulule et al., 1999), which demonstrated that inhibition of detoxification enzymes could partially restore insecticide susceptibility, even in mosquito populations carrying target-site mutations. In particular, the use of piperonyl butoxide (PBO) to inhibit cytochrome P450 monooxygenases, S,S,S-tributyl phosphorotrithioate (DEF) to inhibit carboxylesterases, diethyl maleate (DEM) to inhibit glutathione S-transferases, and sulfinpyrazone (SULF) or 5-nitrouracil to inhibit UDP-glycosyltransferases provided functional evidence that enhanced metabolic activity contributes to resistance phenotypes.

Although synergists are not universally specific inhibitors and may incompletely inhibit resistance-associated enzymes or exhibit cross-pathway effects, underscoring the need for gene-specific and functional validation (Feyereisen, 2005; Liu et al., 2015), synergist bioassays using compounds such as PBO, DEF, and DEM have been invaluable for implicating metabolic detoxification in insecticide resistance (Chouaïbou et al., 2014; Brown et al., 2026; Li et al., 2023; Liu et al., 2015; Wang et al., 2023). Collectively, these observations indicate metabolic detoxification is a major driver of resistance intensity and may contribute to reduced effectiveness of vector control interventions in some settings, independent of target-site insensitivity. Multiple enzyme superfamilies contribute to metabolic resistance, most notably cytochrome P450 monooxygenases, carboxylesterases, and glutathione S-transferases, each with distinct biochemical properties, substrate specificities, and evolutionary trajectories. A summary of metabolic detoxification genes implicated in mosquito insecticide resistance is provided in Table 1. Overlapping gene entries reflect multiple and often complementary lines of evidence, including transcriptional association, synergist-based inference, and functional or biochemical validation, across independent studies and mosquito populations. The following subsections examine these enzyme families in detail, highlighting their physiological roles, genetic architecture, and contributions to resistance phenotypes across major mosquito vectors.

2.1.1 Cytochrome P450 monooxygenases

Cytochrome P450 monooxygenases (P450s) constitute the most extensively characterized and functionally versatile enzyme family involved in metabolic detoxification of insecticides in mosquitoes. Members of this supergene family catalyze diverse Phase I reactions, most commonly oxidative transformations, that modify a broad range of insecticidal compounds, frequently reducing toxicity or, in some cases, generating bioactivated intermediates that are subsequently metabolized or excreted

Table 1 Key metabolic detoxification genes across three major enzyme families implicated in mosquito insecticide resistance, with representative supporting evidence.

Gene (s)	Mosquito species	Insecticide class	Primary evidence supporting resistance association	Key reference (s)
Cytochrome P450 monooxygenases (P450s)				
CYP6M2, CYP6P1, CYP6P3, CYP9K1, CYP4G16	*An. gambiae*	Pyrethroids	Overexpression, synergist (PBO) assays, qRT-PCR	Owusu-Asenso et al. (2026), Vontas et al. (2018)
(An. gambiae) CYP6M2, CYP6Z3, CYP6P3, CYP6P4, CYP6AA1, CYP9K1; (An. funestus) CYP6N1, CYP6M7, CYP6M1 CYP6Z1	*An. gambiae; An. funestus*	Pyrethroids	Transcriptomic profiling, synergist (PBO) assays, qRT-PCR	Matowo et al. (2022)
(An. gambiae) CYP6M2, CYP6P1, CYP6P2, CYP6P3, CYP6P4, CYP6P5, CYP9J5, CYP6Z2; (An. funestus) CYP6P9a, CYP6P9b, CYP6Z1, CYP6M7, CYP9K1; (An. coluzzii) CYP9K1; (An. arabiensis) CYP6P4	*An. gambiae;* An. funestus; An. coluzzii; An. arabiensis	Pyrethroids; carbamates (*CYP6P3*)	Functional metabolism (heterologous *E. coli* expression), inhibition, transgenic overexpression (GAL4/UAS)	Yunta et al. (2016), (2019), Ibrahim et al. (2016), Vontas et al. (2018), Adolfi et al. (2019), Mitchell et al. (2014)

(continued)

Table 1 Key metabolic detoxification genes across three major enzyme families implicated in mosquito insecticide resistance, with representative supporting evidence. (*cont'd*)

Gene (s)	Mosquito species	Insecticide class	Primary evidence supporting resistance association	Key reference (s)
CYP6Z1	*An. funestus*	Carbamate; pyrethroid	Functional metabolism (heterologous *E. coli* expression)	
CYP6AA1	*An. gambiae;* An. coluzzii	Pyrethroid; organophosphates	Copy number variation (CNV)	Njoroge et al. (2022), Lucas et al. (2023)
CYP6P9a/b	*An. funestus*	Pyrethroids; mitochondrial complex I inhibitors	Gene duplication (CNV) with regulatory insertion and overexpression	Tekoh et al. (2025)
CYP9K1	*An. funestus*	Pyrethroids	Coding-sequence SNP (G454A) with functional validation (transgenic *Drosophila*; functional metabolism assays)	Djoko Tagne et al. (2025)
CYP6AA7, CYP9J40, CYP9J34, CYP9M10	*Cx. quinquefasciatus*	pyrethroids	cDNA microarray, qRT-PCR, RNAi, functional metabolism (baculovirus expression system)	Liu et al. (2007), (2011), Li et al. (2015), Gong et al. (2017)

CYP6P14, CYP6BZ2, CYP9J33, CYP9J34, CYP9J40, CYP9J45	*Cx. quinquefasciatus*	Pyrethroids	Overexpression, functional metabolism (baculovirus expression system)	Gong et al. (2022a)
CYP9AL1, CYP9J35, CYP9J45, CYP4C52v1, CYP4D42V1, CYP6BZ2, CYP6P14, CYP6325Y6	*Cx. quinquefasciatus*	pyrethroids	Overexpression, RNAi, homology modeling/ substrate docking	Yang et al. (2021b)
CYP6AA7, CYP9J3, CYP9J40, CYP9J45	*Cx. quinquefasciatus*	pyrethroids	qRT-PCR	Chamnanya et al. (2025), Saha et al. (2025)
CYP325G4, CYP6AA9	*Cx. pipiens pallens*	pyrethroids	Proteomic and transcriptomic profiling, RNAi, qRT-PCR	Xu et al. (2024)
CYP9M10, CYP4H34, CYP6Z10	*Cx. quinquefasciatus*	Pyrethroids	Microarray analysis, qRT-PCR	Komagata et al. (2010)
CYP9M10	*Cx. quinquefasciatus*	Pyrethroids	Targeted mutagenesis (TALEN), CRISPR/ Cas9	Itokawa et al. (2016)
CYP325BC1, CYP9M12	*Cx. quinquefasciatus*	Organophosphates (malathion)	Transcriptomic profiling, RNAi	Huang et al. (2023)

(continued)

Table 1 Key metabolic detoxification genes across three major enzyme families implicated in mosquito insecticide resistance, with representative supporting evidence. (*cont'd*)

Gene (s)	Mosquito species	Insecticide class	Primary evidence supporting resistance association	Key reference (s)
CYP6AA7, CYP4C52v1	*Cx. quinquefasciatus*	pyrethroids	qRT-PCR	Yang and Liu (2011)
CYP9J34	*Cx. pipiens pallens*	Pyrethroids	Functional metabolism, luciferase reporter assays, RNAi	Du et al. (2024)
CYP9J24, CYP9J26, CYP9J28, CYP9J32	*Ae. aegypti*	Pyrethroids	Functional metabolism (in vitro assays)	Stevenson et al. (2012)
CYP6BB2, CYP9J26, CYP9J28, CYP9J32	*Ae. aegypti*	DDT; pyrethroids; carbamates	Field overexpression, functional metabolism assays	Vontas et al. (2020), Sene et al. (2021)
Cyp9J28, Cyp9J32, Cyp9M6	*Ae. aegypti;*	Pyrethroids,	Comparative transcriptomics / field overexpression	*Djiappi-Tchamen et al. (2021)
CYP6BB, CYP9J28, CYP9J32, (and others)	*Ae. aegypti*	Pyrethroids	Targeted deep sequencing, CNV analysis (genomic association)	Faucon et al. (2015)

Genomic duplication of P450s, including CYP6BB2	*Ae. aegypti*	Insecticides not specified	Transcriptomic profiling, Structural variation/gene duplication	Bacot et al. (2024)
Carboxyl/cholinesterases (CCEs)				
Coeae6g	*An. gambiae*	Organophosphates; Pyrethroids; Carbmates	Activity-based chemoproteomics (ABPP), functional validation	Balaska et al. (2025)
estα21, estβ2	*Cx. quinquefasciatus*	Organophosphates	Gene co-amplification, recombinant genomic cloning	Vaughan et al. (1997)
EsterA11, EsterB11	*Cx. pipiens complex*	Organophosphates,	Gene amplification, qRT-PCR	Cui et al. (2007)
AsAe9, AsAe10, AsAce1, AsAce2, AsBe4	*An. sinensis*	Pyrethroids	Transcriptomic profiling, overexpression, functional RNAi	Wei et al. (2025)
CPIJ018231, CPIJ018232, CPIJ018233	*Cx. quinquefasciatus*	Pyrethroids	Transcriptomic overexpression, functional metabolism (Sf9 cells), enzyme activity assays, docking analysis	Reid et al. (2012), Gong et al. (2022b)

(continued)

Table 1 Key metabolic detoxification genes across three major enzyme families implicated in mosquito insecticide resistance, with representative supporting evidence. *(cont'd)*

Gene (s)	Mosquito species	Insecticide class	Primary evidence supporting resistance association	Key reference (s)
Ace−1	*An. gambiae; An. albimanus; Cx. pipiens complex*	Carbamates; organophosphates	Target-site mutation (G119S), sequencing, structural modeling, qPCR genotyping	Weill et al. (2003), (2004), Djogbenou et al. (2007), Essandoh et al. (2013), Oumbouke et al. (2020)
COEAE1F	*An. gambiae*	Pyrethroids	Whole-genome microarray, qRT-PCR	Oumbouke et al. (2020)
CCEae3a	*Ae. aegypti*	Carbamates; organophosphates	GAL4/UAS expression system, transgenic line of *An. gambiae*	Poulton et al. (2024)
AsAe10, AsAce1	*An. gambiae*	Pyrethroids	RNA-seq, qRT-PCR	Wei et al. (2025)
Glutathione S-transferases (GSTs)				
(An. gambiae) AaGSTe4, AgGSTe4; (An. arabiensis) AaGSTe4, AgGSTe4;	*An. gambiae;* An. arabiensis	Pyrethroids	Overexpression, transcriptomic association	Wilding et al. (2015)

(An. gambiae) AgGSTe2; (An. funestus) GSTS1-2, GSTE2	*An. gambiae;* An. funestus	DDT; Pyrethroids	Microarray analysis, qRT-PCR, *Drosophila expression,* I114T mutation in *AgGSTe2*	Mitchell et al. (2014), Nardini et al. (2017)
AfGSTe2	*An. funestus*	DDT; Pyrethroids	Transcriptomic association, coding-sequence mutation (L119F), recombinant enzyme metabolism, transgenic overexpression (*Drosophila*)	Riveron et al. (2014)
CPGSTd1, CPGSTd2	*Cx. pipiens complex*	Pyrethroids	qRT-PCR overexpression	Xu et al. (2017)
GSTd2, GSTe2	*An. sinensis*	Pyrethroids	Protein database annotation, phylogenetic analysis, transcriptomics, qRT-PCR, RNAi, functional metabolism	Tao et al. (2022)
Gste2	*An. gambiae;* Ae. aegypti	Organophosphate;DDT	Functional validation (GAL4/UAS-mediated transgenic overexpression)	Adolfi et al. (2019)

(continued)

Table 1 Key metabolic detoxification genes across three major enzyme families implicated in mosquito insecticide resistance, with representative supporting evidence. *(cont'd)*

Gene (s)	Mosquito species	Insecticide class	Primary evidence supporting resistance association	Key reference (s)
GSTe3, *GSTe4*, *GSTe6*	*An. funestus*	DDT; pyrethroids	Transcriptomics, allelic variation, structural modeling, recombinant metabolism, transgenic expression	Kouamo et al. (2025)
UDP-Glucuronosyltransferases (UGTs)				
GT310B2, *UGT301C2*, *UGT302A3*, *UGT306C2*	*An. funestus*	Pyrethroids	Transcriptomic association, GWAS-PoolSeq, population-genomic differentiation (mutations)	Al-Yazeedi et al. (2024)
UGT308G1, *UGT306A2*, *UGT302A1*	*An. gambiae*; *An. coluzzii*	Pyrethroids; DDT	Transcriptomic and multi-omics association, overexpression, RNAi (no susceptibility restoration)	Logan et al. (2024)
AsUGT308D3, *UGT302A3*	*An. sinensis*	Pyrethroids	Bioinformatics methods, RNA-seq and RT-qPCR, Variant screening	Zhou et al. (2019)

Multiple UGT genes	*An. coluzzii;* An. gambiae; An. arabiensis; An. funestus; Aedes spp; Culex sp	DDT; Pyrethroids	Transcriptomic and genomic surveys (correlative; reviewed in Ellis Logan et al., 2024)	**Logan et al. (2024)

Genes listed represent selected examples from four major detoxification enzyme families—cytochrome P450 monooxygenases (P450s), carboxyl/cholinesterases (CCEs), glutathione S-transferases (GSTs), and UDP-glucuronosyltransferases (UGTs)—associated with insecticide resistance in mosquitoes.

Only genes supported by experimental or population-level evidence are included, such as functional validation (e.g., heterologous expression, transgenic or genome-editing approaches, RNA interference), biochemical assays, synergist-based inference, or consistent associations with resistance phenotypes in field populations (e.g., overexpression or copy number variation). Genes identified solely through bioinformatic or transcriptomic analyses without validation were excluded.

When multiple genes, species, and references are listed within a single row, these represent collective evidence across studies, and specific gene–species relationships may vary among references. Genes may appear in multiple rows to reflect distinct lines of evidence.

Data availability varies among species. In particular, evidence for *Aedes albopictus* remains limited and is largely restricted to transcriptomic observations, with comparatively few validated studies for other detoxification mechanisms. This table highlights well-supported examples and is not intended to be exhaustive.

**Logan et al. 2024. Transcriptomic and genomic surveys are cited via recent syntheses where appropriate and represent correlative evidence unless functional validation is explicitly indicated.

*Djiappi-Tchamen et al., 2021. This study provides rare transcriptomic evidence for resistance-associated P450 expression in Aedes albopictus.

(Feyereisen, 2005; Li et al., 2007). The recurrent recruitment of related P450 families across mosquito populations likely reflects this intrinsic biochemical versatility. Their broad substrate specificity enables metabolism of multiple insecticide classes, including pyrethroids, organochlorines, carbamates, and newer chemistries, thereby facilitating parallel and convergent adaptive responses under insecticide selection (Hemingway et al., 2004; Liu, 2015; Moyes et al., 2017).

Early lines of evidence implicating P450-mediated resistance arose from synergist assays, in which inhibition of monooxygenase activity, most notably via PBO (Li et al., 2023), reduced resistance levels. This approach continues to be widely applied in contemporary field and laboratory studies (Matowo et al., 2022; Wang et al., 2023). Such PBO sensitivity suggested that resistance often reflects elevated P450 activity arising from transcriptional up-regulation, gene amplification/copy number variation, or coding changes that affect catalytic efficiency or substrate binding. As summarized in Fig. 2, synergist-based inference provided an initial functional signal that subsequently guided transcriptomic discovery and multi-level validation of resistance-associated P450 genes. Molecular studies then moved beyond enzyme-level inference to identify specific resistance-associated loci. Early candidate-gene analyses established links between individual P450 genes and resistance phenotypes in particular populations (Amenya et al., 2008; Gong et al., 2005; Liu et al., 2011; Rongnoparut et al., 2003). With the advent of transcriptomic and genomic approaches, it has become clear that metabolic resistance often involves multiple detoxification genes through mechanisms such as transcriptional upregulation and gene copy number variation. Transcriptomic analyses have revealed differential expression of detoxification genes, including cytochrome P450s, in resistant *Culex* mosquitoes (Liu et al., 2007; Reid et al., 2012, 2018), while genome-wide studies have identified extensive copy number variation at insecticide resistance loci (Lucas et al., 2019). These patterns highlight the polygenic and population-specific nature of metabolic resistance and reflect repeated recruitment of functionally related P450 genes within expanded detoxification gene families across diverse mosquito species.

Extending these observations across broader evolutionary scales, comparative genomic and transcriptomic analyses provide further insight into the genetic architecture of metabolic resistance across mosquito genera. Studies in *Anopheles*, *Aedes*, and *Culex* show that resistance is frequently associated with expansion and diversification of particular P450 subfamilies and coordinated overexpression of multiple P450 genes.

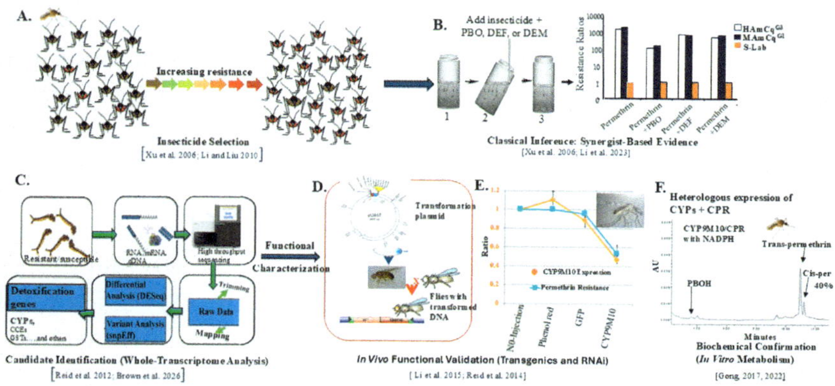

Fig. 2 Conceptual progression from synergist-based inference to molecular and functional validation of metabolic detoxification–mediated insecticide resistance in mosquitoes. (A) Insecticide selection in field populations drives the emergence of resistance populations. (B) Synergism bioassays are used to infer metabolic mechanisms underlying resistance using enzyme inhibitors such as piperonyl butoxide (PBO), diethyl maleate (DEM), or S,S,S-tributyl phosphorotrithioate (DEF). Synergist bioassays are illustrated as a three-step experimental workflow: 1. mosquito larvae are first introduced into a bottle containing tap water; 2 insecticide is then added alone or in combination with a sublethal concentration of a synergist e.g., PBO, DEM, or DEF; 3 mortality is subsequently assessed after exposure, as described by Xu et al. (2006) and Li et al. (2023). Increased sensitivity in the presence of synergists provided early functional evidence linking resistance to metabolic detoxification through P450s, esterases, and/or GSTs. (C) Transcriptomic and genomic analyses enabled further identification of candidate resistance-associated genes, often characterized by transcriptional upregulation, copy-number variation, or coding-sequence mutations. Candidates are typically prioritized based on recurrence across resistant populations and membership in detoxification gene families. (D-F) Direct causal validation is achieved using complementary approaches, including (D) heterologous expression, (E) RNA interference RNAi-mediated knockdown, and (F) in vitro metabolism assays. Heterologous expression systems may include insect cells, yeast, or bacterial systems, as schematically represented. Together, these integrated approaches establish cytochrome metabolic enzymes as mechanistic drivers of insecticide resistance rather than correlative markers of selection.

Gene amplification and copy number variation (CNVs) represent one of key mechanisms contributing to these patterns in some populations (Bacot et al., 2024; Faucon et al., 2015; Hearn et al., 2022; Lucas et al., 2019; Lucas et al., 2023; Nagi and Ingham, 2025; Weetman et al., 2018). Population genomic studies further demonstrate that CNVs, including amplification of detoxification loci, can elevate gene expression and contribute to resistance phenotypes, alongside resistance-associated haplotypes shaped by linked SNPs and regulatory variants (Lucas et al., 2023; Njoroge et al., 2022). Importantly, elevated detoxification gene expression can arise through multiple, non-mutually exclusive mechanisms, including regulatory variation, resistance-associated haplotypes, and copy number variation, which should not should not be considered interchangeable... In addition, coding-sequence mutations can directly enhance metabolic capacity, as exemplified by the resistance-conferring G454A substitution in *CYP9K1* in *An. funestus* (Djoko Tagne et al., 2025). Notably, resistance-associated P450s from the CYP6 and CYP9 families have been repeatedly identified in resistant populations across diverse insect taxa and geographic regions, illustrating recurrent and non-random recruitment of functionally related detoxification enzymes under insecticide selection (Nauen et al., 2022; Vontas et al., 2020). Multiple CYP6- and CYP9-family genes, including *CYP9K1* and *CYP6M2*, have been consistently overexpressed and independently validated in pyrethroid-resistant *Anopheles* populations from geographically distinct regions, including Tanzania and Ghana (Matowo et al., 2022; Owusu-Asenso et al., 2026). Consistent with this pattern, surveys across several African countries repeatedly identify overlapping subsets of CYP6 and CYP9 genes as major contributors to metabolic resistance (Hearn et al., 2022; Ibrahim et al., 2016; Vontas et al., 2020).

This recurrent recruitment extends beyond *Anopheles* to other Culicidae genera. In *Ae. aegypti*, multiple studies implicate CYP6 and CYP9 family members, including *CYP6BB2* and several CYP9J gene subfamily, as key contributors to pyrethroid resistance across geographically distinct populations, with functional validation confirming roles in insecticide metabolism (Sene et al., 2021; Stevenson et al., 2012; Vontas et al., 2020). Similarly, in *Culex* mosquitoes, *CYP9M10* and related CYP9 family genes have been repeatedly identified as resistance-associated and functionally validated in populations from the United States and Japan, indicating that similar detoxification loci are recurrently involved in resistance across geographically distinct populations (Gong et al., 2017, 2022a; Itokawa et al., 2016; Xu et al., 2024; Yang et al., 2021b). Collectively, these findings indicate that P450-mediated resistance is best understood not as the action of single resistance genes, but as a polygenic detoxification system involving multiple functionally related genes that can be differentially recruited under insecticide selection. This architecture may contribute to the rapid emergence and broad distribution of metabolic resistance across mosquito taxa.

As illustrated in Fig. 2, the field has progressed from synergist-based inference and correlative expression analyses to functional approaches capable of establishing causal links between specific P450 genes and resistance. Heterologous expression systems, including baculovirus-infected insect cells (Gong et al., 2022a, 2022b), dual-plasmid co-expression in *Escherichia coli* (Hearn et al., 2022; Vontas et al. 2018; Yunta et al. 2016, 2019), and complementary in vitro metabolism assays, have been widely used to demonstrate that individual mosquito P450 enzymes can metabolize specific insecticides (Gong et al., 2022a; Liu, 2015; Stevenson et al., 2012). Complementary in vivo approaches, including RNA interference–mediated knockdown in mosquitoes and transgenic *Drosophila melanogaster* expression systems, further demonstrate that suppression or ectopic expression of individual P450s can significantly alter resistance levels, providing functional validation within an intact physiological context (Djoko Tagne et al., 2025; Li et al., 2015; Reid et al., 2014). Emerging applications of CRISPR/Cas9 genome editing and copy-number manipulation offer even more definitive tests of P450 function by directly linking gene dosage or sequence variation to resistance phenotypes.

Genome editing has transformed functional genetics by enabling direct tests of causality for both metabolic and target-site resistance mechanisms. Although such approaches are increasingly established for resistance gene functional characterization in insects more broadly (Xu et al., 2025), their application to mosquito insecticide resistance has, until recently, remained comparatively limited. Nevertheless, several studies demonstrate the power of genome editing in mosquitoes. In *Cx. quinquefasciatus*, TALEN- and CRISPR/Cas9-mediated disruption of the overexpressed P450 gene *CYP9M10* produced a dramatic reduction in larval pyrethroid resistance, providing definitive reverse-genetic validation of a metabolic resistance locus (Itokawa et al., 2016). Technical advances in non-model mosquito systems, including optimized CRISPR tools and site-directed transgenesis in mosquitoes, are expected to accelerate genome-editing-based validation of resistance loci (Feng et al., 2021).

In parallel with genome-editing approaches, complementary in vivo functional validation strategies based on transgenic overexpression have also been developed in mosquitoes. Notably, GAL4/UAS-based expression systems in *An. gambiae* enable controlled overexpression of individual detoxification genes, demonstrating that elevated expression of specific P450s, and, in some cases, GSTs, can be sufficient to confer insecticide resistance in a defined genetic background (Adolfi et al., 2019). Together, these approaches firmly establish P450s as mechanistic drivers of insecticide resistance rather than passive correlates of selection, integrating P450-mediated detoxification into an evolutionary genomics framework in which gene regulation, structural variation, and enzymatic capacity collectively shape resistance outcomes under insecticide selection pressure (Weetman et al., 2018). These functional advances now enable resistance-associated P450s to be evaluated not only as biomarkers, but also as experimentally testable targets for resistance management and the development of novel intervention strategies.

2.1.2 Carboxylesterases

Although P450s represent the most extensively characterized metabolic resistance system in mosquitoes, carboxylesterase-mediated resistance has been investigated using a broadly comparable conceptual and experimental framework (Fig. 2). Early evidence for esterase involvement emerged from synergist-based assays, most commonly using DEF (Li et al., 2023), together with biochemical measurements demonstrating elevated esterase activity in resistant mosquito populations (Ketterman et al., 1992). Gene amplification was subsequently recognized as a key mechanism underlying elevated esterase expression in insecticide-resistant insects (Devonshire and Field, 1991). These observations were subsequently reinforced by molecular and genomic analyses identifying resistance-associated esterase loci, followed by functional validation through expression studies, knockdown approaches, and enzymatic assays. Carboxylesterases contribute to metabolic insecticide resistance primarily through hydrolysis or sequestration of ester-containing compounds (Ketterman et al., 1992; Wei et al., 2025). By accelerating chemical degradation or binding and sequestering toxicants away from neuronal targets, esterases reduce the bioavailable dose reaching sensitive tissues. Esterase-mediated resistance is most consistently associated with organophosphates and carbamates, although associations with other insecticide classes have also been reported (Hemingway et al., 2004; Liu, 2015).

In contrast to P450-mediated resistance, which most commonly involves transcriptional up-regulation and, in some cases, moderate CNVs, esterase-mediated resistance has historically been strongly associated with high-level gene amplification, typically involving very high copy number CNVs (Cui et al., 2007; Raymond et al., 1998). Nevertheless, esterase-based resistance can also arise through transcriptional up-regulation and coding-sequence changes that alter catalytic efficiency or binding capacity (Essandoh et al., 2013; Gong et al., 2022b; Reid et al., 2012; Wei et al., 2025). In resistant populations, elevated esterase activity therefore most often reflects enzyme overproduction driven by amplification, although increased transcriptional output or structural variation enhancing catalytic or binding capacity has also been documented (Raymond et al., 1998). Classic studies in *Culex* demonstrated that massive overproduction of specific esterases can confer extremely high resistance levels, establishing esterase amplification as a canonical mechanism. In addition, resistance within the carboxyl/cholinesterase superfamily can also arise via target-site modification of acetylcholinesterase (AChE), most notably the G119S substitution in *ace-1*, which confers high-level resistance to organophosphates and carbamates and has been repeatedly documented across mosquito species (Weill et al., 2003, 2004).

Structural analyses have provided critical insight into how overproduced esterases contribute to resistance. The first crystal structure of a resistance-associated mosquito carboxylesterase, Cqestβ21 from *Cx. quinquefasciatus*, showed that sequestration-type esterases retain the canonical α/β-hydrolase fold but possess distinctive active-site and binding-pocket features that favor high-affinity ligand binding rather than rapid catalytic turnover (Hopkins et al., 2017). Comparative structural and sequence analyses suggest that resistance-associated esterases form a functionally specialized subgroup within the broader carboxyl/cholinesterase family. In many cases, resistance is associated with high-level expression of these enzymes, which can bind and sequester insecticide molecules through rapid binding and slow turnover, or catalytically metabolize them, depending on the specific esterase and system (Cui et al., 2007; Hemingway et al., 1998; Wei et al., 2025). Although esterases generally exhibit narrower substrate specificity than P450s, their capacity for efficient detoxification or sequestration renders them an important and durable component of metabolic resistance, particularly under sustained selection pressure from ester-based insecticides. As summarized in Fig. 2, esterase-mediated resistance has frequently been inferred from synergist-based assays and subsequently supported by transcriptomic profiling and functional characterization of candidate esterase genes, including biochemical metabolism assays and RNA interference, together with complementary structural modeling approaches such as molecular docking (Gong et al., 2022b; Wei et al., 2025; Yang et al., 2021b).

More recently, predictive chemoproteomic approaches have been applied to identify esterase-mediated resistance mechanisms prior to their emergence in field populations. Using competitive activity-based protein profiling with fluorophosphonate probes, Balaska et al. (2025) identified active serine hydrolases in a susceptible *An. gambiae* strain that directly interact with insecticide metabolites. Functional validation showed that the carboxylesterase Coeae6g confers resistance to the organophosphate pirimiphos-methyl and mediates cross-resistance to multiple insecticide classes, including organophosphates, carbamates, and pyrethroids. This study provides a framework for linking enzyme activity to resistance risk and illustrates how chemoproteomic screening can complement genomic surveillance in resistance management. Collectively, these studies establish esterases as a complementary metabolic resistance system that, while less diverse and less frequently implicated than P450s, can play a decisive role in shaping resistance phenotypes. Compared with P450-mediated resistance, esterase-based mechanisms are often more chemically constrained and typically involve a limited set of loci, often with very high copy number amplification, often with very high copy number amplification, but nevertheless illustrate how repeated selection can favor amplification and functional optimization of detoxification genes across diverse mosquito species under intense insecticide pressure.

2.1.3 Glutathione S-transferases

Consistent with the conceptual framework summarized in Fig. 2, studies of glutathione S-transferase (GST)-mediated insecticide resistance have progressed from functional inference to molecular identification and causal validation. Early evidence from synergist assays, most commonly using diethyl maleate (DEM), together with biochemical measurements, implicated GST activity in resistance phenotypes; these findings were subsequently supported by transcriptomic analyses and functional assays linking specific GST genes to detoxification capacity and resistance. GSTs constitute a major detoxification enzyme family in mosquitoes, functioning primarily through glutathione-dependent conjugation of electrophilic insecticides or their oxidative metabolites (Enayati et al., 2005; Ranson and Hemingway, 2005). This conjugation increases compound solubility and facilitates detoxification and sequestration, thereby limiting insecticide accumulation and cellular toxicity. GST-mediated resistance has been most prominently associated with organochlorine detoxification, particularly DDT, although broader roles in mitigating insecticide-induced oxidative stress are increasingly recognized (Enayati et al., 2005). Beyond direct conjugation, GSTs can protect cellular macromolecules from oxidative damage generated during insecticide exposure, enhancing survival under toxic stress.

A particularly compelling example that integrates transcriptional upregulation, structural variation, and population-level selection is provided by *GSTe2* in *Anopheles funestus*. Riveron et al. (2014) demonstrated that *GSTe2* is strongly overexpressed in DDT- and pyrethroid-resistant populations and carries a resistance-associated coding mutation (L119F) that enhances detoxification capacity. Transgenic expression and recombinant enzyme assays further showed that overexpression of *GSTe2* alone can confer resistance, while the L119F substitution increases catalytic efficiency by enlarging the DDT-binding pocket and facilitating substrate access and metabolism. Population genetic analyses revealed strong directional selection on the resistant allele and a close correspondence between *GSTe2* haplotype distribution and regional resistance patterns across Africa.

More recently, this framework has been extended beyond *GSTe2* to encompass allelic variation across the epsilon GST gene cluster in *An. funestus*. Kouamo et al. (2025) showed that *GSTe3*, *GSTe4*, and *GSTe6* exhibit both elevated expression and resistance-associated allelic variants under geographic selection. Functional assays indicated that these allelic variants enhance detoxification of DDT and pyrethroids, while structural modeling suggested that substitutions expand the active-site binding cavity and increase substrate affinity. Transgenic expression further confirmed that selected GSTe alleles alone can reduce insecticide susceptibility in vivo. Collectively, these studies establish that GST-mediated resistance, like P450- and esterase-based resistance, can arise through coordinated regulatory and structural evolution within detoxification gene clusters, shaping both the magnitude and evolutionary trajectory of metabolic resistance.

2.1.4 UDP-glucuronosyltransferases (UGTs)

Growing evidence indicates that UDP-glycosyltransferases (UGTs) represent an increasingly important, yet historically underappreciated, component of metabolic insecticide resistance in mosquitoes, with support derived primarily from transcriptomic and genomic surveys in which UGTs are repeatedly detected among genes upregulated in resistant mosquito populations, despite rarely being the primary focus of targeted resistance studies (Logan et al., 2024). Large-scale transcriptomic, genomic, and integrative multi-omics analyses have consistently identified elevated UGT expression in insecticide-resistant populations, particularly in *Anopheles* species across diverse African regions, with associations reported for multiple insecticide classes, including pyrethroids, carbamates, and organochlorines (Al-Yazeedi et al., 2024, Logan et al., 2024, Zhou et al., 2019).

In several cases, UGT expression is both constitutively elevated in resistant strains and inducible following insecticide exposure, suggesting a flexible regulatory response to chemical stress. Similar expression patterns have also been reported in resistant *Aedes* and *Culex* mosquitoes (Logan et al., 2024), indicating that UGT-associated detoxification responses may represent a conserved feature of resistance across vector taxa.

Beyond transcriptional regulation, emerging population-genomic evidence suggests that coding-sequence variation in UGT genes may also contribute to insecticide resistance in mosquitoes. A recent genome-wide association and targeted sequencing study in *An. funestus* identified consistent Africa-wide overexpression of specific UGTs, together with significant genetic differentiation at multiple UGT loci between pyrethroid-resistant and susceptible populations. Several UGT genes exhibited elevated gene-wise FST values and nonsynonymous substitutions, suggesting local adaptation under insecticide selection pressure (Al-Yazeedi et al., 2024). While these findings implicate both regulatory and coding variation in UGT-mediated resistance, direct functional validation of individual mutations remains to be established.

Functional interrogation of UGTs in mosquitoes has so far yielded mixed results. RNAi-mediated knockdown of highly overexpressed UGT candidates in pyrethroid-resistant *An. coluzzii* achieved substantial transcript suppression, yet failed to restore susceptibility to deltamethrin, even when multiple UGTs were silenced simultaneously (Logan et al., 2024). These findings suggest a high degree of functional redundancy within the UGT family and indicate that individual UGTs are unlikely to act as dominant resistance determinants. Instead, UGTs likely contribute modestly and additively within a broader detoxification framework, potentially acting downstream of phase I metabolism. Notably, evidence from chemical inhibition further supports a functional role for UGTs in mosquito insecticide resistance. In contrast to gene-specific RNAi, treatment with the UGT inhibitor sulfinpyrazone restored susceptibility to pyrethroids and DDT in multiple African malaria vectors, including *An. gambiae*, *An. coluzzii*, *An. arabiensis*, and *An. funestus* (Logan et al., 2024). This apparent discrepancy between RNAi- and inhibitor-based assays is consistent with extensive functional redundancy within the UGT family and suggests that collective UGT activity, rather than individual genes, contributes to resistance phenotypes.

Insight from other insect species reinforces this interpretation. Functional studies employing RNA interference, transgenic expression, heterologous metabolism assays, or CRISPR-based gene disruption across diverse taxa have demonstrated that manipulation of individual UGTs can alter susceptibility to multiple insecticides, typically producing modest but reproducible effects (Scott, 2025). These effects are generally smaller than those associated with major phase I enzymes, supporting a model in which UGTs act primarily as phase II conjugation enzymes that facilitate excretion of insecticides or their oxidative metabolites. Such contributions may become biologically meaningful through interaction with other resistance mechanisms, including enhanced phase I metabolism, reduced penetration, or target-site insensitivity.

2.1.5 Integrated metabolic detoxification mechanisms

Taken together, studies across mosquito species indicate that metabolic insecticide resistance is rarely driven by a single enzyme family but instead reflects the coordinated action of multiple detoxification systems (Hemingway et al., 2004; Liu et al., 2015; Ranson and Lissenden, 2016). Cytochrome P450 monooxygenases often provide the initial detoxification step by oxidizing insecticides or generating reactive intermediates, which are subsequently processed by carboxyl/cholinesterases through hydrolysis or by glutathione S-transferases via conjugation. UGTs, although less extensively characterized in mosquitoes, likely function downstream within this metabolic cascade by glycosylating oxidized or conjugated metabolites, thereby enhancing solubility and promoting excretion. The recurrent co-overexpression of P450s, esterases, GSTs, and, in some cases, UGTs underscores a modular and additive detoxification architecture rather than isolated pathways, conferring both robustness and flexibility for the evolution of high-level and broad-spectrum resistance under sustained insecticide selection pressure.

2.2 Target-site insensitivity as a mechanism of resistance

Target-site insensitivity represents an another major molecular basis of insecticide resistance across insects and arises through structural or functional modifications of proteins targeted by insecticidal compounds (Ranson and Lissenden, 2016; Scott, 2019). Such modifications reduce insecticide binding affinity or disrupt toxic action at the site of interaction, thereby diminishing physiological efficacy (Dong et al., 2014). In contrast to metabolic detoxification, which alters insecticide fate, target-site resistance directly affects toxicodynamics at neuronal target proteins. Mutations affecting voltage-gated sodium channels (Davies et al., 2007), acetylcholinesterase (Weill et al., 2004), acetylcholine receptor (Fouet et al., 2024), and γ-aminobutyric acid (GABA) receptors (Ffrench-Constant and Roush, 1991) constitute the most extensively documented forms of target-site resistance in mosquito vectors and have repeatedly evolved under strong and sustained insecticide selection pressure.

2.2.1 Voltage-gated sodium channels

Alterations in the voltage-gated sodium channel (VGSC) constitute the most widespread and operationally significant form of target-site resistance in mosquitoes, particularly in response to pyrethroids and organochlorines. At the molecular level, pyrethroids and dichlorodiphenyltrichloroethane (DDT) exert their insecticidal effects primarily through direct interaction with the VGSC, where they disrupt normal channel gating and prolong channel opening, leading to sustained neuronal depolarization and neurotoxicity (Dong et al., 2014; Scott, 2019; Soderlund, 2012). Structural modifications of the sodium channel, most commonly arising from single-nucleotide polymorphisms that generate amino-acid substitutions, reduce or abolish insecticide binding affinity and may secondarily alter channel gating properties, thereby diminishing toxic action and conferring resistance (Dong et al., 2014; Silver et al., 2014).

Such substitutions, collectively referred to as knockdown resistance (*kdr*) mutations (Fig. 3), have been documented across multiple mosquito genera and are strongly associated with resistance phenotypes in field populations (Kaur et al., 2025; Kawada et al., 2023; Smith et al., 2018; Tognarelli et al., 2025; Uemura et al., 2024). To date, a large number of VGSC amino-acid substitutions have been identified in mosquito populations resistant to pyrethroids and DDT. Rather than cataloguing all reported variants, the following sections highlight representative and well-characterized mutations in the three major mosquito genera, *Anopheles*, *Aedes*, and *Culex*, that illustrate common patterns of target-site adaptation and convergent evolution. A schematic overview of representative VGSC mutations and their locations within the channel protein is provided in Fig. 3. VGSC-mediated resistance has arisen independently in diverse geographic regions, frequently involving parallel or convergent substitutions at homologous positions within the channel. Individual mutations can confer varying resistance levels depending on allelic state, genetic background, and interaction with other resistance mechanisms. The global dissemination and persistence of VGSC mutations underscore their evolutionary success under sustained pyrethroid-based selection pressure and highlight their central role in shaping contemporary mosquito resistance landscapes (Moyes et al., 2017; Scott, 2019).

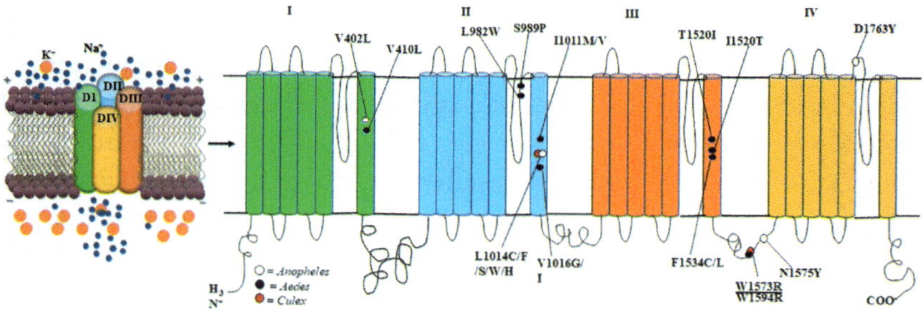

Fig. 3 Representative amino-acid substitutions in the voltage-gated sodium channel (VGSC) associated with resistance to pyrethroids and dichlorodiphenyltrichloroethane (DDT) in mosquitoes. Mutations are mapped onto a generalized insect VGSC topology comprising four homologous domains (I–IV), each containing six transmembrane segments (S1–S6). Substitutions identified in *Anopheles* (open circles), *Aedes* (black circles), and *Culex* (red circles) are overlaid onto a canonical channel schematic to illustrate shared resistance hotspots and recurrent patterns across vector genera. Only selected substitutions are shown, based on prior evidence from (i) functional studies, including electrophysiological assays (e.g., heterologous expression in systems such as *Xenopus* oocytes) or transgenic approaches demonstrating altered channel sensitivity to insecticides, and/or (ii) population-level association studies, in which mutations are consistently associated with resistance phenotypes in field populations. This schematic is not intended to provide an exhaustive catalogue of all reported VGSC variants. Mutation positions are indicated using representative numbering based on a canonical insect VGSC; equivalent substitutions may occur at homologous positions but differ in absolute amino-acid numbering among mosquito species.

In *Ae. aegypti*, pyrethroid resistance is frequently associated with the co-occurrence of multiple VGSC mutations rather than single substitutions alone. Field surveys across Asia, the Americas, and Africa have repeatedly documented double-mutation genotypes associated with elevated resistance, most commonly involving combinations of V1016G or V1016I with F1534C, and, in parts of Southeast Asia, the additional inclusion of S989P (Kawada et al., 2023; Saavedra-Rodriguez et al., 2018; Smith et al., 2018). Experimental and electrophysiological studies further demonstrate that combinations of these mutations generally confer higher levels of resistance than single-site variants, although the magnitude of this effect varies depending on genetic background and experimental system (Du et al., 2013, 2016; Smith et al., 2018; Uemura et al., 2024). For example, S989P enhances V1016G-dependent insensitivity, T1520I reduces F1534C-dependent sensitivity, and the combined S989P–V1016G–F1534C haplotype confers extremely high levels of resistance, illustrating the epistatic architecture underlying *kdr* evolution in this species (Du et al., 2013; Kaur et al., 2025; Uemura et al., 2024). Field-based genotype–phenotype association studies in Vietnamese *Ae. aegypti* further demonstrated that the L982W substitution shows a strong and independent correlation with permethrin resistance, even in populations where canonical *kdr* markers such as F1534C occur at low frequency, highlighting the importance of population-specific mutation profiles and haplotype structure in shaping resistance phenotypes (Kawada et al., 2023).

In addition to the canonical *kdr* substitutions at S989P, V1016G/I, and F1534C/L, a more recently identified voltage-gated sodium channel mutation, V410L, has been repeatedly reported in global *Ae. aegypti* populations and is associated with elevated levels of pyrethroid resistance, particularly when occurring in combination with other *kdr* alleles (Haddi et al., 2017; Enayati et al., 2024). Secondary variants, such as T1520I, have also been documented in field populations, frequently co-occurring with major *kdr* substitutions; however, available evidence suggests that their individual contributions to resistance are generally modest relative to primary *kdr* mutations (Fan and Scott, 2020; Naw et al., 2022).

Importantly, functional studies and population-genetic observations address distinct but complementary questions. Electrophysiological and mutagenesis-based analyses, which artificially combine mutations on a single VGSC haplotype, have shown that the S989P–V1016G–F1534C triple combination can reduce pyrethroid sensitivity by several orders of magnitude, illustrating the potential for extreme resistance if such a haplotype were to arise (Du et al., 2013; Hirata et al., 2014). In contrast, recent high-resolution population-genetic analyses indicate that these mutations often segregate into distinct, non-recombining haplogroups in natural mosquito populations (Kawada et al., 2023; Smith et al., 2018). For example, studies of Indian *Ae. aegypti* populations reveal that the highly resistant triple-*kdr* haplotype is currently absent, consistent with continued operational efficacy of pyrethroids in these regions. Together, these findings highlight that resistance severity in *Ae. aegypti* is shaped not only by the presence of multiple *kdr* mutations, but also by their haplotype structure, recombination dynamics, and genomic context. While functional studies define the upper limits of resistance that VGSC mutations can confer, population-level processes determine how and where such extreme resistance phenotypes emerge and spread in the field. Recent large-scale population genomic analyses further demonstrate the global distribution and ongoing gene flow of insecticide resistance–associated variants across

Ae. aegypti populations, highlighting the dynamic and interconnected nature of resistance evolution worldwide (Crawford et al., 2025).

In contrast to *Ae. aegypti*, where multiple VGSC substitutions contribute to complex *kdr* haplotypes, VGSC-mediated resistance in *Anopheles* mosquitoes is largely dominated by substitutions at a single locus, codon 1014, most commonly L1014F and L1014S, with L1014H reported less frequently (Davies et al., 2007). These alleles together define the classical *kdr* mechanism in *Anopheles* and account for the majority of VGSC-mediated resistance observed across species and geographic regions. Although rare alternative VGSC haplotypes have recently been identified, codon 1014 remains the central and most widely distributed target-site resistance locus in this genus. Importantly, the functional significance of this locus has been definitively established through genome-editing approaches. Functional studies have demonstrated that the L1014F substitution alone can confer substantial resistance to pyrethroids and DDT in vitro, even in the absence of additional VGSC mutations (Burton et al., 2011). This causal role has been directly confirmed using CRISPR/Cas9 genome editing in *An. gambiae* (Grigoraki et al., 2021), where targeted introduction of L1014F into an otherwise fully susceptible genetic background resulted in elevated resistance, reduced mortality following exposure to pyrethroid-treated bed nets, and further enhancement of resistance when combined with over-expression of detoxification enzymes, while also revealing fitness costs that likely constrain fixation in natural populations (Grigoraki et al., 2021).

Although codon 1014 remains the primary VGSC resistance locus in *Anopheles*, recent work has identified an alternative VGSC haplotype that also confers pyrethroid resistance. A linked V402L–I1527T haplotype, mutually exclusive with classical L1014F/S alleles, has been detected at high frequencies in *Anopheles coluzzii* populations from West Africa. CRISPR/Cas9-based functional validation demonstrated that V402L alone confers resistance to pyrethroids and DDT, albeit at lower levels than L1014F and without detectable fitness costs under laboratory conditions, indicating distinct resistance–fitness trade-offs among VGSC alleles (Williams et al., 2022). Together, these genome-editing studies establish that, in contrast to *Ae. aegypti*, where high-level resistance typically arises from the co-occurrence of multiple VGSC mutations, VGSC-mediated resistance in *Anopheles* is characterized by a limited number of alternative haplotypes, with resistance intensity in field populations commonly shaped by interactions between *kdr* mutations and metabolic resistance mechanisms. These genome-editing studies establish CRISPR/Cas9 as a powerful and precise tool for resolving the causal roles, effect sizes, and fitness consequences of VGSC target-site mutations in mosquito vectors (Feng et al., 2021).

Similar to *Anopheles* mosquitoes, VGSC-mediated resistance in *Culex* mosquitoes has historically been considered largely dominated by substitutions at codon 1014. The L1014F mutation is the most widely reported and consistently associated with resistance to pyrethroids and DDT across members of the *Cx. pipiens* complex and *Cx. quinquefasciatus*, while L1014S occurs less frequently and L1014H has been reported only sporadically. Compared with *Ae. aegypti*, where multiple distinct VGSC substitutions frequently co-occur and contribute to high-level resistance, target-site resistance in *Culex* has traditionally been viewed as more genetically constrained. However, more detailed analyses of the sodium channel gene in *Cx. quinquefasciatus* have revealed a more complex picture. Comprehensive sequencing identified multiple nonsynonymous and synonymous mutations distributed across the VGSC coding region, the frequencies of which increased markedly following permethrin selection and were strongly correlated with resistance intensity (Li et al., 2012; Xu et al., 2012). Importantly, specific combinations of co-occurring mutations, including the simultaneous presence of multiple substitutions within individual mosquitoes, were repeatedly associated with high levels of resistance across independent field and laboratory-selected populations. These findings indicate that, although codon 1014 remains a central determinant, the accumulation of multi-site VGSC mutations under selection pressure is a common feature of permethrin-resistant *Culex* populations.

Despite these advances, functional validation of VGSC mutations in *Culex* remains comparatively limited relative to *Anopheles* and *Aedes*, largely due to historical constraints in molecular and genome-editing tools. In field populations, high levels of resistance are frequently observed in mosquitoes carrying both VGSC mutations and elevated metabolic detoxification capacity, suggesting that combined target-site and metabolic mechanisms contribute to resistance intensity (e.g., Liu, 2015, Mitchell et al., 2014). Available evidence indicates that while VGSC-mediated resistance in *Culex* may be less extensively characterized than in *Aedes*, it nonetheless involves coordinated mutation accumulation (Li et al., 2012, Xu et al., 2012) and interaction with metabolic mechanisms under sustained insecticide selection pressure. Across mosquito genera, VGSC-mediated target-site resistance provides a critical foundation for resistance to pyrethroids and DDT, but rarely operates in isolation under field conditions (Hemingway et al., 2004;

Ranson and Lissenden, 2016). The magnitude and operational significance of resistance are shaped by species-specific genetic architectures, ranging from complex, multi-mutation VGSC haplotypes in *Aedes*, to a limited set of dominant loci in *Anopheles* and *Culex*, as well as by interactions with metabolic detoxification mechanisms. Functional validation and field observations together underscore that target-site mutations define baseline susceptibility thresholds, while additional resistance mechanisms modulate resistance intensity and control outcomes. Notably, *kdr*-mediated resistance is not uniformly distributed across mosquito species; for example, it has historically been rare or absent in *Anopheles funestus*, where resistance is primarily driven by metabolic mechanisms, although recent studies suggest that target-site mutations may be emerging in some populations (Mulamba et al., 2014; Irving and Wondji, 2017). Collectively, these patterns highlight both shared principles and distinct genus-specific trajectories in the evolution of VGSC-mediated resistance across mosquito vectors.

2.2.2 Acetylcholinesterase

Alterations in acetylcholinesterase (AChE), the molecular target of organophosphate and carbamate insecticides, represent a well-characterized but relatively limited form of target-site resistance in mosquitoes. Organophosphates and carbamates exert toxicity by inhibiting AChE, leading to the accumulation of acetylcholine at synapses and neuromuscular junctions and resulting in sustained neuronal excitation and paralysis (Weill et al., 2004). Resistance to these compounds arises primarily through amino-acid substitutions within the active site or surrounding regions of the enzyme that reduce insecticide binding affinity while preserving sufficient catalytic activity for neuronal function.

The most extensively documented and functionally validated mutation conferring AChE-based resistance is G119S in the *ace-1* gene. This substitution, which replaces glycine with serine at position 119, is located within the oxyanion hole, a key structural element involved in substrate stabilization. Structural and biochemical studies have demonstrated that G119S markedly reduces the binding affinity of organophosphates and carbamates to AChE, thereby conferring high levels of resistance and cross-resistance to both insecticide classes (Weill et al., 2003, 2004). The G119S mutation has been repeatedly identified in resistant populations of *Cx. pipiens*, *An. gambiae*, and *An. albimanus*, and is sufficient on its own to produce strong target-site insensitivity. Despite its effectiveness, the G119S substitution imposes substantial fitness costs, primarily due to reduced enzymatic efficiency, which can limit its persistence in the absence of insecticide pressure. In several mosquito species, this constraint has been partially alleviated through duplication of the *ace-1* gene, resulting in the coexistence of resistant and susceptible AChE copies within the same genome. This duplicated architecture allows mosquitoes to retain resistance while maintaining adequate cholinergic function, representing a distinct evolutionary solution compared with VGSC-mediated resistance (Djogbénou et al., 2009, Labbé et al., 2007). In addition, increased copy number variances of the resistant allele can further enhance resistance by increasing allele dosage, thereby contributing both to fitness compensation and elevated resistance levels (Grau-Bové et al., 2021).

Population genetic studies further indicate that both the *ace-1* G119S resistance allele and associated gene duplications can spread via introgression among *Anopheles* populations, facilitating the regional dissemination and long-term maintenance of acetylcholinesterase-based resistance despite its inherent fitness costs (Djogbenou et al., 2008). In contrast to voltage-gated sodium channel–mediated resistance, which is characterized by numerous mutations and frequent combinatorial haplotypes conferring graded resistance phenotypes, acetylcholinesterase-based resistance in mosquitoes is largely driven by a single substitution, G119S. This striking difference likely reflects the strong structural and functional constraints acting on AChE, where most amino-acid changes severely compromise enzymatic activity and impose prohibitive fitness costs. Although AChE-mediated resistance is functionally constrained relative to VGSC-mediated resistance, its prevalence in field populations likely reflects a combination of factors, including fitness costs, genetic architecture, and differences in insecticide use and selection pressure. Notably, *ace-1* resistance alleles can reach high frequencies in populations under strong selection, indicating that functional constraints do not necessarily limit their spread.

2.2.3 GABA-gated chloride channels

The γ-aminobutyric acid (GABA)-gated chloride channel represents another established neuronal target site for insecticide action and resistance in mosquitoes. Cyclodiene insecticides and phenylpyrazoles, most notably dieldrin and fipronil, exert their toxic effects by antagonizing GABA receptor function, thereby disrupting inhibitory neurotransmission and causing uncontrolled neuronal excitation (Ffrench-Constant and Roush, 1991; Ffrench-Constant et al., 1993). Target-site resistance at the GABA receptor is primarily mediated by amino-acid substitutions in the receptor subunit encoded by the *Rdl* (resistance to dieldrin) gene. The most widely documented and functionally validated mutation is A296G (or A296S), which reduces insecticide binding and confers high levels of resistance to cyclodienes, as well as cross-resistance to other GABA-targeting compounds (Ffrench-Constant et al., 1993). This substitution has been identified across multiple mosquito genera, including *Anopheles*, *Aedes*, and *Culex*, and represents a classic example of convergent evolution under insecticide selection pressure. Similar to acetylcholinesterase-based resistance, the mutational landscape of GABA receptor resistance in mosquitoes appears to be highly constrained, with few viable substitutions capable of conferring resistance without severe fitness penalties. Consequently, unlike VGSC-mediated resistance, which involves numerous mutations and combinatorial haplotypes, *Rdl*-based resistance is dominated by a single recurrent substitution.

Despite its clear mechanistic basis, the contemporary operational importance of GABA receptor–mediated resistance in mosquitoes is more limited than that of VGSC or metabolic mechanisms. The restricted use of cyclodienes due to environmental persistence and regulatory withdrawal has reduced selective pressure on *Rdl* variants in many regions (Xie et al., 2024). Nevertheless, the continued use of phenylpyrazoles and related compounds in specific control contexts underscores the ongoing relevance of GABA receptor mutations for resistance surveillance and management. Together with alterations in VGSCs and acetylcholinesterase, mutations in GABA-gated chloride channels illustrate the diversity of molecular strategies by which mosquitoes reduce insecticide sensitivity at neuronal target sites, complementing metabolic detoxification pathways in shaping resistance phenotypes.

3. Regulatory and evolutionary integration of insecticide resistance in mosquitoes

The preceding sections have described the principal molecular mechanisms underlying insecticide resistance in mosquitoes. In natural populations, however, these mechanisms rarely function independently. Instead, resistance phenotypes represent emergent adaptive outcomes arising from coordinated, multilayered interactions among metabolic detoxification systems, target-site modifications, and the regulatory architectures that govern gene expression and physiological responsiveness (Liu, 2015; Al Naggar et al., 2025).

Evidence from both field and laboratory studies across mosquito genera demonstrates that high-intensity resistance consistently reflects the co-occurrence of multiple mechanisms within shared genetic backgrounds (Tognarelli et al., 2025; Xu et al., 2017; Yewhalaw et al., 2011). In *Anopheles*, elevated pyrethroid resistance often involves the combination of VGSC target-site mutations (e.g., L1014F) with increased expression of detoxification enzymes, including P450s as well as carboxylesterases (Al Naggar et al., 2025; Oumbouke et al., 2020). Transcriptomic analyses have identified overexpression of multiple detoxification-related genes in resistant field populations, however, direct associations between specific genes and resistance to individual insecticides are not always clearly resolved (Oumbouke et al., 2020). In *Ae. aegypti*, high resistance intensity is likewise associated with the presence of complex VGSC haplotypes together with increased detoxification gene expression or copy numbers at detoxification loci (Brown et al., 2025; Al Naggar et al., 2025; Saavedra-Rodriguez et al., 2018; Wang et al., 2023). Comparable multilayered resistance architectures have also been documented in *Culex* mosquitoes (Chamnanya et al., 2025; Xu et al., 2005; Yang and Liu, 2011; Yang et al., 2021b). Collectively, these patterns indicate that extreme resistance phenotypes emerge not from single mutational events but from integrated, systems-level adaptive architectures shaped by sustained insecticide selection. Target-site mutations may establish baseline insensitivity, while metabolic and regulatory mechanisms amplify resistance magnitude, broaden chemical tolerance, and mitigate physiological costs (Scott, 2017; Walsh et al., 2022), however, the relative contributions of these mechanisms vary among species, with resistance in *An. funestus* being predominantly driven by metabolic processes. Such multilayered configurations enhance robustness and may facilitate the recurrent evolution of high-level resistance across taxa. Elucidating how these components interact, and how their expression is coordinated across tissues, developmental stages, and ecological contexts, is therefore essential for explaining resistance intensity and predicting evolutionary trajectories under persistent selection.

At the molecular scale, coordinated resistance phenotypes are increasingly understood as products of regulatory network evolution that modulate both the expression and integration of multiple resistance mechanisms, particularly detoxification pathways (Liu, 2015). Genomic (Hearn et al., 2022; Lucas et al., 2019), transcriptomic (Reid et al., 2012, 2014), and functional genetic (Adolfi et al., 2019, Li and Liu, 2017, 2018, 2019) studies demonstrate that resistance evolution is shaped not only by structural variation in resistance genes, but also by regulatory processes influencing resistance magnitude, tissue specificity, inducibility, and developmental expression;. Consistent with this view, overexpression of resistance-associated cytochrome P450 genes frequently reflects dynamic transcriptional responses to insecticide exposure and environmental stress. Emerging evidence further implicates conserved signaling pathways, including G protein–coupled receptor–linked cascades, second-messenger signaling, redox-responsive pathways involving reactive oxygen species, and downstream transcriptional regulators, in the coordinated and context-dependent control of detoxification gene expression and resistance phenotypes (Averill-Bates, 2024; Bottino-Rojas et al., 2015; Li and Liu, 2017, 2018; Liu et al., 2021a, 2021b).

3.1 GPCR-mediated signaling and cAMP–PKA regulation of metabolic resistance

Accumulating evidence indicates that regulatory signaling pathways, rather than structural variation in detoxification enzymes alone, play an important role in shaping metabolic insecticide resistance in mosquitoes. Among these, G protein–coupled receptor (GPCR)–mediated signaling has emerged as a key upstream regulatory mechanism capable of modulating cytochrome P450 gene expression and detoxification capacity, with the most direct experimental support coming from studies in *Culex* mosquitoes (Li and Liu, 2017, 2018, 2019; Liu et al., 2007, 2011). GPCRs constitute a large and diverse class of sensory and regulatory receptors that transduce extracellular cues, including physiological stress signals and endogenous ligands, into intracellular second-messenger cascades, providing a mechanistic framework through which environmental exposures can influence gene expression programs (Liu et al., 2021a, 2021b).

In *Cx. quinquefasciatus*, early systems-level transcriptomic analyses comparing resistant and susceptible strains revealed coordinated upregulation of genes involved in both detoxification and signal transduction, including GPCRs, suggesting that resistance involves regulatory interactions among multiple genes rather than isolated enzymatic changes (Liu et al., 2007). Subsequent tissue-specific expression studies identified a resistance-associated GPCR preferentially overexpressed in Malpighian tubules, implicating GPCR signaling in post-prandial diuresis and xenobiotic excretion as contributors to resistance (Liu et al., 2011). Functional validation using RNA interference demonstrated that suppression of resistance-associated GPCR genes resulted in reduced expression of multiple P450 genes and increased insecticide susceptibility, providing direct evidence that GPCRs act upstream of metabolic detoxification pathways in *Culex* mosquitoes (Li and Liu, 2017, 2018, 2019; Li et al., 2015).

Further dissection of this regulatory architecture revealed that downstream components of the GPCR signaling cascade, including the G-protein αs subunit, adenylyl cyclase, cyclic AMP (cAMP), and protein kinase A (PKA), are required for maintenance of resistance phenotypes. RNAi-mediated knockdown of these intracellular effectors led to decreased P450 expression and significantly enhanced insecticide sensitivity in both resistant laboratory strains and field populations, while complementary in vitro experiments showed that activation of this pathway elevated cAMP levels, induced P450 expression, and increased tolerance to insecticides (Li and Liu, 2017, 2018, 2019). Together, these findings support a GPCR–Gαs–adenylyl cyclase–cAMP–PKA signaling cascade that transcriptionally regulates detoxification enzymes and amplifies metabolic resistance in *Culex* mosquitoes (Fig. 4). Notably, many of the cytochrome P450 genes regulated downstream of this GPCR–cAMP–PKA pathway have been independently validated as resistance determinants through complementary in vivo approaches, including RNA interference–mediated knockdown in mosquitoes and transgenic *Drosophila melanogaster* expression systems, in which suppression or ectopic expression of individual P450s significantly alters insecticide susceptibility (Li et al., 2015; Yang et al., 2021b).

While the GPCR–cAMP–PKA pathway represents a trans-acting regulatory mechanism that modulates detoxification gene expression through signaling cascades, resistance evolution can also involve cis-regulatory variation within promoter regions of detoxification genes. In *Cx. quinquefasciatus*, a regulatory region upstream of *CYP9M10*, a partial CuRE1 (*Culex* repetitive element 1) transposable element, was shown to drive markedly elevated expression relative to susceptible alleles, with reporter assays demonstrating enhanced promoter activity and field studies revealing strong associations between this allele and permethrin resistance (Wilding et al., 2012). Although a transposable element CuRE1 initially appeared to contribute to regulatory function, subsequent functional analyses indicated that it likely serves as a marker for an adjacent cis-regulatory motif under positive selection. These findings highlight the importance of promoter evolution and regulatory sequence variation in shaping resistance phenotypes in natural mosquito populations. Notably, *CYP9M10* has also been identified as a downstream target of GPCR–cAMP–PKA signaling in *Culex*, indicating that resistance-associated P450 expression can be shaped simultaneously by

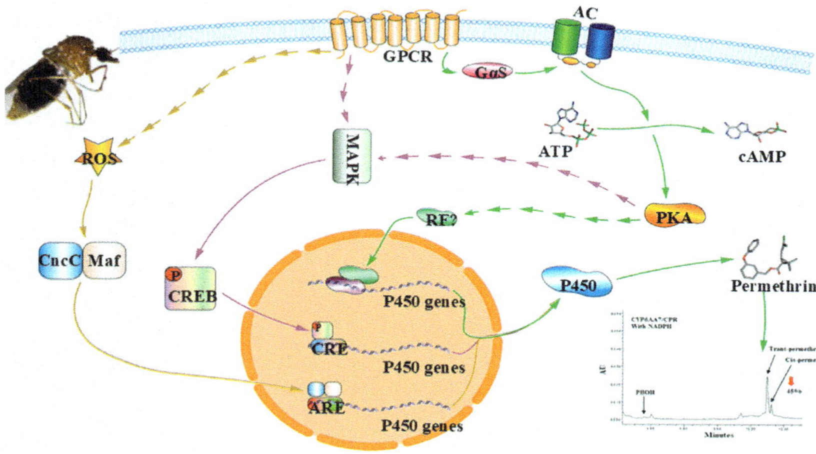

Fig. 4 Conceptual model of regulatory signaling pathways modulating cytochrome P450–mediated metabolic insecticide resistance in mosquitoes. The GPCR–Gαs–adenylyl cyclase–cAMP–protein kinase A (PKA) signaling cascade has been functionally validated in *Culex* mosquitoes and shown to transcriptionally regulate resistance-associated P450 genes, thereby amplifying detoxification capacity and resistance phenotypes (Section 4.1). Activation of GPCR signaling stimulates cAMP production and PKA activity, leading to phosphorylation of downstream transcriptional regulators and increased expression of detoxification enzymes. Parallel stress-responsive branches, including reactive oxygen species (ROS) dependent on redox signaling and mitogen-activated protein kinase (MAPK) pathways, are depicted as conserved regulatory modules. Components of redox-responsive regulation (e.g., CncC/Maf) have functional support in mosquitoes, whereas MAPK-dependent transcriptional regulation has been more extensively characterized in other insect systems. The precise mechanistic integration of these pathways in mosquito insecticide resistance remains incompletely resolved. ROS signaling is proposed to activate redox-sensitive transcription factors such as CncC/Maf, while MAPK cascades may converge on regulators including CREB, providing parallel and potentially synergistic routes for inducible detoxification gene expression. These signaling modules are further proposed to converge at the level of transcriptional and epigenetic control of P450 gene networks, enabling context-dependent modulation of resistance intensity across tissues, developmental stages, and environmental conditions. Solid arrows indicate experimentally validated regulatory relationships in mosquitoes; dashed arrows denote conserved or hypothesized interactions supported primarily by evidence from other insect systems and partial characterization in mosquitoes.

trans-acting signaling cascades and cis-regulatory promoter variation. This convergence highlights the multilayered regulatory architecture underlying metabolic resistance evolution.

Although comparable GPCR-centered regulatory pathways have not yet been mechanistically resolved in *Anopheles* or *Aedes*, coordinated overexpression of detoxification genes and inducible metabolic responses suggest the presence of upstream regulatory processes. Whether such regulation involves GPCR-mediated signaling similar to that described in other systems remains to be determined. Such signaling-based regulationcoud provide mosquitoes with the flexibility to modulate resistance intensity across tissues and developmental stages, enabling adaptive responses to fluctuating insecticide exposure while minimizing fitness costs. Determining whether conserved GPCR-linked signaling pathways contirbute to metabolic resistance across mosquito taxa remains an important direction for future research.

3.2 ROS- and redox-mediated stress signaling in insecticide resistance

In addition to GPCR–cAMP–PKA–mediated signaling, oxidative stress and reactive oxygen species (ROS)–responsive pathways (Averill-Bates, 2024) have been associated with changes in detoxification gene expression and physiological responses in mosquitoes. Insecticide exposure and detoxification processes can perturb redox homeostasis, and resistant mosquitoes often exhibit altered antioxidant capacity and oxidative stress profiles relative to susceptible counterparts, consistent with a close coupling between stress physiology and resistance expression (Oliver and Brooke, 2016). Rather than acting solely as damaging byproducts, ROS function as signaling molecules that activate conserved stress-response pathways and reprogram transcription (Averill-Bates, 2024). A key redox-responsive regulatory axis in insects is the Nrf2/CncC pathway and its partner Maf, which together coordinate antioxidant defenses and detoxification gene expression (Ingham et al., 2017; Misra et al., 2011; Zhang et al., 2016).

These redox-responsive processes are incorporated into the conceptual framework shown in Fig. 4, which illustrates how ROS-mediated signaling may act in parallel with GPCR-driven pathways to modulate detoxification of gene expression in mosquitoes. In mosquitoes, functional evidence supports regulatory involvement of components of this pathway in detoxification gene expression: Maf-S has been shown to regulate detoxification gene expression in *An. gambiae*, and the Nrf2 ortholog in *Ae. aegypti* plays roles in redox biology and detoxification-related physiology (Ingham et al., 2017). In addition to redox-responsive Nrf2/CncC signaling, other stress-activated kinase cascades may contribute to regulation of detoxification gene expression. In the whitefly *Bemisia tabaci*, MAPK–CREB signaling has been implicated in regulating cytochrome P450 expression in response to xenobiotic stress (Yang et al., 2020).

Although their precise contributions to mosquito insecticide resistance remain incompletely resolved, such conserved stress-responsive regulatory architectures provide plausible mechanisms by which signaling pathways may modulate resistance intensity across tissues and developmental stages.

Although the mechanistic linkage between GPCR activation and ROS generation has not yet been directly demonstrated in mosquitoes, analogous regulatory interactions are well characterized in other eukaryotic systems. In mammalian systems, GPCR stimulation can activate NADPH oxidases to generate reactive oxygen species (ROS), and these ROS act as second messengers that modulate kinase cascades and redox-sensitive transcription factors, including MAPKs and Nrf2 pathways (Finkel, 2011; Lassegue and Griendling, 2010; Petry and Gorlach, 2019; Schieber and Chandel, 2014; Ushio-Fukai, 2009). This conserved interplay between receptor-mediated signaling and redox regulation provides a plausible mechanistic basis for crosstalk between GPCR and ROS pathways in insects. Accordingly, the dashed connection between GPCR signaling and ROS modules in Fig. 4 reflects a conserved but as yet unvalidated regulatory linkage in mosquito insecticide resistance.

At the population level, redox-responsive regulation provides a plausible mechanism for the plasticity and context dependence of metabolic resistance observed in the field, including variation in resistance intensity across ecological conditions and physiological states. Redox-mediated regulation of detoxification gene expression has been functionally characterized in multiple non-mosquito insect species, and oxidative stress–linked resistance phenotypes have been repeatedly documented in mosquito vectors, providing independent evidence that ROS-mediated signaling can influence cytochrome P450 expression and stress adaptation across insects (Ingham et al., 2017; Misra et al., 2011; Oliver and Brooke, 2016). However, a recent critical synthesis highlights that although overactivation of the CncC/Keap1 pathway is frequently associated with elevated detoxification enzyme expression, definitive causal links between this pathway and insecticide resistance remain unresolved across arthropods, underscoring the need for functional genomic validation in resistant populations (Amezian and Van Leeuwen, 2025).

4. Future directions and perspectives

Together, these findings indicate that metabolic resistance reflects multilayered regulatory evolution operating across promoters, transcription factor, and post-transcriptional -level control. Resolving how these regulatory layers interact with upstream signaling pathways will be essential for defining the network architectures that shape adaptive resistance phenotypes in mosquito populations. Despite substantial advances in identifying resistance-associated genes and pathways, major knowledge gaps remain in understanding how regulatory networks coordinate insecticide resistance across molecular, physiological, and ecological scales. Although transcriptional and signaling pathways controlling detoxification enzymes, particularly cytochrome P450s, have been increasingly implicated, the upstream regulatory architecture, including the identity of regulatory variants, network connectivity, and context-dependent modulation, remains incompletely resolved. Emerging functional genomics tools, especially CRISPR/Cas-based genome editing (Xu et al., 2025) and targeted regulatory perturbation, now provide unprecedented opportunities to move beyond correlative expression analyses and directly test the causal roles of regulatory elements, signaling components, and transcriptional regulators in adaptive resistance phenotypes.

Future work integrating genome editing with population genomics, epigenomics, and systems-level modeling will be essential for dissecting how regulatory variation shapes resistance magnitude, fitness trade-offs, and evolutionary trajectories in natural mosquito populations. A central challenge will be distinguishing stable, heritable regulatory adaptations that contribute to long-term evolutionary change from inducible or reversible stress responses that reflect short-term physiological plasticity. Clarifying this distinction will be critical for predicting resistance persistence, anticipating the evolutionary durability of resistance alleles, and designing intervention strategies that minimize selection for stable, high-level resistance. By shifting emphasis from individual resistance genes to the regulatory networks that govern their expression and evolutionary dynamics, the field can move toward a more predictive and mechanistically grounded framework for adaptive resistance evolution, one that better informs sustainable, evolution-aware vector control strategies.

5. Conclusion

In summary, insecticide resistance in mosquitoes is best understood not as the consequence of isolated mutations or individual detoxification genes, but as a multilayered adaptive phenotype involving structural variation, gene expression changes, and environmental selection. As one of the most rapid and recurrent examples of contemporary adaptive evolution in natural populations, resistance illustrates how strong anthropogenic selection can repeatedly reshape genomic and regulatory architectures within constrained molecular systems. The recurrent emergence of similar resistance configurations across mosquito taxa reflects both the predictability of intense insecticide-driven selection and the limited set of viable molecular solutions available within neuronal targets and detoxification pathways. Yet the intensity of resistance is influenced by the combined effects of gene identity and expression, whereas the durability and geographic spread of resistance are shaped by factors such as fitness trade-offs, ecological selection, and population dynamics. These multilayered interactions shape not only resistance magnitude, but also its evolutionary stability and persistence in heterogeneous ecological landscapes. Moving forward, a predictive framework for resistance evolution will require resolving how cis- and trans-regulatory variation, signaling networks, redox-responsive pathways, and broader gene regulatory mechanisms collectively structure adaptive trajectories in natural mosquito populations. By integrating mechanistic insect physiology with evolutionary genomics, systems biology, and functional validation, the field is poised to transition from retrospective documentation of resistance to anticipatory, evolution-informed management strategies capable of extending the operational lifespan of vector control interventions.

Acknowledgments

Authors thank the support from College of Agriculture, Auburn University.

Data availability

No new data was generated or analyzed in support of this research.

Funding

The project was supported by the following awards: AAES Hatch/Multistate Grants ALA015–1−16009, ALA015–1−19148 to NL.

CRediT authorship contribution statement

NL: Conceptualization, Writing – Original Draft, Writing – Review and Editing, Visualization (figure preparation).

Declaration of competing interest

The authors declare that they have no known competing financial interests or personal relationships that could have appeared to influence the work reported in this paper.

References

Achee, N.L., Grieco, J.P., Vatandoost, H., Seixas, G., Pinto, J., Ching-Ng, L., et al., 2019. Alternative strategies for mosquito-borne arbovirus control. PLoS Negl. Trop. Dis. 13, e0006822. https://doi.org/10.1371/journal.pntd.0006822.

Adolfi, A., Poulton, B., Anthousi, A., Macilwee, S., Ranson, H., Lycett, G.J., 2019. Functional genetic validation of key genes conferring insecticide resistance in the major African malaria vector, *Anopheles gambiae*. Proc. Natl. Acad. Sci. U. S. A. 116, 25764–25772. https://doi.org/10.1073/pnas.1914633116.

Al Naggar, Y., Fahmy, N.M., Alkhaibari, A.M., Al-Akeel, R.K., Alharbi, H.M., Mohamed, A., et al., 2025. Mechanisms and genetic drivers of resistance of insect pests to insecticides and approaches to its control. Toxics 13, 681. https://doi.org/10.3390/toxics13080681.

Al-Yazeedi, T., Muhammad, A., Irving, H., Ahn, S.J., Hearn, J., Wondji, C.S., 2024. Overexpression and nonsynonymous mutations of UDP-glycosyltransferases are potentially associated with pyrethroid resistance in *Anopheles funestus*. Genomics 116, 110798. https://doi.org/10.1016/j.ygeno.2024.110798.

Amenya, D.A., Naguran, R., Lo, T.-C.M., Ranson, H., Spillings, B.L., Wood, O.R., et al., 2008. Overexpression of a cytochrome P450 (CYP6P9) in a major African malaria vector, *Anopheles funestus*, resistant to pyrethroids. Insect Mol. Biol. 17, 19–25. https://doi.org/10.1111/j.1365-2583.2008.00766.x.

Amezian, D., Van Leeuwen, T., 2025. Is the CncC/Keap1 complex a major factor in conferring pesticide resistance in arthropods? A critical review. Insect Biochem. Mol. Biol. 184, 104400. https://doi.org/10.1016/j.ibmb.2025.104400.

Averill-Bates, D., 2024. Reactive oxygen species and cell signaling. Review. Biochim. Biophys. Acta. Mol. Cell Res. 1871, 119573. https://doi.org/10.1016/j.bbamcr.2023.119573.

Bacot, T., Haberkorn, C., Guilliet, J., Cattel, J., Kefi, M., Nadalin, L., et al., 2024. A genomic duplication spanning multiple P450s contributes to insecticide resistance in the dengue mosquito *Aedes aegypti*. Peer Community J. 4, e110. https://doi.org/10.24072/pcjournal.497.

Balaska, S., Grigoraki, L., Lycett, G., Weetman, D., Oladepo, F., Colman, F., et al., 2025. Predictive chemoproteomics and functional validation reveal Coeae6g-mediated insecticide cross-resistance in the malaria vector *Anopheles gambiae*. Nat. Commun. 16, 10772. https://doi.org/10.1038/s41467-025-65827-4.

Bhatt, S., Gething, P.W., Brady, O.J., Messina, J.P., Farlow, A.W., Moyes, C.L., et al., 2015. The global distribution and burden of dengue. Nature 496, 504–507. https://doi.org/10.1038/nature12060.

Bottino-Rojas, V., Talyuli, O.A.C., Jupatanakul, N., Sim, S., Dimopoulos, G., Venancio, T.M., et al., 2015. Heme signaling impacts global gene expression, immunity and dengue virus infectivity in *Aedes aegypti*. PLoS ONE 10, e0135985. https://doi.org/10.1371/journal.pone.0135985.

Brown, D., Dub, M., Manganaris, J., Liu, N., 2025. Challenges in insecticide resistance management: implications of alternating selection in the yellow fever mosquito, *Aedes aegypti*. Pestic. Biochem. Physiol. 217, 106832. https://doi.org/10.1016/j.pestbp.2025.106832.

Brown, D., Houck, R., Liu, N., 2026. Integrated transcriptomic and variant analysis reveals molecular mechanisms of pyrethroid resistance in a genetically homogenized Cas9 strain of Aedes aegypti. Arch. Toxicol. 100, 2543–2558. https://doi.org/10.1007/s00204-026-04326-x.

Burton, M.J., Mellor, I.R., Duce, I.R., Davies, T.G.E., Field, L.M., Williamson, M.S., 2011. Differential resistance of insect sodium channels with *kdr* mutations to deltamethrin, permethrin and DDT. Insect Biochem. Mol. Biol. 41, 723–732. https://doi.org/10.1016/j.ibmb.2011.05.004.

Cano, J., Rebollo, M.P., Golding, N., Pullan, R.L., Crellen, T., Soler, A., et al., 2014. The global distribution and transmission limits of lymphatic filariasis: past and present. Parasites Vectors 7, 466. https://doi.org/10.1186/s13071-014-0466-x.

Chamnanya, S., Kiddela, B., Saingamsook, J., Nachaiwieng, W., Lumjuan, N., Somboon, P., et al., 2025. Overexpression of multiple cytochrome P450 genes with and without knockdown resistance mutations confers high resistance to deltamethrin in *Culex quinquefasciatus*. Infect. Dis. Poverty 14, 2. https://doi.org/10.1186/s40249-024-01269-2.

Chouaïbou, M., Zivanovic, G.B., Knox, T.B., Jamet, H.P., Bonfoh, B., 2014. Synergist bioassays: a simple method for initial metabolic resistance investigation of field *Anopheles gambiae* s.l. populations. Acta Trop 130, 108–111.

Crawford, J.E., Balcazar, D., Redmond, S., Rose, N.H., Youd, H.A., Lucas, E.R., et al., 2025. 1206 genomes reveal origin and movement of *Aedes aegypti* driving increased dengue risk. Science 389, eads3732. https://doi.org/10.1126/science.ads3732.

Cui, F., Weill, M., Berthomieu, A., Raymond, M., Qiao, C.-L., 2007. Characterization of novel esterases in insecticide-resistant mosquitoes. Insect Biochem. Mol. Biol. 37, 1131–1137. https://doi.org/10.1016/j.ibmb.2007.07.002.

Davies, T.G.E., Field, L.M., Usherwood, P.N.R., Williamson, M.S., 2007. A comparative study of voltage-gated sodium channels in the Insecta: implications for pyrethroid resistance in Anopheline and other Neopteran species. Insect Mol. Biol. 16, 361–375. https://doi.org/10.1111/j.1365-2583.2007.00733.x.

Devonshire, A.L., Field, L.M., 1991. Gene amplification and insecticide resistance. Annu. Rev. Entomol. 36, 1–23. https://doi.org/10.1146/annurev.en.36.010191.000245.

Djiappi-Tchamen, B., Nana-Ndjangwo, M.S., Mavridis, K., Talipouo, A., Nchoutpouen, E., Makoudjou, I., et al., 2021. Analyses of insecticide resistance genes in *Aedes aegypti* and *Aedes albopictus* mosquito populations from Cameroon. Genes. (Basel) 12, 828. https://doi.org/10.3390/genes12060828.

Djogbenou, L., Chandre, F., Berthomieu, A., Dabire, R., Koffi, A., Alout, H., et al., 2008. Evidence of introgression of the ace-1R mutation and of the *ace-1* duplication in West African *Anopheles gambiae* s.s. PLoS ONE 3, e2172. https://doi.org/10.1371/journal.pone.0002172.

Djogbenou, L., Weill, M., Hougard, J.-M., Raymond, M., Akogbeto, M., Chandre, F., 2007. Characterization of insensitive acetylcholinesterase (*ace-1R*) in *Anopheles gambiae* (Diptera: Culicidae): resistance levels and dominance. J. Med. Entomol. 44, 805–810. https://doi.org/10.1093/jmedent/44.5.805.

Djogbénou, L., Labbé, P., Chandre, F., Pasteur, N., Weill, M., 2009. Ace-1 duplication in *Anopheles gambiae*: a challenge for malaria control. Malar. J. 8, 70. https://doi.org/10.1186/1475-2875-8-70.

Djoko Tagne, C.S., Kouamo, M.F.M., Tchouakui, M., Muhammad, A., Mugenzi, L.J.L., Tatchou-Nebangwa, N.M.T., et al., 2025. A single mutation G454A in the P450 *CYP9K1* drives pyrethroid resistance in the major malaria vector *Anopheles funestus* reducing bed net efficacy. Genetics 229, iyae181. https://doi.org/10.1093/genetics/iyae181.

Dong, K., Du, Y., Rinkevich, F., Nomura, Y., Xu, P., Wang, L., et al., 2014. Molecular biology of insect sodium channels and pyrethroid resistance. Insect Biochem. Mol. Biol. 50, 1–17. https://doi.org/10.1016/j.ibmb.2014.03.012.

Du, Y., Nomura, Y., Satar, G., Hu, Z., Nauen, R., He, S.Y., et al., 2013. Molecular evidence for dual pyrethroid-receptor sites on a mosquito sodium channel. Proc. Natl. Acad. Sci. U. S. A 110, 11785–11790. https://doi.org/10.1073/pnas.1305118110.

Du, Y., Nomura, Y., Zhorov, B.S., Dong, K., 2016. Sodium channel mutations and pyrethroid resistance in *Aedes aegypti*. Insects 7, 60. https://doi.org/10.3390/insects7040060.

Du, J., Yin, H., Li, J., Zhang, W., Ding, G., Zhou, D., et al., 2024. Transcription factor B-H2 regulates *CYP9J34* expression conveying deltamethrin resistance in *Culex pipiens pallens*. Pest. Manag. Sci. 80, 1991–2000. https://doi.org/10.1002/ps.7934.

Enayati, A.A., Ranson, H., Hemingway, J., 2005. Insect glutathione S-transferases and insecticide resistance. Insect Mol. Biol. 14, 3–8. https://doi.org/10.1111/j.1365-2583.2004.00529.x.

Enayati, A., Valadan, R., Bagherzadeh, M., Cheraghpour, M., Nikookar, S.H., Fazeli-Dinan, M., et al., 2024. *Kdr* genotyping and the first report of V410L and V1016I *kdr* mutations in voltage-gated sodium channel gene in *Aedes aegypti* (Diptera: Culicidae) from Iran. Parasit. Vectors 17, 34. https://doi.org/10.1186/s13071-024-06123-w.

Essandoh, J., Yawson, A.E., Weetman, D., 2013. Acetylcholinesterase (*Ace-1*) target site mutation 119S is strongly diagnostic of carbamate and organophosphate resistance in *Anopheles gambiae s.s.* and *Anopheles coluzzii* across southern Ghana. Malar. J. 12, 404. https://doi.org/10.1186/1475-2875-12-404.

Fan, Y., Scott, J.G., 2020. The F1534C *voltage-sensitive sodium channel* mutation confers 7- to 16-fold resistance to pyrethroid insecticides in *Aedes aegypti*. Pest. Manag. Sci. 76, 2251–2259. https://doi.org/10.1002/ps.5763.

Faucon, F., Dusfour, I., Gaude, T., Navratil, V., Boyer, F., Chandre, F., et al., 2015. Identifying genomic changes associated with insecticide resistance in the dengue mosquito Aedes aegypti by deep targeted sequencing. Genome Res. 25, 1347–1359. https://doi.org/10.1101/gr.189225.115.

Feng, X., López Del Amo, V., Mameli, E., Lee, M., Bishop, A.L., Perrimon, N., et al., 2021. Optimized CRISPR tools and site-directed transgenesis towards gene drive development in *Culex quinquefasciatus* mosquitoes. Nat. Commun. 12, 2960. https://doi.org/10.1038/s41467-021-23239-0.

Feyereisen, R., 2005. Insect cytochrome P450. In: Gilbert, L., Iatrou, K., Gill, S. (Eds.), Comprehensive Molecular Insect Science. vol. 4. Elsevier, Oxford, pp. 1–77.

Ffrench-Constant, R.H., Rocheleau, T.A., Steichen, J.C., Chalmers, A.E., 1993. A point mutation in a *Drosophila* GABA receptor confers insecticide resistance. Nature 363, 449–451. https://doi.org/10.1038/363449a0.

Ffrench-Constant, R.H., Roush, R.T., 1991. Gene mapping and cross-resistance in cyclodiene insecticide-resistant *Drosophila melanogaster* (Mg.). Genet. Res. 57, 17–21. https://doi.org/10.1017/S0016672300028986.

Finkel, T., 2011. Signal transduction by reactive oxygen species. J. Cell Biol. 194, 7–15. https://doi.org/10.1083/jcb.201102095.

Fouet, C., Pinch, M.J., Ashu, F.A., Ambadiang, M.M., Bouaka, C., Batronie, A.J., et al., 2024. Field-evolved resistance to neonicotinoids in the mosquito *Anopheles gambiae*, is associated with downregulation and mutations of nicotinic acetylcholine receptor subunits combined with cytochrome P450-mediated detoxification. Pestic. Biochem. Physiol. 206, 106205. https://doi.org/10.1016/j.pestbp.2024.106205.

Gao, H., Hu, W., Cui, C., Wang, Y., Zheng, Y., Jacobs-Lorena, M., et al., 2025. Emerging challenges for mosquito-borne disease control and the promise of symbiont-based transmission-blocking strategies. PLoS Pathog. 21, e1013431. https://doi.org/10.1371/journal.ppat.1013431.

Gong, M.Q., Gu, Y., Hu, X.B., Sun, Y., Ma, L., Li, X.L., et al., 2005. Cloning and overexpression of *CYP6F1*, a cytochrome P450 gene, from deltamethrin-resistant *Culex pipiens pallens*. Acta Biochim. Biophys. Sin. (Shanghai). 37, 317–326. https://doi.org/10.1111/j.1745-7270.2005.00042.x.

Gong, Y., Li, T., Feng, Y., Liu, N., 2017. The function of two P450s, CYP9M10 and CYP6AA7, in the permethrin resistance of *Culex quinquefasciatus*. Sci. Rep. 7, 587. https://doi.org/10.1038/s41598-017-00486-0.

Gong, Y., Li, T., Li, Q., Liu, S., Liu, N., 2022a. The central role of multiple P450 genes and their co-factor CPR in the development of permethrin resistance in the mosquito *Culex quinquefasciatus*. Front. Physiol. 12, 802584. https://doi.org/10.3389/fphys.2021.802584.

Gong, Y., Li, M., Li, T., Liu, N., 2022b. Molecular and functional characterization of three novel carboxylesterases in the detoxification of permethrin in the mosquito, *Culex quinquefasciatus*. Insect Sci. 29, 199–214. https://doi.org/10.1111/1744-7917.12927.

Grau-Bové, X., Lucas, E., Pipini, D., Rippon, E., van 't Hof, A.E., Constant, E., et al., 2021. Resistance to pirimiphos-methyl in West African *Anopheles* is spreading via duplication and introgression of the Ace1 locus. PLoS Genet. 17, e1009253. https://doi.org/10.1371/journal.pgen.1009253.

Grigoraki, L., Cowlishaw, R., Nolan, T., Donnelly, M., Lycett, G., Ranson, H., 2021. CRISPR/Cas9-modified Anopheles gambiae carrying *kdr* mutation L1014F functionally validate its contribution to insecticide resistance and combined effect with metabolic enzymes. PLoS Genet. 17, e1009556. https://doi.org/10.1371/journal.pgen.1009556.

Gubler, D.J., 1998. Resurgent vector-borne diseases as a global health problem. Emerg. Infect. Dis. 4, 442–450. https://doi.org/10.3201/eid0403.980326.

Haddi, K., Tomé, H.V.V., Du, Y., Valbon, W.R., Nomura, Y., Martins, G.F., et al., 2017. Detection of a new pyrethroid resistance mutation (V410L) in the sodium channel of Aedes aegypti: a potential challenge for mosquito control. Sci. Rep. 7, 46549. https://doi.org/10.1038/srep46549.

Hancock, P.A., Ochomo, E., Messenger, L.A., 2024. Genetic surveillance of insecticide resistance in African *Anopheles* populations to inform malaria vector control. Trends Parasitol. 40, 604–618. https://doi.org/10.1016/j.pt.2024.04.016.

Hearn, J., Djoko Tagne, C.S., Ibrahim, S.S., Tene-Fossog, B., Mugenzi, L.M.J., Irving, H., et al., 2022. Multi-omics analysis identifies a *CYP9K1* haplotype conferring pyrethroid resistance in the malaria vector *Anopheles funestus* in East Africa. Mol. Ecol. 31, 3642–3657. https://doi.org/10.1111/mec.16497.

Hemingway, J., Hawkes, N.J., McCarroll, L., Ranson, H., 2004. The molecular basis of insecticide resistance in mosquitoes. Insect Biochem. Mol. Biol. 34, 653–665. https://doi.org/10.1016/j.ibmb.2004.03.018.

Hemingway, J., Hawkes, N., Prapanthadara, L., Jayawardenal, K.G., Ranson, H., 1998. The role of gene splicing, gene amplification and regulation in mosquito insecticide resistance. Philos. Trans. R. Soc. Lond. B Biol. Sci. 353, 1695–1699. https://doi.org/10.1098/rstb.1998.0320.

Hirata, K., Komagata, O., Itokawa, K., Yamamoto, A., Tomita, T., Kasai, S., 2014. A single crossing-over event in voltage-sensitive Na+ channel genes may cause critical failure of dengue mosquito control by insecticides. PLoS Negl. Trop. Dis. 8, e3085. https://doi.org/10.1371/journal.pntd.0003085.

Hopkins, D.H., Fraser, N.J., Mabbitt, P.D., Carr, P.D., Oakeshott, J.G., Jackson, C.J., 2017. Structure of an insecticide-sequestering carboxylesterase from the disease vector *Culex quinquefasciatus*: what makes an enzyme a good insecticide sponge? Biochemistry 56, 5512–5525. https://doi.org/10.1021/acs.biochem.7b00774.

Huang, X., Kaufman, P.E., Athrey, G.N., Fredregill, C., Alvarez, C., Shetty, V., et al., 2023. Potential key genes involved in metabolic resistance to malathion in the southern house mosquito, *Culex quinquefasciatus*, and functional validation of *CYP325BC1* and *CYP9M12* as candidate genes using RNA interference. BMC Genomics 24, 160. https://doi.org/10.1186/s12864-023-09241-4.

Huang, Q., Zhou, Y., Xu, J., Zhou, G., Guo, Q., Duan, Y., et al., 2025. Mechanisms of pyrethroid resistance in Culicidae mosquitoes from Hainan Island, China. Parasit. Vectors 18, 411. https://doi.org/10.1186/s13071-025-07052-y.

Ibrahim, S.S., Ndula, M., Riveron, J.M., Irving, H., Wondji, C.S., 2016. The P450 *CYP6Z1* confers carbamate/pyrethroid cross-resistance in a major African malaria vector beside a novel carbamate-insensitive N485I *acetylcholinesterase-1* mutation. Mol. Ecol. 25, 3436–3452. https://doi.org/10.1111/mec.13673.

Ingham, V.A., Pignatelli, P., Moore, J.D., Wagstaff, S., Ranson, H., 2017. The transcription factor Maf-S regulates metabolic resistance to insecticides in the malaria vector *Anopheles gambiae*. BMC Genomics 18, 669. https://doi.org/10.1186/s12864-017-4086-7.

Irving, H., Wondji, C.S., 2017. Investigating knockdown resistance *(kdr)* mechanism against pyrethroids/DDT in the malaria vector *Anopheles funestus* across Africa. BMC Genet. 18, 76. https://doi.org/10.1186/s12863-017-0539-x.

Itokawa, K., Komagata, O., Kasai, S., Ogawa, K., Tomita, T., 2016. Testing the causality between *CYP9M10* and pyrethroid resistance using TALEN and CRISPR/Cas9 technologies. Sci. Rep. 6, 24652. https://doi.org/10.1038/srep24652.

Kasai, S., Weerashinghe, I.S., Shono, T., 1998. P450 monooxygenases are an important mechanism of permethrin resistance in *Culex quinquefasciatus* Say larvae. Arch. Insect Biochem. Physiol. 37, 47–56. https://doi.org/10.1002/(SICI)1520-6327(1998)37:13.0.CO;2-S.

Kaur, T., Kushwah, R.S., Pradhan, S., Das, M.K., Kona, M.P., Anushrita, Mittal, et al., 2025. Knockdown-resistance (*kdr*) mutations in Indian *Aedes aegypti* populations: lack of recombination among haplotypes bearing V1016G, F1534C, and F1534L *kdr* alleles. PLoS Negl. Trop. Dis. 19, e0013126. https://doi.org/10.1371/journal.pntd.0013126.

Kawada, H., Higa, Y., Kasai, S., 2023. Reconsideration of importance of the point mutation L982W in the voltage-sensitive sodium channel of the pyrethroid-resistant *Aedes aegypti* (L.) (Diptera: Culicidae) in Vietnam. PLoS ONE 18, e0285883. https://doi.org/10.1371/journal.pone.0285883.

Ketterman, A.J., Jayawardena, K.G., Hemingway, J., 1992. Purification and characterization of a carboxylesterase involved in insecticide resistance from the mosquito *Culex quinquefasciatus*. Biochem. J. 287, 355–360. https://doi.org/10.1042/bj2870355.

Komagata, O., Kasai, S., Tomita, T., 2010. Overexpression of cytochrome P450 genes in pyrethroid-resistant *Culex quinquefasciatus*. Insect Biochem. Mol. Biol. 40, 146–152. https://doi.org/10.1016/j.ibmb.2010.01.006.

Kouamo, M.F.M., Ibrahim, S.S., Muhammad, A., Gadji, M., Hearn, J., Wondji, C.S., 2025. Allelic variation in a cluster of epsilon glutathione S-transferase genes contributes to DDT and pyrethroid resistance in the major African malaria vector *Anopheles funestus*. BMC Genomics 26, 452. https://doi.org/10.1186/s12864-025-11637-3.

Labbé, P., Berthomieu, A., Berticat, C., Alout, H., Raymond, M., Lenormand, T., et al., 2007. Independent duplications of the acetylcholinesterase gene conferring insecticide resistance in the mosquito *Culex pipiens*. Mol. Biol. Evol. 24, 1056–1067. https://doi.org/10.1093/molbev/msm025.

Lassegue, B., Griendling, K.K., 2010. NADPH oxidases: functions and pathologies in the vasculature. Arterioscler. Thromb. Vasc. Biol. 30, 653–661. https://doi.org/10.1161/ATVBAHA.108.181610.

Li, T., Cao, C., Yang, T., Zhang, L., He, L., Xi, Z., et al., 2015. A G-protein-coupled receptor regulation pathway in cytochrome P450-mediated permethrin resistance in mosquitoes, Culex quinquefasciatus. Sci. Rep. 5, 17772. https://doi.org/10.1038/srep17772.

Li, T., Liu, N., 2010. Inheritance of permethrin resistance in Culex quinquefasciatus J. Med. Entomol 47 2010 1127 1134 10.1603/me10142Li, T., Liu, N., 2010. Inheritance of permethrin resistance in Culex quinquefasciatus. J. Med. Entomol. 47, 1127–1134. https://doi.org/10.1603/me10142.

Li, T., Liu, N., 2017. Regulation of P450-mediated permethrin resistance in *Culex quinquefasciatus* by the GPCR/Gαs/AC/cAMP/PKA signaling cascade. Biochem. Biophys. Rep. 12, 12–19. https://doi.org/10.1016/j.bbrep.2017.08.010.

Li, T., Liu, N., 2018. The function of G-protein-coupled receptor regulatory cascade in southern house mosquitoes (Diptera: Culicidae). J. Med. Entomol. 55, 1–9. https://doi.org/10.1093/jme/tjy022.

Li, T., Liu, N., 2019. Role of the G-protein-coupled receptor signaling pathway in insecticide resistance. Int. J. Mol. Sci. 20, 4300. https://doi.org/10.3390/ijms20174300.

Li, X., Schuler, M.A., Berenbaum, M.R., 2007. Molecular mechanisms of metabolic resistance to synthetic and natural xenobiotics. Annu. Rev. Entomol. 52, 231–253. https://doi.org/10.1146/annurev.ento.51.110104.151104.

Li, T., Wang, Y., Liu, N., 2023. Synergism study for investigating possible mechanisms of insecticide resistance in mosquitoes. Cold Spring Harb. Protoc, pdb.prot108042. https://doi.org/10.1101/pdb.prot108042.

Li, T., Zhang, L., Reid, W.R., Xu, Q., Dong, K., Liu, N., 2012. Multiple mutations and mutation combinations in the sodium channel of permethrin-resistant mosquitoes, *Culex quinquefasciatus*. Sci. Rep. 2, 781. https://doi.org/10.1038/srep00781.

Liu, N., 2015. Insecticide resistance in mosquitoes: impact, mechanisms, and research directions. Annu. Rev. Entomol. 60, 537–559. https://doi.org/10.1146/annurev-ento-010814-020828.

Liu, N., Brown, D.J., 2026. Insecticide resistance in the house fly (Musca domestica): metabolic detoxification, key genes, and regulatory characterization. Curr. Opin. Insect Sci. 76, 101510. https://doi.org/10.1016/j.cois.2026.101510.

Liu, N., Li, M., Gong, Y., Liu, F., Li, T., 2015. Cytochrome P450s – their expression, regulation, and role in insecticide resistance. Pestic. Biochem. Physiol. 120, 77–81. https://doi.org/10.1016/j.pestbp.2015.01.006.

Liu, N., Li, T., Reid, W.R., Yang, T., Zhang, L., 2011. Multiple cytochrome P450 genes: their constitutive overexpression and permethrin induction in insecticide-resistant mosquitoes, *Culex quinquefasciatus*. PLoS ONE 6, e23403. https://doi.org/10.1371/journal.pone.0023403.

Liu, N., Li, T., Wang, Y., Liu, S., 2021a. G-protein coupled receptors (GPCRs) in insects-a potential target for new insecticide development. Molecules 26, 2993. https://doi.org/10.3390/molecules26102993.

Liu, N., Liu, H., Zhu, F., Zhang, L., 2007. Differential expression of genes in pyrethroid-resistant and susceptible mosquitoes, *Culex quinquefasciatus*. Gene 394, 61–68. https://doi.org/10.1016/j.gene.2007.01.032.

Liu, N., Liu, H., Zhu, F., Zhang, L., 2007. Differential expression of genes in pyrethroid resistant and susceptible mosquitoes, Culex quinquefasciatus (S.). Gene 394, 61–68. https://doi.org/10.1016/j.gene.2007.01.032.

Liu, N., Wang, Y., Li, T., Feng, X., 2021b. G-protein coupled receptors (GPCRs): signaling pathways, characterization, and functions in insect physiology and toxicology. Int. J. Mol. Sci. 22, 5260. https://doi.org/10.3390/ijms22105260.

Logan, R.A.E., Mäurer, J.B., Wapler, C., Ingham, V.A., 2024. Uridine diphosphate (UDP)-glycosyltransferases (UGTs) are associated with insecticide resistance in the major malaria vectors *Anopheles gambiae s.l.* and *Anopheles funestus*. Sci. Rep. 14, 19821. https://doi.org/10.1038/s41598-024-70713-y.

Lu, K., Cheng, Y., Li, W., Li, Y., Zeng, R., Song, Y., 2020. Activation of CncC pathway by ROS burst regulates cytochrome P450 CYP6AB12 responsible for λ-cyhalothrin tolerance in *Spodoptera litura*. J. Hazard. Mater. 387, 121698. https://doi.org/10.1016/j.jhazmat.2019.121698.

Lucas, E.R., Miles, A., Harding, N.J., Clarkson, C.S., Lawniczak, M.K.N., Kwiatkowski, D.P., et al., 2019. Whole-genome sequencing reveals high complexity of copy number variation at insecticide resistance loci in malaria mosquitoes. Genome Res. 29, 1250–1261. https://doi.org/10.1101/gr.245795.118.

Lucas, E.R., Nagi, S.C., Egyir-Yawson, A., Essandoh, J., Dadzie, S., Chabi, J., et al., 2023. Genome-wide association studies reveal novel loci associated with pyrethroid and organophosphate resistance in *Anopheles gambiae* and *Anopheles coluzzii*. Nat. Commun. 14, 4946. https://doi.org/10.1038/s41467-023-40693-0.

Matowo, J., Weetman, D., Pignatelli, P., Wright, A., Charlwood, J.D., Kaaya, R., et al., 2022. Expression of pyrethroid metabolizing P450 enzymes characterizes highly resistant *Anopheles* vector species targeted by successful deployment of PBO-treated bednets in Tanzania. PLoS ONE 17, e0249440. https://doi.org/10.1371/journal.pone.0249440.

Misra, J.R., Horner, M.A., Lam, G., Thummel, C.S., 2011. Transcriptional regulation of xenobiotic detoxification in *Drosophila*. Genes. Dev. 25, 1796–1806. https://doi.org/10.1101/gad.17280911.

Mitchell, S.N., Rigden, D.J., Dowd, A.J., Lu, F., Wilding, C.S., Weetman, D., et al., 2014. Metabolic and target-site mechanisms combine to confer strong DDT resistance in *Anopheles gambiae*. PLoS ONE 9, e92662. https://doi.org/10.1371/journal.pone.0092662.

Moyes, C.L., Vontas, J., Martins, A.J., Ng, L.C., Koou, S.Y., Dusfour, I., et al., 2017. Contemporary status of insecticide resistance in the major *Aedes* vectors of arboviruses infecting humans. PLoS Negl. Trop. Dis. 11, e0005625. https://doi.org/10.1371/journal.pntd.0005625.

Mulamba, C., Riveron, J.M., Ibrahim, S.S., Irving, H., Barnes, K.G., Mukwaya, L.G., et al., 2014. Widespread pyrethroid and DDT resistance in the major malaria vector *Anopheles funestus* in East Africa is driven by metabolic resistance mechanisms. PLOS ONE 9, e110058. https://doi.org/10.1371/journal.pone.0110058.

Nagi, S.C., Ingham, V.A., 2025. A multi-omic meta-analysis reveals novel mechanisms of insecticide resistance in malaria vectors. Commun. Biol. 8, 790. https://doi.org/10.1038/s42003-025-08221-6.

Nardini, L., Hunt, R.H., Dahan-Moss, Y.L., Christie, N., Christian, R.N., Coetzee, M., et al., 2017. Malaria vectors in the democratic republic of the Congo: the mechanisms that confer insecticide resistance in *Anopheles gambiae* and *Anopheles funestus*. Malar. J. 16, 448. https://doi.org/10.1186/s12936-017-2099-y.

Nauen, R., Bass, C., Feyereisen, R., Vontas, J., 2022. The role of cytochrome P450s in insect toxicology and resistance. Annu. Rev. Entomol. 67, 105–124. https://doi.org/10.1146/annurev-ento-070621-061328.

Naw, H., Vo, T.C., Le, H.G., Kang, J.-M., Mya, Y.Y., Myint, M.K., et al., 2022. Knockdown resistance mutations in the voltage-gated sodium channel of *Aedes aegypti* (Diptera: Culicidae) in Myanmar. Insects 13, 322. https://doi.org/10.3390/insects13040322.

Njoroge, H., Vant Hof, A., Oruni, A., Pipini, D., Nagi, S.C., Lynd, A., et al., 2022. Identification of a rapidly spreading triple mutant for high-level metabolic insecticide resistance in *Anopheles gambiae* provides a real-time molecular diagnostic for antimalarial intervention deployment. Mol. Ecol. 31, 4307–4318. https://doi.org/10.1111/mec.16591.

Oliver, S.V., Brooke, B.D., 2016. The role of oxidative stress in the longevity and insecticide resistance phenotype of the major malaria vectors *Anopheles arabiensis* and *Anopheles funestus*. PLoS ONE 11, e0151049. https://doi.org/10.1371/journal.pone.0151049.

Onen, H., Luzala, M.M., Kigozi, S., Sikumbili, R.M., Muanga, C.-J.K., Zola, E.N., et al., 2023. Mosquito-borne diseases and their control strategies: an overview focused on green synthesized plant-based metallic nanoparticles. Insects 14, 221. https://doi.org/10.3390/insects14030221.

Oumbouke, W.A., Pignatelli, P., Barreaux, A.M.G., Tia, I.Z., Koffi, A.A., Alou, L.P.A., et al., 2020. Fine-scale spatial investigation of multiple insecticide resistance and underlying target-site and metabolic mechanisms in *Anopheles gambiae* in central Cote d'Ivoire. Sci. Rep. 10, 15066. https://doi.org/10.1038/s41598-020-71933-8.

Owusu-Asenso, C.M., Abdulai, A., Sraku, I.K., Attah, S.K., Aboagye-Antwi, F., Afrane, Y.A., 2026. Comparative analysis reveals concentration-dependent and site-specific detoxification gene expression patterns in pyrethroid-resistant Anopheles gambiae s.l. from Accra, Ghana. Sci. Rep. 16, 3717. https://doi.org/10.1038/s41598-025-33720-1.

Pelloquin, B., Agossa, F., Acford-Palmer, H., Clark, T., Ogoma, S.B., Williams, M., et al., 2025. Novel insecticide resistance mutations associated with variable PBO synergy in *Anopheles gambiae s.l.* from the Democratic Republic of Congo. Sci. Rep. 15, 27618. https://doi.org/10.1038/s41598-025-09016-9.

Petry, A., Gorlach, A., 2019. Regulation of NADPH oxidases by G protein-coupled receptors. Antioxid. Redox Signal. 30, 74–94. https://doi.org/10.1089/ars.2018.7525.

Poulton, B.C., Colman, F., Anthousi, A., Sattelle, D.B., Lycett, G.J., 2024. *Aedes aegypti* CCEae3A carboxylesterase expression confers carbamate, organophosphate and limited pyrethroid resistance in a model transgenic mosquito. PLoS Negl. Trop. Dis. 18, e0011595. https://doi.org/10.1371/journal.pntd.0011595.

Ranson, H., Hemingway, J., 2005. Mosquito glutathione transferases. Methods Enzymol 401, 226–241. https://doi.org/10.1016/S0076-6879(05)01014-1.

Ranson, H., Lissenden, N., 2016. Insecticide resistance in African *Anopheles mosquitoes*: a worsening situation that needs urgent action to maintain malaria control. Trends Parasitol 32, 187–196. https://doi.org/10.1016/j.pt.2015.11.010.

Raymond, M., Chevillon, C., Guillemaud, T., Lenormand, T., Pasteur, N., 1998. An overview of the evolution of overproduced esterases in the mosquito *Culex pipiens*. Philos. Trans. R. Soc. Lond. B Biol. Sci. 353, 1707–1711. https://doi.org/10.1098/rstb.1998.0322.

Reid, W.R., Thornton, A., Pridgeon, J.W., Becnel, J.J., Tang, F., Estep, A., et al., 2014. Transcriptional analysis of four family 4 P450s in a Puerto Rico strain of *Aedes aegypti* (Diptera: Culicidae) compared with an Orlando strain and their possible functional roles in permethrin resistance. J. Med. Entomol. 51, 605–615. https://doi.org/10.1603/me13228.

Reid, W.R., Zhang, L., Gong, Y., Li, T., Liu, N., 2018. Gene expression profiles of the Southern house mosquito *Culex quinquefasciatus* during exposure to permethrin. Insect Sci. 25, 439–453. https://doi.org/10.1111/1744-7917.12438.

Reid, W.R., Zhang, L., Liu, F., Liu, N., 2012. The transcriptome profile of the mosquito *Culex quinquefasciatus* following permethrin selection. PLoS ONE 7, e47163. https://doi.org/10.1371/journal.pone.0047163.

Riveron, J.M., Yunta, C., Ibrahim, S.S., Djouaka, R., Irving, H., Menze, B.D., et al., 2014. A single mutation in the *GSTe2* gene allows tracking of metabolically based insecticide resistance in a major malaria vector. Genome Biol. 15, R27. https://doi.org/10.1186/gb-2014-15-2-r27.

Rongnoparut, P., Boonsuepsakul, S., Chareonviriyaphap, T., Thanomsing, N., 2003. Cloning of cytochrome P450 *CYP6P5* and *CYP6AA2* from *Anopheles minimus* resistant to deltamethrin. J. Vector Ecol. 28, 150–158.

Saavedra-Rodriguez, K., Maloof, F.V., Campbell, C.L., Garcia-Rejon, J., Lenhart, A., Penilla, P., et al., 2018. Parallel evolution of *vgsc* mutations at domains IS6, IIS6 and IIIS6 in pyrethroid resistant *Aedes aegypti* from Mexico. Sci. Rep. 8, 6747. https://doi.org/10.1038/s41598-018-25222-0.

Saha, A., Das, P., Das, S., Saha, D., 2025. Overexpression of cytochrome P450 and esterase genes in pyrethroid-resistant wild *Culex quinquefasciatus* populations from sub-Himalayan West Bengal, India. Parasitol. Res. 124, 68. https://doi.org/10.1007/s00436-025-08487-z.

Schieber, M., Chandel, N.S., 2014. ROS function in redox signaling and oxidative stress. Curr. Biol. 24, R453–R462. https://doi.org/10.1016/j.cub.2014.03.034.

Scott, J.G., 2017. Evolution of resistance to pyrethroid insecticides in *Musca domestica*. Pest. Manag. Sci. 73, 716–720. https://doi.org/10.1002/ps.4328.

Scott, J.G., 2019. Life and death at the voltage-sensitive sodium channel: evolution in response to insecticide use. Annu. Rev. Entomol. 64, 243–257. https://doi.org/10.1146/annurev-ento-011118-112420.

Scott, J.G., 2025. Insect UDP-glycosyltransferases and xenobiotic metabolism. Pestic. Biochem. Physiol. 216, 106731. https://doi.org/10.1016/j.pestbp.2025.106731.

Sene, N.M., Mavridis, K., Ndiaye, E.H., Diagne, C.T., Gaye, A., Ngom, E.H.M., et al., 2021. Insecticide resistance status and mechanisms in *Aedes aegypti* populations from Senegal. PLoS Negl. Trop. Dis. 15, e0009393. https://doi.org/10.1371/journal.pntd.0009393.

Silver, K.S., Du, Y., Nomura, Y., Oliveira, E.E., Salgado, V.L., Zhorov, B.S., et al., 2014. Voltage-gated sodium channels as insecticide targets. Adv. Insect Physiol. 46, 389–433. https://doi.org/10.1016/B978-0-12-417010-0.00005-7.

Smith, L.B., Kasai, S., Scott, J.G., 2018. Voltage-sensitive sodium channel mutations S989P + V1016G in Aedes aegypti confer variable resistance to pyrethroids, DDT and oxadiazines. Pest. Manag. Sci. 74, 737–745. https://doi.org/10.1002/ps.4771.

Soderlund, D.M., 2012. Molecular mechanisms of pyrethroid insecticide neurotoxicity: recent advances. Arch. Toxicol. 86, 165–181. https://doi.org/10.1007/s00204-011-0726-x.

Stevenson, B.J., Pignatelli, P., Nikou, D., Paine, M.J., 2012. Pinpointing P450s associated with pyrethroid metabolism in the dengue vector, *Aedes aegypti*: developing new tools to combat insecticide resistance. PLoS Negl. Trop. Dis. 6, e1595. https://doi.org/10.1371/journal.pntd.0001595.

Tao, F., Si, F.L., Hong, R., He, X., Li, X.Y., Qiao, L., et al., 2022. Glutathione S-transferase (GST) genes and their function associated with pyrethroid resistance in the malaria vector *Anopheles sinensis*. Pest. Manag. Sci. 78, 4127–4139. https://doi.org/10.1002/ps.7031.

Tekoh, T.A., Mugenzi, L.M.J., Menze, B., Tchapga, W., Wondji, M., Tchouakui, M., et al., 2025. The duplicated cytochrome P450 CYP6P9a/b confers cross-resistance to a mitochondrial complex I inhibitor in the African malaria vector *Anopheles funestus*. BMC Genomics 26, 837. https://doi.org/10.1186/s12864-025-11984-1.

Tognarelli, J., Moya, P.R., González, C.R., Collao-Ferrada, X., 2025. Global distribution and impact of knockdown resistance mutations in *Aedes aegypti* on pyrethroid resistance. Parasit. Vectors 18, 382. https://doi.org/10.1186/s13071-025-06817-9.

Uemura, N., Itokawa, K., Komagata, O., Kasai, S., 2024. Recent advances in the study of knockdown resistance mutations in *Aedes* mosquitoes with a focus on several remarkable mutations. Curr. Opin. Insect Sci. 63, 101178. https://doi.org/10.1016/j.cois.2024.101178.

Ushio-Fukai, M., 2009. Vascular signaling through G protein-coupled receptors: new concepts. Curr. Opin. Nephrol. Hypertens 18, 153–159. https://doi.org/10.1097/MNH.0b013e3283252efe.

Vaughan, A., Hawkes, N., Hemingway, J., 1997. Co-amplification explains linkage disequilibrium of two mosquito esterase genes in insecticide-resistant *Culex quinquefasciatus*. Biochem. J. 325, 359–365. https://doi.org/10.1042/bj3250359.

Venkatesan, P., 2024. Global upsurge in dengue in. Lancet Infect. Dis. 24, e620. https://doi.org/10.1016/S1473-3099(24)00609-1.

Vontas, J., Grigoraki, L., Morgan, J., Tsakireli, D., Fuseini, G., Segura, L., et al., 2018. Rapid selection of a pyrethroid metabolic enzyme CYP9K1 by operational malaria control activities. Proc. Natl. Acad. Sci. U. S. A. 115, 4619–4624. https://doi.org/10.1073/pnas.1719663115.

Vontas, J., Katsavou, E., Mavridis, K., 2020. Cytochrome P450-based metabolic insecticide resistance in Anopheles and Aedes mosquito vectors: muddying the waters. Pestic. Biochem. Physiol. 170, 104666. https://doi.org/10.1016/j.pestbp.2020.104666.

Vulule, J.M., Beach, R.F., Atieli, F.K., McAllister, J.C., Brogdon, W.G., Roberts, J.M., et al., 1999. Elevated oxidase and esterase levels associated with permethrin tolerance in *Anopheles gambiae* from Kenyan villages using permethrin-impregnated nets. Med. Vet. Entomol. 13, 239–244. https://doi.org/10.1046/j.1365-2915.1999.00177.x.

Walsh, T.K., Heckel, D.G., Wu, Y., Downes, S., Gordon, K.H.J., Oakeshott, J.G., 2022. Determinants of insecticide resistance evolution: comparative analysis among *Heliothines*. Annu. Rev. Entomol. 67, 387–406. https://doi.org/10.1146/annurev-ento-080421-071655.

Wang, Y., Wang, X., Brown, D.J., An, M., Xue, R.D., Liu, N., 2023. Insecticide resistance: status and potential mechanisms in *Aedes aegypti*. Pestic. Biochem. Physiol. 195, 105577. https://doi.org/10.1016/j.pestbp.2023.105577.

Weetman, D., Djogbenou, L., Lucas, E., 2018. Copy number variation (CNV) and insecticide resistance in mosquitoes: evolving knowledge or an evolving problem? Curr. Opin. Insect Sci. 27, 82–88. https://doi.org/10.1016/j.cois.2018.04.005.

Wei, Y., Gu, X., Si, F., Chen, X., Qiao, L., Yan, H., et al., 2025. Transcriptome and functional analyses reveal carboxylesterase genes involved in pyrethroid resistance in *Anopheles sinensis* (Diptera: Culicidae). Insects 16, 938. https://doi.org/10.3390/insects16090938.

Weill, M., Lutfalla, G., Mogensen, K., Chandre, F., Berthomieu, A., Berticat, C., et al., 2003. Comparative genomics: insecticide resistance in mosquito vectors. Nature 423, 136–137. https://doi.org/10.1038/423136b.

Weill, M., Malcolm, C., Chandre, F., Mogensen, K., Berthomieu, A., Marquine, M., et al., 2004. The unique mutation in *ace-1* giving high insecticide resistance is easily detectable in mosquito vectors. Insect Mol. Biol. 13, 1–7. https://doi.org/10.1111/j.1365-2583.2004.00452.x.

Wilding, C.S., Smith, I., Lynd, A., Yawson, A.E., Weetman, D., Paine, M.J.I., et al., 2012. A *cis*-regulatory sequence driving metabolic insecticide resistance in mosquitoes: functional characterisation and signatures of selection. Insect Biochem. Mol. Biol. 42, 699–707. https://doi.org/10.1016/j.ibmb.2012.06.003.

Wilding, C.S., Weetman, D., Rippon, E.J., Steen, K., Mawejje, H.D., Barsukov, I., et al., 2015. Parallel evolution or purifying selection, not introgression, explains similarity in the pyrethroid detoxification-linked $GSTE_4$ of *Anopheles gambiae* and *Anopheles arabiensis*. Mol. Genet. Genomics 290, 201–215. https://doi.org/10.1007/s00438-014-0910-9.

Williams, J., Cowlishaw, R., Sanou, A., Ranson, H., Grigoraki, L., 2022. *In vivo* functional validation of the V402L voltage-gated sodium channel mutation in the malaria vector *Anopheles gambiae*. Pest. Manag. Sci. 78, 1155–1163. https://doi.org/10.1002/ps.6731.

World Health Organization, 2017. Global Vector Control Response 2017–2030World Health Organization, Geneva. https://www.who.int/publications/i/item/9789241512978.

World Health Organization, 2021. Global Malaria Programme – Insecticide ResistanceWorld Health Organization, Geneva. https://www.who.int/teams/global-malaria-programme/prevention/vector-control/insecticide-resistance.

World Health Organization, 2025. Dengue: global situation, surveillance and progress – 2024 update. Wkly. Epidemiol. Rec. 100. pp. 665–678. https://www.who.int/publications/i/item/who-wer10052-665-678.

World Health Organization, 2024. Vector-borne diseases. https://www.who.int/news-room/fact-sheets/detail/vector-borne-diseases.

Xie, N., Bickley, B.A., Gross, A.D., 2024. GABA-gated chloride channel mutation (*Rdl*) induces cholinergic physiological compensation resulting in cross resistance in *Drosophila melanogaster.* Pestic. Biochem. Physiol. 203, 105972. https://doi.org/10.1016/j.pestbp.2024.105972.

Xu, Y., Du, J., Zhang, K., Li, J., Zou, F., Li, X., et al., 2024. The dual resistance mechanism of *CYP325G4* and *CYP6AA9* in *Culex pipiens pallens* legs according to transcriptome and proteome analysis. J. Agric. Food Chem. 72, 27150–27162. https://doi.org/10.1021/acs.jafc.4c05708.

Xu, Q., Liu, H., Zhang, L., Liu, N., 2005. Resistance in the mosquito *Culex quinquefasciatus*, and possible mechanisms for resistance. Pest. Manag. Sci. 61, 1096–1102. https://doi.org/10.1002/ps.1090.

Xu, W., Liu, S., Zhang, Y., Gao, J., Yang, M., Liu, X., et al., 2017. Cypermethrin resistance conferred by increased target insensitivity and metabolic detoxification in *Culex pipiens pallens* Coq. Pestic. Biochem. Physiol. 142, 77–82. https://doi.org/10.1016/j.pestbp.2017.01.008.

Xu, Q., Wang, M., Zeng, J., Sun, H., Wei, X., Jiang, H., et al., 2025. CRISPR/Cas technology in insect insecticide resistance. Insects 16, 345. https://doi.org/10.3390/insects16040345.

Xu, Q., Wang, H., Zhang, L., Liu, N., 2006. *Kdr* allelic variation in pyrethroid-resistant mosquitoes, *Culex quinquefasciatus* (S). Biochem. Biophys. Res. Commun. 345, 774–780. https://doi.org/10.1016/j.bbrc.2006.04.155.

Xu, Q., Zhang, L., Li, T., Zhang, L., He, L., Dong, K., et al., 2012. Evolutionary adaptation of the amino acid and codon usage of the mosquito sodium channel following insecticide selection in field mosquitoes. PLoS One 7, e47609. https://doi.org/10.1371/journal.pone.0047609.

Yang, X., Deng, S., Wei, X., Yang, J., Zhao, Q., Yin, C., et al., 2020. MAPK-directed activation of the whitefly transcription factor *CREB* leads to P450-mediated imidacloprid resistance. Proc. Natl. Acad. Sci. U. S. A. 117, 10246–10253. https://doi.org/10.1073/pnas.1913603117.

Yang, T., Li, T., Feng, X., Li, M., Liu, S., Liu, N., 2021b. Multiple cytochrome P450 genes: conferring high levels of permethrin resistance in mosquitoes, *Culex quinquefasciatus*. Sci. Rep. 11, 9041. https://doi.org/10.1038/s41598-021-88121-x.

Yang, T., Liu, N., 2011. Genome analysis of cytochrome P450s and their expression profiles in insecticide resistant mosquitoes, *Culex quinquefasciatus*. PLoS One 6, e29418. https://doi.org/10.1371/journal.pone.0029418.

Yang, X., Wei, X., Yang, J., Du, T., Yin, C., Fu, B., et al., 2021a. Epitranscriptomic regulation of insecticide resistance. Sci. Adv. 7, eabe5903. https://doi.org/10.1126/sciadv.abe5903.

Yewhalaw, D., Wassie, F., Steurbaut, W., Spanoghe, P., Van Bortel, W., Denis, L., et al., 2011. Multiple insecticide resistance: an impediment to insecticide-based malaria vector control program. PLoS ONE 2011, e16066. https://doi.org/10.1371/journal.pone.0016066.

Yunta, C., Grisales, N., Nász, S., Hemmings, K., Pignatelli, P., Voice, M., et al., 2016. Pyriproxyfen is metabolized by P450s associated with pyrethroid resistance in *An. gambiae*. Insect Biochem. Mol. Biol. 78, 50–57. https://doi.org/10.1016/j.ibmb.2016.09.001.

Yunta, C., Hemmings, K., Stevenson, B., Koekemoer, L.L., Matambo, T., Pignatelli, P., et al., 2019. Cross-resistance profiles of malaria mosquito P450s associated with pyrethroid resistance against WHO insecticides. Pestic. Biochem. Physiol 161, 61–67. https://doi.org/10.1016/j.pestbp.2019.06.007.

Zhang, J., Wang, X., Vikash, V., Ye, Q., Wu, D., Liu, Y., et al., 2016. ROS and ROS-mediated cellular signaling. Oxid. Med. Cell. Longev 2016, 4350965. https://doi.org/10.1155/2016/4350965.

Zhou, Y., Fu, W.B., Si, F.L., Yan, Z.T., Zhang, Y.J., He, Q.Y., et al., 2019. UDP-glycosyltransferase genes and their association and mutations associated with pyrethroid resistance in *Anopheles sinensis* (Diptera: Culicidae). Malar. J. 18, 62. https://doi.org/10.1186/s12936-019-2705-2.

CHAPTER TWO

Thermal and desiccation tolerance in mosquitoes: physiological, ecological, evolutionary perspectives

Joshua B. Benoit[a,*] and Chloé Lahondère[b]
[a]Department of Biological Sciences, University of Cincinnati, Cincinnati, OH, United States
[b]Department of Biochemistry, Virginia Polytechnic Institute and State University, Blacksburg, VA, United States
*Corresponding author. e-mail address: joshua.benoit@uc.edu

Contents

Advances in Insect Physiology, Volume 70
ISSN 0065-2806, https://doi.org/10.1016/bs.aiip.2026.05.006

Summary

Mosquitoes occupy an extraordinary range of ecological contexts despite their strong dependence on environmental temperature and water availability, making tolerance to thermal extremes and desiccation stress central to mosquito survival and evolution. As ectotherms with relatively permeable cuticles and complex aquatic–terrestrial life cycles, mosquitoes are acutely sensitive to heat, cold, and water limitation, and variation in these tolerances shapes their geographic distributions, seasonal dynamics, invasion success, and capacity to transmit pathogens. This chapter synthesizes current knowledge on the physiology, ecology, and evolution of thermal and desiccation tolerance in mosquitoes by integrating evidence across life stages, taxa, and levels of biological organization, from eggs and larvae developing in thermally and hydrologically variable habitats to adults facing extreme conditions during host seeking, dispersal, and reproduction. We review conceptual frameworks for measuring and interpreting stress tolerance, including thermal performance curves, acute versus chronic stress responses, phenotypic plasticity, and evolutionary trade-offs, and examine how these traits are constrained or facilitated by life-history strategies. We further summarize the molecular, physiological, and genomic mechanisms underlying stress tolerance, including heat shock responses, membrane remodeling, regulation of metabolic and oxidative stress, cuticular and osmoregulatory adaptations, and the role of gene regulatory networks revealed through population genomics, transcriptomics, and experimental evolution. Thermal and desiccation stresses are frequently co-occurring and share underlying pathways, generating physiological conflicts that shape correlated or decoupled evolutionary trajectories across taxa and environments. Finally, we discuss how stress tolerance feeds back on mosquito longevity, immunity, pathogen development, and vectorial capacity, and consider the implications of rapid environmental and climate change for predictive modeling and vector control. By linking physiological limits, plasticity, ecological shifts and evolutionary potential, this chapter highlights the importance of integrative, trait-based approaches for forecasting mosquito responses to novel thermal and hydric regimes and for mitigating the public health impacts of mosquito-borne diseases.

1. Introduction: environmental stress and the evolution of mosquito biology

Mosquitoes are among the most widely distributed insects on Earth, occupying habitats that span tropical rainforests, temperate wetlands, arid landscapes, and densely populated urban environments (Haba et al., 2025; Kraemer et al., 2019; Sinka and M, 2013). This ecological breadth of mosquitoes is striking, given their strong dependence on specific environmental temperatures and water availability necessary for larval development (Chu et al., 2022; Holmes and Benoit, 2019; Brown et al., 2023). As ectotherms with relatively permeable cuticles and complex aquatic–terrestrial life cycles, mosquitoes are sensitive to thermal extremes and desiccation stress. Consequently, tolerance to heat, cold, and water limitation has played a central role in shaping mosquito evolution, ecology, and their capacity to act as vectors for pathogens of human and animal disease (Chandrasegaran et al., 2020; Mordecai et al., 2019).

Thermal and desiccation stresses operate across multiple spatial and temporal scales. For example, immediate changes can occur within minutes during blood ingestion (Benoit et al., 2011; Lahondère and Lazzari, 2012). Daily fluctuations in temperature and humidity influence activity patterns, feeding success, and survival, while seasonal extremes shape population dynamics, phenology, and geographic range limits over periods of weeks to months (Chen et al., 2024; Kingsolver, 1989; Kerr et al., 2025; Terblanche and Lehmann, 2025). Over evolutionary time, historical climatic variability has driven divergence among mosquito lineages, favoring distinct physiological strategies for coping with heat, cold, and dryness (Lahondère and Bonizzoni, 2022; Couper et al., 2025; Abbasi, 2025). In the modern era, anthropogenic climate change and urbanization are intensifying these selective pressures, exposing mosquitoes to novel combinations of high temperature, low humidity, and altered microclimates such as urban heat islands (Rulli et al., 2025; Rose et al., 2020; Chakraborty et al., 2026; Chakraborty et al., 2025).

From ecological and evolutionary perspectives, thermal and desiccation tolerance are not independent traits but integrated phenotypes emerging from molecular, physiological, behavioral, and life-history processes that interact to allow the survival and proliferation of mosquitoes (Yilmaz et al., 2021; Holmes and Benoit, 2019; González-Tokman et al., 2020; Brown et al., 2023). These traits influence key fitness components, including development rate, fecundity, longevity, and dispersal, and they interact with immunity and pathogen transmission. In this chapter, we summarize current knowledge on the biological processes and evolutionary aspects of thermal and desiccation tolerance in mosquitoes, integrating evidence across life stages, taxa, and levels of biological organization. We emphasize how these stress tolerances shape mosquito survival, diversification, invasion success, and vectorial capacity. Lastly, we highlight key gaps and future research directions that will be critical to understanding the dynamic roles that thermal and dehydration stress play in mosquito biology.

2. Conceptual frameworks for studying mosquito stress tolerance

2.1 Defining and measuring stress tolerance

Thermal tolerance in mosquitoes, along with other animals, is most commonly quantified using metrics such as critical thermal maxima (CT_{max}) and minima (CT_{min}), lethal temperature thresholds (e.g., LT_{50} or LT_{100}), and performance-based traits including development rate, adult longevity, fecundity, blood-feeding success, and flight activity measured across temperature gradients (Couper et al., 2024; Oliveira et al., 2021; Ørsted et al., 2022). CT_{max} and CT_{min} assays have been widely applied in mosquitoes and other insects to estimate acute upper and lower thermal limits and to compare thermal sensitivity among species, populations, and life stages (Paaijmans, Blanford, et al., 2010; Angilletta, 2009; Couret et al., 2014; Couret and Benedict, 2014; Carlassara et al., 2024; Couper et al., 2024). Lethal temperature assays provide complementary information on mortality risk under extreme heat or cold exposure and are particularly relevant for understanding survival during episodic thermal extremes such as heat waves or cold snaps (Bayoh and Lindsay, 2004; Carrington et al., 2013). In contrast, performance-based thermal response curves quantify sublethal effects of temperature on traits directly linked to fitness and vectorial capacity (Shocket et al., 2025; Kellermann et al., 2019; Shapiro et al., 2017), revealing how temperature shapes population growth rates, transmission potential, and geographic distribution under ecologically realistic conditions (Mordecai et al., 2019; Tesla et al., 2018).

Desiccation tolerance in mosquitoes is typically assessed by measuring survival time at low relative humidity, rates of water loss, cuticular permeability, and estimates of critical water content (Gray et al., 2009; Benoit, Patrick, et al., 2010; Accoti et al., 2025). Empirical studies across mosquito taxa demonstrate substantial interspecific and intraspecific variation in dehydration resistance, often associated with habitat aridity, seasonality, and life-history strategy, where some species, such as many within the lineage of *Aedes*, exhibit higher dehydration resistance (Benoit and Denlinger, 2007; Benoit, Patrick, et al., 2010; Hagan et al., 2018; Reidenbach et al., 2014; Gray et al., 2009). Gravimetric measurements of water loss rates have been used extensively in mosquitoes to link desiccation resistance to cuticular hydrocarbons, body size, and metabolic rate (Gray and Bradley, 2005; Fouet et al., 2012; Benoit and Denlinger, 2007; Benoit, Lopez-Martinez, et al., 2010). These traits strongly influence adult activity patterns, host-seeking success, and survival in dry or urban environments where access to free water is limited (Holmes et al., 2025; Hagan et al., 2018; Manzano-Alvarez et al., 2023; Wahaab et al., 2025; Bozic et al., 2024; Abu et al., 2024).

Thermal and dehydration stress in mosquitoes are tightly coupled, with water balance profoundly influencing thermal tolerance and vice versa (Holmes and Benoit, 2019; Brown et al., 2023). First, increasing temperature and decreasing relative humidity lead to higher saturation deficits, resulting in greater water loss (Holmes and Benoit, 2019; Chakraborty et al., 2026). Elevated temperatures accelerate metabolic rates and respiratory water loss (Benoit and Denlinger, 2010; Hadley, 1994; Benoit et al., 2023; Sinclair et al., 2024), increasing the rate at which mosquitoes dehydrate during both resting and active periods (i.e., host seeking and flight), thereby reducing survival under hot and dry conditions (Benoit, Lopez-Martinez, et al., 2010; Gray and Bradley, 2005). Conversely, dehydration can directly constrain thermal tolerance by impairing neuromuscular function, disrupting ion balance, and limiting the capacity for heat shock protein expression (Harrison et al., 2012), thereby reducing CT_{max} and increasing mortality at high temperatures. Lastly, the process of evaporative cooling is critical to prevent overheating in mosquitoes (Lahondère and Lazzari, 2012), but this may be limited if the humidity is too high or the water content of the mosquito is too low to tolerate the losses. These interactions are significant in natural environments, where high temperatures and low humidity frequently co-occur during dry seasons (Heino et al., 2023; Tavakol et al., 2020), intensifying selection on integrated physiological traits such as lower cuticular permeability, metabolic suppression, and behavioral avoidance. As a result, the evolution of mosquito stress tolerance is best understood as a multidimensional process shaped by the combined effects of thermal and dehydration stress, rather than by either stressor in isolation. Thus, modeling and experimental studies can treat humidity and temperature as separate environmental variables, but must consider their combined influence—usually captured as saturation deficit —since interactive effects can drastically increase dehydration and thermal stress beyond what would be predicted from either factor alone.

Distinguishing between acute and chronic stress responses is essential for interpreting the evolutionary significance of thermal and desiccation tolerance in mosquitoes (Hidalgo et al., 2016; Rose et al., 2020; Chakraborty et al., 2026). Acute thermal limits often determine survival during rare but intense events such as heat waves or cold snaps, which can impose strong episodic selection and drive rapid population declines (Bayoh and Lindsay, 2004; Reinhold et al., 2018; Paaijmans, Blanford, et al., 2010). In contrast, chronic exposure to moderately stressful temperatures influences development time, adult size, reproductive output, immune competence, and lifespan, shaping long-term population persistence and vector competence (Covey et al., 2025; Villena et al., 2022; Murdock et al., 2012; Alto and Bettinardi, 2013; Carrington et al., 2013). Similarly, short-term desiccation resistance during host-seeking flights may rely on behavioral avoidance or transient physiological adjustments to move to areas with higher humidity (Laursen et al., 2023; Kessler and Guerin, 2008), whereas long-term survival in arid or seasonal environments reflects evolved differences in water balance regulation, cuticular properties, shifted feeding patterns, and metabolic suppression (Holmes et al., 2025; Hagan et al., 2018; Reidenbach et al., 2014; Huxley et al., 2025). As different experimental metrics capture distinct components of stress response, evolutionary and ecological analyses of mosquito stress tolerance must explicitly consider the environmental context and temporal scale of exposure. Measurements focused solely on acute tolerance may underestimate the importance of chronic sublethal stress in shaping population dynamics and species distributions. Integrative approaches that combine acute thermal and desiccation limits with shifts that occur during prolonged, sublethal stress exposures are essential for understanding how mosquitoes adapt to heterogeneous and changing environments, particularly in the context of climate change and urbanization (Rose et al., 2020; Murdock et al., 2016; Couret and Benedict, 2014).

2.2 Thermal and dehydration performance and fitness landscapes

Thermal performance curves (TPCs) and, to a lesser extent, dehydration performance curves (DPCs) provide a robust framework for quantifying how temperature and dehydration shape organismal performance and fitness-related traits for mosquitoes (Fig. 1). A wide range of studies has examined TPCs across numerous insect systems (Pawar and Kontopoulos, 2025; Abarca et al., 2024; Shocket et al., 2025; Dennington et al., 2025). DPC-based studies are limited to a single study that assessed blood-feeding propensity in relation to dehydration levels (Hagan et al., 2018). As such, we will restrict most of our discussion to TPCs and emphasize that the absence of studies on DPCs affects our understanding of dehydration-induced shifts in insect biological parameters. The lack of DPCs is likely a by-product of linking specific levels of water loss with traits, which is more challenging to accomplish than thermal stress, as only temperature exposure is necessary to assess TPCs for ectotherms (Hagan et al., 2018; Benoit et al., 2009; Rosendale et al., 2016), while DPCs require measuring water loss at specific time points under dry conditions.

TPCs typically exhibit a unimodal shape in which performance increases gradually with temperature as biochemical reaction rates accelerate, reaches a peak at an optimal temperature (T_{opt}), and then declines steeply at higher temperatures due to protein denaturation, membrane dysfunction, and failure of physiological homeostasis (Angilletta, 2009; Huey and Kingsolver, 1989; Sinclair et al., 2016). Evolutionary divergence in TPCs can occur through multiple, non-mutually exclusive mechanisms, including shifts in T_{opt}, changes in thermal breadth (the range of temperatures over which high performance is maintained), and variation in maximum performance. These dimensions reflect underlying trade-offs in enzyme kinetics, metabolic stability, and resource allocation, making TPCs a direct link between physiology, ecology, and evolutionary fitness (Schulte et al., 2011; Huey and Kingsolver, 1989). In mosquitoes, TPCs vary substantially among species, populations, and between trait types, reflecting adaptation to heterogeneous thermal environments across latitudinal and elevational gradients (Dennington et al., 2025; Sinclair et al., 2016;

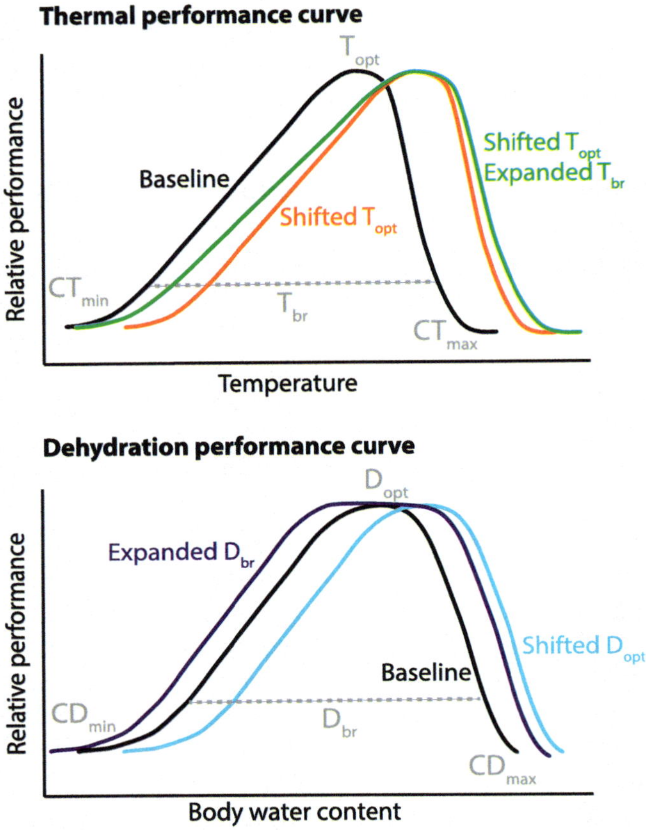

Fig. 1 Thermal and dehydration performance curves and evolutionary trade-offs in mosquitoes. Conceptual thermal performance curves (TPCs) and dehydration performance curves (DPCs). Curves illustrate variation in optimal temperature, performance breadth, and maximal performance, as well as plastic shifts following acclimation. TPCs and DPCS shape fitness components such as development rate, fecundity, and survival. Optimal temperature (T_{opt}) is the temperature at which performance is maximized. The thermal breadth (T_{Br}) range of temperatures allows for sustained performance. Critical limits (CT_{min} and CT_{max}) are temperatures where performance ceases. Abbreviations for DPC are similar except based on an internal effect (body water content) rather than an external stress (temperature).

Dennington et al., 2024; Shocket et al., 2025). Outside optimal ranges, potential impacts on mosquito biology span a wide spectrum, from minor to lethal, as summarized in (Fig. 2).

Similar to other insects, subtropical and tropical mosquitoes often exhibit higher thermal optima but relatively narrow performance breadths for traits

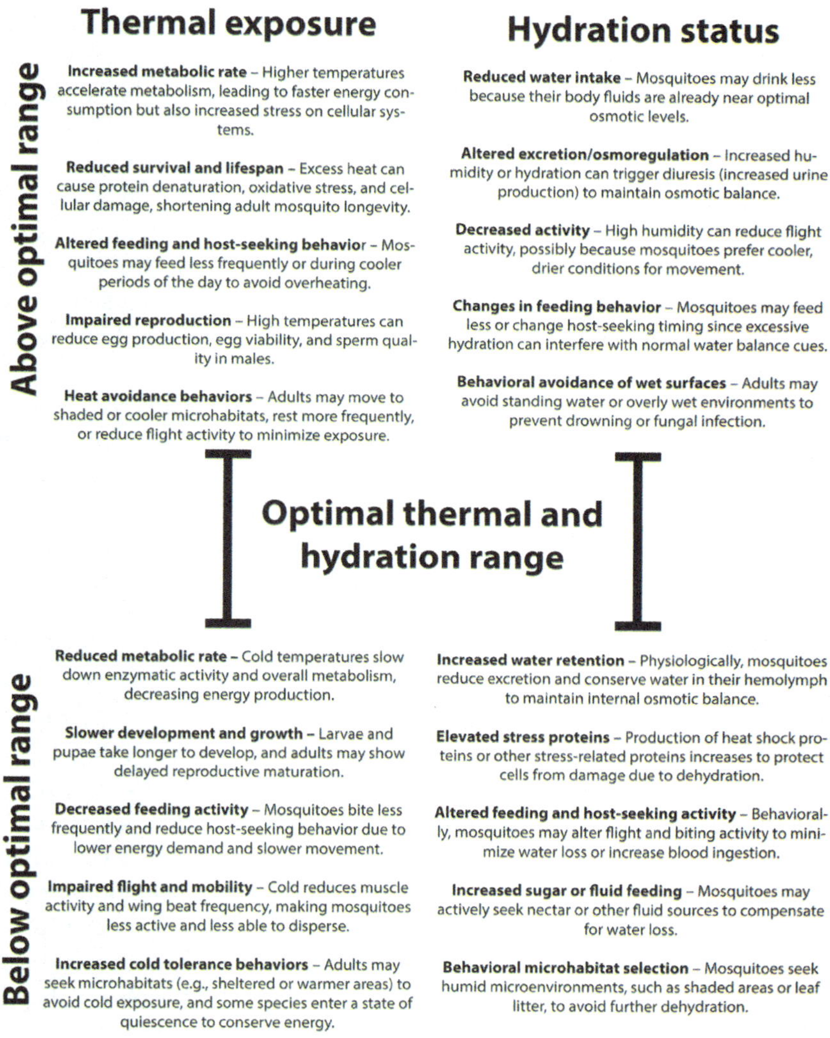

Fig. 2 Mosquito responses to deviations from thermal and hydration optima. Physiological and behavioral changes occur when mosquitoes experience suboptimal conditions.

such as development rate, survival, and biting frequency, consistent with specialization on warm, thermally stable environments (Mordecai et al., 2019; Couret et al., 2014). Temperate species and populations, including *Culex* spp. and high-latitude *Aedes* populations, tend to display broader TPCs that maintain moderate performance across wider temperature ranges, albeit sometimes with lower peak performance (Farjana et al., 2012; Alto and Juliano, 2001; Ciota et al., 2014). Notably, some mosquito stages show minimal variation in thermal resistance across large geographic ranges or habitats, whereas others show significant variation even across relatively small areas (Couper et al., 2024; Chakraborty et al., 2025).

These differences are thought to reflect evolutionary trade-offs, as selection for broad thermal tolerance can reduce maximal performance or energetic efficiency at any single temperature, a pattern consistent with generalist–specialist trade-offs observed across ectotherms (Angilletta et al., 2010; Sinclair et al., 2016; Pawar and Kontopoulos, 2025). Importantly, mosquito TPCs are trait-specific and vary across life-history stages, indicating that thermal adaptation is multifaceted and may generate conflicting selective pressures across development, reproduction, and vectorial capacity. Lastly, there are likely critical dynamics between TPCs and DPCs that have been explored, yet humidity-thermal stress dynamics have yet to be sufficiently explored (Brown et al., 2023; Holmes and Benoit, 2019), except for recent studies (Huxley et al., 2025; Chakraborty et al., 2026; Johnson et al., 2026).

2.3 Phenotypic plasticity in thermal and desiccation stress responses

Phenotypic plasticity plays a central role in mosquito responses to thermal and desiccation stress by allowing individuals to adjust physiological and life-history traits in response to environmental variation (Vinauger and Chandrasegaran, 2024; Dennington et al., 2024, 2025). Across insect species, plastic responses to temperature influence survival, development, and performance (Sgrò et al., 2016; Somero, 2010; Chakraborty et al., 2025), and these mechanisms are particularly relevant for mosquitoes exposed to thermal fluctuations. Thermal conditions experienced during larval development can have persistent effects on adult phenotype, shaping traits such as body size, longevity, immunity, and reproductive output (Mackay et al., 2023; Vantaux et al., 2016; Ezeakacha and Yee, 2019), with downstream consequences for population dynamics and vectorial capacity. In *Ae. aegypti*, fluctuating developmental temperatures alter survival and key life-history traits, demonstrating how early thermal environments can propagate across life stages to influence adult performance (Carrington et al., 2013; Alto and Juliano, 2001).

Short-term thermal acclimation further contributes to plasticity in adult mosquitoes and other insects (Warusawithana et al., 2025; Sørensen et al., 2019). Experimental studies across insects, including mosquitoes, show that recent thermal history can shift physiological performance curves and tolerance limits, allowing individuals to maintain function under variable or stressful temperatures. Such acclimatory responses can modify metabolic rates, activity, and survival, highlighting the capacity for rapid physiological adjustment in changing environments (Somero, 2010; Sinclair et al., 2016; Warusawithana et al., 2025; Bowler, 2005; Hoffmann et al., 2003). These responses are especially relevant under natural conditions, where mosquitoes experience large daily temperature fluctuations that differ substantially from constant laboratory conditions (Paaijmans, Blanford, et al., 2010; Shocket et al., 2025; Beck-Johnson et al., 2017).

Although desiccation plasticity has received less attention than thermal plasticity in mosquitoes, available evidence indicates that desiccation resistance is closely linked to physiological traits governing water balance that could be plastic, including cuticular permeability and hydrocarbon composition (Chown and Nicolson, 2004; Gibbs and Rajpurohit, 2010). In *Anopheles coluzzii,* environmentally driven variation in cuticular hydrocarbons and water loss rates explains substantial differences in desiccation tolerance across humidity regimes, and seasonal shifts in these traits are consistent with plastic responses rather than fixed genetic differences (Arcaz et al., 2016). Such plasticity likely enhances persistence in habitats characterized by substantial temporal variation in humidity.

Plastic responses can buffer mosquito populations against short-term environmental variability by allowing individuals to maintain performance without immediate genetic change (Kellermann et al., 2012; Hidalgo et al., 2016; Chandrasegaran et al., 2025). Thermal acclimation, for example, may reduce the strength of directional selection on stress-tolerance traits by broadening the range of conditions over which mosquitoes can function effectively (Somero, 2010; Sgrò et al., 2016). At the same time, spatial and seasonal variation in thermal and desiccation tolerance across populations underscores the importance of environmentally induced phenotypic variation in shaping ecological and evolutionary responses to climate variability (Arcaz et al., 2016).

Plasticity interacts with evolutionary processes in complex ways. By altering trait expression across environments, plastic responses can mask underlying genetic variation from selection or reshape the selective landscape itself, complicating efforts to distinguish plasticity from local adaptation (Stevens et al., 2025; Sommer, 2020). In heterogeneous environments, selection may favor high plasticity despite associated energetic or regulatory costs, whereas in more stable environments, reduced plasticity and more canalized stress-tolerance traits may be favored (Somero, 2010; Sgrò et al., 2016). This may be even more critical during climate change, as extreme weather events are expected to increase (Heino et al., 2023; Herring et al., 2014; Newman and Noy, 2023). Disentangling the relative roles of phenotypic plasticity and genetic adaptation remains a key challenge in evolutionary physiology and mosquito vector biology.

2.4 Trade-offs and constraints of increased stress resistance

Traits that enhance stress tolerance in mosquitoes—such as elevated expression of heat shock proteins (HSPs), changes in cuticular hydrocarbon (CHC) profiles, or shifts in metabolic output—often have costs because they require energetic and biochemical resources that might otherwise support growth, reproduction, or immune defense (Savvidou et al., 2025; Garland et al., 2022; Zera and Harshman, 2001). For example, HSPs are widely recognized as molecular chaperones that stabilize proteins under stress and help maintain cellular homeostasis (King & MacRae, 2015). In mosquitoes, dehydration stress induces the expression of Hsp70 and Hsp90 transcripts, and reducing the expression of these proteins via RNA interference significantly lowers dehydration tolerance, indicating their functional importance in stress resistance (Benoit et al., 2010). However, synthesizing and maintaining elevated levels of HSPs is energetically costly and can divert limited resources from other vital processes, such as reproduction (Alagar et al., 2022). Desiccation resistance in mosquitoes is also linked to variation in cuticular hydrocarbon composition. In *An. coluzzii*, specific CHCs account for a large proportion of the variation in desiccation tolerance, suggesting that investment in particular hydrocarbon profiles enhances water balance but may entail a metabolic cost (Arcaz et al., 2016). Such structural and biochemical investments reflect there will likely be allocation trade-offs: energy allocated to desiccation resistance via CHC biosynthesis is energy not available for reproduction or other life-history functions.

In addition to physiological constraints, genetic correlations among traits can limit the independent evolution of thermal and desiccation tolerance (Logan and Christian, 2020; Gerken et al., 2015; Schou et al., 2022). When the same genes or regulatory networks influence multiple traits (pleiotropy), natural selection for greater stress tolerance may unintentionally shift correlated traits in directions detrimental to fitness. Such genetic correlations reduce the ability of natural selection to optimize each trait independently, constraining evolutionary responses under changing environmental conditions. For example, selection on stress response pathways that enhance HSP expression may be genetically linked to variation in developmental time or fecundity, illustrating how adaptation to stress can entail developmental costs. Understanding these physiological and genetic trade-offs is critical for predicting how mosquito populations will respond to climate change, warming temperatures, and increasing drought frequency. Ignoring the costs associated with enhanced stress tolerance could lead to overestimating the adaptive capacity of mosquitoes and misinterpreting vector dynamics under future environmental change scenarios.

3. Thermal tolerance across mosquito life stages

3.1 Egg thermal tolerance

Mosquito eggs experience some of the most extreme and variable thermal conditions encountered across the mosquito life cycle, particularly in species that oviposit in exposed, terrestrial, or ephemeral aquatic habitats. Unlike larvae and adults, eggs are immobile and unable to behaviorally thermoregulate, rendering them especially vulnerable to fluctuations in temperature and humidity (Clements, 1992; Chakraborty et al., 2026; Dias et al., 2025). Importantly, there are distinct differences among egg stages within major mosquito groups, with most eggs hatching within a few days. In contrast, some lineages, such as those within the Aedini tribe, have prolonged embryonic stages lasting weeks to months or even years - only emerging when submerged in water (Clements, 1992; Sota and Mogi 1992b, 1992a). Elevated temperatures generally accelerate embryonic development but can increase mortality through protein denaturation, oxidative stress, metabolic reserve depletions, or disruption of membrane integrity, whereas low temperatures slow development and may cause chilling or freezing injury when physiological thermal tolerance limits are exceeded (Chakraborty et al., 2025; Ajayi, Oyen, et al., 2024; Doeurk et al., 2025; Dias et al., 2025).

In temperate- and high-latitude species, egg cold tolerance is a critical adaptation that enables persistence of specific mosquito species through winter (Denlinger and Armbruster, 2014; Diniz et al., 2017). Many mosquitoes exhibit facultative or obligate egg diapause, characterized by developmental arrest, metabolic suppression, and shifts in gene expression that enhance resistance to prolonged cold exposure and the dry conditions of winter (Denlinger and Armbruster, 2014; Diniz et al., 2017). Diapausing eggs can remain viable for months under fluctuating cold temperatures (Fischer et al., 2025; Won and Choi, 2025; Sota and Mogi 1992b), allowing hatching to be synchronized with favorable spring conditions (i.e., higher temperatures and longer days). In contrast, tropical and subtropical mosquitoes rarely experience freezing stress but are frequently exposed to extreme heat, particularly in sunlit artificial containers such as tires, buckets, and cisterns. More details are provided in Section 9 on mosquito cold tolerance and overwintering. Temperatures within these microhabitats can substantially exceed ambient air temperatures, making heat tolerance a significant determinant of egg survival and subsequent population growth (Juliano et al., 2002; Couret et al., 2014).

Thermal tolerance varies among populations of *Ae. aegypti* eggs, which are associated with urbanization and human-associated ecological traits (Chakraborty et al., 2025). Specifically, eggs from populations collected in areas with higher human population density and stronger human host preference exhibited higher viability following high-temperature stress. This suggests that egg thermotolerance may be a critical factor in the persistence of urban *Ae. aegypti* populations under heat stress (Chakraborty et al., 2025). Overall, evolutionary divergence in egg thermal tolerance reflects habitat use and broader life-history strategies. Species that exploit transient or unpredictable environments often evolve broader thermal tolerance ranges, buffering embryos against rapid temperature fluctuations. Consequently, egg-level tolerance to thermal stress has likely been a key factor underlying the ecological flexibility, global spread, and invasive success of specific mosquito species, linking embryonic stress resistance to patterns of urban adaptation, climate resilience, and vector persistence.

3.2 Larval thermal ecology

Larval mosquitoes develop in aquatic habitats that differ widely in thermal stability, from large, permanent water bodies to small, ephemeral containers. Large habitats buffer temperature fluctuations, whereas small containers can experience rapid, extreme diel changes, sometimes exceeding 15–20 °C within a day (Sota et al., 1994; Kumar et al., 2016; Sukiato et al., 2019). Temperature strongly influences larval growth rate, development time, survival, and adult body size (Sukiato et al., 2019; Barreaux et al., 2018). Warmer conditions generally accelerate development but can reduce survival and adult size when thermal optima are exceeded (Atkinson, 1994; Couret et al., 2014; Sukiato et al., 2019; Barreaux et al., 2018). Larval rearing temperature also affects adult traits, including longevity, fecundity, and vector competence, with consequences for cohort synchrony and disease transmission potential (Alto and Bettinardi, 2013; Carrington et al., 2013). Together, these physiological and behavioral responses highlight the critical role of larval thermal ecology in shaping mosquito life-history traits and pathogen transmission. Importantly, factors including elevation have a direct impact on species' abilities to develop and thus on abundance. For example, *Cx. erethyzonfer* larvae were found in higher numbers at higher elevation but only on one of the mountain slopes where they were collected in Costa Rica (Amundrud and Srivastava, 2020), emphasizing the species low heat tolerance.

3.3 Pupal vulnerability to heat stress

The pupal stage in mosquitoes represents a potential thermal bottleneck because pupae are non-feeding, have limited behavioral thermoregulatory capacity, and high stress is likely to prevent adult tissue development. There are few studies that have directly assessed thermal tolerance of the pupal stage when compared to other stages (Reinhold et al., 2018; Couper et al., 2024; Chakraborty et al., 2025). In most cases, larval and pupal development are combined and not treated separately. Short periods of high temperature (42 °C) result in a significant stress response (Zhao et al., 2010), but tracing how this impacts adult emergence and survival is not usually conducted (Mourya et al. 2004). Fluctuating asymmetry (FA) of traits, such as wing length, and differences in adult sizes have been known to be impacted by thermal stress experienced during the larval and pupal stages (Mourya et al., 2004; Mpho et al., 2002; Rueda et al., 1990; Loetti et al., 2011). A major highlight is that targeted studies of pupal stages will be necessary to assess stage-specific mosquito responses to thermal stress and climate variability.

3.4 Adult thermal tolerance

Adult mosquitoes encounter significant thermal stress throughout their life cycle, as their small bodies make them highly sensitive to environmental temperatures. High temperatures can impair flight performance, increase metabolic costs, and elevate mortality, while low temperatures constrain flight, host-seeking, and mating activities (Murdock et al., 2012; Rivas et al., 2018; Upshur et al., 2019; Lahondère et al., 2023). Thermal tolerance of adults influences geographic distribution, seasonal activity, and diel behavior (Reinhold et al., 2018; Lahondère and Bonizzoni, 2022; Mordecai et al., 2019). Differences in thermal limits among species contribute to their ability to colonize temperate versus tropical regions, with cold-tolerant species emerging earlier in spring and heat-tolerant species thriving in warmer months (Reinhold et al., 2018).

Sex-specific differences in thermal physiology are particularly pronounced in females due to additional demands associated with blood feeding, egg development, and repeated gonotrophic cycles. Ingesting a warm blood meal imposes rapid internal thermal stress, triggering physiological heat-shock responses (Pawar et al., 2024; Benoit et al., 2011). While males generally exhibit lower thermal tolerance, sex-specific selection on thermal physiology and stress-response pathways arises from these reproductive and blood-feeding challenges, influencing survival, reproductive timing, and population dynamics. Age also has a general impact on thermal tolerance (Lyons et al., 2014; Martin and Hillyer, 2024), in which declining thermal tolerance accelerates aging and reduces longevity.

Several studies indicate that factors such as latitude or elevation affect mosquito thermal tolerance (Lyberger et al., 2025), which has also been shown in multiple insect species, including *Drosophila* (e.g., (Sørensen et al., 2005)). In most cases, insects collected at higher latitudes or elevations tend to have lower heat tolerance than those from lower, i.e., warmer, elevations. However, it is worth noting that responses may differ depending on the stress resistance traits measured (e.g., CT_{max}, CT_{min}, T_{opt}, or HSP70 synthesis) and climate (e.g., temperate vs. tropical) (Chen and Lewis, 2024). For *Cx. tarsalis*, CT_{max} positively correlated with maximum temperatures recorded at each collection site along a collection gradient, and the population from higher elevation (i.e., cooler) had higher metabolic rates overall (i.e., thermal compensation, (Vorhees et al., 2013)). Finally, as climatic conditions vary throughout the year, mosquitoes may exhibit seasonal variations in their thermal breadth. Shifts in thermal tolerance have indeed been observed in adults from multiple species, including *Ae. vexans* and *Cx. quinquefasciatus* and *An. crucians* (Oliveira et al., 2021) Yet again, not all thermal traits are affected in the same way at different times of year, and there are significant stage and species differences, suggesting that heat tolerance is not always highest during warm seasons.

3.5 Behavioral changes in relation to thermal stress

Mosquitoes often employ behavioral thermoregulation, such as seeking shaded areas during heat stress, to mitigate effects on survival and metabolism (Murdock et al., 2012). Despite strong evidence that temperature strongly shapes mosquito behavior, thermal avoidance behaviors are not often framed as a distinct adaptive category. Studies on the aquatic stages are less common, where larvae will move to a specific optimum temperature well above lower temperatures that prevent growth and below higher temperatures that cause stress (Hug et al., 2025; Liu and Zwiebel, 2013).

Adult mosquitoes actively exploit fine-scale microhabitat heterogeneity by selecting shaded vegetation, cooler resting surfaces, or sheltered indoor environments that buffer against extreme temperatures (Sauer et al., 2021; Murdock et al., 2017; Pincebourde et al., 2016). Mosquitoes may also shift activity to crepuscular or nocturnal periods, when thermal conditions are more favorable, and adjust host-seeking, feeding, and oviposition behaviors to minimize heat exposure (Meuti, 2025; Abbasi, 2025; Souza and Weaver. 2024). In some cases, mosquitoes will likely exhibit short-distance movements to track optimal thermal niches within a landscape (Trewin et al., 2020; Pincebourde and Woods, 2020), effectively maintaining body temperatures within tolerable ranges. These behavioral adjustments can substantially buffer individuals from both lethal and sublethal thermal conditions, helping preserve physiological function, reproductive capacity, and fitness across fluctuating environments. Incorporating behavioral thermoregulation into studies of mosquito ecology is therefore essential for accurately predicting vector persistence, distribution, and pathogen transmission potential under increasingly variable and warming climates.

3.6 Blood as a source of heat stress

Blood feeding represents a major and biologically relevant source of acute heat stress in mosquitoes (Benoit et al., 2011; Lahondère and Lazzari, 2012; Benoit et al., 2019; Reinhold et al., 2021). During feeding, mosquitoes rapidly ingest warm vertebrate blood at temperatures often exceeding ambient conditions by 15–25 °C, leading to a rapid rise in temperature. Because mosquitoes have limited thermal inertia (Benoit et al., 2011), this sudden heat load poses a risk of protein denaturation and cellular damage. In *Ae. aegypti*, a warm blood meal elicits a strong heat shock response, including rapid upregulation of HSP70. This response is essential for maintaining gut integrity and supporting digestion and egg production, indicating that blood-meal–induced heat stress is a significant constraint rather than a passive consequence of feeding (Benoit et al., 2011; Lahondère and Lazzari, 2012; Benoit et al., 2019). In addition to molecular responses, mosquitoes employ behavioral and physiological mechanisms to mitigate thermal stress during blood feeding. One prominent strategy is evaporative cooling, in which mosquitoes excrete and retain fluid droplets at the abdomen during feeding, lowering body temperature through evaporation (Lahondère and Lazzari, 2012). Thermal imaging studies show that this behavior substantially reduces abdominal temperature and helps prevent overheating during feeding on warm-blooded hosts (Lahondère and Lazzari, 2012; Reinhold et al., 2021).

3.7 Comparative and evolutionary patterns of thermal resistance

Aedes species often exhibit high heat tolerance and rapid development, traits that facilitate colonization of warm, variable environments (Reinhold et al., 2018). Experimental studies have shown that rapid shifts to extreme temperatures dramatically reduce survival in *Aedes* species; however, when rearing and exposure temperatures match, survival and physiological performance improve, indicating thermal acclimation.

Within *Ae. aegypti*, research has revealed phenotypic variation in thermal tolerance across geographically diverse populations (Chakraborty et al., 2025; Chakraborty et al., 2026; Dennington et al., 2025, 2024). Such local adaptation suggests that thermal stress responses are dynamic, with implications for population resilience under climate change. Comparative work between *Ae. aegypti* and *Culex quinquefasciatus* demonstrate distinct thermal tolerance patterns across genera. *Cx. quinquefasciatus* exhibits inherently greater cold tolerance, while *Ae. aegypti* is better adapted to withstand heat stress, indicating divergent strategies likely shaped by evolutionary pressures in their respective ecological niches (Chura et al., 2023).

Beyond species-level physiological differences, environmental temperature is a major driver of mosquito life history and performance. Both *Ae. aegypti* and *Ae. albopictus* show temperature-dependent variation in survival, development, and reproductive traits, and thermal stress can influence geographic distribution and seasonal activity (Reinhold et al., 2018; Chakraborty et al., 2025). Variations in egg-stage thermotolerance, for example, are associated with urbanization and human interactions, highlighting ecological nuances in thermal adaptation (Chakraborty et al., 2025).

These comparative patterns of stress resistance have important evolutionary implications. Thermal adaptation may influence not only short-term physiological responses but long-term population differentiation and diversification. Genotypic variation underlying heat tolerance suggests that evolutionary adaptation to thermal environments can occur over relatively short timescales, potentially reshaping species' ecological niches and distributions (Dennington et al., 2025, 2024). Meanwhile, the divergence in cold versus heat tolerance among genera reflects historical selection pressures that have shaped life-history strategies, reproductive timing, and overwintering behavior to allow survival across longitudinal or elevation gradients (Chura et al., 2023). Together, these patterns highlight how thermal adaptation underpins mosquito diversification and ecological success, enabling different genera to thrive across a wide range of climates and habitat types.

4. Molecular and physiological mechanisms of thermal tolerance

Thermal tolerance in mosquitoes is maintained by a suite of conserved and lineage-specific molecular and physiological mechanisms that collectively determine cellular stability, metabolic performance, and organismal survival under temperature stress (González-Tokman et al., 2020). These mechanisms operate across hierarchical levels—from protein folding and membrane structure to whole-organism energy balance—as highlighted in comparative insect physiology and thermal biology literature (Angilletta, 2009; Somero, 2010). Mosquitoes mitigate thermal stress across life stages through coordinated structural, physiological, and behavioral mechanisms that buffer acute and prolonged heat exposure (Fig. 3). Eggs rely on protective chorion properties, rapid induction of HSPs, and developmental arrest or delayed hatching to limit damage and preserve viability under high temperatures. Larvae and pupae employ behavioral thermoregulation by moving to cooler, deeper, or shaded aquatic microhabitats, coupled with physiological responses such as reduced feeding, accelerated development, and activation of antioxidant and heat shock pathways. In adults, heat stress is alleviated through behavioral avoidance of extreme temperatures, altered activity patterns, and physiological buffering during energetically demanding processes like blood feeding and reproduction. Together, these stage-specific responses promote survival and reproductive success while also shaping life-history trade-offs and long-term adaptation to warming environments. Specific aspects are discussed in more detail in 4.1–4.4.

4.1 Heat shock protein response

Among the most extensively studied mechanisms of thermal tolerance is the heat shock response. HSPs, including members of the HSP70, HSP90, and small HSP families, act as molecular chaperones that stabilize nascent

Fig. 3 Mechanisms of thermal resistance across mosquito life stages. Schematic overview of structural, physiological, and behavioral adaptations that reduce thermal stress in mosquito eggs, larvae, pupae, and adults (before and after blood feeding).

proteins, prevent denaturation, and facilitate refolding under thermal stress (Banfi et al., 2025; Mack and Attardo, 2024). In mosquitoes, constitutive expression levels and inducibility of HSPs vary among species and populations, reflecting differences in thermal regimes and evolutionary histories (González-Tokman et al., 2020; Banfi et al., 2025). For example, *Ae. aegypti* and *Ae. albopictus* larvae show significant upregulation of HSP genes, such as HSP26, HSP83, and Hsc70, in response to heat stress (Arun et al., 2017; Zhao et al., 2010). Experimental work in other mosquito species shows that HSP expression correlates with survival after thermal challenge (Carlassara et al., 2024; Mack and Attardo, 2024), indicating a direct physiological role for these proteins in buffering heat stress.

4.2 Membrane composition and homeoviscous adaptation

Thermal stress challenges membrane integrity and function. Homeoviscous adaptation—the adjustment of membrane lipid composition to maintain optimal fluidity—is a conserved response among ectotherms, helping preserve function across temperatures (Hazel, 1995). Changes in lipid saturation and membrane composition have been documented across various arthropods as a thermal acclimation response, although specific lipidomic studies in mosquitoes remain limited and are predominantly associated with overwintering (see Section 8; (Reynolds et al., 2012; Batz and Armbruster, 2018)).

4.3 Metabolic regulation and energy allocation during thermal stress

Temperature exerts profound effects on metabolic rate, typically increasing metabolic demand and energy turnover at higher temperatures (González-Tokman et al., 2020). Elevated temperatures often necessitate the suppression of non-essential processes and reallocation of resources toward essential cellular maintenance and stress mitigation. In insects, transcriptomic and metabolic profiling studies reveal coordinated regulation of glycolysis, oxidative phosphorylation, lipid metabolism, and other central metabolic pathways in response to thermal stress (Singh et al., 2022; Pathak et al., 2025). Under cold exposure, metabolic suppression is common, and expression of genes involved in energy conservation and mitochondrial efficiency increases, reflecting a shift toward energy-saving phenotypes during thermal challenges (González-Tokman et al. 2020; Lv et al. 2023).

4.4 Oxidative stress and cellular damage

Thermal stress frequently increases the production of reactive oxygen species (ROS), which can damage proteins, lipids, and nucleic acids (Mack and Attardo, 2024; Slimen et al., 2014; Zhang et al., 2019). Elevated ROS levels result from accelerated metabolic rates and mitochondrial dysfunction under temperature extremes. Antioxidant defenses—including superoxide dismutase (SOD), catalase, glutathione-dependent enzymes, and other enzymatic and non-enzymatic antioxidants—are integral to mitigating oxidative damage and maintaining cellular integrity. Variation in antioxidant regulation can influence aging, longevity, and life-history trade-offs (Barr et al., 2024; Martin et al., 2026), with implications for population fitness and vectorial capacity in disease transmission contexts.

5. Desiccation stress in mosquito ecology and evolution

Desiccation stress is a major selective force shaping mosquito physiology, behavior, and life history, particularly in arid, seasonal, and urban environments (Holmes and Benoit, 2019; Brown et al., 2023; Benoit and Denlinger, 2010). As with thermal tolerance, shifts in desiccation resistance result from structural, physiological, and behavioral traits that collectively maintain water balance.

Mosquitoes deploy stage-specific structural, physiological, and behavioral mechanisms to limit water loss across eggs, larvae, and adults (Fig. 4). Eggs exhibit the highest intrinsic desiccation resistance, relying on chorion impermeability, wax layer deposition, and accumulation of protective molecules (e.g., trehalose and heat shock proteins), with prolonged dry exposure inducing developmental arrest and long-term viability under arid conditions. Importantly, the egg stages, particularly for some species that are dehydration resistant, such as *Aedes* (Sota and Mogi 1992b, 1992a; Chakraborty et al., 2026), are likely critical to the distribution and movement of mosquitoes due to trade and travel (Kraemer et al., 2019; Pabst et al., 2025). Larvae, which are constrained to aquatic habitats, respond to short-term dehydration risk by reducing activity and rapidly adjusting internal osmolyte levels. In contrast, prolonged stress can trigger metabolic suppression, altered growth trajectories, and even death in relatively short periods if the larval habitat dries. Adults mitigate water loss behaviorally by seeking humid refugia and reducing activity, and physiologically by remodeling cuticular lipids, increasing hemolymph concentration, and activating

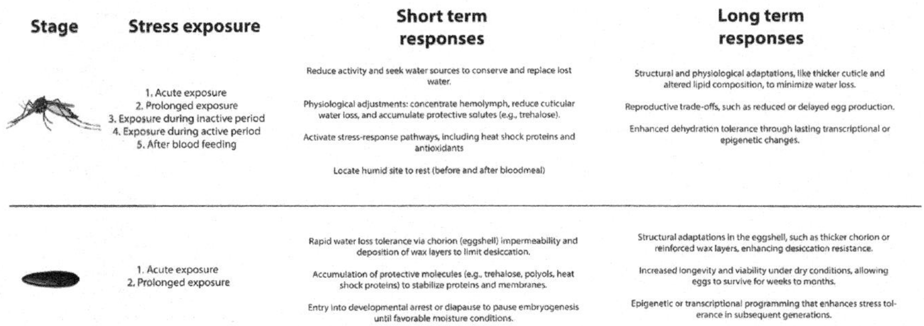

Fig. 4 Mechanisms of desiccation resistance across mosquito life stages. Schematic overview of structural, physiological, and behavioral adaptations that reduce water loss in mosquito eggs and adults.

stress-response pathways. Across life stages, repeated or prolonged dehydration can lead to longer-term transcriptional or epigenetic changes that enhance future stress tolerance but may impose reproductive or developmental trade-offs. Specific aspects are discussed in more detail in 5.1–5.5.

5.1 Environmental sources of desiccation stress

Desiccation stress results from a combination of low humidity, high temperatures, and limited access to free water (Hagan et al., 2018; Benoit and Denlinger, 2010; Benoit et al., 2023). Adult mosquitoes are vulnerable during host-seeking, dispersal, and resting in exposed environments, while eggs and larvae may experience desiccation in drying or ephemeral habitats (Chu et al., 2022). Urban environments often exacerbate water stress through reduced vegetation, increased impervious surfaces, and elevated temperatures (Rose et al., 2020; Chakraborty et al., 2026), creating stresses that are distinct or elevated above those in natural ecosystems. In some cases, such a cities in deserts, the alternative can occur where artificial water sources associated with humans can provide resources for mosquito survival.

5.2 Egg and larval desiccation resistance

Egg desiccation resistance represents a significant evolutionary innovation in mosquitoes, particularly in the *Aedes* and closely related species (Sota and Mogi 1992a; Sota, 1993a). Following oviposition, mosquito embryos secrete a serosal cuticle beneath the chorion, which functions as a critical barrier to water loss and confers dramatic increases in egg survival under dry conditions. Experimental work across insects has demonstrated that formation of the serosal cuticle reduces trans-eggshell water flux by orders of magnitude, effectively decoupling embryonic survival from ambient humidity (Rezende et al., 2008; Vargas et al., 2014; Prasad et al., 2023). In mosquitoes, this adaptation allows eggs to withstand prolonged periods of desiccation, facilitating persistence through seasonal droughts and unpredictable rainfall regimes. Desiccation-resistant eggs have been particularly important in the evolutionary success of *Aedes* mosquitoes (Rezende et al., 2008; Vargas et al., 2014; Sota and Mogi 1992a), but other lineages (e.g., *Culex* and *Anopheles*) lack this adaptation. Species such as *Ae. aegypti* and *Ae. albopictus* lay eggs above the water line in natural and artificial containers (Chakraborty et al., 2025; Sota and Mogi, 1992a; Urbanski et al., 2010), where they are frequently exposed to drying conditions. These eggs can remain viable for months, and in some cases years, enabling populations to bridge unfavorable seasons and rapidly exploit newly available water following rainfall (Sota and Mogi, 1992a; Faull and Williams, 2015; Chakraborty et al., 2026; Westby et al., 2025).

Comparative and experimental studies indicate that enhanced egg desiccation resistance in *Aedes* is tightly linked to their invasive capacity and global spread via human-mediated transport, including trade in used tires and other water-holding containers (Vargas et al., 2014; Juliano and Philip Lounibos, 2005; Juliano et al., 2002). In contrast to eggs, larval stages are generally far less tolerant of desiccation (Benoit and Denlinger, 2007; Benoit, Lopez-Martinez, et al., 2010; Chu et al., 2022), reflecting their obligate aquatic requirements. Nonetheless, larvae exhibit limited physiological and behavioral mechanisms to mitigate the dehydration risk during transient habitat drying (Koenraadt et al., 2003; Bick and Penn, 1947). These include rapid development, behavioral avoidance of shallow or warming microhabitats, and in some species, the ability to survive brief periods of water loss within damp substrates (Benoit and Denlinger, 2007; Benoit, Lopez-Martinez, et al., 2010; Chu et al., 2022).

5.3 Adult desiccation tolerance mechanisms

Adult mosquitoes mitigate water loss through an integrated suite of physiological, morphological, and behavioral adaptations that collectively maintain water balance under dry conditions (Gibbs and Rajpurohit, 2010; Benoit et al., 2023; Benoit and Denlinger, 2010). One of the most critical barriers to desiccation is the insect cuticle, whose permeability is primarily determined by the quantity and composition of cuticular hydrocarbons (CHCs) (Gibbs and Rajpurohit, 2010; Blomquist and Ginzel, 2021; Arcaz et al., 2016). These long-chain lipids form a hydrophobic layer on the epicuticle that restricts transcuticular water loss. In mosquitoes, variation in CHC chain length, degree of saturation, and branching patterns is strongly associated with desiccation resistance, with longer-chain and more saturated hydrocarbons generally conferring lower cuticular permeability (Gibbs and Rajpurohit, 2010; Gibbs, 1998; Reidenbach et al., 2014). Comparative studies among mosquito species, along with insects, and populations demonstrate that desert- or dry-adapted taxa, as well as individuals reared under low humidity, often exhibit increased total CHC abundance or altered CHC profiles that enhance resistance to dehydration (Foley and Telonis-Scott, 2011; Ferveur et al., 2018; Wang et al., 2022; Nayal et al., 2024).

Respiratory water loss represents another challenge for adult mosquitoes, particularly given their high mass-specific metabolic rates (Benoit and Denlinger, 2007; Brown et al., 2023; Holmes and Benoit, 2019). Mosquitoes reduce water loss during respiration through tight regulation of spiracular opening, including reduced spiracular conductance and temporal clustering of gas exchange events (Mamai et al., 2016; Arcaz et al., 2016; Wagoner et al., 2014). This form of spiracular control limits evaporative loss but may constrain oxygen uptake and carbon dioxide release, potentially imposing metabolic trade-offs during prolonged desiccation stress (Chown et al., 2011). Beyond restricting water loss, adult mosquitoes also enhance desiccation tolerance by maximizing internal water stores and maintaining hemolymph homeostasis. Hemolymph volume, ion balance, and osmotic pressure are actively regulated during dehydration, with the Malpighian tubules and hindgut playing key roles in minimizing water excretion while preserving ionic balance (Coast, 2009; Beyenbach, 2003). Accumulation of compatible osmolytes such as sugars and amino acids can further stabilize cells during dehydration, although this mechanism has been less extensively characterized in mosquitoes than in other insects (Yancey, 2005).

5.4 Behavioral and life-history strategies to avoid dehydration

Behavioral adaptations play a crucial role in mitigating desiccation stress in mosquitoes, often acting synergistically with physiological mechanisms to reduce water loss. Many mosquito species actively select humid resting sites such as vegetation, tree holes, animal burrows, or human dwellings, where microclimatic conditions buffer against extreme heat and low humidity (Benoit et al., 2023; Chu et al., 2022; Kessler and Guerin, 2008). Shifts in daily activity patterns also reduce exposure to desiccating conditions: host-seeking, mating, and oviposition behaviors are frequently concentrated during crepuscular or nocturnal periods when temperatures are lower and relative humidity is higher (Ajayi, Susanto, et al., 2024; Ajayi et al., 2022; Wynne et al., 2024). Experimental and field studies demonstrate that access to shaded or humid refugia can substantially increase adult survival under arid or seasonal dry conditions (Holmes et al., 2025; Kessler and Guerin, 2008), highlighting the importance of fine-scale habitat selection in determining mosquito persistence and distribution.

Life-history strategies further influence desiccation tolerance through trade-offs among growth, reproduction, and survival. Body size is a particularly important determinant: larger mosquitoes generally exhibit lower surface-area-to-volume ratios, which reduces relative rates of water loss and increases survival under low-humidity conditions (Briegel, 1990; Benoit and Denlinger, 2007; Benoit, Lopez-Martinez, et al., 2010; Gray and Bradley, 2005). As a result, larval environmental conditions that influence adult size, such as temperature, nutrition, and density, indirectly shape adult desiccation tolerance. However, increased body size often comes at the cost of longer larval development times and higher resource requirements, which may be disadvantageous in ephemeral or resource-limited habitats (Kawecki and Stearns, 1993; Telang and Wells, 2004). Blood feeding can be used as a behavioral method to replenish water lost (Manzano-Alvarez et al., 2023; Hagan et al., 2018; Holmes et al., 2025), but frequent host-seeking exposes mosquitoes to desiccating environments, and the risk of death increases with greater host interactions. This increased blood feeding is critical for mosquitoes to survive periods when free water is unavailable (Holmes et al., 2025). Seasonal avoidance of dry periods can occur through aestivation, migration, or diapause (Benoit, 2010; Dao et al., 2014; Diallo et al., 2022; Lehmann et al., 2010). Lastly, as with thermal tolerance, age has a distinct effect on dehydration tolerance (Lyons et al., 2014; Benoit and Denlinger, 2007).

5.5 Sugar feeding reduces hydric stress and extends lifespan

Adult mosquitoes rely on plant-derived sugars as a central component of their energy budget, and accumulating evidence demonstrates that sugar feeding influences survival, thermal tolerance, and resistance to desiccation in species-specific ways. Access to sugar influences activity and survival under various thermal conditions, with sugar-fed mosquitoes of both sexes living longer than water-fed individuals (Upshur et al., 2019). This demonstrates that access to carbohydrates may help buffer the effects of potential environmental stressors such as high temperatures and risk of desiccation. Field investigations of *An. gambiae* showed that males feed on sugar at night at high frequency, whereas females modulate sugar intake according to reproductive state and host availability (Gary and Foster, 2006). Earlier work on *Cx. tarsalis* similarly documented frequent sugar feeding in both sexes at various stages of their life (e.g., entering diapause, diapausing, and terminating diapause), highlighting that regular carbohydrate acquisition is fundamental to energy homeostasis in this species and likely contributes to resilience in fluctuating environmental conditions (Reisen et al., 1986). For *An. sergentii*, populations exhibited significantly higher adult survival in sugar-rich areas compared to sugar-poor areas (Gu et al., 2011). *Ae. aegypti* exhibited lower sugar feeding rates compared to *Cx. quinquefasciatus* and that rates were correlated with environmental conditions (Olson et al., 2020). In contrast, *Ae. albopictus*, showed high sugar feeding rates in the wild and dry conditions, implying a compensatory role of carbohydrates in mitigating desiccation stress (Fikrig et al., 2020). Floral resource availability in the wild significantly affects mosquito survival under stressful environmental conditions (i.e., arid), reinforcing the link between nectar access/quality and stress resilience (Branham et al., 2025).

Together, these studies demonstrate that sugar feeding is not merely a supplementary behavior but a key ecological determinant of mosquito performance under abiotic stress. Species/individuals that routinely acquire sugar appear to maintain broader tolerance to cold or hydric challenges. In contrast, species/individuals with lower or more opportunistic sugar feeding show survival benefit when sugar is available but may be more sensitive to environmental variability in its absence. Overall, it appears clear that sugar-feeding ecology fundamentally shapes mosquito survival trajectories and physiological resilience to temperature extremes and desiccation stress.

5.6 Blood as a source of osmotic stress

In mosquitoes, blood feeding represents a profound osmotic challenge due to the rapid ingestion of a large volume of vertebrate blood, which is hypoosmotic relative to mosquito hemolymph (Lehane, 2005). A single blood meal can exceed the mosquito's unfed body mass, causing immediate dilution of hemolymph solutes and increased hydrostatic pressure within the gut and circulatory system. Without rapid water removal, this imbalance can disrupt neuromuscular function, metabolic processes, and general homeostasis. Consequently, osmotic stress following blood ingestion is among the most acute challenges faced by adult female mosquitoes (Beyenbach, 2003; Piermarini, 2016).

Mosquitoes mitigate post–blood meal osmotic stress through an exceptionally rapid diuresis driven by the Malpighian tubules and hindgut (Halberg and Denholm, 2024). Within minutes of feeding, diuretic hormones stimulate ion transport processes that promote the excretion of excess sodium, potassium, and chloride, with water following osmotically. This postprandial diuresis can eliminate a substantial fraction of the ingested fluid within the first hour after feeding, restoring hemolymph osmolarity while retaining proteins and other nutrients required for egg production (Beyenbach, 2003; Piermarini, 2016; Coast, 2009). Key molecular components of this response include V-type H^+-ATPases, cation exchangers, and aquaporins, which together enable high rates of fluid secretion and precise regulation of ion balance.

The osmotic stress associated with blood feeding is not only an acute challenge but also a repeated, energetically costly process throughout a mosquito's lifespan (Lehane, 2005). Recurrent cycles of rapid diuresis and osmotic adjustment may impose trade-offs with other physiological functions, including immunity and longevity, particularly under conditions of environmental dehydration or thermal stress. Thus, blood-meal–induced osmotic stress is a central selective pressure shaping mosquito renal physiology and contributes to the tight integration of feeding, reproduction, and stress tolerance in these vectors (Beyenbach, 2003; Piermarini, 2016). Importantly, even though there is a substantial osmotic stress, a blood meal represents a rapid mechanism to replenish water stress, as mentioned in 5.5, where hemolymph osmolality can return to the pre-dehydration levels within a short time (Holmes et al., 2023).

6. Integration of thermal and desiccation tolerance

Thermal and desiccation stresses frequently co-occur in nature, particularly in hot, arid, or seasonally dry environments, and selection often acts on these stressors simultaneously rather than in isolation (Juliano et al., 2002; Holmes and Benoit, 2019; Brown et al., 2023). For mosquitoes, which are small-bodied ectotherms with high surface-area-to-volume ratios, elevated temperatures coincide with low relative humidity, intensifying water-loss rates while simultaneously challenging cellular and enzymatic stability. As a result, many of the physiological mechanisms underlying tolerance to heat and dehydration overlap. These include the induction of HSPs synthesis and other molecular chaperones that stabilize proteins under stress, metabolic suppression that reduces endogenous heat production and respiratory water loss, and enhanced antioxidant defenses that mitigate oxidative damage generated during both thermal stress and dehydration (Hagan et al., 2018; Benoit and Denlinger, 2010; Benoit et al., 2023). From an evolutionary perspective, this shared physiological architecture creates the potential for correlated responses to selection, such that adaptation to one stressor may indirectly alter tolerance to the other. As an example, a recent study focusing on *An. stephensi* across five relative humidities (30–90%) and seven temperatures (16–38 °C) shows that humidity modifies thermal performance of key traits (Johnson et al., 2026). When incorporated into distribution models, humidity reshapes predication of population fitness and areas of mosquito distribution (Johnson et al., 2026), highlighting the need to move beyond temperature-only predictions.

Despite this overlap, physiological conflicts between thermal and desiccation tolerance are common and may constrain simultaneous optimization of both traits. One prominent example is evaporative cooling, which can be an effective short-term response to heat stress but necessarily increases water loss and accelerates dehydration, particularly in low-humidity environments (Chown et al., 2011). Similarly, increased respiratory rates at high temperatures can enhance oxygen delivery but also increase transpiration throughout the tracheal system, further exacerbating the risk of desiccation. Consequently, selection for high thermal tolerance may come at the cost of reduced desiccation resistance, or vice versa, depending on local environmental conditions and the temporal structure of stress exposure.

Empirical studies in mosquitoes reveal mixed evolutionary patterns consistent with these competing pressures. In some taxa and populations, tolerance to heat and desiccation is positively correlated, suggesting either pleiotropy or parallel selection in environments where heat and dryness reliably co-occur (Juliano et al., 2002; Reidenbach et al., 2014). For example, species or populations occupying arid or urban heat-island environments often exhibit enhanced tolerance to both stressors, potentially reflecting selection on shared traits such as cuticular hydrocarbon composition or metabolic suppression (Brown et al., 2023; Chakraborty et al., 2025; Rose et al., 2020). In contrast, other studies demonstrate weak or absent correlations between thermal and desiccation tolerance, indicating that these traits can evolve independently when selection pressures are decoupled or when genetic correlations constrain joint responses (Kellermann et al., 2012; Hoffmann et al., 2003). Such decoupling may occur in habitats where high temperatures coincide with high humidity, or where life-history traits such as larval habitat choice or adult resting behavior buffer individuals from simultaneous exposure to both stressors.

7. Cross-stage and transgenerational effects of thermal and desiccation stress in mosquitoes

Stressors experienced during larval or pupal development can persistently shape adult phenotypes, influencing body size, metabolic reserves, cuticular properties, and tolerance to subsequent environmental challenges (Barreaux et al., 2018; Telang and Wells, 2004; Vinauger and Chandrasegaran, 2024; Evans et al., 2020). For example, larvae exposed to elevated temperatures or reduced water availability could emerge as adults with enhanced resistance to heat or dehydration (Aboagye-Antwi and Tripet, 2010; Ezeakacha and Yee, 2019; Gray, 2013; Ukubuiwe et al., 2018), but these gains frequently come with trade-offs such as reduced longevity, smaller body, or diminished fecundity. These carryover effects indicate that stress responses are integrated across life stages rather than reset at metamorphosis. This suggests that interpretations based solely on adult-stage assays will be limited as optimal larval conditions are rare.

Beyond within-generation carryover, transgenerational effects of thermal and desiccation stress are increasingly recognized as important components of mosquito stress biology (Chakraborty et al., 2025; Sgrò et al., 2016; Heilig et al., 2026; Foo et al., 2019). Mosquito lines maintained at elevated temperatures showed increased thermal resistance, providing evidence that *Aedes* populations can adapt their thermal performance in response to prolonged exposure to higher temperatures (Dennington et al., 2024). Parental exposure to environmental stress can alter egg characteristics, including size, chorion structure, and desiccation tolerance, as well as offspring development rates and survival (Chakraborty et al., 2025; Westby et al., 2025). Such effects may arise through changes in maternal provisioning, altered hormonal signaling, or epigenetic mechanisms that modulate gene expression in the next generation (Carvalho, 2023; Glastad et al., 2019). While these parental effects can enhance offspring tolerance to harsh conditions, trade-offs often occur, including reduced reproductive output, multiple blood meals required for egg deposition, delayed development, or altered population growth trajectories.

These cross-stage and transgenerational effects have significant implications for predicting mosquito responses to environmental change. Reliance on single-stage measurements of stress tolerance risks oversimplifying population-level outcomes, particularly under climate scenarios characterized by increased temperature extremes and episodic drought. Carryover and parental effects can decouple immediate stress resistance from longer-term fitness, leading to nonlinear or counterintuitive population responses (Harrison et al., 2011). Explicit incorporation of these processes into experimental designs, reviews, and predictive models will be essential for forecasting mosquito population dynamics, range expansion, and vectorial capacity under future climatic conditions.

8. Thermal and humidity detection

Mosquitoes rely on highly specialized sensory systems to detect heat as a critical short-range cue during host seeking. Female mosquitoes, particularly species such as *Ae. aegypti* are attracted to thermal stimuli that approximate vertebrate skin temperatures, thereby helping them locate hosts as they approach. Thermosensation is mediated by sensory neurons in the antennae and other head appendages that express temperature-sensitive channels like those in the transient receptor potential (TRP) family; for example, the TRPA1 channel, in conjunction with two opsins, tunes mosquito heat-seeking behavior to host-relevant warmth (Corfas and Vosshall, 2015; Chandel et al., 2024). In addition to these thermoreceptors, ionotropic receptors such as Ir93a and Ir68a contribute to combined heat and humidity detection, with specific sensory neurons playing roles in both thermosensation and short-range humidity gradients that guide host approach and blood feeding (Laursen et al., 2023; Tang et al., 2024). Disrupting these receptors impairs mosquitoes' ability to orient to heat and humidity cues, underscoring the central role of these sensory pathways in feeding behavior.

Humidity detection, or hygroreception, provides mosquitoes with complementary information about host proximity, water availability, and desiccation risk. Elevated humidity gradients emanating from the skin, breath, or moist surfaces near hosts enhance mosquito responsiveness to thermal and olfactory cues, favoring landing and feeding behaviors. At the molecular level, hygrosensation in mosquitoes involves ionotropic receptor pathways similar to those characterized in *Drosophila*, with Ir93a acting as a coreceptor and more specific receptors, such as Ir68a, mediating responses to humid air (Laursen et al., 2023; Tang et al., 2024). Functional studies have shown that distinct humidity-responsive neurons—Dry Cells and Moist Cells—require different IR family members, such as Ir40a and Ir68a, with Ir68a-dependent Moist Cells being especially important for behaviors such as oviposition site seeking in *Ae. aegypti* (Tang et al., 2024). Together, these heat- and humidity-sensing systems allow mosquitoes to fine-tune host-seeking and reproductive behaviors in response to complex environmental cues.

9. Mosquito exposure to cold

9.1 Exposure to cold

Cold tolerance varies substantially among species and life stages for mosquitoes (Moser et al., 2023; Tippelt et al., 2020). Broadly, mosquitoes exhibit either freeze avoidance or chill tolerance rather than true freeze tolerance, except for some specific species (Copeland and Craig, 1990), relying on physiological mechanisms that prevent internal ice formation or mitigate cellular damage at low but nonfreezing temperatures. In general, most mosquitoes will undergo cold knockdown at temperatures near or slightly below freezing (Couper et al., 2024; Reinhold et al., 2018; Jass et al., 2019). Temperate species such as *C. pipiens* and *A. albopictus* demonstrate pronounced seasonal plasticity, entering diapause or producing cold-resistant eggs that enable survival during winter (discussed in more detail below). In contrast, many tropical mosquito species remain highly cold-sensitive, with drastic declines in function at temperatures as high as 5 °C (Mordecai et al., 2019; Reinhold et al., 2018), limiting their geographic distribution and shaping patterns of seasonal abundance and range expansion.

At the physiological level, mosquito cold tolerance involves coordinated biochemical and cellular adjustments that stabilize membranes, maintain ion balance, and protect macromolecules during cold exposure (Hayward and Colinet, 2023; Denlinger and Lee, 2010; Clark and Worland, 2008). Accumulation of cryoprotective metabolites—including sugars, polyols, and certain amino acids—reduces cellular dehydration and membrane phase transitions, while alterations in lipid composition help maintain membrane fluidity at low temperatures. Distinct changes in cytoskeletal structures have been documented as critical for tolerance to cold exposure (Kim et al., 2006). Cold exposure also induces expression of stress-response pathways such as HSPs, antioxidant defenses, and metabolic suppression, which collectively reduce damage associated with chilling injury and oxidative stress.

9.2 Overwintering – dehydration and cold exposure

Overwintering represents a bottleneck in the life cycle of many mosquito species, particularly in temperate and continental climates where individuals must survive prolonged exposure to low temperatures and limited water availability (Denlinger, 2022; Denlinger and Armbruster, 2014). Depending on species and latitude, mosquitoes overwinter as eggs, larvae, or adults, and each strategy entails distinct physiological challenges. Eggs of container-breeding species such as *Ae. albopictus* frequently experiences winter desiccation in addition to cold stress, as breeding habitats dry or freeze (Urbanski et al., 2010; Denlinger and Armbruster, 2014). In contrast, adult overwintering species, including *Cx. pipiens*, enter reproductive diapause in sheltered microhabitats such as caves, basements, or culverts, where low temperatures reduce metabolic demands, but water loss remains a persistent threat (Robich and Denlinger, 2005; Denlinger and Armbruster, 2014). Thus, overwintering success depends on integrating cold-tolerance mechanisms with strategies that limit dehydration during extended periods of inactivity. Importantly, as food sources are limited for adults and eggs have finite energy reserves, the depletion of energy reserves due to chronic bouts of stress or overwintering periods that are too warm can be a limiting factor in the overwintering of mosquitoes (Rozsypal et al., 2021; Benoit, Patrick, et al., 2010).

Physiologically, overwintering mosquitoes rely on a suite of overlapping mechanisms that mitigate both freezing injury and water loss. Diapause is central to this process, involving metabolic suppression, accumulation of energy reserves (primarily lipids), and enhanced stress tolerance (Denlinger and Armbruster, 2014). Cold exposure can exacerbate dehydration through chilling-induced disruption of ion and water balance, a phenomenon well documented in insects more broadly (Sinclair et al., 2015). In mosquito eggs, enhanced desiccation resistance is achieved through specialized chorionic structures and changes in cuticular permeability that slow water loss and confer resistance to cold-induced damage (Urbanski et al., 2010; Rezende et al., 2008). In adults, increased levels of cryoprotective molecules such as glycerol and sugars may function in two ways: stabilizing membranes at low temperatures and reducing osmotic stress associated with dehydration (Benoit and Denlinger, 2007; Urbanski et al., 2010). These shared protective pathways highlight the strong physiological coupling between cold and desiccation stress during overwintering.

From an ecological and evolutionary perspective, the interaction between dehydration and cold exposure during overwintering has essential consequences for mosquito distribution, population dynamics, and disease risk (Benoit, 2010; Denlinger, 2022). Survival through winter determines spring population size and, consequently, the timing and intensity of disease transmission (Kraemer et al., 2019). Selection imposed by harsh winters can favor genotypes with enhanced diapause depth, reduced water loss rates, or improved tolerance to combined cold–dry stress, potentially constraining or facilitating range expansion under climate change. For example, the spread of *Ae. albopictus* into temperate regions has been linked to the evolution of eggs that are more cold- and desiccation-resistant (Thomas et al., 2012). As winters become shorter but more variable, understanding how dehydration and cold jointly shape overwintering success will be essential for predicting future mosquito persistence and the emergence of vector-borne diseases.

10. Evolutionary genetics of stress tolerance

10.1 Standing genetic variation and local adaptation

Population genetic and genomic studies demonstrate that mosquito species harbor substantial standing genetic variation for traits related to thermal and desiccation tolerance, providing the evolutionary raw material necessary for rapid adaptive responses to spatially and temporally heterogeneous environments (Anopheles gambiae 1000 Genomes Consortium, 2020; Crawford et al., 2025; Gloria-Soria et al., 2016; Haba et al., 2025; Kotsakiozi et al., 2018). Experimental and comparative work across mosquito taxa indicates that populations exhibit substantial genetic differences due to heritable variation (Chakraborty et al., 2025; Dennington et al., 2024; Batz et al., 2020; Zani et al., 2005). This standing variation is significant under rapid climate change, as it allows populations to respond to selection without requiring novel mutations.

Strong evidence for local adaptation comes from clinal patterns observed along latitudinal, altitudinal, and aridity gradients. In *An gambiae*, the frequency of chromosomal inversions such as 2La varies predictably across moisture gradients in Africa and is closely associated with desiccation resistance and water balance physiology, implicating selection imposed by arid environments (Gray et al., 2009; Ayala et al., 2019). Similarly, *Ae. aegypti* exhibit differentiation in stress tolerance and life-history traits between populations (Chakraborty et al., 2025; Dennington et al., 2025; Soudi et al., 2023; Bennett et al., 2021), consistent with adaptation to local thermal regimes despite ongoing gene flow.

Urbanization represents an additional axis of environmental heterogeneity that can drive local adaptation in mosquito populations (Rose et al., 2020; Haba et al., 2025). Urban mosquitoes experience elevated temperatures, altered humidity, and novel larval habitats relative to rural populations, creating intense selective pressures on thermal and desiccation tolerance traits

(Chakraborty et al., 2025; Fyie et al., 2023; Murdock et al., 2017). Landscape genomic analyses of *Ae. aegypti* have identified allele frequency shifts and candidate loci associated with temperature and precipitation variables (Rose et al., 2020), suggesting that standing genetic variation facilitates adaptation to urban microclimates. Although explicit urban–rural genomic contrasts remain limited, available evidence suggests that mosquitoes have the capacity to adapt to anthropogenic environments. These patterns suggest that local adaptation to environmental stress commonly proceeds through modulation of gene expression and physiological tuning, enabling flexible yet heritable responses to local thermal and hydric conditions.

10.2 Experimental evolution and functional validation to establish stress-responsive mechanisms

Experimental evolution provides some of the clearest evidence that specific insects can rapidly adapt to stress, but a major focus of these studies has been on *Drosophila* (Gefen et al., 2006; Sota, 1993a; Telonis-Scott et al., 2006; Mesas et al., 2021). Laboratory passage experiments in mosquitoes demonstrate that tolerance to high temperature can shift over relatively few generations (Dennington et al., 2024). Artificial selection for desiccation resistance has only been examined for mosquitoes in a single study without replicated lines (Sota and Mogi, 1992b), but has been assessed in detail for *Drosophila* (Gefen et al., 2006; Telonis-Scott et al., 2006; Gibbs et al., 1997). Combined selection experiments targeting increased (or decreased) thermal and dehydration tolerance could provide complementary insights into the adaptive potential of mosquitoes in relation to these stresses. In mosquitoes, behavioral thermoregulation and humidity seeking are most often thought of as immediate, plastic responses that enable individuals to mitigate acute risks of overheating or dehydration by selecting favorable microhabitats. These stress-avoiding traits are likely evolvable as substantial differences in behavioral aspects has been noted between population (Ajayi, Susanto, et al., 2024; Wynne et al., 2024; Rose et al., 2020), as interspecific and intraspecific variation in resting-site selection, activity timing, and humidity preference could be subject to selection and play a critical role in shaping mosquito persistence, which could be examined under targeted selection based studies.

Functional validation approaches increasingly complement experimental evolution by establishing causal links between candidate genes and stress tolerance phenotypes in mosquitoes (Airs and Bartholomay, 2017; Yadav et al., 2023; Palli, 2023).

RNA interference (RNAi) studies in mosquitoes have identified heat shock proteins (particularly HSP70), antioxidant enzymes, aquaporins, and genes involved in lipid metabolism and cuticle biosynthesis as contributors to thermal and desiccation resistance (Benoit et al., 2010). Targeting trehalose, either through RNAi to suppress enzyme activity or transporters, directly impacts dehydration tolerance and response to dry conditions in mosquitoes (Liu et al., 2013; Hagan et al., 2018). The application of CRISPR/Cas9 genome editing in *Ae. aegypti* has further enabled precise gene manipulation, allowing direct tests of gene function under stress (Venkataraman et al., 2023; Laursen et al., 2023; Raji et al., 2019). Together, these approaches are likely to confirm that many stress-response genes exhibit pleiotropic effects on survival, development, and reproduction. Ultimately, integrating experimental evolution and functional validation is essential for resolving the genetic architecture of thermal and desiccation tolerance in mosquitoes.

11. Role of the microbiome in tolerance to thermal and dehydration stress

The role of the microbiome as a modulator of thermal and dehydration tolerance remains underrepresented in the study of mosquito stress physiology, despite growing evidence that microbial associates shape mosquito metabolism, cuticle formation, and overall stress resilience (Onyango et al., 2020; Ware-Gilmore et al., 2021). In mosquitoes, larval and adult microbiomes influence nutrient acquisition, energy storage, and lipid metabolism (Dickson et al., 2017; Didion et al., 2021; Valzania et al., 2018), processes that are tightly linked to heat tolerance and water balance. Microbially derived metabolites and micronutrients can affect cuticular composition and general metabolite profiles (Kelley et al., 2025; Duplais et al., 2021; Graber and Corrie, 2026), including cuticle thickness/darkening, hydrocarbon profiles, and accumulation of protective osmolytes that regulate water loss and can prevent dehydration-induced damage. The presence or absence of *Wolbachia* is known to have distinct effects of survival of mosquitoes during sub-optimal conditions (Ross et al., 2020, 2023). This is critical as *Wolbachia*-based control protocols are important, but may be impaired under periods of stress (Ross et al., 2023; Hien et al., 2021; Ross et al., 2020).

Thermal and dehydration stress can reshape the mosquito microbiome, generating feedback that further influences host performance (Lv et al., 2023; Onyango et al., 2020; Guégan et al., 2018). Exposure to elevated temperatures or low-humidity conditions has been shown to alter microbial diversity and community structure, with downstream consequences for digestion, oxidative stress management, and immune function. In contrast to thermal tolerance, the role of the microbiome in dehydration resistance is comparatively poorly characterized, with few studies directly testing how microbial manipulation affects water loss rates, hemolymph osmolality, or survival under desiccating conditions. The indirect evidence from studies linking microbiota to cuticle integrity, energy reserves, and ion transport suggests that microbial effects on dehydration tolerance are likely, but remain largely untested in mosquitoes. Recognizing that the microbiome likely contributes to dehydration and thermal tolerance, highlights a critical gap in mosquito biology and underscores the need for studies that jointly examine microbiome composition and environmental variability.

12. Implications of thermal and dehydration stress for vectorial capacity and pathogen transmission

Mosquito vectorial capacity is susceptible to environmental stressors (Fig. 5), including temperature extremes and dehydration, because these factors simultaneously affect adult survival, feeding behavior, immune competence, and pathogen development. Thermal stress has nonlinear and species-specific effects on vector competence. Moderate warming generally accelerates pathogen replication and shortens the extrinsic incubation period (EIP) for malaria parasites, dengue, Zika, and other arboviruses, potentially increasing transmission rates (Mordecai et al., 2019; Alto and Bettinardi, 2013). However, extreme temperature (high and low) can sharply reduce mosquito survival, suppress immune function, and disrupt feeding behavior, often counterbalancing the pathogen-facilitating effects of moderate warming (Murdock et al., 2012; Paaijmans et al., 2010). For example, in *Anopheles* mosquitoes, exposure to high thermal stress impairs melanization and antimicrobial peptide activity, increasing susceptibility to *Plasmodium* (Martin and Hillyer, 2024; Condé et al., 2021). Yet concomitant reductions in lifespan may limit the proportion of vectors that survive long enough to transmit malaria effectively.

Thermal stress

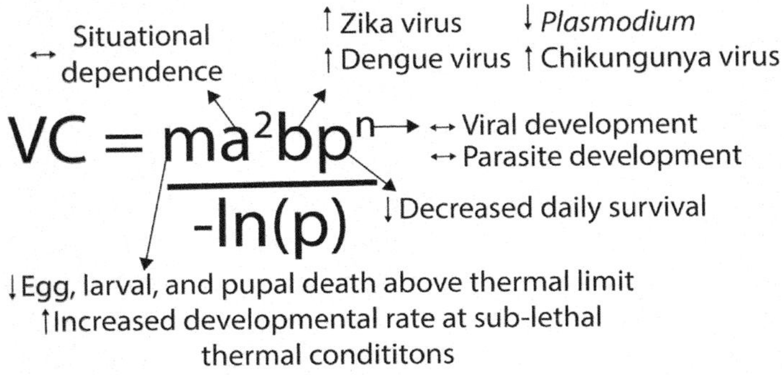

Key Takeaways of Thermal Stress

1. The "Inverted U" Effect: Transmission does not increase indefinitely with heat. It follows a curve where it rises until the optimum is reached, then crashes once the temperature causes physiological stress or death to the mosquito.
2. The Lifespan-Incubation Race: For transmission to occur, a mosquito must live longer than the Extrinsic Incubation Period (EIP). Thermal stress often shortens the mosquito's life so much that it dies before it ever becomes infectious, even if the pathogen is replicating faster.
3. Asymmetric Response: Research indicates that Dengue (and other viruses like Zika) generally has a higher thermal tolerance than Malaria. This means that as the planet warms, we may see a "disease swap" in some regions where it becomes too hot for Malaria but perfectly suited for Dengue.

Dehydration stress

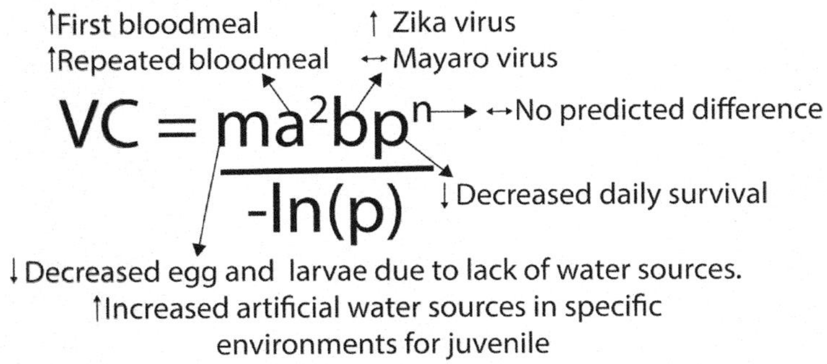

Key Takeaways of Dehydration Stress

1. Dry conditions increase mosquito biting and host switching, dramatically raising opportunities for disease transmission.
2. Dehydration makes mosquitoes more susceptible to infection by weakening internal barriers, even without increasing viral load.
3. Disease risk is highest at intermediate dryness, where dehydration boosts biting but still allows mosquitoes to survive long enough to transmit pathogens.

Fig. 5 Links between thermal/dehydration stress tolerance, pathogen transmission, and vectorial capacity. Conceptual framework linking thermal and desiccation tolerance to mosquito longevity, immunity, pathogen development, and

Interestingly, infection with certain pathogens has been shown to alter mosquito thermal biology, thereby increasing their potential exposure to thermal stress and, subsequently, dehydration. For example, *Ae. aegypti* infected with Sindbis virus exhibited a higher thermopreference than non-infected mosquitoes, while *Ae. japonicus* had a lower thermopreference when infected with *Dirofilaria immitis* (Hug et al., 2024). Another study in *An. stephensi* showed that infection with *Plasmodium yoelii* and/or various entomopathogenic fungi, while affecting the mosquito thermal behavior, did not directly impact their thermal preference (Blanford et al., 2009). This suggests a pathogen-mosquito species-specific effect on the insects' thermal behavior, but more studies with other models are needed before generalizing to other mosquito and pathogen species.

Temperature alterspathogen development in complex ways. Arboviruses such as dengue virus and Zika virus replicate more rapidly at elevated temperatures, reducing the EIP and, theoretically, increasing transmission potential (Perdomo et al., 2025); however, excessive heat can destabilize viral replication in the mosquito midgut or salivary glands (Tesla et al., 2018). Similarly, *Plasmodium* develops more rapidly within optimal thermal ranges. Still, both heat shock and cold stress can lead to aberrant parasite development or vector mortality, generating highly nonlinear relationships between environmental temperature and transmission risk (Murdock et al., 2012; Paaijmans et al., 2010). Daily temperature fluctuations add further complexity, as brief periods of thermal extremes can disproportionately affect pathogen growth relative to mosquito physiology, highlighting the importance of realistic environmental modeling in predicting disease dynamics.

Dehydration stress interacts with thermal effects in ways that can either amplify or mitigate transmission potential. High temperatures increase evaporative water loss, which may accelerate dehydration, reduce adult longevity, and alter host-seeking and blood-feeding behavior (Benoit et al., 2010; Hagan et al., 2018; Holmes et al., 2025). Dehydrated mosquitoes, however, often display increased blood-feeding propensity and activity,

transmission potential. The figure emphasizes nonlinear and context-dependent effects of environmental stress on vector competence and disease dynamics. m represents vector density relative to the host, a is the daily probability of host feeding (squared to account for the two feeding events required for pathogen transmission), b is transmission efficiency or vector competence, p is the vector's daily survival probability, and n is the number of days required for the pathogen to disseminate from the mosquito midgut to the site of exit or extrinsic incubation period (EIP).

potentially offsetting reduced lifespan and enhancing transmission opportunities (Hagan et al., 2018; Manzano-Alvarez et al., 2023; Bozic et al., 2024), even causing male mosquitoes to ingest blood (Bozic et al., 2024). Furthermore, dehydration appears to directly affect viral transmission (Abu et al., 2024; Accoti et al., 2025; Wahaab et al., 2025). This behavioral response is particularly relevant under drought or urban heat conditions, where dehydration is common. Moreover, dehydration, combined with elevated temperature, can alter mosquito immunity, modulating susceptibility to pathogens in ways that depend on both the magnitude and duration of the stressor (Murdock et al., 2012; Alto and Bettinardi, 2013).

From an evolutionary perspective, selection for thermal tolerance may extend transmission seasons and facilitate vector range expansion under climate change. Stress-tolerant genotypes are more likely to survive extreme temperatures, maintain feeding activity, and sustain pathogen development within optimal windows, potentially reshaping mosquito–pathogen coevolutionary dynamics (Kraemer et al., 2019; Ryan et al., 2019). However, trade-offs among stress tolerance, longevity, immunity, and reproduction can yield complex, context-dependent outcomes for disease transmission. Understanding these interactions is critical for predicting vector-borne disease dynamics in an era of increasing thermal variability and extreme weather events.

13. Climate change, predictive models, and vector control

Climate change is altering global thermal and hydric environments, increasing variability in temperature, humidity, and precipitation patterns that directly influence mosquito biology, distribution, and disease-transmission potential (Mordecai et al., 2019; Ryan et al., 2019). Mosquitoes are ectothermic organisms whose physiological rates—such as development, mortality, fecundity, and vector competence—are all strongly temperature-dependent, and can respond nonlinearly to changing climatic conditions. Higher temperatures and altered rainfall regimes can both expand suitable habitat in some regions and impose stress in others, affecting mosquito survival and disease risk.

To forecast these responses, mechanistic predictive models that incorporate the biological processes underlying mosquito life cycles are more robust than simple statistical correlations (Johnson et al., 2015; Ryan et al., 2019).

Trait-based models link environmental drivers, such as temperature, to experimentally derived thermal performance curves for key biological traits, enabling the prediction of how vector populations and transmission potential change across climatic gradients (Mordecai et al., 2019; Ryan et al., 2019; Dennington et al., 2024). Integrating hydric stress (e.g., effects of desiccation and drought) further refines these models, as water availability affects breeding habitat and larval survival.

Climate change can exert selective pressure, driving evolutionary and phenotypic responses in mosquito populations. Experimental studies have demonstrated that populations of *Ae. aegypti* exhibit significant variation in thermal tolerance, and that thermal performance traits can shift over multiple generations under different temperature regimes, indicating potential adaptive responses to warming climates (Dennington et al., 2024, 2025). Genetic variation in heat tolerance has been documented in other species (Hoffmann et al., 2003; Kellermann et al., 2012), illustrating the capacity for mosquitoes to adapt to climate change in ways that may alter disease risk and range expansions (Couper et al., 2024, 2025).

In the context of vector control, understanding stress responses is critical because environmental stress interacts with the efficacy of control measures. Thermal and desiccation stress can influence mosquito behavior and physiology, thereby affecting exposure to control tools, such as insecticide-treated nets and indoor residual spraying. For example, insecticide-resistant populations may retain thermal tolerance, and phenotypic plasticity can buffer mosquitoes against environmental extremes without reducing resistance traits, complicating control efforts (Ioannou et al. 2024), but there have been documented trade-offs with starvation and desiccation tolerance (Savvidou et al., 2025). This underscores the need to integrate ecological and evolutionary dynamics into control strategies.

Importantly, macroclimatic variables derived from weather stations used in modeling often poorly reflect the microclimates actually experienced by mosquitoes and other insects (Lembrechts et al., 2019; Pincebourde and Woods, 2020; Murdock et al., 2017; Paaijmans, Imbahale, et al., 2010), which typically occupy buffered environments such as shaded vegetation, leaf litter, wall and roof crevices, indoor resting sites, and the interiors of natural and artificial containers. These microhabitats can differ substantially from ambient conditions, exhibiting reduced temperature extremes, higher and more stable humidity, and altered daily thermal cycles. As a result, mosquitoes may avoid acute thermal or desiccation stress by moving to these favorable microclimates even when regional weather data indicate conditions that exceed their experimentally defined physiological limits. This growing recognition of macro–microclimate decoupling has important implications for predicting mosquito survival, seasonal dynamics, and geographic range limits, particularly under climate change scenarios (Murdock et al., 2017; Faridah et al., 2022; Wimberly et al., 2020). Overall, laboratory measurements of thermal tolerance, desiccation resistance, or performance curves may overestimate, or in some cases underestimate, stress exposure if mosquitoes routinely exploit microhabitats that buffer environmental extremes. Ignoring microclimate use can therefore lead to inaccurate projections of population persistence, disease transmission risk, and range expansion or contraction (Murdock et al., 2017; Faridah et al., 2022; Wimberly et al., 2020; Kerr et al., 2025), underscoring the need to integrate fine-scale habitat conditions into mosquito physiology and ecological models.

Process-based models can simulate the impacts of novel control approaches, such as *Wolbachia* releases and incompatible insect techniques (Ross et al., 2023; Hien et al., 2021; Ross et al., 2020; Foo et al., 2019). These interventions depend on climatic suitability for *Wolbachia*-infected mosquitoes and vector population dynamics; thus, models that include climate effects on life history traits and mosquito metapopulation structure can help optimize release strategies (e.g., (Lim et al., 2024)). Climate-informed predictive frameworks can identify periods of environmental vulnerability, improve the timing of interventions, and anticipate evolutionary responses that may erode the effectiveness of control measures. Overall, integrating mechanistic climate models, trait variation, adaptive potential, and intervention timing enhances our capacity to forecast vector responses and design sustainable vector management strategies in a changing climate.

14. Male stress tolerance and its role in control

Males have received less attention in the literature with lower efforts to synthesize male-specific thermal and desiccation constraints, as males were thought to be less critical to disease transmission (Benoit and Denlinger, 2007; Chakraborty et al., 2025; Savvidou et al., 2025; Mirieri et al., 2025), despite their clear relevance to mating and insect control strategies such as the sterile insect technique (SIT) and incompatible insect technique (IIT). Available studies indicate that males are often more sensitive to environmental stressors than females (Chakraborty et al., 2025; Mirieri et al., 2025). Thermal stress can impair male flight performance, courtship behavior, and sperm production or viability, leading to reduced mating success even when female survival and fecundity remain largely unaffected (Bader and Williams, 2012; Pekľanská et al., 2025). Likewise, desiccation stress may disproportionately limit male activity, shorten the effective mating window, and accelerate depletion of energetic reserves required for sustained mate searching (Benoit and Denlinger, 2007; Benoit, Lopez-Martinez, et al., 2010; Savvidou et al., 2025). A more explicit synthesis of male-focused thermal and desiccation constraints is therefore essential in review treatments to fully capture how environmental stress shapes population dynamics and to improve predictions relevant to both ecology and applied control programs.

15. Conclusions and future directions

Thermal and desiccation tolerance are central axes of mosquito evolution, shaping geographic distributions, seasonal abundance, and vectorial capacity. Variations in these traits influence survival under extreme environmental conditions, reproductive success, and interactions with predators and pathogens. Recent advances in genomics, transcriptomics, and proteomics have begun to elucidate the molecular and cellular mechanisms underpinning these tolerances, including heat shock proteins, metabolic regulation, cuticular adaptations, and osmoregulatory processes. Linking these physiological traits to ecological and evolutionary dynamics provides critical insights into how mosquito populations adapt to local environmental pressures and how these adaptations influence pathogen transmission.

Looking ahead, a multidisciplinary approach will be essential for predicting mosquito responses to global change. Integrative multi-omics approaches, spanning transcriptomics, genomics, metabolomics, and proteomics across life stages and ecologically relevant stress regimes, can uncover regulatory plasticity often undetected under constant laboratory conditions (Carrington et al., 2013; Paaijmans et al., 2010). Extending these studies to chromatin accessibility and DNA methylation will clarify how short-term plastic responses may become stabilized or constrained over evolutionary time. This is critical, as acute climatic extremes, such as heat waves and episodic drought pulses, can impose rapid and severe physiological stress that overwhelms organismal tolerance limits, driving immediate population collapse, compared with gradual, long-term warming trends do not. Population genomics and experimental evolution offer complementary insights into rapid adaptation (Anopheles gambiae 1000 Genomes Consortium, 2020; Crawford et al., 2025; Gloria-Soria et al., 2016; Haba et al., 2025; Kotsakiozi et al., 2018), but functional validation remains limited to linking to environmental conditions. Experimental evolution under fluctuating thermal and humidity regimes better captures natural selective pressures than constant extreme conditions, and coupling these experiments with time-series RNA-seq and whole-genome resequencing can reveal the tempo and mode of adaptation. Integration with CRISPR-based perturbations will be essential for establishing causal links between genotype and phenotype.

There are also several critical gaps that limit our ability to predict mosquito population persistence and disease risk under rapid climate change. Most thermal data are derived from constant laboratory conditions that poorly reflect the microclimates mosquitoes actually experience, such as container walls, indoor resting sites, and shaded vegetation, creating uncertainty in linking physiological limits to real exposure. Moreover, research emphasizes plasticity and adaptation while largely neglecting upper physiological ceilings and irreversible failure thresholds that may drive abrupt population collapse during extreme heat events. Thermal biology is insufficiently integrated across the entire life cycle and gonotrophic cycle, despite evidence that heat stress during development or blood feeding can impose lasting constraints on reproduction and survival. Male thermal biology remains a blind spot, despite the fact that male fertility and survival strongly influence population growth and control strategies. In addition, carry-over and transgenerational effects of thermal stress are rarely examined, potentially obscuring how resilience or vulnerability propagates across generations. Behavioral thermoregulation and avoidance are typically treated as proximate responses rather than evolvable traits that may strongly mediate selection. Lastly, thermal tolerance is rarely studied as interacting in concert with dehydration, nutrition, and microbiome composition, even though these stressors interact to shape survival and thermal tolerance. Addressing these gaps will extend our understanding of how thermal and dehydration stress is experienced, constrained, and integrated across biological scales.

A final critical area remains the link between stress tolerance and disease transmission, and vectorial capacity, which has been studied only for specific mosquitoes in relation to particular stresses (Accoti et al., 2025;

Abu et al., 2024; Villena et al., 2022; Manzano-Alvarez et al., 2023). Evolutionary changes in thermal and desiccation tolerance may alter immunity, metabolism, and microbiome composition, producing non-linear effects on pathogen development and transmission. Addressing this challenge requires interdisciplinary approaches combining evolutionary physiology, vector biology, and epidemiological modeling. Developing predictive, trait-based models that integrate physiological limits, plasticity, genetic variation, and evolutionary constraints is essential for forecasting mosquito distributions and disease risk under climate change. Such models will bridge basic research and applied vector control, informing targeted interventions for populations most likely to survive extreme conditions. Overall, a holistic understanding of thermal and desiccation tolerance will be indispensable to understanding the evolutionary biology and public health of mosquitoes.

Acknowledgments

This development of this review was supported by the National Institute of Allergy and Infectious Diseases of the National Institutes of Health under Award (R01AI148551 and R21AI166633) and the Department of Biological Sciences at the University of Cincinnati to J. B. B. and National Science Foundation (NSF-IOS Award #2114127), USDA NIFA VA-160160, the Global Change Center, and the Department of Biochemistry at Virginia Tech (VA, USA) to C. L. We apologize if specific references or topics were omitted, as this is an expansive review of thermal and dehydration in mosquitoes that required we limit discussion of specific areas.

References

Abarca, M., Parker, A.L., Larsen, E.A., et al., 2024. How development and survival combine to determine the thermal sensitivity of insects. PLoS One 19 (1), e0291393.

Abbasi, E., 2025. Climate change and vector-borne disease transmission: the role of insect behavioral and physiological adaptations. Integr. Organismal Biol. (Oxford, Engl.) 7 (1) obaf011.

Aboagye-Antwi, F., Tripet, F., 2010. Effects of larval growth condition and water availability on desiccation resistance and its physiological basis in adult *Anopheles gambiae* sensu stricto. Malar. J. 9 (1), 225.

Abu, A.E.I., Margaret, B., Accoti, A., Sylla, M., Dickson, L.B., 2024. Low humidity enhances Zika virus infection and dissemination in *Aedes aegypti* mosquitoes. mSphere 9 (8), e0040124.

Accoti, Margaret Becker, A., Abu, A.E.I., et al., 2025. Dehydration-induced Ae-Aper50 regulates midgut infection in *Aedes aegypti* mosquitoes. mBio 16 (3), e0120724.

Airs, P.M., Bartholomay, L.C., 2017. RNA interference for mosquito and mosquito-borne disease control. Insects 8 (1), 4.

Ajayi, O.M., Marlman, J.M., Gleitz, L.A., et al., 2022. Behavioral and postural analyses establish sleep-like states for mosquitoes that can impact host landing and blood feeding. J. Exp. Biol. 225 (11). https://doi.org/10.1242/jeb.244032.

Ajayi, O.M., Oyen, K.J., Davies, B., et al., 2024. Egg hatching success is influenced by the time of thermal stress in four hard tick species. J. Med. Entomol. 61 (1), 110–120.

Ajayi, O.M., Susanto, E.E., Wang, L., et al., 2024. Intra-species quantification reveals differences in activity and sleep levels in the yellow fever mosquito, *Aedes aegypti*. Med. Vet. Entomol. 38 (4), 482–494.

Alagar Boopathy, L.R., Suleima, J.-T., Alecki, C., Vera, M., 2022. Mechanisms tailoring the expression of heat shock proteins to proteostasis challenges. J. Biol. Chem. 298 (5), 101796.

Alto, B.W., Bettinardi, D., 2013. Temperature and dengue virus infection in mosquitoes: independent effects on the immature and adult stages. Am. J. Tropical Med. Hyg. 88 (3), 497–505.

Alto, B.W., Juliano, S.A., 2001. Temperature effects on the dynamics of *Aedes albopictus* (Diptera: Culicidae) populations in the laboratory. J. Med. Entomology 38 (4), 548–556.

Amundrud, S.L., Srivastava, D.S., 2020. Thermal tolerances and species interactions determine the elevational distributions of insects. Glob. Ecol. Biogeogr.: A J. Macroecol. 29 (8), 1315–1327.

Angilletta Jr., M.J., 2009. Thermal Adaptation: A Theoretical and Empirical Synthesis. Oxford University Press.

Angilletta Jr, M.J., Huey, R.B., Frazier, M.R., 2010. Thermodynamic effects on organismal performance: is hotter better? Physiol. Biochem. Zool. 83 (2), 197–206.

Anopheles gambiae 1000 Genomes Consortium, 2020. Genome variation and population structure among 1142 mosquitoes of the African Malaria vector species *Anopheles gambiae* and *Anopheles coluzzii*. Genome Res. 30 (10), 1533–1546.

Arcaz, A.C., Huestis, D.L., Dao, A., et al., 2016. Desiccation tolerance in *Anopheles coluzzii*: the effects of spiracle size and cuticular hydrocarbons. J. Exp. Biol. 219 (Pt 11), 1675–1688.

Arun, S., Shriram, A.N., Muruganandam, N., Thamizhmani, R., 2017. Expression of heat shock proteins (HSPs) in *Aedes aegypti* (L) and *Aedes albopictus* (Skuse) (Diptera: Culicidae) larvae in response to thermal stress. Acta Tropica 167 (March), 121–127.

Atkinson, D., 1994. Temperature and organism size—A biological law for ectotherms? Advances in Ecological Research. Advances in Ecological Research. Elsevier.

Ayala, Simo Zhang, D., Chateau, M., et al., 2019. Association mapping desiccation resistance within chromosomal inversions in the African Malaria vector *Anopheles gambiae*. Mol. Ecol. 28 (6), 1333–1342.

Bader, C.A., Williams, C.R., 2012. Mating, ovariole number and sperm production of the dengue vector mosquito *Aedes aegypti* (L.) in Australia: broad thermal optima provide the capacity for survival in a changing climate. Physiol. Entomol. 37 (2), 136–144.

Banfi, D., Bianchi, T., Mastore, M., Brivio, M.F., 2025. The role of heat shock proteins in insect stress response, immunity, and climate adaptation. Insects 16 (7), 741.

Barreaux, A.M.G., Stone, C.M., Barreaux, P., Koella, J.C., 2018. The relationship between size and longevity of the malaria vector *Anopheles gambiae* (s.s.) depends on the larval environment. Parasites & Vectors 11 (1), 485.

Barr, J.S., Lindsay, E.M., Tate, A.T., Hillyer, J.F., 2024. Warmer environmental temperature accelerates aging in mosquitoes, decreasing longevity and worsening infection outcomes. Immun. & Ageing: I & A 21 (1), 61.

Batz, Z.A., Armbruster, P.A., 2018. Diapause-associated changes in the lipid and metabolite profiles of the Asian Tiger mosquito, *Aedes albopictus*. J. Exp. Biol. 221 (Pt 24) jeb189480.

Batz, Z.A., Clemento, A.J., Fritzenwanker, J., Ring, T.J., Garza, J.C., Armbruster, P.A., 2020. Rapid adaptive evolution of the diapause program during range expansion of an invasive mosquito. Evolution; Int. J. Org. Evolution 74 (7), 1451–1465.

Bayoh, M.N., Lindsay, S.W., 2004. Temperature-related duration of aquatic stages of the afrotropical malaria vector mosquito *Anopheles gambiae* in the laboratory. Med. Vet. Entomol. 18 (2), 174–179.

Beck-Johnson, L.M., Nelson, W.A., Paaijmans, K.P., Read, A.F., Thomas, M.B., Bjørnstad, O.N., 2017. The importance of temperature fluctuations in understanding mosquito population dynamics and malaria risk. R. Soc. Open. Sci. 4 (3), 160969.

Bennett, K.L., McMillan, W.O., Loaiza, J.R., 2021. The genomic signal of local environmental adaptation in *Aedes aegypti* mosquitoes. Evolut. Appl. 14 (5), 1301–1313.

Benoit, J.B., 2010. Water management by dormant insects: comparisons between dehydration resistance during summer aestivation and winter diapause. Prog. Mol. Subcell. Biol. 49, 209–229.

Benoit, J.B., Denlinger, D.L., 2007. Suppression of water loss during adult diapause in the northern house mosquito, *Culex pipiens*. J. Exp. Biol. 210 (Pt 2), 217–226.

Benoit, J.B., Denlinger, D.L., 2010. Meeting the challenges of on-host and off-host water balance in blood-feeding arthropods. J. Insect Physiol. 56 (10), 1366–1376.

Benoit, J.B., Lazzari, C.R., Denlinger, D.L., Lahondère, C., 2019. Thermoprotective adaptations are critical for arthropods feeding on warm-blooded hosts. Curr. Opin. Insect Sci. 34 (August), 7–11.

Benoit, J.B., Lopez-Martinez, G., Elnitsky, M.A., Lee Jr, R.E., Denlinger, D.L., 2009. Dehydration-induced cross tolerance of *Belgica antarctica* larvae to cold and heat is facilitated by trehalose accumulation. Comp. Biochem. Physiol. Part. A, Mol. & Integr. Physiol. 152 (4), 518–523.

Benoit, J.B., Lopez-Martinez, G., Patrick, K.R., Phillips, Z.P., Krause, T.B., Denlinger, D.L., 2011. Drinking a hot blood meal elicits a protective heat shock response in mosquitoes. Proc. Natl Acad. Sci. U S Am. 108 (19), 8026–8029.

Benoit, J.B., Lopez-Martinez, G., Phillips, Z.P., Patrick, K.R., Denlinger, D.L., 2010. Heat shock proteins contribute to mosquito dehydration tolerance. J. Insect Physiol. 56 (2), 151–156.

Benoit, J.B., McCluney, K.E., DeGennaro, M.J., Dow, J.A.T., 2023. Dehydration dynamics in terrestrial arthropods: from water sensing to trophic interactions. Annu. Rev. Entomol. 68 (January), 129–149.

Benoit, J.B., Patrick, K.R., Desai, K., Hardesty, J.J., Krause, T.B., Denlinger, D.L., 2010. Repeated bouts of dehydration deplete nutrient reserves and reduce egg production in the mosquito *Culex pipiens*. J. Exp. Biol. 213 (Pt 16), 2763–2769.

Beyenbach, K.W., 2003. Transport mechanisms of diuresis in malpighian tubules of insects. J. Exp. Biol 206 (Pt 21), 3845–3856.

Bick, G.H., Penn, G.H., 1947. Resistance of mosquito larvae and pupae to experimental drought. Ann. Entomological Soc. Am. 40 (1), 82–86.

Blanford, S., Read, A.F., Thomas, M.B., 2009. Thermal behaviour of *Anopheles stephensi* in response to infection with malaria and fungal entomopathogens. Malar. J. 8 (April), 72.

Blomquist, G.J., Ginzel, M.D., 2021. Chemical ecology, biochemistry, and molecular biology of insect hydrocarbons. Annu. Rev. Entomol. 66 (1), 45–60.

Bowler, K., 2005. Acclimation, heat shock and hardening. J. Therm. Biol. 30 (2), 125–130.

Bozic, J., Joseph, R.E., Krizek, R.S., et al., 2024. Revisiting the paradigm of anhematophagy in male mosquitoes. bioRxiv. https://doi.org/10.1101/2024.10.08.617226. October 8.

Branham, E.J., Nelson, I.E., Rochlin, I., et al., 2025. Fantastic feasts and where to find them: mosquito (Diptera: Culicidae) sugar feeding and survivorship on endemic flowers of arid scrublands. Arthropod-Plant Interact 19 (3). https://doi.org/10.1007/s11829-025-10151-3.

Briegel, H., 1990. Metabolic relationship between female body size, reserves, and fecundity of *Aedes aegypti*. J. Insect Physiol. 36 (3), 165–172.

Brown, J.J., Pascual, M., Wimberly, M.C., Johnson, L.R., Murdock, C.C., 2023. Humidity - the overlooked variable in the thermal biology of mosquito-borne disease. Ecol. Lett. 26 (7), 1029–1049.

Carlassara, M., Khorramnejad, A., Oker, H., et al., 2024. Population-specific responses to developmental temperature in the arboviral vector *Aedes albopictus*: implications for climate change. Glob. Change Biol. 30 (3), e17226.

Carrington, L.B., Armijos, M.V., Lambrechts, L., Barker, C.M., Scott, T.W., 2013. Effects of fluctuating daily temperatures at critical thermal extremes on *Aedes aegypti* life-history traits. PLoS One 8 (3), e58824.

Carvalho, C.F., de, 2023. Epigenetic effects of climate change on insects. Curr. Opin. Insect Sci. 57 (101029), 101029.

Chakraborty, A., Sgrò, C.M., Mirth, C.K., 2025. Untangling plastic responses to combined thermal and dietary stress in insects. Curr. Opin. Insect Sci. 68 (101328), 101328.

Chakraborty, S., Shah, D., Dayal, D., et al., 2025. Parental exposure to wet and dry conditions shapes the viability and thermotolerance of eggs in *Aedes aegypti*. bioRxiv. https://doi.org/10.1101/2025.06.16.658957.

Chakraborty, S., Zigmond, E., Levan, K., et al., 2026. Drought tolerance of *Aedes aegypti* mosquito eggs is influenced by adaptation to local climate conditions and associations with humans. BioRxiv. https://doi.org/10.64898/2026.01.15.699549.

Chakraborty, Souvik, Zigmond, E., Shah, S., et al., 2025. Thermal tolerance of *Aedes aegypti* mosquito eggs is associated with urban adaptation and human interactions. J. Therm. Biol no. 104167 (June): 104167.

Chandel, A., DeBeaubien, N.A., Ganguly, A., et al., 2024. Thermal infrared directs host-seeking behaviour in *Aedes aegypti* mosquitoes. Nature 633 (8030), 615–623.

Chandrasegaran, K., Lahondère, C., Escobar, L.E., Vinauger, C., 2020. Linking mosquito ecology, traits, behavior, and disease transmission. Trends Parasitol. 36 (4), 393–403.

Chandrasegaran, K., Walker, M., Marano, J.M., et al., 2025. Larval environment reshapes mosquito disease risk via phenotypic and molecular plasticity. bioRxiv. https://doi.org/10.1101/2025.06.16.659972.

Chen, J., Lewis, O.T., 2024. Limits to species distributions on tropical mountains shift from high temperature to competition as elevation increases. Ecol. Monogr. 94 (1). https://doi.org/10.1002/ecm.1597.

Chen, Shyh-Chi, Goodhart, G., Eaton, D., et al., 2024. Humidity as a zeitgeber for circadian entrainment of insect systems. bioRxivorg. https://doi.org/10.1101/2024.09.04.611265.

Chown, S.L., Nicolson, S.W., 2004. Water Balance Physiology. Insect Physiological Ecology: Mechanisms And Patterns. pp. 87–114.

Chown, S.L., Sørensen, J.G., Terblanche, J.S., 2011. Water loss in insects: an environmental change perspective. J. Insect Physiol. 57 (8), 1070–1084.

Chu, E., Chakraborty, S., Benoit, J.B., DeGennaro, M.M., 2022. Water homeostasis and hygrosensation in mosquitoes. In: Hill, S., Ignell, R., Lazzari, C., Lorenzo, M. (Eds.), Sensory Ecology of Disease Vectors. Wageningen Academic Publishers.

Chura, M., Healy, K., Diaz, R., Kaller, M., 2023. Effects of species, sex, and diet on thermal tolerance of *Aedes aegypti* and *Culex quinquefasciatus* (Diptera: Culicidae). J. Med. Entomology 60 (4), 637–643.

Ciota, A.T., Matacchiero, A.C., Kilpatrick, A.M., Kramer, L.D., 2014. The effect of temperature on life history traits of *Culex* mosquitoes. J. Med. Entomology 51 (1), 55–62.

Clark, M.S., Worland, M.R., 2008. How insects survive the cold: molecular mechanisms-a review. J. Comp. Physiol. B, Biochemical, Systemic, Environ. Physiol. 178 (8), 917–933.

Clements, A.N., 1992. The Biology of Mosquitoes: Volume 1. Development, Nutrition and Reproduction. Chapman & Hall.

Coast, G.M., 2009. Neuroendocrine control of ionic homeostasis in blood-sucking insects. J. Exp. Biol. 212 (Pt 3), 378–386.

Condé, Erika Hernandez-Torres, R., Claudio-Piedras, F., et al., 2021. Heat shock causes lower plasmodium infection rates in *Anopheles albimanus*. Front. Immunol. 12 (June), 584660.

Copeland, R.S., Craig Jr., G.B., 1990. Cold hardiness of tree-hole mosquitoes in the great lakes region of the United States. Can. J. Zool. 68 (6), 1307–1314.

Corfas, R.A., Vosshall, L.B., 2015. The cation channel TRPA1 tunes mosquito thermotaxis to host temperatures. eLife 4, e11750. https://doi.org/10.7554/eLife.11750.

Couper, L.I., Dodge, T.O., Hemker, J.A., et al., 2025. Evolutionary adaptation under climate change: *Aedes* sp. demonstrates potential to adapt to warming. Proc. Natl Acad. Sci. U S A. 122 (2), e2418199122.

Couper, L.I., Farner, J.E., Lyberger, K.P., Lee, A.S., Mordecai, E.A., 2024. Mosquito thermal tolerance is remarkably constrained across a large climatic range. Proceedings. Biol. Sci. 291 (2015), 20232457.

Couret, J., Benedict, M.Q., 2014. A meta-analysis of the factors influencing development rate variation in *Aedes aegypti* (Diptera: Culicidae). BMC Ecol. Preprint, pp. 14. https://doi.org/10.1186/1472-6785-14-3.

Couret, J., Dotson, E., Benedict, M.Q., 2014. Temperature, larval diet, and density effects on development rate and survival of *Aedes aegypti* (Diptera: Culicidae). PLoS ONE. Preprint, pp. 9. https://doi.org/10.1371/journal.pone.0087468.

Covey, H.O., Wilson, R., Gurrola-Mares, Y., McMillan, J.R., Brelsfoard, C.L., 2025. The thermal stress response of *Aedes aegypti* and *Aedes albopictus* when exposed to rapid temperature changes. Parasites & Vectors 18 (1), 300.

Crawford, J.E., Balcazar, D., Redmond, S., et al., 2025. 1206 Genomes Reveal Origin and Movement of Aedes aegypti Driving Increased Dengue Risk. Science (New York, N.Y.).

Dao, A., Yaro, A.S., Diallo, M., Timbiné, S., Huestis, D.L., Kassogué, Y., Traoré, A.I., Sanogo, Z.L., Samaké, D., Lehmann, T., 2014. Signatures of aestivation and migration in Sahelian malaria mosquito populations. Nature 516 (7531), 387–390. https://doi.org/10.1038/nature13987.

Denlinger, D.L., 2022. Insect Diapause. Cambridge University Press.

Denlinger, D.L., Armbruster, P.A., 2014. Mosquito diapause. Annu. Rev. Entomol. 59, 73–93.

Denlinger, D.L., Jr, R.E.L., 2010. Low Temperature Biology of Insects. Cambridge University Press.

Dennington, N.L., Grossman, M.K., Teeple, J.L., et al., 2025. Phenotypic variation in populations of the mosquito vector, *Aedes aegypti*, and implications for predicting the effects of temperature and climate change on dengue transmission. PLoS Neglected Tropical Dis. 19 (11), e0013623.

Dennington, N.L., Grossman, M.K., Ware-Gilmore, F., et al., 2024. Phenotypic adaptation to temperature in the mosquito vector, *Aedes aegypti*. Glob. Change Biol. 30 (1), e17041.

Diallo, M., Samake, D., Yossi, O., Veru, L.M., Graber, L.C., Conte, A.R., Kouam, C., Krajacich, B.J., Lehmann, T., 2022. Isotopic evidence that aestivation allows malaria mosquitoes to persist through the dry season in the Sahel. Nat. Ecol. Evol. 6 (11), 1687–1699.

Dias, R., Leite, M.P.C., Corrêa-do-Nascimento, G.S., et al., 2025. Impact of thermal variation on egg hatching and the life cycle of *Aedes* (Protomacleaya) *terrens* (Diptera: Culicidae) in a laboratory environment. Life (Basel, Switz.) 15 (7), 1038.

Dickson, Davy Jiolle, L.B., Minard, G., et al., 2017. Carryover effects of larval exposure to different environmental bacteria drive adult trait variation in a mosquito vector. Sci. Adv. 3 (8), e1700585.

Didion, E.M., Sabree, Z.L., Kenyon, L., et al., 2021. Microbiome reduction prevents lipid accumulation during early diapause in the northern house mosquito, *Culex pipiens Pipiens*. J. Insect Physiol. 134 (October), 104295.

Diniz, Araujo, D.F., Ribeiro de Albuquerque, C.M., Oliva, L.O., Varjal de Melo-Santos, M.A., Ayres, C.F.J., 2017. Diapause and quiescence: dormancy mechanisms that contribute to the geographical expansion of mosquitoes and their evolutionary success. Parasites & Vectors 10 (1), 310.

Doeurk, B., Leng, S., Long, Z., Maquart, P.-O., Boyer, S., 2025. Impact of temperature on survival, development and longevity of *Aedes aegypti* and *Aedes albopictus* (Diptera: Culicidae) in Phnom Penh, Cambodia. Parasites & Vectors 18 (1), 362.

Duplais, C., Sarou-Kanian, V., Massiot, D., et al., 2021. Gut bacteria are essential for normal cuticle development in herbivorous turtle ants. Nat. Commun. 12 (1), 676.

Evans, M.V., Newberry, P.M., Murdock, C.C., 2020. Carry-over effects of the larval environment in mosquito-borne disease systems. Population Biology of Vector-Borne Diseases. Oxford University Press.

Ezeakacha, N.F., Yee, D.A., 2019. The role of temperature in affecting carry-over effects and larval competition in the globally invasive mosquito *Aedes albopictus*. Parasites & Vectors 12 (1), 123.

Faridah, L., Fauziah, N., Agustian, D., et al., 2022. Temporal correlation between urban microclimate, vector mosquito abundance, and dengue cases. J. Med. Entomol. 59 (3), 1008–1018.

Farjana, T., Tuno, N., Higa, Y., 2012. Effects of temperature and diet on development and interspecies competition in *Aedes aegypti* and *Aedes albopictus*. Med. Vet. Entomol. 26 (2), 210–217.

Faull, K.J., Williams, C.R., 2015. Intraspecific variation in desiccation survival time of *Aedes aegypti* (L.) mosquito eggs of Australian origin. J. Vector Ecol.: J. Soc. Vector Ecol. 40 (2), 292–300.

Ferveur, J.-F., Cortot, J., Rihani, K., Cobb, M., Everaerts, C., 2018. Desiccation resistance: effect of cuticular hydrocarbons and water content in *Drosophila melanogaster* adults. PeerJ 6 (February), e4318.

Fikrig, K., Peck, S., Deckerman, P., et al., 2020. Sugar feeding patterns of New York *Aedes albopictus* mosquitoes are affected by saturation deficit, flowers, and host seeking. PLoS Neglected Tropical Dis. 14 (10), e0008244.

Fischer, S., De Majo, M.S., Di Battista, C., Campos, R.E., 2025. Effects of temperature and humidity on the survival and hatching response of diapausing and non-diapausing *Aedes aegypti* eggs. J. Insect Physiol. 161 (104726), 104726.

Foley, B.R., Telonis-Scott, M., 2011. Quantitative genetic analysis suggests causal association between cuticular hydrocarbon composition and desiccation survival in *Drosophila melanogaster*. Heredity 106 (1), 68–77.

Foo, I.J.-H., Hoffmann, A.A., Ross, P.A., 2019. Cross-generational effects of heat stress on fitness and *Wolbachia* density in *Aedes aegypti* mosquitoes. Tropical Med. Infect. Dis. 4 (1), 13.

Fouet, C., Gray, E., Besansky, N.J., Costantini, C., 2012. Adaptation to aridity in the Malaria mosquito *Anopheles gambiae*: chromosomal inversion polymorphism and body size influence resistance to desiccation. PLoS One 7 (4), e34841.

Fyie, L.R., Tronetti, H.R., Gardiner, M.M., Meuti, M.E., 2023. Potential for urban warming to postpone overwintering dormancy of temperate mosquitoes. J. Therm. Biol. 115 (103594), 103594.

Garland Jr, T., Downs, C.J., Ives, A.R., 2022. Trade-offs (and constraints) in organismal biology. Physiol. Biochem. Zool. 95 (1), 82–112.

Gary Jr, R.E., Foster, W.A., 2006. Diel timing and frequency of sugar feeding in the mosquito *Anopheles gambiae*, depending on sex, gonotrophic state and resource availability. Med. Vet. Entomol. 20 (3), 308–316.

Gefen, E., Marlon, A.J., Gibbs, A.G., 2006. Selection for desiccation resistance in adult *Drosophila melanogaster* affects larval development and metabolite accumulation. J. Exp. Biol. 209 (Pt 17), 3293–3300.

Gerken, A.R., Eller, O.C., Hahn, D.A., Morgan, T.J., 2015. Constraints, independence, and evolution of thermal plasticity: probing genetic architecture of long- and short-term thermal acclimation. Proc. Natl Acad. Sci. U S Am. 112 (14), 4399–4404.

Gibbs, A.G., Chippindale, A.K., Rose, M.R., 1997. Physiological mechanisms of evolved desiccation resistance in *Drosophila melanogaster*. J. Exp. Biol. 200 (Pt 12), 1821–1832.

Gibbs, A.G., 1998. Water-proofing properties of cuticular lipids. Am. Zool. 38 (3), 471–482.

Gibbs, A.G., Rajpurohit, S., 2010. Cuticular lipids and water balance. In: Blomquist, G.J., A-Genevieve Bagneres. (Eds.), Insect Hydrocarbons. Cambridge University Press.

Glastad, K.M., Hunt, B.G., Goodisman, M.A.D., 2019. Epigenetics in insects: genome regulation and the generation of phenotypic diversity. Annu. Rev. Entomology 64 (1), 185–203.

Gloria-Soria, Diego Ayala, A., Bheecarry, A., et al., 2016. Global genetic diversity of *Aedes aegypti*. Mol. Ecol. 25 (21), 5377–5395.

González-Tokman, D., C-Aguilar, A., Dáttilo, W., Lira-Noriega, A., Sánchez-Guillén, R.A., Villalobos, F., 2020. Insect responses to heat: physiological mechanisms, evolution and ecological implications in a warming world. Biol. Rev. Camb. Philos. Soc. 95 (3), 802–821.

Graber, L.C., Corrie, S.M., 2026. Insect-microbiome interactions in a changing world. Curr. Opin. Insect Sci. 4, 101495.

Gray, E.M., 2013. Thermal acclimation in a complex life cycle: the effects of larval and adult thermal conditions on metabolic rate and heat resistance in *Culex pipiens* (Diptera: Culicidae). J. Insect Physiol. 59 (10), 1001–1007.

Gray, E.M., Bradley, T.J., 2005. Physiology of desiccation resistance in *Anopheles gambiae* and *Anopheles arabiensis*. Am. J. Tropical Med. Hyg. 73 (3), 553–559.

Gray, E.M., Rocca, K.A.C., Costantini, C., Besansky, N.J., 2009. Inversion 2La Is associated with enhanced desiccation resistance in *Anopheles gambiae*. Malar. J. 8 (September), 215.

Guégan, M., Minard, G., Tran, F.-H., Tran Van, V., Dubost, A., Valiente Moro, C., 2018. Short-term impacts of anthropogenic stressors on *Aedes albopictus* mosquito vector microbiota. FEMS Microbiology Ecol 94 (12). https://doi.org/10.1093/femsec/fiy188.

Gu, W., Müller, G., Schlein, Y., Novak, R.J., Beier, J.C., 2011. Natural plant sugar sources of anopheles mosquitoes strongly impact malaria transmission potential. PLoS One 6 (1), e15996.

Haba, Y., Consortium, P., Korlević, P., et al., 2025. Ancient origin of an urban underground mosquito. Science 390 (6771) eady4515.

Hadley, N.F., 1994. Water Relations of Terrestrial Arthropods. Cambridge at the University Press, pp. 356–pages.

Hagan, R.W., Didion, E.M., Rosselot, A.E., et al., 2018. Dehydration prompts increased activity and blood feeding by mosquitoes. Sci. Rep. 8 (1), 6804.

Halberg, K.V., Denholm, B., 2024. Mechanisms of systemic osmoregulation in insects. Annu. Rev. Entomol. 69 (January), 415–438.

Harrison, J.F., Woods, H.A., Roberts, S.P., 2012. Ecological and Environmental Physiology of Insects. Ecological and Environmental Physiology Series. Oxford University Press.

Harrison, X.A., Blount, J.D., Inger, R., Norris, D.R., Bearhop, S., 2011. Carry-over effects as drivers of fitness differences in animals. J. Anim. Ecol. 80 (1), 4–18.

Hayward, S.A., Colinet, H., 2023. Metabolomics as a tool to elucidate biochemical cold adaptation in insects. Curr. Opin. Insect Sci. 58 (101061), 101061.

Hazel, J.R., 1995. Thermal adaptation in biological membranes: is homeoviscous adaptation the explanation? Annu. Rev. Physiol. 57 (1), 19–42.

Heilig, M., Edwards, M.J., Armbruster, P.A., 2026. Suppression of transgenerational lipid provisioning inhibits desiccation resistance, but not diapause, in the vector mosquito, *Aedes albopictus*. Res. Sq. https://doi.org/10.21203/rs.3.rs-8302219/v1.

Heino, M., Kinnunen, P., Anderson, W., et al., 2023. Increased probability of hot and dry weather extremes during the growing season threatens global crop yields. Sci. Rep. 13 (1), 3583.

Herring, S.C., Hoerling, M.P., Peterson, T.C., Stott, P.A., 2014. Explaining extreme events of 2013 from a climate perspective. Bull. Am. Meteorol. Soc. 95 (9), S1–S104.

Hidalgo, K., Siaussat, D., Braman, V., et al., 2016. Comparative physiological plasticity to desiccation in distinct populations of the malarial mosquito *Anopheles coluzzii*. Parasites & Vectors 9 (1), 565.

Hien, N.T., Anh, D.D., Le, N.H., et al., 2021. Environmental factors influence the local establishment of *Wolbachia* in *Aedes aegypti* mosquitoes in two small communities in Central Vietnam. Gates Open. Res. 5, 147.

Hoffmann, A.A., Sørensen, J.G., Loeschcke, V., 2003. Adaptation of *Drosophila* to temperature extremes: bringing together quantitative and molecular approaches. J. Therm. Biol. 28 (3), 175–216.

Holmes, C.J., Benoit, J.B., 2019. Biological adaptations associated with dehydration in mosquitoes. Insects 10 (11). https://doi.org/10.3390/insects10110375.

Holmes, C.J., Brown, E.S., Sharma, D., et al., 2023. Dehydration alters transcript levels in the mosquito midgut, likely facilitating rapid rehydration following a bloodmeal. Insects 14 (3). https://doi.org/10.3390/insects14030274.

Holmes, C.J., Chakraborty, S., Ajayi, O.M., et al., 2025. Multiple blood feeding bouts in Mosquitoes allow for prolonged survival and are predicted to increase viral transmission during dry periods. iScience 28 (2), 111760.

Huey, R.B., Kingsolver, J.G., 1989. Evolution of thermal sensitivity of ectotherm performance. Trends Ecol. & Evolution 4 (5), 131–135.

Hug, D.O.H., Gretener-Ziegler, R., Stegmayer, R.I., Mathis, A., Verhulst, N.O., 2024. Altered thermal preferences of infected or immune-challenged *Aedes aegypti* and *Aedes japonicus* mosquitoes. Sci. Rep. 14 (1), 12959.

Hug, D.O.H., Hugo, E., Verhulst, N.O., 2025. Avoid the light to avoid the heat? Thermal and light preferences of *Aedes aegypti* mosquito larvae. Anim. Behav. 219 (123011), 123011.

Huxley, P.J., Brown, J.J., Laurent, B.S., et al., 2025. Beyond temperature: relative humidity systematically shifts the temperature dependence of population growth in a malaria vector. bioRxiv. https://doi.org/10.1101/2025.05.30.656372.

Ioannou, C.S., Savvidou, E.C., Apocha, L., Terblanche, J.S., Papadopoulos, N.T., 2024. Insecticide resistant mosquitoes remain thermal stress resistant, without loss of thermal plasticity. Sci. Total. Environ. 912 (169443), 169443.

Jass, Amanda, Yerushalmi, G.Y., Davis, H.E., Donini, A., MacMillan, H.A., 2019. An impressive capacity for cold tolerance plasticity protects against ionoregulatory collapse in the disease vector *Aedes aegypti*. J. Exp. Biol. 222 (Pt 24) jeb214056.

Johnson, B., Huxley, P.J., Brown, J.J., et al., 2026. Humidity shapes the thermal niche of *Anopheles Stephensi*, an invasive malaria vector. bioRxiv. https://doi.org/10.64898/2026.03.28.715035.

Johnson, Tal Ben-Horin, L.R., Lafferty, K.D., et al., 2015. Understanding uncertainty in temperature effects on vector-borne disease: a Bayesian approach. Ecology 96 (1), 203–213.
Juliano, S.A., Philip Lounibos, L., 2005. Ecology of invasive mosquitoes: effects on resident species and on human health. Ecol. Lett. 8 (5), 558–574.
Juliano, S.A., O'Meara, G.F., Morrill, J.R., Cutwa, M.M., 2002. Desiccation and thermal tolerance of eggs and the coexistence of competing mosquitoes. Oecologia 130 (3), 458–469.
Kawecki, T.J., Stearns, S.C., 1993. The evolution of life histories in spatially heterogeneous environments: optimal reaction norms revisited. Evolut. Ecol. 7 (2), 155–174.
Kellermann, V., Chown, S.L., Schou, M.F., et al., 2019. Comparing thermal performance curves across traits: how consistent are they? J. Exp. Biol. 222 (11) jeb193433.
Kellermann, V., Overgaard, J., Hoffmann, A.A., Fløjgaard, C., Svenning, J.-C., Loeschcke, V., 2012. Upper thermal limits of *Drosophila* are linked to species distributions and strongly constrained phylogenetically. Proc. Natl Acad. Sci. U S Am. 109 (40), 16228–16233.
Kelley, M., Rathore, S., Chandrasegaran, K., et al., 2025. A microbiome-derived nutrient underlies tyrosine metabolism and predator avoidance in mosquito larvae. bioRxiv. https://doi.org/10.1101/2025.04.02.646203.
Kerr, J.T., Gordon, S.C.C., Chen, I.-C., et al., 2025. Effects of microclimate variation on insect persistence under global change. Nat. Rev. Biodivers. 1 (8), 532–542.
Kessler, S., Guerin, P.M., 2008. Responses of *Anopheles gambiae*, *Anopheles stephensi*, *Aedes aegypti*, and *Culex pipiens* mosquitoes (Diptera: Culicidae) to cool and humid refugium conditions. J. Vector Ecol.: J. Soc. Vector Ecol. 33 (1), 145–149.
Kim, M., Robich, R.M., Rinehart, J.P., Denlinger, D.L., 2006. Upregulation of two actin genes and redistribution of actin during diapause and cold stress in the northern house mosquito, *Culex pipiens*. J. Insect Physiol. 52 (11-12), 1226–1233.
Kingsolver, J.G., 1989. Weather and the population dynamics of insects: integrating physiological and population ecology. Physiol. Zool. 62 (2), 314–334.
Koenraadt, C.J.M., Paaijmans, K.P., Githeko, A.K., Knols, B.G.J., Takken, W., 2003. Egg hatching, larval movement and larval survival of the Malaria vector Anopheles gambiae in desiccating habitats. Malar. J. 2 (1), 20.
Kotsakiozi, P., Evans, B.R., Gloria-Soria, A., et al., 2018. Population structure of a vector of human diseases: *Aedes aegypti* in its ancestral range, Africa. Ecol. Evolution 8 (16), 7835–7848.
Kraemer, M.U.G., Reiner Jr, R.C., Brady, O.J., et al., 2019. Past and future spread of the arbovirus vectors *Aedes aegypti* and *Aedes albopictus*. Nat. Microbiol. 4 (5), 854–863.
Kumar, G., Singh, R.K., Pande, V., Dhiman, R.C., 2016. Impact of container material on the development of *Aedes aegypti* larvae at different temperatures. J. Vector Borne Dis. 53 (2), 144–148.
Lahondère, C., Bonizzoni, M., 2022. Thermal biology of invasive aedes mosquitoes in the context of climate change. Curr. Opin. Insect Sci. 51 (100920), 100920.
Lahondère, C., Lazzari, C.R., 2012. Mosquitoes cool down during blood feeding to avoid overheating. Curr. Biology: CB 22 (1), 40–45.
Lahondère, C., Vinauger, C., Liaw, J.E., Tobin, K.K.S., Joiner, J.M., Riffell, J.A., 2023. Effect of temperature on mosquito olfaction. Integr. Comp. Biol. 63 (2), 356–367.
Laursen, W.J., Budelli, G., Tang, R., et al., 2023. Humidity sensors that alert mosquitoes to nearby hosts and egg-laying sites. Neuron ahead of print. https://doi.org/10.1016/j.neuron.2022.12.025.
Lehane, M.J., 2005. The Biology of Blood-Sucking in Insects. Cambridge University Press.
Lehmann, T., Dao, A., Yaro, A.S., Adamou, A., Kassogue, Y., Diallo, M., Sékou, T., Coscaron-Arias, C., 2010. Aestivation of the African malaria mosquito, Anopheles gambiae in the Sahel. Am. J. Trop. Med. Hyg. 83 (3), 601–606.

Lembrechts, J.J., Nijs, I., Lenoir, J., 2019. Incorporating microclimate into species distribution models. Ecography 42 (7), 1267–1279.
Lim, P.L.J., Cook, A.R., Bansal, S., Chow, J.Y., Lim, J.T., 2024. *Wolbachia* incompatible insect technique program optimization over large spatial scales using a process-based model of mosquito metapopulation dynamics. BMC Biol. 22 (1), 269.
Liu, C., Zwiebel, L.J., 2013. Molecular characterization of larval peripheral thermosensory responses of the malaria vector mosquito *Anopheles gambiae*. PLoS One 8 (8), e72595.
Liu, K., Dong, Y., Huang, Y., Rasgon, J.L., Agre, P., 2013. Impact of trehalose transporter knockdown on *Anopheles gambiae* stress adaptation and susceptibility to *Plasmodium Falciparum* infection. Proc. Natl Acad. Sci. U S Am. 110 (43), 17504–17509.
Loetti, Verónica, Schweigmann, N., Burroni, N., 2011. Development rates, larval survivorship and wing length of *Culex pipiens* (Diptera: Culicidae) at constant temperatures. J. Nat. History 45 (35-36), 2203–2213.
Logan, M.L., Christian, L.C., 2020. Genetic constraints, transcriptome plasticity, and the evolutionary response to climate change. Front. Genet 11 (September), 538226.
Lv, Wen-Xiang, Cheng, P., Lei, J.-J., et al., 2023. Interactions between the gut micro-community and transcriptome of *Culex pipiens Pallens* under low-temperature stress. Parasites & Vectors 16 (1), 12.
Lyberger, K., Farner, J.E., Couper, L., Mordecai, E.A., 2025. Plasticity in mosquito size and thermal tolerance across a latitudinal climate gradient. J. Anim. Ecol. 94 (3), 330–339.
Lyons, C.L., Coetzee, M., Terblanche, J.S., Chown, S.L., 2014. Desiccation tolerance as a function of age, sex, humidity and temperature in adults of the African Malaria vectors *Anopheles arabiensis* and *Anopheles funestus*. J. Exp. Biol. 217 (Pt 21), 3823–3833.
Mack, L.K., Attardo, G.M., 2024. Heat shock proteins, thermotolerance, and insecticide resistance in mosquitoes. Front. Insect Sci. 4 (January), 1309941.
Mackay, A.J., Yan, J., Kim, C.-H., Barreaux, A.M.G., Stone, C.M., 2023. Larval diet and temperature alter mosquito immunity and development: using body size and developmental traits to track carry-over effects on longevity. Parasites & Vectors 16 (1), 434.
Mamai, W., Mouline, K., Parvy, J.-P., et al., 2016. Morphological changes in the spiracles of *Anopheles gambiae* S.l (Diptera) as a response to the dry season conditions in Burkina Faso (West Africa). Parasites & Vectors 9 (1), 11.
Manzano-Alvarez, J., Terradas, G., Holmes, C.J., Benoit, J.B., Rasgon, J.L., 2023. Dehydration stress and mayaro virus vector competence in *Aedes aegypti*. J. Virol. 97 (12), e0069523.
Martin, L.E., Barr, J.S., Cartailler, J.-P., Shrestha, S., Estévez-Lao, T.Y., Hillyer, J.F., 2026. Warmer temperature accelerates senescence by modifying the aging-dependent changes in the mosquito transcriptome, altering immunity, metabolism, and DNA repair. Immun. Ageing 23 (1).
Martin, L.E., Hillyer, J.F., 2024. Higher temperature accelerates the aging-dependent weakening of the melanization immune response in mosquitoes. PLoS Pathog. 20 (1), e1011935.
Mesas, A., Jaramillo, A., Castañeda, L.E., 2021. Experimental evolution on heat tolerance and thermal performance curves under contrasting thermal selection in *Drosophila subobscura*. J. Evolut. Biol. 34 (5), 767–778.
Meuti, M.E., 2025. Environmental factors that regulate mosquito physiology and behavior. Annu. Rev. Entomol. https://doi.org/10.1146/annurev-ento-121423-013620. ahead of print.
Mirieri, C.K., Ros, V.I.D., Abd-Alla, A.M.M., van Oers, M.M., 2025. Stress in dipteran insects mass-reared for sterile insect technique applications. Insect Sci. 32 (5), 1495–1523.
Mordecai, E.A., Caldwell, J.M., Grossman, M.K., et al., 2019. Thermal biology of mosquito-borne disease. Ecol. Lett. 22 (10), 1690–1708.

Moser, Kane, S., Barnard, M., Frantz, R.M., et al., 2023. Scoping review of *Culex* mosquito life history trait heterogeneity in response to temperature. Parasites & Vectors 16 (1), 200.

Mourya, D.T., Yadav, P., Mishra, A.C., 2004. Effect of temperature stress on immature stages and susceptibility of *Aedes aegypti* mosquitoes to Chikungunya virus. Am. J. Tropical Med. Hyg. 70 (4), 346–350.

Mpho, M., Callaghan, A., Holloway, G.J., 2002. Effects of temperature and genetic stress on life history and fluctuating wing asymmetry in *Culex pipiens* mosquitoes. Eur. J. Entomol. 99 (4), 405–412.

Murdock, C.C., Paaijmans, K.P., Bell, A.S., et al., 2012. Complex effects of temperature on mosquito immune function. Proceedings. Biol. Sci. R. Soc. 279 (1741), 3357–3366.

Murdock, C.C., Sternberg, E.D., Thomas, M.B., 2016. Malaria transmission potential could be reduced with current and future climate change. Sci. Rep. 6 (1), 27771.

Murdock, C.C., Evans, M.V., McClanahan, T.D., Miazgowicz, K.L., Tesla, B., 2017. Fine-scale variation in microclimate across an urban landscape shapes variation in mosquito population dynamics and the potential of *Aedes albopictus* to transmit arboviral disease. PLoS Neglected Tropical Dis. 11 (5), e0005640.

Nayal, K., Krupp, J.J., Abdalla, O.H.M.H., Levine, J.D., 2024. Cuticular hydrocarbons promote desiccation resistance by preventing transpiration in *Drosophila melanogaster*. J. Exp. Biol. 227 (23). https://doi.org/10.1242/jeb.247752.

Newman, R., Noy, I., 2023. The global costs of extreme weather that are attributable to climate change. Nat. Commun. 14 (1), 6103.

Oliveira, B.F., Yogo, W.I.G., Hahn, D.A., Yongxing, J., Scheffers, B.R., 2021. Community-wide seasonal shifts in thermal tolerances of mosquitoes. Ecology 102 (7), e03368.

Olson, M.F., Garcia-Luna, S., Juarez, J.G., et al., 2020. Sugar feeding patterns for *Aedes aegypti* and *Culex quinquefasciatus* (Diptera: Culicidae) mosquitoes in South Texas. J. Med. Entomol. 57 (4), 1111–1119.

Onyango, G.M., Bialosuknia, M.S., Payne, F.A., Mathias, N., Ciota, T.A., Kramer, D.L., 2020. Increase in temperature enriches heat tolerant taxa in *Aedes aegypti* midguts. Sci. Rep. 10 (1), 19135.

Ørsted, M., Jørgensen, L.B., Overgaard, J., 2022. Finding the right thermal limit: a framework to reconcile ecological, physiological and methodological aspects of CTmax in ectotherms. J. Exp. Biol. 225 (19). https://doi.org/10.1242/jeb.244514.

Paaijmans, K.P., Blanford, S., Bell, A.S., Blanford, J.I., Read, A.F., Thomas, M.B., 2010. Influence of climate on Malaria transmission depends on daily temperature variation. Proc. Natl Acad. Sci. U S Am. 107 (34), 15135–15139.

Paaijmans, K.P., Imbahale, S.S., Thomas, M.B., Takken, W., 2010. Relevant microclimate for determining the development rate of malaria mosquitoes and possible implications of climate change. Malar. J. 9 (1), 196.

Pabst, R., Sousa, C.A., Essl, F., et al., 2025. Global invasion patterns and dynamics of disease vector mosquitoes. Nat. Commun. 16 (1), 9127.

Palli, S.R., 2023. RNAi Turns 25: contributions and challenges in insect science. Front. Insect Sci. 3 (October), 1209478.

Pathak, A.K., Quek, S., Sharma, R., et al., 2025. Thermal variation influences the transcriptome of the major malaria vector *Anopheles Stephensi*. Commun. Biol. 8 (1), 112.

Pawar, Samraat, Huxley, P.J., Smallwood, T.R.C., et al., 2024. Variation in temperature of peak trait performance constrains adaptation of arthropod populations to climatic warming. Nat. Ecol. & Evol. 8 (3), 500–510.

Pawar, S., Kontopoulos, D.-G., 2025. Toward a general understanding of thermal performance curves in biology. Proc. Natl. Acad. Sci. U S Am. 122 (51), e2528528122.

Pekľanská, M., van Heerwaarden, B., Hoffmann, A.A., Nouzová, M., Šíma, R., Ross, P.A., 2025. Elevated developmental temperatures below the lethal limit reduce Aedes aegypti fertility. J. Exp. Biol. 228 (3), JEB249803.

Perdomo, H.D., Khorramnejad, A., Cham, N.M., Kropf, A., Sogliani, D., Bonizzoni, M., 2025. Prolonged exposure to heat enhances mosquito tolerance to viral infection. Commun. Biol. 8 (1), 168.

Piermarini, P.M., 2016. Renal excretory processes in mosquitoes. In: Advances in Insect Physiology. (Vol. 51). Academic Press, pp. 393–433.

Pincebourde, S., Murdock, C.C., Vickers, M., Sears, M.W., 2016. Fine-scale microclimatic variation can shape the responses of organisms to global change in both natural and urban environments. Integr. Comp. Biol. 56 (1), 45–61.

Pincebourde, S., Woods, H.A., 2020. There is plenty of room at the bottom: microclimates drive insect vulnerability to climate change. Curr. Opin. Insect Sci. 41 (October), 63–70.

Prasad, A., Sreedharan, S., Bakthavachalu, B., Laxman, S., 2023. Eggs of the mosquito *Aedes aegypti* survive desiccation by rewiring their polyamine and lipid metabolism. PLoS Biol. 21 (10), e3002342.

Raji, J.I., Melo, N., Castillo, J.S., et al., 2019. *Aedes aegypti* mosquitoes detect acidic volatiles found in human odor using the IR8a pathway. Curr. Biol.: CB 29 (8), 1253–1262 e7.

Reidenbach, K.R., Cheng, C., Liu, F., Liu, C., Besansky, N.J., Zainulabeuddin, S., 2014. Cuticular differences associated with aridity acclimation in African Malaria vectors carrying alternative arrangements of inversion 2La. Parasites & Vectors 7 (April), 176.

Reinhold, J.M., Lazzari, C.R., Lahondère, C., 2018. Effects of the environmental temperature on *Aedes aegypti* and *Aedes albopictus* mosquitoes: a review. Insects 9 (4), 158.

Reinhold, J.M., Shaw, R., Lahondère, C., 2021. Beat the heat: *Culex quinquefasciatus* regulates its body temperature during blood feeding. J. Therm. Biol. 96 (102826), 102826.

Reisen, W.K., Meyer, R.P., Milby, M.M., 1986. Patterns of fructose feeding by culex tarsalis (Diptera: Culicidae). J. Med. Entomol. 23 (4), 366–373.

Reynolds, J.A., Poelchau, M.F., Rahman, Z., Armbruster, P.A., Denlinger, D.L., 2012. Transcript profiling reveals mechanisms for lipid conservation during diapause in the Mosquito, *Aedes albopictus*. J. Insect Physiol. 58 (7), 966–973.

Rezende, G.L., Martins, A.J., Gentile, C., et al., 2008. Embryonic desiccation resistance in *Aedes aegypti*: presumptive role of the chitinized serosal cuticle. BMC Dev. Biol 8 (1), 82.

Rivas, G.B.S., T-de-Freitas, R., Pavan, M.G., Lima, J.B.P., Peixoto, A.A., Bruno, R.V., 2018. Effects of light and temperature on daily activity and clock gene expression in two mosquito disease vectors. J. Biol. Rhythm 33 (3), 272–288.

Robich, R.M., Denlinger, D.L., 2005. Diapause in the mosquito *Culex pipiens* evokes a metabolic switch from blood feeding to sugar gluttony. Proc. Natl Acad. Sci. U S Am. 102 (44), 15912–15917.

Rosendale, A.J., Romick-Rosendale, L.E., Watanabe, M., Dunlevy, M.E., Benoit, J.B., 2016. Mechanistic underpinnings of dehydration stress in the American Dog tick revealed through RNA-Seq and metabolomics. J. Exp. Biol. 219 (12), 1808–1819.

Rose, N.H., Sylla, M., Badolo, A., et al., 2020. Climate and urbanization drive Mosquito preference for humans. Curr. Biol.: CB 30 (18), 3570–3579.e6.

Ross, P.A., Axford, J.K., Yang, Q., et al., 2020. Heatwaves cause fluctuations in wMel *Wolbachia* densities and frequencies in *Aedes aegypti*. PLoS Neglected Tropical Dis. 14 (1), e0007958.

Ross, P.A., Elfekih, S., Collier, S., et al., 2023. Developing *Wolbachia*-based disease interventions for an extreme environment. PLoS Pathog. 19 (1), e1011117.

Rozsypal, J., Moos, M., Ivo, R., Košťál, V., 2021. Do energy reserves and cold hardiness limit winter survival of *Culex pipiens*? Comp. Biochem. Physiol. Part. A, Mol. & Integr. Physiol. 255 (110912), 110912.

Rueda, L.M., Patel, K.J., Axtell, R.C., Stinner, R.E., 1990. Temperature-dependent development and survival rates of *Culex quinquefasciatus* and *Aedes aegypti* (Diptera: Culicidae). J. Med. Entomol. 27 (5), 892–898.

Rulli, M.C., D'Odorico, P., Galli, N., et al., 2025. Land use change and infectious disease emergence. Rev. Geophys. (Washington, D. C.: 1985). 63 (2). https://doi.org/10.1029/2022rg000785.

Ryan, S.J., Carlson, C.J., Mordecai, E.A., Johnson, L.R., 2019. Global expansion and redistribution of *Aedes*-borne virus transmission risk with climate change. PLoS Neglected Tropical Dis. 13 (3), e0007213.

Sauer, F.G., Grave, J., Lühken, R., Kiel, E., 2021. Habitat and microclimate affect the resting site selection of mosquitoes. Med. Vet. Entomol. 35 (3), 379–388.

Savvidou, E.C., Ioannou, C.S., Apocha, L., Terblanche, J.S., Papadopoulos, N.T., 2025. Desiccation tolerance and possible starvation trade-offs in larvicide resistant *Culex* and *Aedes* mosquitoes. iScience 28 (6), 112521.

Schou, M.F., Engelbrecht, A., Brand, Z., Svensson, E.I., Cloete, S., Cornwallis, C.K., 2022. Evolutionary trade-offs between heat and cold tolerance limit responses to fluctuating climates. Sci. Adv. 8 (21), eabn9580.

Schulte, P.M., Healy, T.M., Fangue, N.A., 2011. Thermal performance curves, phenotypic plasticity, and the time scales of temperature exposure. Integr. Comp. Biol. 51 (5), 691–702.

Sgrò, C.M., Terblanche, J.S., Hoffmann, A.A., 2016. What can plasticity contribute to insect responses to climate change? Annu. Rev. Entomol. 61 (1), 433–451.

Shapiro, L.L.M., Whitehead, S.A., Thomas, M.B., 2017. Quantifying the effects of temperature on mosquito and parasite traits that determine the transmission potential of human malaria. PLoS Biol. 15 (10), e2003489.

Shocket, M.S., Bernhardt, J.R., Miazgowicz, K.L., et al., 2025. Mean daily temperatures predict the thermal limits of malaria transmission better than hourly rate summation. Nat. Commun. 16 (1), 3441.

Sinclair, B.J., Coello Alvarado, L.E., Ferguson, L.V., 2015. An invitation to measure insect cold tolerance: methods, approaches, and workflow. J. Therm. Biol. 53 (October), 180–197.

Sinclair, B.J., Marshall, K.E., Sewell, M.A., et al., 2016. Can we predict ectotherm responses to climate change using thermal performance curves and body temperatures? Ecol. Lett. 19 (11), 1372–1385.

Sinclair, B.J., Saruhashi, S., Terblanche, J.S., 2024. Integrating water balance mechanisms into predictions of insect responses to climate change. J. Exp. Biol. 227 (10). https://doi.org/10.1242/jeb.247167.

Singh, P., Kumar, P., Pande, V., Kumar, V., Dhiman, R.C., 2022. Untargeted metabolomics-based response analysis of temperature and insecticide exposure in *Aedes aegypti*. Sci. Rep. 12 (1), 2066.

Sinka, M.E., 2013. Global distribution of the dominant vector species of malaria. In: Manguin, S. (Ed.), *Anopheles* Mosquitoes - New Insights into Malaria Vectors. InTech.

Slimen, I.B., Najar, T., Ghram, A., Dabbebi, H., Ben Mrad, M., et al., 2014. Reactive oxygen species, heat stress and oxidative-induced mitochondrial damage. A review. Int. J. Hyperth.: Off. J. Eur. Soc. Hyperthermic Oncology, North. Am. Hyperthermia Group 30 (7), 513–523.

Somero, G.N., 2010. The physiology of climate change: how potentials for acclimatization and genetic adaptation will determine 'Winners' and 'Losers'. J. Exp. Biol. 213 (6), 912–920.

Sommer, R.J., 2020. Phenotypic plasticity: from theory and genetics to current and future challenges. Genetics 215 (1), 1–13.

Sørensen, J.G., Norry, F.M., Scannapieco, A.C., Loeschcke, V., 2005. Altitudinal variation for stress resistance traits and thermal adaptation in adult *Drosophila buzzatii* from the new world. J. Evolut. Biol. 18 (4), 829–837.

Sørensen, J.G., Hamann, M., Kristensen, T.N., Lauritzen, J.M.S., Noer, N.K., Høye, T.T., et al., 2019. Rapid Induction of the heat hardening response in an arctic insect. Biol. Lett 15 (10), 20190613.

Sota, T., 1993a. Response to selection for desiccation resistance in *Aedes albopictus* eggs (Diptera: Culicidae). Appl. Entomol. Zool. 28 (2), 161–168.

Sota, T., Mogi, M., 1992a. Interspecific variation in desiccation survival time of *Aedes* (Stegomyia) mosquito eggs is correlated with habitat and egg size. Oecologia 90 (3), 353–358.

Sota, T., Mogi, M., 1992b. Survival time and resistance to desiccation of diapause and non-diapause eggs of temperate *Aedes (Stegomyia)* mosquitoes. Entomol. Experimentalis et. Applicata 63 (2), 155–161.

Sota, T., Mogi, M., Hayamizu, E., 1994. Habitat stability and the larval mosquito community in treeholes and other containers on a temperate island. Popul. Ecol. 36 (1), 93–104.

Soudi, S., Crepeau, M., Collier, T.C., Lee, Y., Cornel, A.J., Lanzaro, G.C., 2023. Genomic signatures of local adaptation in recent invasive *Aedes aegypti* populations in California. BMC Genomics 24 (1), 311.

Souza, W.M.D., Weaver, S.C., 2024. Effects of climate change and human activities on vector-borne diseases. Nat. Reviews. Microbiol. 22 (8), 476–491.

Stevens, D.R., Susan, A.F., Mathis, K.A., Meyer, N.P., 2025. Importance of considering the plastic expression of phenotypes when determining evolutionary relationships. BioEssays: N. Rev. Mol. Cell. Dev. Biol. 47 (8), e70033.

Sukiato, F., Wasserman, R.J., Foo, S.C., Wilson, R.F., Cuthbert, R.N., 2019. The effects of temperature and shading on mortality and development rates of *Aedes aegypti* (Diptera: Culicidae). J. Vector Ecol.: J. Soc. Vector Ecol. 44 (2), 264–270.

Tang, R., Busby, R., Laursen, W.J., Keane, G.T., Garrity, P.A., 2024. Functional dissection of mosquito humidity sensing reveals distinct dry and moist cell contributions to blood feeding and oviposition. Proc. Natl Acad. Sci. U S Am. 121 (35) e2407394121.

Tavakol, A., Rahmani, V., Harrington Jr., J., 2020. Probability of compound climate extremes in a changing climate: a copula-based study of hot, dry, and windy events in the Central United States. Environ. Res. Lett. 15 (10), 104058.

Telang, A., Wells, M.A., 2004. The effect of larval and adult nutrition on successful autogenous egg production by a mosquito. J. Insect Physiol. 50 (7), 677–685.

Telonis-Scott, M., Guthridge, K.M., Hoffmann, A.A., 2006. A new set of laboratory-selected *Drosophila melanogaster* lines for the analysis of desiccation resistance: response to selection, physiology and correlated responses. J. Exp. Biol. 209 (Pt 10), 1837–1847.

Terblanche, J.S., Lehmann, P., 2025. Metabolic flexibility in insects: patterns, mechanisms, and implications. Annu. Rev. Entomol. https://doi.org/10.1146/annurev-ento-120523-051033. ahead of print.

Tesla, B., Demakovsky, L.R., Mordecai, E.A., et al., 2018. Temperature drives Zika virus transmission: evidence from empirical and mathematical models. Proceedings. Biol. Sci. 285 (1884), 20180795.

Thomas, S.M., Obermayr, U., Fischer, D., Kreyling, J., Beierkuhnlein, C., 2012. Low-temperature threshold for egg survival of a post-diapause and non-diapause European aedine strain, *Aedes albopictus* (Diptera: Culicidae). Parasites & Vectors 5 (1), 100.

Tippelt, L., Werner, D., Kampen, H., 2020. Low temperature tolerance of three *Aedes albopictus* strains (Diptera: Culicidae) under constant and fluctuating temperature scenarios. Parasites & Vectors 13 (1), 587.

Trewin, B.J., Pagendam, D.E., Zalucki, M.P., et al., 2020. Urban landscape features influence the movement and distribution of the Australian container-inhabiting mosquito vectors *Aedes aegypti* (Diptera: Culicidae) and *Aedes Notoscriptus* (Diptera: Culicidae). J. Med. Entomol. 57 (2), 443–453.

Ukubuiwe, A.C., Olayemi, I.K., Arimoro, F.O., et al., 2018. Influence of rearing-water temperature on life stages' vector attributes, distribution and utilisation of metabolic reserves in *Culex quinquefasciatus* (Diptera: Culicidae): implications for disease transmission and vector control. J. Basic. & Appl. Zool. 79 (1). https://doi.org/10.1186/s41936-018-0045-3.

Upshur, I.F., Bose, E.A., Hart, C., Lahondère, C., 2019. Temperature and sugar feeding effects on the activity of a laboratory strain of *Aedes aegypti*. Insects 10 (10), 347.

Urbanski, J.M., Benoit, J.B., Michaud, M.R., Denlinger, D.L., Armbruster, P., 2010. The molecular physiology of increased egg desiccation resistance during diapause in the invasive mosquito, *Aedes albopictus*. Proc. R. Soc. B: Biol. Sci. 277 (1694), 2683–2692.

Valzania, L., Coon, K.L., Vogel, K.J., Brown, M.R., Strand, M.R., 2018. Hypoxia-induced transcription factor signaling is essential for larval growth of the mosquito *Aedes aegypti*. Proc. Natl Acad. Sci. U S Am. 115 (3), 457–465.

Vantaux, A., Lefèvre, T., Cohuet, A., Dabiré, K.R., Roche, B., Roux, O., 2016. Larval nutritional stress affects vector life history traits and human malaria transmission. Sci. Rep. 6 (1), 36778.

Vargas, H.C.M., Farnesi, L.C., Martins, A.J., Valle, D., Gustavo Lazzaro, R., 2014. Serosal cuticle formation and distinct degrees of desiccation resistance in embryos of the mosquito vectors *Aedes aegypti, Anopheles aquasalis and Culex quinquefasciatu*. J. Insect Physiol. 62, 54–60.

Venkataraman, K., Shai, N., Lakhiani, P., et al., 2023. Two novel, tightly linked, and rapidly evolving genes underlie *Aedes aegypti* mosquito reproductive resilience during drought. eLife 12. https://doi.org/10.7554/eLife.80489.

Villena, O.C., Ryan, S.J., Murdock, C.C., Johnson, L.R., 2022. Temperature impacts the environmental suitability for malaria transmission by *Anopheles gambiae* and *Anopheles Stephensi*. Ecology 103 (8), e3685.

Vinauger, C., Chandrasegaran, K., 2024. Context-specific variation in life history traits and behavior of *Aedes aegypti* mosquitoes. Front. Insect Sci. 4 (September), 1426715.

Vorhees, A.S., Gray, E.M., Bradley, T.J., 2013. Thermal resistance and performance correlate with climate in populations of a widespread mosquito. Physiological Biochemical Zool. 86 (1), 73–81.

Wagoner, K.M., Lehmann, T., Huestis, D.L., Ehrmann, B.M., Cech, N.B., Wasserberg, G., 2014. Identification of morphological and chemical markers of dry- and wet-season conditions in female *Anopheles gambiae* mosquitoes. Parasites & Vectors 7 (1), 294.

Wahaab, A., Alvarez, J.M., Minh, H.T.N., Benoit, J.B., Rasgon, J.L., 2025. Effect of prolonged dehydration stress on the vector competence of *Aedes aegypti* for Mayaro virus. bioRxiv. https://doi.org/10.1101/2025.06.16.660000.

Wang, Z., Receveur, J.P., Pu, J., et al., 2022. Desiccation resistance differences in *Drosophila* species can be largely explained by variations in cuticular hydrocarbons. eLife 11 (December). https://doi.org/10.7554/elife.80859.

Ware-Gilmore, F., Sgrò, C.M., Xi, Z., et al., 2021. Microbes increase thermal sensitivity in the mosquito *Aedes aegypti*, with the potential to change disease distributions. PLoS Neglected Tropical Dis. 15 (7), e0009548.

Warusawithana, A.L., van Heerwaarden, B., Hoffmann, A.A., Perran, A.R., 2025. Heat hardening enhances mosquito heat tolerance in a species-specific and trait-specific manner. bioRxiv. https://doi.org/10.1101/2025.09.14.676167.

Westby, K., Tayon, L., Orlinick, B., Smith, A., Kim, M., 2025. Interpopulation variation in egg mortality in response to hot and dry conditions in North American populations of the invasive tiger mosquito, *Aedes albopictus*. Authorea Inc. https://doi.org/10.22541/au.176225263.33248262/v1.

Wimberly, M.C., Davis, J.K., Evans, M.V., et al., 2020. Land cover affects microclimate and temperature suitability for arbovirus transmission in an urban landscape. PLoS Neglected Tropical Dis. 14 (9), e0008614.

Won, M.H., Choi, K.S., 2025. Study of *Aedes albopictus* hatching rate by low-temperature stress. Entomol. Res. 55 (2). https://doi.org/10.1111/1748-5967.70018.

Wynne, N.E., Applebach, E., Chandrasegaran, K., et al., 2024. *Aedes albopictus* colonies from different geographic origins differ in their sleep and activity levels but not in the time of peak activity. Med. Vet. Entomol. 38 (4), 495–507.

Yadav, M., Dahiya, N., Sehrawat, N., 2023. Mosquito gene targeted RNAi studies for vector control. Funct. & Integr. Genomics 23 (2), 180.

Yancey, P.H., 2005. Organic osmolytes as compatible, metabolic and counteracting cytoprotectants in high osmolarity and other stresses. J. Exp. Biol. 208 (15), 2819–2830.

Yilmaz, A.R., Diamond, S.E., Martin, R.A., 2021. Evidence for the evolution of thermal tolerance, but not desiccation tolerance, in response to hotter, drier city conditions in a cosmopolitan, terrestrial isopod. Evolut. Appl. 14 (1), 12–23.

Zani, P.A., Swanson, S.E.T., Corbin, D., et al., 2005. Geographic variation in tolerance of transient thermal stress in the mosquito *Wyeomyia smithii*. Ecology 86 (5), 1206–1211.

Zera, A.J., Harshman, L.G., 2001. The physiology of life history trade-offs in animals. Annu. Rev. Ecol. Syst. 32 (1), 95–126.

Zhang, J., Ahmad, S., Wang, L.-Y., Han, Q., Jian-Chun, Zhang, Yan-Ping, L., 2019. Cell death induced by α-terthienyl via reactive oxygen species-mediated mitochondrial dysfunction and oxidative stress in the midgut of *Aedes aegypti* larvae. Free. Radic. Biol. & Med. 137 (June), 87–98.

Zhao, L., James, J.B., Clark, G.G., Kenneth, J.L., 2010. Expression of *AeaHsp26* and *AeaHsp83* in *Aedes aegypti* (Diptera: Culicidae) larvae and pupae in response to heat shock stress. J. Med. Entomol. 47 (3), 367–375.

CHAPTER THREE

Tracing the evolutionary history of insect-specific orthoflavivirus through phylogenetics

Cecília Artico Banho[a], Victoria Bernardi[a], Igor da Silva Teixeira[a], Beatriz de Carvalho Marques[a], Mauricio Lacerda Nogueira[a,b,*], and Nikos Vasilakis[b,c,d,*]

[a]Laboratório de Pesquisas em Virologia, Faculdade de Medicina de São José do Rio Preto, São José do Rio Preto, São Paulo, Brazil
[b]Department of Pathology, University of Texas Medical Branch, Galveston, TX, United States
[c]Center for Vector-Borne and Zoonotic Diseases, University of Texas Medical Branch, Galveston, TX, United States
[d]Institute for Human Infection and Immunity, University of Texas Medical Branch, Galveston, TX, United States
*Corresponding authors. e-mail address: mauricio.nogueira@edu.famerp.br; nivasila@utmb.edu

Contents

Abstract

In this chapter, we trace the evolutionary history of insect-specific orthoflaviviruses (ISOFs) through phylogenetics, integrating current knowledge about their discovery, diversity, evolution, and ecological dynamics. We discuss the distinction between classical insect-specific orthoflaviviruses (cISOFs) and dual-host affiliated insect-specific orthoflaviviruses (dISOFs), evaluating their phylogenetic relationships with mosquito-borne orthoflaviviruses (MBOFs) and their evolutionary and biological

Advances in Insect Physiology, Volume 70
ISSN 0065-2806, https://doi.org/10.1016/bs.aiip.2026.05.004

implications. The chapter further explores viral maintenance mechanisms, host restriction, and virus-virus interactions in mosquitoes, emphasizing their ecological and epidemiological relevance. Additionally, we discuss applied perspectives, highlighting the potential of ISOFs as platforms for vaccine development, antiviral research, and innovative biological control strategies against arboviruses.

1. Introduction

Arboviruses are a diverse group of vector-borne viruses that cause significant human morbidity and mortality (Huang et al., 2023; Young, 2018). Over the recent decades, their emergence and re-emergence have intensified, representing an increasing global public health concern (Huang et al., 2023; Young, 2018). According to Huang et al., arboviruses comprise 13 families, 34 genera, and 460 viral species (Huang et al., 2023). By definition, arboviruses are viruses maintained primarily, or to a significant extent, in nature through transmission cycles between susceptible vertebrate hosts mediated by hematophagous arthropods (Casals, 1971). This concept highlights the central role of the transmission cycle, involving the virus, vertebrate host, and arthropod vector. Arboviral classification requires evidence of a transmission cycle between arthropods and vertebrates that reflects natural transmission. Thus, viral properties unrelated to this cycle, while sometimes informative, are not defining criteria (Casals, 1971). Currently, recognized vectors include mosquitoes, ticks, sandflies (*Phlebotomus*), and biting midges (*Culicoides*) (Casals, 1971; Huang et al., 2023; Young, 2018). Despite this definition, certain viruses are classified within arboviral groups based on their genetic relatedness to known arboviruses, even if arthropod transmission has not been confirmed. Notably, members of the *Flaviviridae* family are of particular interest, as they include several pathogens of significant global health importance (Young, 2018).

The family *Flaviviridae* comprises enveloped, positive-sense single-stranded RNA (ssRNA+) viruses infecting a wide range of vertebrate and invertebrate hosts (Lindenbach, 2013). Following the recent proposed taxonomic reorganization that is under consideration for ratification by the International Committee on Taxonomy of Viruses (ICTV), the family currently includes five genera: *Orthoflavivirus*, *Tamanavirus*, *Termitovirus*, *Guaicovirus*, and *Jingmenvirus* (Simmonds et al., 2025). Members of this family exhibit great genomic diversity, including both non-segmented and segmented RNA genomes, and are associated with a broad spectrum of infections in humans and animals, ranging from hepatotropic diseases to arthropod-borne illnesses of substantial medical and veterinary relevance (Lindenbach, 2013; Simmonds et al., 2025, 2017).

Among these genera, *Orthoflavivirus* deserves particular attention because it includes most viral species classically recognized as arboviruses (arthropod-borne viruses), i.e., viruses transmitted by hematophagous arthropods, primarily mosquitoes and ticks (Holbrook, 2017; Lindenbach, 2013; Westaway et al., 1985). This genus encompasses some of the most medically and veterinary important pathogens worldwide, including yellow fever virus (YFV – *Orthoflavivirus flavi*), dengue virus (DENV – *Orthoflavivirus denguei*), West Nile virus (WNV – *Orthoflavivirus nilense*), Japanese encephalitis virus (JEV – *Orthoflavivirus japonicum*), and tick-borne encephalitis virus (TBEV – *Orthoflavivirus encephalitidis*). These viruses have a wide geographic distribution and are frequently associated with outbreaks and epidemics with significant public health consequences (Brinton, 2002; Chancey et al., 2015; Gao et al., 2019; Gritsun et al., 2003; Im et al., 2020; Lim et al., 2025).

The orthoflavivirus genome encodes a single open reading frame (ORF) that is translated into a single polyprotein, which is subsequently processed by co- and post-translational cleavages into structural proteins [capsid (C), pre-membrane (prM), and envelope (E)] and non-structural proteins (NS1, NS2A, NS2B, NS3, NS4A, NS4B, and NS5) (Lindenbach, 2013; Liu et al., 2010; Villordo and Gamarnik, 2009). In addition, the genomic RNA contains two untranslated regions flanking the ORF (5′ UTR and 3′ UTR), whose sizes may vary, typically around ~100 nt and ~400–800 nt, respectively. A type I cap is present at the 5′ end (m 7 GpppA m p) and is followed by a conserved AG dinucleotide sequence. The 3′ end does not terminate in a poly (A) tail, but rather in a conserved CU dinucleotide sequence (Liu et al., 2010; Markoff, 2003; Villordo and Gamarnik, 2009).

According to the official taxonomy of the ICTV, the genus *Orthoflavivirus* currently comprises more than 50 viral species recognized in the Master Species List (International Committee on Taxonomy of Viruses (ICTV), 2023; Simmonds et al., 2017). However, recent studies have revealed an increasing number of orthoflavivirus-related lineages, particularly in arthropods, many of which still lack formal classification or an ICTV-recognized species name and remain provisional or unclassified (International Committee on Taxonomy of Viruses (ICTV), 2023). This scenario largely reflects advances in sequencing technologies and the widespread application of metagenomic approaches, which have uncovered substantial viral diversity previously unknown (Bolling et al., 2015b;

Charles et al., 2018; da Silva Ferreira et al., 2020; Orba et al., 2021). Thus, the officially recognized taxonomic richness of the genus represents only the ratified fraction and does not necessarily encompass the full extent of circulating diversity.

From an ecological perspective, orthoflaviviruses are horizontally transmitted to vertebrate hosts, primarily mammals and birds, which often serve as amplifying hosts and/or reservoirs (Blitvich and Firth, 2017, 2015; International Committee on Taxonomy of Viruses (ICTV), 2023; Kuno and Chang, 2005). In vertebrates, the clinical severity of infection varies widely, ranging from asymptomatic cases to severe and potentially fatal disease, including hemorrhagic syndromes and neurological disorders (Gould and Solomon, 2008; International Committee on Taxonomy of Viruses (ICTV), 2023). The vast majority of orthoflavivirus species can replicate actively in both vertebrate and invertebrate hosts. However, a subset of viruses exhibits a distinct ecology: some orthoflaviviruses are associated with rodents or bats, lacking confirmed arthropod vectors, and are thus classified as no known vector (NKV) (Blitvich and Firth, 2017). Examples include Entebbe bat virus (*Orthoflavivirus entebbeense*), Modoc virus (*Orthoflavivirus modocense*), and Rio Bravo virus (*Orthoflavivirus bravoense*), among others (Blitvich and Firth, 2017; International Committee on Taxonomy of Viruses (ICTV), 2023).

Robust phylogenetic inferences consistently demonstrate segregation between mosquito-borne orthoflavivirus (MBOFs) and tick-borne orthoflaviviruses (TBOFs), and the formation of additional groups by NKV viruses (Blitvich and Firth, 2017, 2015; Grard et al., 2010). Recurrently, the MBOF and TBOF groups are well-supported monophyletic clades in phylogenetic trees, reflecting distinct evolutionary histories and specific adaptations to vectors and hosts (Fig. 1) (Blitvich and Firth, 2015; Kuno et al., 1998). However, NKV are split into two groups: one clustering within the MBOF clade, composed of the Entebbe bat virus group, and another, closely related to the TBOF clade, which encompasses the Modoc and Rio Bravo virus groups (Fig. 1) (Blitvich and Firth, 2015; Charles et al., 2017; Cook et al., 2012; Kholodilov et al., 2024; Rathore and St. John, 2020).

In addition to phylogenetic classification, orthoflavivirus species have historically been categorized into antigenic groups (serocomplexes) based on the ability of polyclonal post-immune sera raised against one orthoflavivirus to neutralize other related viruses, reflecting cross-reactivity patterns primarily mediated by epitopes in the E protein, the main determinant of

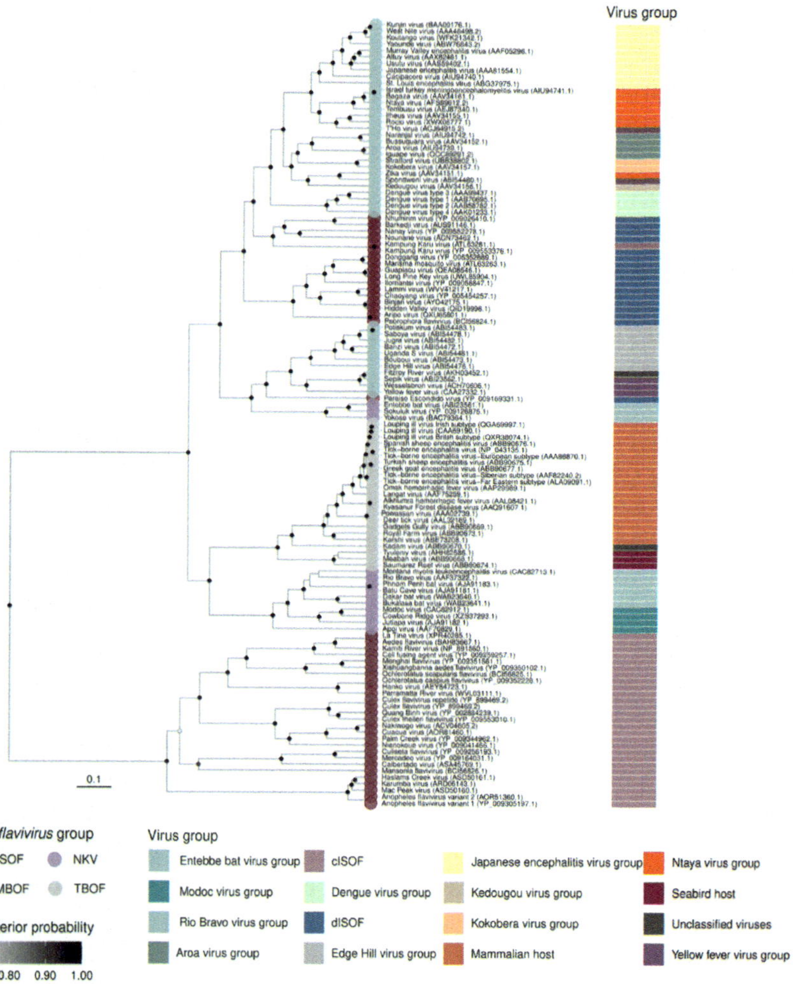

Fig. 1 Phylogenetic tree of the *Orthoflavivirus* genus. Phylogenetic tree inferred from polyprotein sequences comprising 2659 amino acid sites. Tip colors represent the major groups of *Orthoflavivirus*: Mosquito-borne orthoflaviviruses (MBOF), Tick-borne orthoflaviviruses (TBOF), viruses with no known vector (NKV), and insect-specific orthoflaviviruses (ISOF). Colored boxes indicate virus groups, including historically recognized orthoflavivirus groups, insect-specific orthoflavivirus clades (cISOF and dISOF), and host-associated groupings. Posterior probabilities are indicated by a gray gradient at the nodes. The phylogenetic tree is midpoint-rooted. Tree scale represents the number of substitutions per site.

neutralization (Fig. 1) (Calisher et al., 1989; Rathore and St. John, 2020). These serocomplexes include classical groups such as dengue, yellow fever, and encephalitis complexes associated with different vectors, among others (Calisher et al., 1989; International Committee on Taxonomy of Viruses (ICTV), 2023; Rathore and St. John, 2020). This antigenic framework has important practical implications for serological diagnosis, the interpretation of cross-protective immunity, and the understanding of shared antigenicity among phylogenetically related species.

However, over the past few decades, the discovery of a fourth genetic lineage within the genus *Orthoflavivirus* has substantially changed our understanding of its evolution. This new lineage, named insect-specific orthoflavivirus (ISOFs), was important for redefining current views on the origin and diversification of the genus (Blitvich and Firth, 2015; Bolling et al., 2015b; Cook et al., 2012; Crabtree et al., 2003; Kuno et al., 1998; Stollar and Thomas, 1975). ISOFs constitute a genetically diverse group within the genus *Orthoflavivirus*, characterized by replication restricted to arthropods, mainly mosquitoes, suggesting biological barriers at cell entry, host factor usage, and/or post-entry steps (Blitvich and Firth, 2015; Cui et al., 2025; de Oliveira Ribeiro et al., 2021; Kuno, 2007; Tangudu et al., 2021; Tree et al., 2016; Wastika et al., 2020; Zhang et al., 2022). Since their identification, ISOFs have been detected worldwide and are frequently associated with medically important vectors, such as *Aedes*, *Culex*, and *Anopheles* (Table 1) (Blitvich and Firth, 2015; Calisher and Higgs, 2018; Calzolari et al., 2016; Cook et al., 2012). Their wide geographic distribution, together with the high levels of genetic diversity observed, suggests that these viruses have a long evolutionary history with their invertebrate hosts (Bolling et al., 2015b).

The evolutionary relevance of ISOFs extends beyond the simple addition of new species to the *Orthoflavivirus* genus. In fact, they represent a lineage that provides critical insights into the processes shaping the transition from insect-restricted to vertebrate-host cycles. Phylogenetic analyses based on complete ORFs or conserved genomic regions indicate that ISOFs are not monophyletic but are instead subdivided into major clades with distinct patterns of divergence and conservation. These clades likely reflect independent evolutionary origins followed by coevolution with specific mosquito vectors, with diversification shaped by environmental factors, host biology, and virus–vector interactions (Fig. 1) (Blitvich and Firth, 2015; Cook et al., 2012; de Oliveira Ribeiro et al., 2021; Huhtamo et al., 2009; Junglen et al., 2009; Kolodziejek et al., 2013; Vasilakis and Tesh, 2015).

Table 1 Taxonomic, phylogenetic, and epidemiological characteristics of insect-specific orthoflaviviruses (ISOFs).

ISOFs	Abbreviation	Phylogenetic classification	Detection host	Year of first report	Country of detection	References
Aedes flavivirus	AEFV	cISOF	*Aedes albopictus and Aedes flavopictus*	2009	Japan	Hoshino et al. (2009)
Anopheles flavivirus	AnFV	cISOF	*Anopheles gambiae*	2016	Liberia	Fauver et al. (2016)
Calbertado virus	CLBOV	cISOF	*Culex tarsalis*	2011	Canada	Tyler et al. (2011)
cell fusing agent virus	CFAV	cISOF	*Aedes aegypti* cell line (Peleg)	1975	USA	Stollar and Thomas (1975)
Cuacua virus	CuCuV	cISOF	*Mansonia* sp.	2014	Mozambique	Cholleti et al. (2016)
Culex flavivirus	CxFV	cISOF	*Culex pipiens*	2007	Japan	Hoshino et al. (2007)
Culex theileri flavivirus	CxthFV	cISOF	*Culex theileri*	2017	Indonesia	Calzolari et al. (2012), Ferreira et al. (2013)
Culiseta flavivirus	CsFV	cISOF	*Culiseta melanura*	2016	USA	Misencik et al. (2016)
Hanko virus	HANKV	cISOF	*Aedes caspius*	2012	Finland	Huhtamo et al. (2012)
Kamiti River virus	KRV	cISOF	*Aedes macintoshi*	2003	Kenya	Crabtree et al. (2003)
Menghai flavivirus	MFV	cISOF	*Aedes albopictus*	2017	China	Zhang et al. (2017)

(continued)

Table 1 Taxonomic, phylogenetic, and epidemiological characteristics of insect-specific orthoflaviviruses (ISOFs). (*cont'd*)

ISOFs	Abbreviation	Phylogenetic classification	Detection host	Year of first report	Country of detection	References
Mercadeo virus	MECDV	cISOF	*Culex* sp.	2015	Panama	Carrera et al. (2015)
Nakiwogo virus	NAKV	cISOF	*Mansonia africana nigerrima*	2009	Uganda	Cook et al. (2009)
Niénokoué virus	NIEV	cISOF	*Culex* sp.	2017	Côte d'Ivoire	Cook et al. (2009)
Ochlerotatus caspius flavivirus	OCFV	cISOF	*Aedes caspius*	2013	Portugal	Ferreira et al. (2013)
Palm Creek virus	PCFV	cISOF	*Coquillettidia xanthogaster*	2013	Australia	Hobson-Peters et al. (2013)
Parramatta River virus	PaRV	cISOF	*Aedes vigilax*	2015	Australia	McLean et al. (2015)
Quang Binh virus	QBV	cISOF	*Culex tritaeniorhynchus*	2009	Vietnam	Crabtree et al. (2009)
Xishuangbanna aedes flavivirus	XFV	cISOF	*Aedes albopictus*	2016	China	Fan et al. (2016)

Ochlerotatus scapularis flavivirus	OSFV	cISOF	*Aedes scapularis*	2021	Bolivia	Orba et al. (2021)
Mansonia flavivirus	MAFV	cISOF	*Mansonia titillans*	2021	Bolivia	Orba et al. (2021)
Karumba virus	KRBV	cISOF	*Anopheles meraukensis*	2017	Australia	Colmant et al. (2017)
Mac Peak virus	McPV	cISOF	*Anopheles farauti*	2017	Australia	Colmant et al. (2017)
Haslams Creek virus	HaCV	cISOF	*Anopheles annulipes*	2017	Australia	Colmant et al. (2017)
Tabanus rufidens flavivirus	TrFV	cISOF	*Tabanus rufidens*	2021	Japan	Kobayashi et al. (2021)
Barkedji virus	BJV	dISOF	*Culex perexiguus*	2013	Senegal/Israel	Kolodziejek et al. (2013)
Chaoyang virus	CHAOV	dISOF	*Aedes vexans nipponii*	2013	Republic of Korea	Lee et al. (2013)
Donggang virus	DONV	dISOF	*Aedes* sp.	2009	China	Fang et al. (2021a)
Guapiacu virus	GUAPV	dISOF	*Aedes terrens* and *Aedes scapularis*	2021	Brazil	de Oliveira Ribeiro et al. (2021)
Ilomantsi virus	ILOV	dISOF	*Aedes riparius*	2014	Finland	Huhtamo et al. (2014)

(continued)

Table 1 Taxonomic, phylogenetic, and epidemiological characteristics of insect-specific orthoflaviviruses (ISOFs). (*cont'd*)

ISOFs	Abbreviation	Phylogenetic classification	Detection host	Year of first report	Country of detection	References
Kampung Karu virus	KPKV	dISOF	*Anopheles tesselatus*	2018	Malaysia	Huhtamo et al. (2009)
La Tina virus	LTNV	cISOF	*Aedes scapularis*	2018	Peru	Guzman et al. (2018)
Lammi virus	LAMV	dISOF	*Aedes cinereus*	2009	Finland	Huhtamo et al. (2009)
Long Pine Key virus	LPKV	dISOF	*Anopheles crucians*	2017	USA	Guzman et al. (2018)
Marisma mosquito virus	MMV	dISOF	*Aedes caspius*	2012	Spain	Huhtamo et al. (2009)
Nanay virus	NANV	dISOF	*Culex (Melanoconion) ocossa*	2013	Peru	Huhtamo et al. (2009)
Nhumirim virus	NHUV	dISOF	*Culex chidesteri*	2015	Brazil	Pauvolid-Corrêa et al. (2015)
Nounane virus	NOUV	dISOF	*Uranotaenia mashonaensis*	2009	Côte d'Ivoire	Junglen et al. (2009)
Paraiso Escondido virus	PEV	dISOF	*Psathyromyia abonnenci*	2015	Ecuador	Alkan et al. (2015)

Psorophora flavivirus	PSFV	dISOF	*Psorophora albigenu*	2018	Bolívia	Orba et al. (2021)
Aripo virus	ARPV	dISOF	*Psorophora albipes*	2021	Trinidad and Tobago	Auguste et al. (2021)
Binjari virus	BinJV	dISOF	*Aedes normanensis*	2019	Australia	Hobson-Peters et al. (2019)
Hidden Valley virus	HVV	dISOF	*Aedeomyia catasticta*	2020	Australia	Harrison et al. (2020)

Notes: Year, host, and country correspond to the earliest published report describing each virus. Although Yamadai flavivirus is classified as an insect-specific orthoflavivirus by the International Committee on Taxonomy of Viruses (ICTV), it was included in our analysis under the designation Quang Binh virus due to its high genetic similarity and its annotation in the National Center for Biotechnology Information (accession: AB981186.1). **Abbreviations:** ISOFs, insect-specific orthoflaviviruses; cISOFs, classical insect-specific orthoflaviviruses; dISOFs, dual-host affiliated insect-specific orthoflaviviruses.

Although ISOFs are detected predominantly in mosquitoes (family *Culicidae*), related complete or partial genomes have also been described in other dipterans, such as sand flies and chironomids, and horse flies (Kobayashi et al., 2021) suggesting that the evolutionary association between orthoflaviviruses and insects may be broader than initially thought (Alkan et al., 2015; Cook et al., 2013; Moureau et al., 2010). This expansion of the invertebrate host spectrum raises hypotheses regarding the evolutionary history of the genus, including an ancestral origin in arthropods followed by the acquisitions of vertebrate infectivity, or secondary loss of vertebrate infectivity in lineages that became strictly insect-adapted (Bolling et al., 2015b; Carvalho and Long, 2021; Öhlund et al., 2019; Vasilakis and Tesh, 2015).

Therefore, ISOFs have become recognized as central components for reconstructing the evolutionary history of orthoflavivirus, as they provide a valuable comparative system to investigate molecular mechanisms associated with vertebrate adaptation, the evolution of antigenic determinants, the plasticity of untranslated regions, and selection patterns across structural and non-structural genes (Blitvich and Firth, 2015; de Oliveira Ribeiro et al., 2021; Tangudu et al., 2021; Tree et al., 2016; Wastika et al., 2020; Zhang et al., 2022). Thus, integrating taxonomy, phylogeny, metagenomics, and viral ecology positions ISOFs as among the most informative groups for understanding the diversification and evolutionary history of the genus *Orthoflavivirus*. Within this framework, this chapter will focus on ISOFs, providing a brief introduction to their history and discovery, an overview of their evolution and natural maintenance, a discussion of their roles in laboratory studies, and the biological characteristics that have driven recent interest.

Currently, several terms are used to describe viruses that infect only invertebrate hosts, including arthropod-specific viruses (ASVs), insect-specific viruses (ISVs), and mosquito-specific viruses (MSVs) (Agboli et al., 2019). Although these terms are related, they represent distinct classification categories. ASVs include viruses associated with a broad range of arthropods, such as insects and ticks, that naturally replicate both in vivo and in vitro within these hosts but are inherently unable to replicate in vertebrates (Huang et al., 2023). ISVs are a subgroup of ASVs, comprising emerging viruses restricted to insects and incapable of replicating in mammalian or avian cells. Within this subgroup, MSVs are further restricted to mosquitoes (*Culicidae* family) (Agboli et al., 2019). These classifications are mainly based on current sampling and experimental systems available and may change as new host associations are uncovered. For the purpose of this chapter, the term insect-specific orthoflavivirus (ISOFs) is used to refer to viruses detected in insects from the *Culicidae* and *Psychodidae* families that belong exclusively to the *Orthoflavivirus* genus and have been shown in recent studies to be unable to replicate in vertebrate cells. The term mosquito-borne orthoflavivirus (MBOF) refers to arboviruses widely recognized as transmitted by mosquitoes (family *Culicidae*), and the term tick-borne orthoflavivirus (TBOF) refers to arboviruses widely recognized as transmitted by ticks.

2. Discovery of insect-specific orthoflavivirus

Understanding viral diversity and transmission cycles that shape the modern virosphere has been built on key historical milestones in virology. Within this context, the study of orthoflavivirus began in the early twentieth century, a period marked by major advances in recognizing the viral nature of infectious diseases (Stokes et al., 1928). The identification of the YFV established it as the first human disease attributed to a viral agent (Stokes et al., 1928) and provided the first experimental demonstration of the transmission of a human virus by mosquitoes (Reed et al., 2001). These findings were central to the development of the arbovirus concept, referring to viruses transmitted by hematophagous arthropods to vertebrate hosts (Reed et al., 2001).

This transmission model has dominated the understanding of the orthoflaviviruses biology, reinforcing the association of these viruses with transmission cycles involving arthropod vectors and vertebrate hosts (Weaver and Barrett, 2004). This conceptual framework began to be challenged with the discovery of viruses that appeared to be restricted exclusively to insect hosts (Weaver and Barrett, 2004). These viruses became known as insect-specific viruses (ISVs), defined as viruses that naturally infect mosquitoes and for which experimental evidence indicates replication restricted to cells of invertebrate origin (Bolling et al., 2015b). Although ISVs have since been identified across multiple viral families, they were first described within the genus *Orthoflavivirus* (Agboli et al., 2019; Cook et al., 2006; Huang et al., 2023; Öhlund et al., 2019; Vasilakis and Tesh, 2015).

The first ISOF to be characterized was the Cell-fusing agent virus (CFAV), which was isolated in 1975 from *Aedes aegypti* cell cultures. Early experimental observations revealed cytopathic effects and provided initial evidence of host restriction, as CFAV failed to infect vertebrate cell lines (Stollar and Thomas, 1975). Only years later did molecular characterization clarify the taxonomic placement of CFAV, confirming its inclusion within the genus *Orthoflavivirus*, despite its marked genetic divergence from previously described members of the group (Cammisa-Parks et al., 1992). Further support for the existence of orthoflaviviruses associated only with insect hosts was provided by the subsequent discovery of Kamiti River virus (KRV), which was isolated directly from mosquitoes collected in natural settings (Crabtree et al., 2003). Unlike CFAV, KRV underwent genomic sequencing soon after its isolation, reflecting significant advances in virological and molecular techniques available at the time (Crabtree et al., 2003).

Since the discovery of CFAV and KRV, the rate of ISV identification has increased substantially (Kraemer et al., 2015). These findings have greatly expanded our understanding of the ecological and evolutionary diversity of the genus *Orthoflavivirus* (Roundy et al., 2017). Currently, ISOFs are recognized as a distinct group and classified into two main categories: (i) classical insect-specific orthoflaviviruses (cISOFs), which form phylogenetically distinct lineages separate from orthoflaviviruses infecting vertebrates (MBOFs and TBOFs), and (ii) dual-host affiliated insect-specific orthoflaviviruses (dISOFs), which are phylogenetically closer to MBOFs (Bolling et al., 2015b). This shift from a vertebrate-centered view of *Orthoflavivirus* transmission to recognition of insect-restricted cycles has promoted new discoveries and broadened our understanding of the diversity, evolution, and ecology of orthoflavivirus and the virosphere as a whole (Bolling et al., 2015b).

3. Diversity, evolution, and ecology of insect-specific orthoflaviviruses

Increasing research on the genus *Orthoflavivirus* has revealed a broad spectrum of viruses with distinct biological characteristics, including arboviruses that infect vertebrates and viruses restricted exclusively to insects (Roundy et al., 2017; Weaver and Barrett, 2004). In this context, entomological surveillance, based on the systematic collection of mosquitoes and other hematophagous insects in the field, has become an essential component of arbovirus monitoring programs. Beyond enhancing the detection of medically important viruses, these efforts have substantially expanded our understanding of insect-associated viral diversity (Calzolari et al., 2016).

Among these viruses, ISOFs have emerged as an increasingly recognized group (Roundy et al., 2017). They have been detected in multiple mosquito genera, including *Aedes* (Hoshino et al., 2009), *Anopheles* (Fauver et al., 2016), *Culex* (Batson et al., 2021), *Coquillettidia* (Hobson-Peters et al., 2013), *Culiseta* (Misencik et al., 2016), *Mansonia* (Cook et al., 2009), *Psorophora* (Orba et al., 2021), and *Uranotaenia* (Junglen et al., 2009), as well as sand flies (Alkan et al., 2015) and horse flies (Kobayashi et al., 2021), reflecting their broad association with diverse mosquito lineages (Table 1; Fig. 2A), suggesting long-term host–virus associations and possible coevolutionary dynamics with different mosquito species (Dudas and Obbard, 2015). Moreover, ISOFs have been identified in natural mosquito populations across multiple regions worldwide, including all continents except Antarctica, highlighting their extensive geographic distribution (Table 1; Fig. 2B).

The continuous discovery of new viruses within this group suggests that their true diversity remains substantially underestimated, reflecting gaps in our knowledge of the mosquito virosphere and the complexity of host-virus interactions. Their widespread presence in mosquito populations also raises important questions about their ecological roles, including potential interactions with pathogenic orthoflaviviruses, effects on host biology, and impacts on arbovirus transmission dynamics.

3.1 Evolution and phylogenetic relationship of insect-specific orthoflavivirus

A defining distinction between MBOFs and ISOFs is host restriction. MBOFs require a dual-host cycle, alternating between mosquitoes and vertebrates, whereas ISOFs cannot infect vertebrates but replicate efficiently in insect-derived cells (Blitvich and Firth, 2015; Vasilakis and Tesh, 2015).

This host restriction is supported by multiple experimental studies since the initial identification of ISOFs, including CFAV (Cook et al., 2006; Stollar and Thomas, 1975), KRV (Crabtree et al., 2003; Sang et al., 2003), Palm Creek virus (PCV) (Hobson-Peters et al., 2013), Culex flavivirus (CxFV) (Hoshino et al., 2007; Kim et al., 2009), Aedes flavivirus (AeFV) (Hoshino et al., 2009), Quang Bihn virus (QBV) (Crabtree et al., 2009), as well as from the ISOFs that have been described recently, such as Nhumirin virus (NHUV) (Pauvolid-Corrêa et al., 2015), Guapiaçu virus

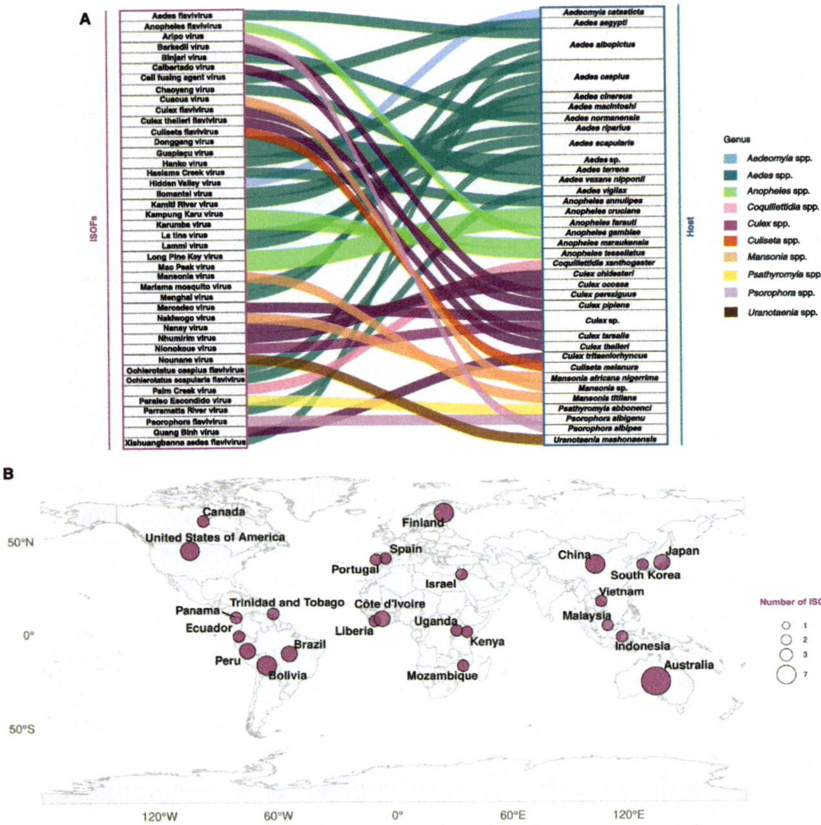

Fig. 2 Global overview of the characterized ISOFs. (A) Sankey diagram illustrating the associations between ISOFs and the host mosquito genus in which they were first identified. Each stream represents a virus-host association and is colored by mosquito genus. (B) Geographic distribution of the characterized ISOFs by country. The map shows the countries where each virus was first reported in the scientific literature. The circles indicate the number of ISOFs characterized in each country. The mosquito genus *Ochlerotatus* is herein referred to as *Aedes*, reflecting current taxonomic classification.

(GUAPV) (de Oliveira Ribeiro et al., 2021), and Aripo virus (ARPV) (Auguste et al., 2021). ISOFs typically replicate efficiently in mosquito-derived cell lines; although, the occurrence of cytopathic effects (CPE) and plaque formation varies with the specific viral species examined (Auguste et al., 2021; Hoshino et al., 2009, 2007). However, these viruses did not replicate in a broad range of vertebrate cell types, such as human, primate, hamster, swine, and avian cells, even under conditions optimized for replication, including temperatures between 27 and 33 °C (Hobson-Peters et al., 2013; Huhtamo et al., 2014; Junglen et al., 2009). Additionally, intracerebral inoculation of young mice showed that certain ISOFs neither caused disease nor demonstrated evidence of replication (Haddow et al., 2013; Huhtamo et al., 2009; Kim et al., 2009).

Nonetheless, some ISOFs present controversial evidence regarding their potential to infect vertebrates. For example, Marisma virus (MMV) induced CPE and detectable viral RNA in Vero and BHK-21 cells between 5 and 7 days post-inoculation, but both viral RNA and CPE were absent after the first subsequent passage, indicating an inability to sustain replication (Vázquez et al., 2012). Similarly, Ribeiro et al. identified Guapiaçu virus (GUAPV) and sequenced its complete genome from the plasma of a patient with febrile syndrome (de Oliveira Ribeiro et al., 2021). In vitro studies, however, showed that the virus did not replicate in mammalian cell types, emphasizing that detection of viral RNA does not necessarily indicate active replication (Blitvich and Firth, 2015; de Oliveira Ribeiro et al., 2021). Functional studies of KRV demonstrate that this virus does not replicate efficiently in vertebrate cells with intact innate immunity. However, turning off key antiviral innate pathways allows KRV to translate its genes, produce viral proteins, and release infectious particles, though at low levels (Tree et al., 2016). These results suggest that the vertebrate innate immune response might be one of the barriers to ISOFs replication. Even with these pathways suppressed, replication remains limited, suggesting additional constraints, possibly related to replication efficiency, virion assembly, or immune evasion (Tree et al., 2016).

These findings establish the central paradigm of ISOF biology: although these viruses retain the fundamental molecular architecture of orthoflaviviruses, current evidence indicates that they cannot complete the replication cycle in vertebrates. This apparent limitation defines a distinct evolutionary niche and underscores their insect-specific nature, although it remains possible that rare or context-dependent permissive interactions have not yet been captured by available experimental systems.

Genomic analysis of ISOFs showed that they display the canonical genomic architecture of the genus *Orthoflavivirus*, characterized by a positive-sense single-stranded RNA genome (ssRNA+) of approximately 10–11 kb, containing a single ORF that encodes a polyprotein subsequently processed into structural (C, prM/M, and E) and non-structural proteins (NS1, NS2A, NS2B, NS3, NS4A, 2K, NS4B, and NS5), flanked by 5′ and 3′ untranslated regions (5′UTR and 3′UTR). Despite this conserved organization, phylogenetic analyses have consistently shown that

ISOFs are not monophyletic. They are composed of the cISOFs, which constitute a highly divergent clade phylogenetically separated from all other orthoflaviviruses (Fig. 1), and the dISOFs, which cluster within the MBOF clade but do not appear to be competent to infect vertebrates (Fig. 1). This pattern suggests that the inability to infect vertebrates may have arisen independently in separate lineages derived from MBOF-like ancestors (Blitvich and Firth, 2015).

This distribution raises key hypotheses regarding the evolutionary origins of ISOFs: they represent ancestral lineages from which MBOFs emerged by acquiring the dual-host cycle, or they have originated from dual-host ancestors that gradually lost vertebrate competence through specialization to invertebrate hosts (Bolling et al., 2015b; Halbach et al., 2017). Currently, approximately 42 ISOFs with full-length polyprotein coding sequence are deposited in GenBank (Table 1), forming the basis for the analyses presented in this review. Most ISOFs have been discovered since the early 2000s, with a marked increase since 2005, driven by advances in next-generation sequencing, particularly metagenomic approaches, which have greatly facilitated the detection and characterization of novel viruses (Fig. 3).

Among the ISOFs described to date (Table 1), 34 are currently listed by the ICTV as "related, unclassified viruses," grouped into two categories: Insect-specific orthoflaviviruses (which include the majority of cISOFs) and *Viruses with no known arthropod vector* (which include most dISOFs) (Table 1) (International Committee on Taxonomy of Viruses (ICTV), 2023). In addition to the viruses recognized by the ICTV, this review includes several ISOFs documented in the literature but not yet formally classified, despite having full genome sequences available at Genbank, such as the GUAPV (de Oliveira Ribeiro et al., 2021), Karumba virus (KRBV), Mac Peak virus (McPV), Haslams Creek virus (HaCV) (Colmant et al., 2017) Psorophora flavivirus (PSFV), Mansonia flavivirus (MAFV), Ochlerotatus scapularis flavivirus (Orba et al., 2021), Aripo virus (ARPV) (Auguste et al., 2021), Binjari virus (BinJV) (Hobson-Peters et al., 2019) and Hidden Valley virus (HVV) (Harrison et al., 2020).

A comprehensive phylogenetic reconstruction including cISOFs, dISOFs, and MBOFs corroborates previous results by placing cISOFs as a basal lineage within the genus *Orthoflavivirus* (Fig. 4) (Blitvich and Firth, 2015; Cook et al., 2012; de Oliveira Ribeiro et al., 2021; Huhtamo et al., 2012). In the phylogenetic analysis, only viruses with complete genomes were used; thus, ISOFs such as Ngoye virus (NGOV) (Grard et al., 2006)

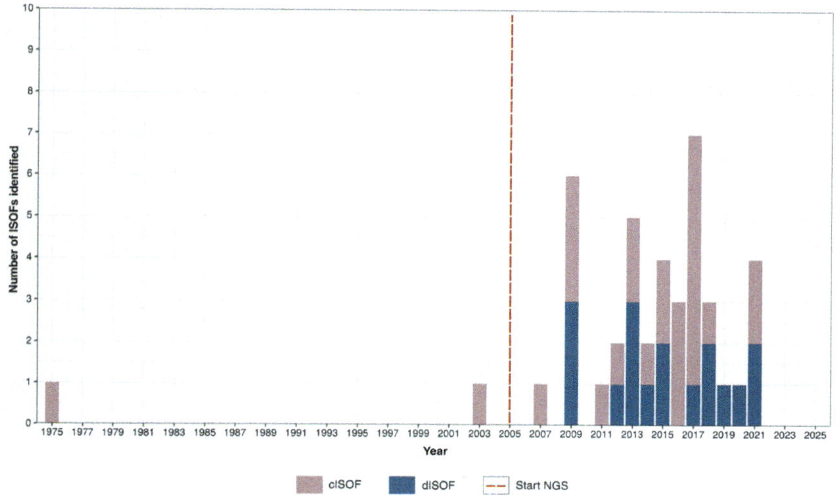

Fig. 3 Temporal distribution of insect-specific orthoflaviviruses (ISOFs) worldwide, by year of first report. Bars represent the total number of viruses described annually, stratified by classification as classical insect-specific orthoflaviviruses (cISOFs) or dual-host-affiliated insect-specific orthoflaviviruses (dISOFs). The year assigned to each virus corresponds to its first description in the scientific literature. The increase in virus discovery over time coincides with the expansion of Next-Generation Sequencing (NGS) technologies.

and Aedes galloisi flavivirus (AGFV) (Hoshino et al., 2012) were not included. The basal clade, composed of cISOFs, can be further subdivided according to host-associated patterns, with viruses detected in *Aedes* spp. tending to cluster more closely with each other than those identified in *Culex* spp. (Fig. 4). Interestingly, viruses such as Nakiwogo virus (NAKV), Cuacua virus, Parramatta River virus (PaRV), and Culiseta flavivirus (CsFV) isolated from *Mansonia, Coquillettidia* and *Culiseta* species, respectively, further expand the host spectrum of this group and seem to be more related to those viruses found in *Culex* species (Fig. 4). Viruses such as Mac Peak virus (McPV), Haslams Creek virus (HaCV), Karumba virus (KRBV), and Anopheles flavivirus (AnFV), all identified in *Anopheles* species, are closely related to one another (Fig. 4).

Although dISOFs are phylogenetically close to MBOFs, phylogenetic analyses indicate that they are not monophyletic, and their phylogenetic structure appears to reflect, at least in part, host associations and geographic distribution (Fig. 4). The most divergent dISOF is Paraiso Escondido virus (PEV), the only ISOF identified in sand flies to date.

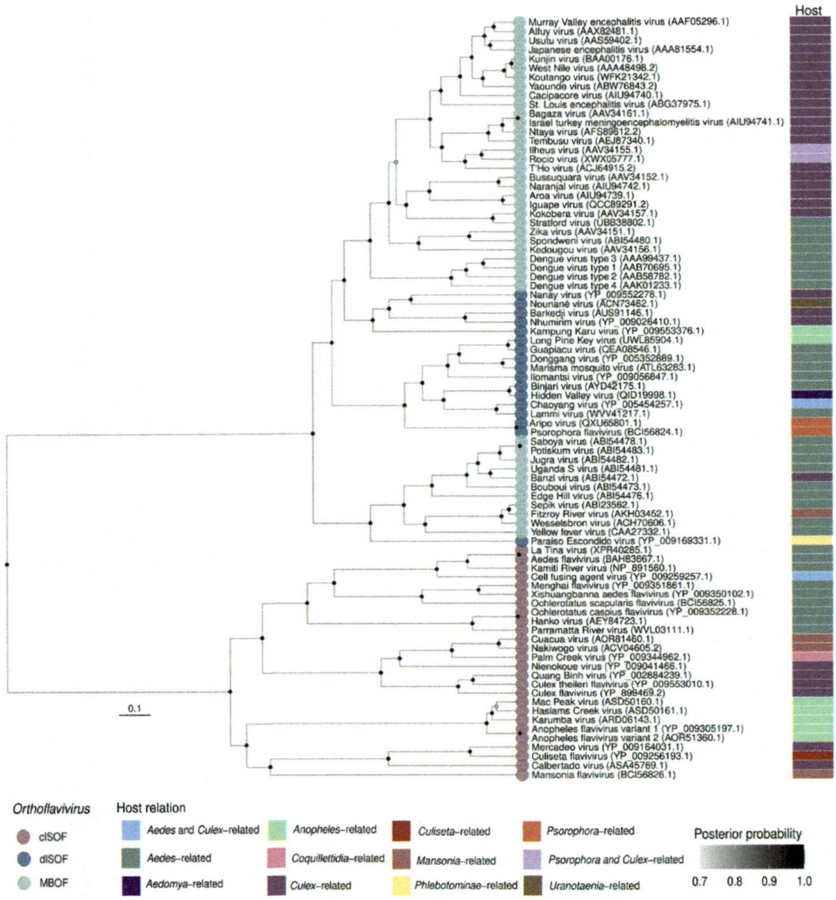

Fig. 4 Phylogenetic tree of orthoflaviviruses associated with mosquito vectors. Phylogenetic tree inferred from polyprotein sequences comprising 2657 amino acid sites. Tip colors represent the major groups of orthoflaviviruses associated with mosquito vectors, including Mosquito-borne orthoflaviviruses (MBOF), classical insect-specific orthoflaviviruses (cISOF), and dual-host affiliated insect-specific orthoflaviviruses. Colored boxes indicate the invertebrate host range of the *Orthoflavivirus*. Posterior probabilities are indicated by a gray gradient at the nodes. The phylogenetic tree is midpoint rooted. Tree scale represents the number of substitutions per site.

PEV is closely related to MBOFs classified within the yellow fever and Edge Hill virus groups, while being distantly related to the remaining dISOFs (Fig. 4), as previously reported by Alkan et al., 2015 (Alkan et al., 2015). The other dISOFs are divided into two main clades. One clade comprises MMV, Donggang virus (DONV), GUAPV, Long Pine Key virus (LPKV), Ilomantsi virus (ILOV), Lammi virus (LAMV), Chaoyang virus (CHAOV), Binjari virus (BinJV), Hidden Valley virus (HVV), Aripo virus (ARPV), and Psorophora flavivirus (PSFV), which display a broad host range, including *Aedes*, *Anopheles*, *Aedomya,* and *Psorophora* species. The second clade encompasses Nanay virus (NANV), Nounané virus (NOUV), Nhumirim virus (NHUV), Barkedji virus (BJV), and Kampung Karu virus (KPKV), which are primarily associated with *Culex* and *Uranotaenia* mosquitoes (Fig. 4).

ISOFs exhibit extensive diversity, often reflecting divergence patterns shaped by species-specific characteristics and geographic distributions, as demonstrated by a more detailed phylogenetic analysis (Fig. 5). It is noteworthy that although many ISOFs are commonly associated with specific mosquito species or genera, exceptions to this pattern have been documented. This has been observed for Culex theileri flavivirus (CxthFV), QBV, and CxFV, which, although predominantly associated with *Culex* species, have also been identified in *Anopheles vagus*, *Culicoides sp.*, and in *Aedes caspius* and *An. sinensis*, respectively (Fig. 5). Within the dISOF clade, a similar pattern is observed for LPKV, which, although more commonly associated with *Aedes* species, has also been detected in *Culex* and *Anopheles* species (Fig. 5).

The ISOFs also present a recurring history of taxonomic redundancy, in which viruses described in different countries were later recognized as closely related variants of the same species (Blitvich and Firth, 2015; Calzolari et al., 2012; Ferreira et al., 2013; Huhtamo et al., 2014, 2012; Vázquez et al., 2012). This scenario underscores the importance of considering previously proposed genetic similarity features in flavivirus taxonomy. An early study of Kuno et al. suggested that nucleotide identity values above 84% within the NS5 region could support classification within the same species (Kuno et al., 1998). However, as this criterion originally used partial gene sequences, it should be interpreted with caution when applied to complete coding sequences. No universal cutoff has been established for complete-genome comparisons in orthoflaviviruses. Instead, species classification relies on a combination of genetic, phylogenetic, and biological criteria. Indeed, according to ICTV guidelines, species demarcation criteria must integrate multiple lines of evidence. These include nucleotide and deduce amino acid sequence relatedness, antigenic characteristics, geographic association, vector association, host association, disease association, and ecological characteristics (International Committee on Taxonomy of Viruses (ICTV), 2023).

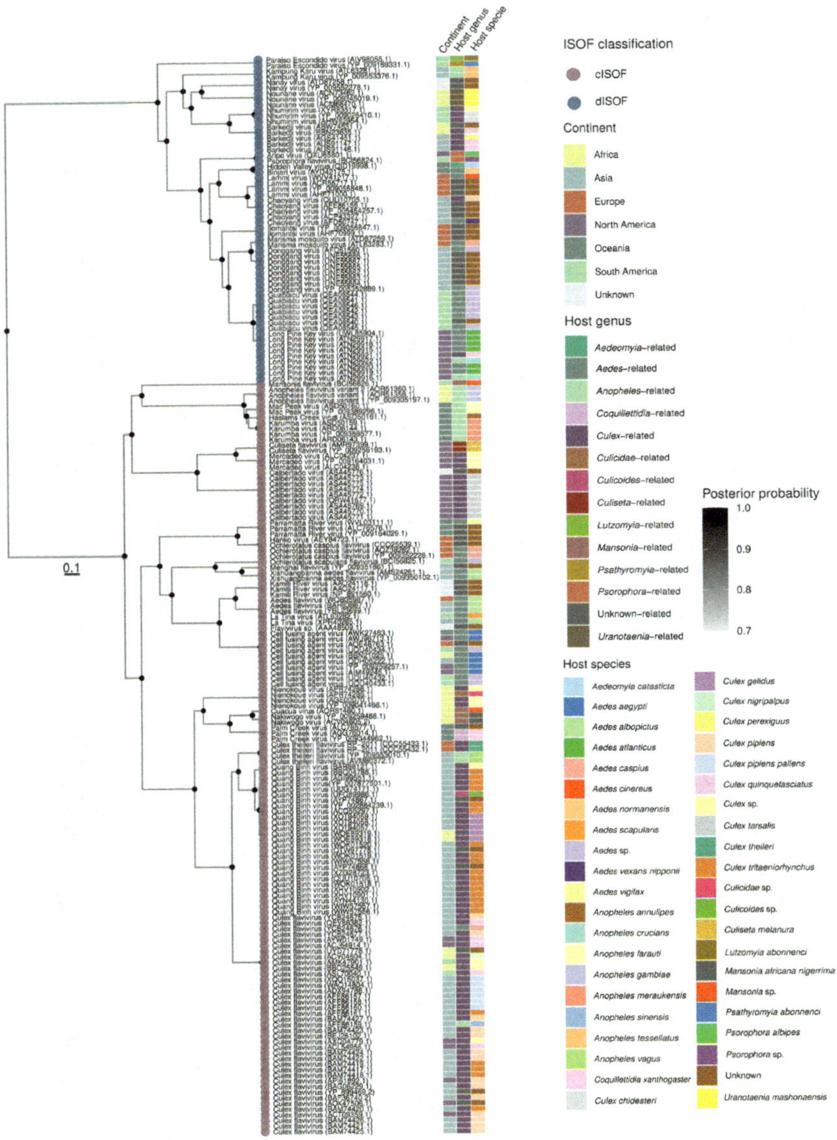

Fig. 5 Phylogenetic tree of insect-specific orthoflaviviruses. Phylogenetic tree inferred from polyprotein sequences comprising 2953 amino acid sites. Tip colors represent the major ISOF groups, including classical insect-specific orthoflaviviruses (cISOFs) and dual-host affiliated insect-specific orthoflaviviruses (dISOFs). Colored boxes indicate the continent where each virus sequence was identified, as well as the invertebrate host range of the ISOFs (genus and species). Posterior probabilities are indicated by a gray gradient at the nodes. The phylogenetic tree is midpoint rooted. Tree scale represents the number of substitutions per site.

In this context, we performed a nucleotide similarity analysis based on whole-genome sequences of cISOFs, dISOFs, and MBOFs to better characterize their identity patterns. Our results showed that La Tina virus (LTV – KY320649.1) and Aedes flavivirus (AeFV – OR098296.1) shared about 91 % of nucleotide identity, while Hanko (HNK) virus and Ochlerotatus caspius flavivirus (OCFV) shared 89 % nucleotide identity, whereas Aripo virus (ARPV - MZ358890.1) and Psorophora flavivirus (PsFV - LC567151.1) shared about 84 % of similarity (Fig. 6). Interestingly, all these viruses, share combination of genetic, phylogenetic, and biological criteria, such as genetic similarity, as proposed by Kuno et al. (Kuno et al., 1998), phylogenetic classification as cISOFs or dISOFs (Table 1, Fig. 5), and host range, that suggests that they should be classified as variants of the same species as s also observed for Anopheles flavivirus variants 1 and 2, which shared 94 % nucleotide identity (Fig. 6). Consistent with the divergence pattern observed in the amino acid phylogenetic tree, the nucleotide similarity analysis showed that dISOFs shared 42–44 % nucleotide identity with cISOFs, whereas they exhibited 55–61 % identity with MBOFs (Fig. 6).

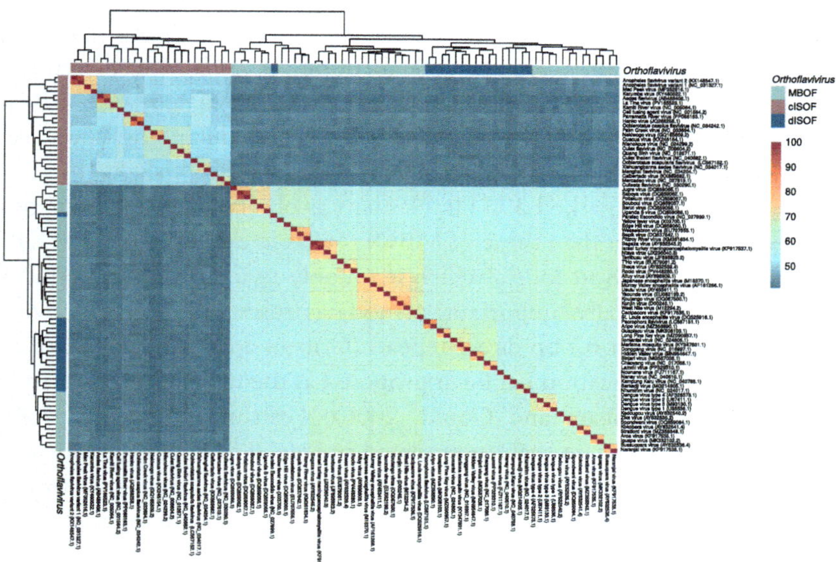

Fig. 6 Nucleotide similarity analysis. Hierarchical clustering of rows and columns reveal similarity patterns among cISOFs, dISOFs, and MBOFs.

The phylogenetic proximity between dISOFs and MBOFs supports the hypothesis that pathogenic orthoflaviviruses may have originated from ancestors exclusively associated with arthropods, with vertebrate tropism acquired during evolution, potentially through independent events involving complex genetic changes (Blitvich and Firth, 2015). Thus, it is possible that dISOFs may represent evolutionarily intermediate forms, resulting from relatively recent transitions in host range (de Oliveira Ribeiro et al., 2021; Guzman et al., 2018). Analogous situations are observed among NKV, such as Entebbe bat virus, Sokuluk virus, and Yokose virus, which lack known arthropod vectors but phylogenetically cluster with MBOFs (Blitvich and Firth, 2015). These examples support the notion that host-tropism shifts have occurred repeatedly throughout the evolutionary history of orthoflaviviruses, involving both gains and losses of transmission competence (Pontremoli et al., 2021). Although there is no evidence that current ISOFs can evolve into vertebrate pathogens, the group's evolutionary trajectory suggests that host tropism can be altered through accumulated mutations and the reorganization of regulatory genomic elements (Elrefaey et al., 2020; Tangudu et al., 2021; Zhang et al., 2022). Thus, identifying the specific changes that enable such transitions remains a major challenge in contemporary evolutionary virology.

Recent studies have sought to elucidate core genomic differences among MBOFs, cISOFs, and dISOFs, focusing on three main dimensions: the functional role of UTRs; sfRNA biogenesis associated with xrRNA structures in the 3′UTR; and compositional biases, including codon and dinucleotide usage.

Among ISOFs, marked 3′UTR plasticity is observed, and these regions are frequently longer than those of MBOFs. Classic examples include KRV (approximately 1208 nt long), which is roughly twice the length typically observed in medically important orthoflaviviruses, and CFAV, which contains two perfect 67-bp direct repeats (Gritsun and Gould, 2006a). For KRV, internal duplication has been proposed as the mechanism underlying this expansion (Gritsun and Gould, 2006a; Gritsun and Gould, 2006). Multiple studies indicate that untranslated regions (UTRs) constitute a key barrier to the transition from insect-specific restriction to dual-host tropism (Elrefaey et al., 2020; Gritsun and Gould, 2006b). Research in this field has demonstrated that short direct repeats within the 3′UTR contribute to the formation of subgenomic orthoflaviviral RNAs (sfRNAs). sfRNAs, which are widely studied in MBOFs, play important roles in viral pathogenicity, immune evasion, and host adaptation. They are generated through incomplete degradation of the viral genomic RNA by the cytoplasmic 5–3′ exoribonuclease XRN1, where degradation is blocked by XRN1-resistant RNA (xrRNA) structures in the 3′UTR (Zhang et al., 2020). In this context, Slonchak et al. demonstrated that both cISOFs and dISOFs generate sfRNAs through the XRN1-resistance mechanism, indicating that sfRNA biogenesis is not restricted to MBOFs (Slonchak et al., 2022). The authors identified multiple XRN1-resistant RNA elements in divergent ISOF lineages and showed that sfRNA production is conserved even in highly divergent viruses, supporting an early evolutionary origin of this mechanism within orthoflaviviruses. Reinforcing these results, broader analyses of 3′UTRs across orthoflaviviruses indicate that, despite substantial sequence variability, these regions maintain conserved functional RNA structures such as xrRNAs (Junglen et al., 2017). Additionally, comparative structural analyses further indicate that cISOFs, such as CxFV, generally exhibit shorter hairpins, occasionally duplicated, but with limited higher-order structural organization. In contrast, dISOFs such as CHAOV and NHUV retain conserved structural blocks resembling the stem-loop (SL) and dumbbell (DB) elements characteristic of MBOFs (Villordo et al., 2016). Consistently, analysis of GUAPV revealed conserved SL-II, SL-IV, and DB elements shared with MBOFs (de Oliveira Ribeiro et al., 2021), indicating a higher degree of structural continuity between dISOFs and dual-host lineages.

In addition to the 3′UTR function, other factors appear to play important roles in ISOF host restrictions (Elrefaey et al., 2020). Zhang et al. (2017) demonstrated that one limitation preventing dISOFs from infecting vertebrates occurs not at the stage of cellular entry, but during post-entry events associated with viral RNA replication (Zhang et al., 2022). Using DONV as a model, the authors showed that this virus can bind to and enter vertebrate cells through mechanisms similar to those of MBOFs, including endocytosis and pH-dependent fusion, indicating that envelope structural proteins are not the critical barrier for tropism. Despite successful entry, DONV fails to establish productive replication, suggesting that the key restriction occurs at the level of viral genome replication and replication complex assembly. Interestingly, replacing the DONV 5′ and 3′ UTRs with those of Zika virus was sufficient to restore replication in vertebrate cells, demonstrating that these regions are essential determinants of vertebrate permissiveness; however, it decreased the E protein expression in mosquito cells, suggesting that temperature is also a key factor for orthoflavivirus dual-cycle host adaptation. Moreover, even when chimeric viruses are generated, such as the YFV/NIEV chimera expressing the Niénokoué virus (NIEV) envelope proteins or the Binjari virus (BinJV) genome backbone containing the structural gene sequences (prM–E) of WNV (BinJ/WNVKUN-prME), replication in vertebrate cells remains restricted (Harrison et al., 2020; Junglen et al., 2017).

Another informative component is the compositional signature related to codon and dinucleotide usage. Vertebrates exhibit pronounced under-representation of CpG and UpA dinucleotides, whereas mosquitoes show strong depletion of UpA but not necessarily CpG (Blitvich and Firth, 2015). Interestingly, cISOFs, such as KRV, CxFV, and CFAV, display genomic profiles more similar to those of mosquitoes, with marked UpA under-representation but without strong CpG depletion (Lobo et al., 2009). However, Blitvich et al. demonstrated that, on average, dual dISOFs exhibit higher CpG usage than dual-host orthoflaviviruses but lower CpG frequencies than cISOFs, thereby reinforcing their intermediate compositional profile (Blitvich and Firth, 2015). Similar findings were reported by Kenney et al., who showed that the codon usage pattern of NHUV (dISOF) is more similar to that of WNV than to that of the CxFV (cISOF) (Kenney et al., 2014). Reinforcing this pattern, Colmant et al. demonstrated that, when comparing the frequencies of CpG dinucleotides, clusters with high CpG frequency were observed for cISOFs (such as PCV), intermediate frequencies for dISOFs (such as BinJV and HVV), whereas MBOFs, such as Usutu virus and WNV, showed a lower frequency of CpG dinucleotides (Colmant et al., 2021). Additionally, normalized Codon Adaptation Index (nCAI) analyses revealed that cISOFs show low adaptation to human codon usage, whereas MBOFs display higher nCAI values consistent with efficient replication in vertebrates. dISOFs exhibited intermediate values, suggesting partial compatibility with human codon preferences (de Oliveira Ribeiro et al., 2021).

Together, these findings indicate that dISOFs share compositional features more closely aligned with dual-host flaviviruses than with classical cISOFs, consistent with their phylogenetic position (Fig. 4).

3.2 Maintenance in nature

Among members of the genus *Orthoflavivirus*, MBOFs are characterized by transmission cycles that require both a vertebrate host and an arthropod vector, mainly mosquitoes (Weaver and Barrett, 2004). The persistence of these viruses in nature depends on the ecological interaction between competent vector populations and susceptible vertebrate hosts. Consequently, their geographic distribution is closely linked to the spatial and ecological overlap between vectors and vertebrate hosts (Kramer et al., 2008; Weaver and Barrett, 2004). In contrast, ISOFs are restricted to invertebrate hosts and do not require vertebrate hosts to complete their transmission cycle, suggesting that they employ distinct ecological strategies for long-term maintenance (Blitvich and Firth, 2015; Roundy et al., 2017).

Among the proposed mechanisms, vertical transmission, particularly transovarial transmission (TOT), is considered one of the principal mechanisms for the maintenance of ISOFs in nature. TOT refers to the infection of developing eggs within the ovaries of infected females, resulting in progeny that emerge already infected and enabling intergenerational viral persistence (Blitvich and Firth, 2017). This mechanism has been experimentally documented for several ISOFs and is further supported by evidence from field studies (Anderson et al., 2020; Bolling et al., 2015a, 2011; Cook et al., 2006; Farfan-Ale et al., 2010; Haddow et al., 2013; Logan et al., 2022; Lutomiah et al., 2007; Saiyasombat et al., 2011; Sang et al., 2003). Venereal transmission can also serve as a potential mechanism for viral maintenance. Infected male mosquitoes can transmit the virus to uninfected females during mating, and this type of transmission has been reported for both MBOFs (Anderson et al., 2020; Logan et al., 2022; Shroyer, 1990) and ISOFs (Bolling et al., 2015a). Beyond vertical transmission, horizontal transmission routes have also been proposed as alternative mechanisms for ISV maintenance. In this context, plants could potentially contribute to the ecological persistence of ISVs. Because adult mosquitoes feed on plant nectar (Herreros-Moya et al., 2025), infected individuals might deposit viral particles onto plant surfaces, enabling indirect transmission to uninfected mosquitoes. However, there is no empirical evidence to support the role of plants in the ISVs maintenance cycle, so their participation in viral ecology remains speculative (Roundy et al., 2017). Another possible route involves exposure to contaminated aquatic environments during the larval stage (Agboli et al., 2019) or interactions with parasites, such as mermithid nematodes, which have been associated with increased ISV infection rates (Agboli et al., 2019; Evangelina et al., 2013; Muttis et al., 2012). Additionally, recent metatranscriptomic evidence suggests potential virus sharing between mosquitoes and ectoparasitic water mites, indicating that parasite-mediated horizontal transmission may occur in natural settings (Ortiz-Baez et al., 2022).

Understanding the maintenance cycles of ISOFs is essential for elucidating their ecological dynamics and for exploring their potential implications for vector competence. The coexistence of ISOFs and MBOFs within natural mosquito populations raises important questions about potential viral interactions and their implications for vector biology and arboviral ecology (Roundy et al., 2017; Vasilakis and Tesh, 2015).

4. Applications

Mosquito-borne diseases, including DENV, Zika virus (ZIKV), and YFV, cause significant global health and economic burdens (Costa et al., 2023; Shepard et al., 2016), driving the search for novel and effective control strategies. ISOFs have emerged as practical tools in this context because they replicate exclusively in insect cells and do not infect vertebrates, thereby minimizing risks to humans and animals (Blitvich and Firth, 2015; Cui et al., 2025; Tree et al., 2016). These characteristics enable their deployment in biotechnological applications such as vector control, vaccine platform development, and paratransgenesis, representing a shift toward more targeted and sustainable approaches to limiting arbovirus transmission (Bolling et al., 2015b; Erasmus and Weaver, 2017; Hall et al., 2025; Öhlund et al., 2019).

ISOFs are investigated as biocontrol agents to suppress the replication and transmission of pathogenic arboviruses within mosquito populations. This suppression may occur through direct viral interference or by enhancing the mosquito's innate immune response to other viruses (Bolling et al., 2015b; Hobson-Peters et al., 2013). Additionally, ISOFs are under investigation as platforms for vaccine development and as vehicles for paratransgenesis, wherein genetically modified bacteria or viruses are introduced into mosquito populations to reduce vector competence (Hobson-Peters et al., 2019). These approaches mark a transition from traditional chemical control methods to more targeted and environmentally sustainable strategies. In the following sections, we will discuss the competitive interactions between ISOFs and arboviruses, mechanisms of coinfection, the phenomenon of superinfection exclusion (SIE), and the development of ISOF-based vaccine platforms, highlighting their implications for mosquito-borne disease control.

4.1 Vector and transmission control

4.1.1 Competitive interactions

Viruses frequently infect hosts or vectors sequentially, leading to complex interactions that can shape viral dynamics within these systems (Ciota, 2019;

Lim et al., 2025; Vogels et al., 2019). Superinfection refers to the sequentially infections of a biological system by more than one virus and may result in three distinct outcomes: (i) synergistic interactions, in which the replication or fitness of one or both viruses enhance; (ii) antagonistic interactions, in which the replication or fitness of one or both viruses inhibit; and (iii) neutral interactions, where no measurable impact on viral fitness occurs (Brustolin et al., 2023; Ciota, 2019). Such superinfection can occur not only at the cellular level but also at the tissue, organismal, or vector level (Auguste et al., 2021; Blackard, 2012; Hobson-Peters et al., 2013; Romo et al., 2018). These interactions are particularly important in the context of arboviruses and ISOFs, which often share ecological niches and can coinfect the same mosquito species.

ISOF prevalence varies across mosquito populations, with detection rates ranging from 10 % to over 80 % depending on species and geographic location (Fang et al., 2021b, 2018; Farfan-Ale et al., 2009; Papa et al., 2014). Overall, prevalence estimates are based on pooled mosquito samples rather than individual analyses, which can introduce bias and potentially misrepresent the true infection rates and biological significance of ISOF prevalence. Despite this methodological limitation, the widespread detection of ISOFs across studies suggests that their coexistence within mosquito populations is common. This frequent coexistence may result from mixed-mode transmission via both horizontal and vertical routes (Agboli et al., 2019). Due to frequent detection and high prevalence of some ISVs during orthoflaviviruses screening, researchers have begun testing whether ISOF superinfection can modulate viral replication dynamics (Olmo et al., 2023). Modulation may occur through direct competition for cellular resources, induction of host antiviral immune responses such as RNA interference pathways (RNAi), or interference with cellular entry and replication machinery. These processes ultimately influence the replication, transmission, and pathogenicity of superinfecting arboviruses (Chan et al., 2026).

Simultaneous coinfection of mosquitoes, in which two or more viruses infect the vector simultaneously, has also been studied as a competitive interaction (Table 2) (Kent et al., 2010). However, researchers consider these events artificial because they rarely occur in nature. In the wild, mosquitoes are more likely to acquire viruses sequentially through multiple blood meals or environmental exposures. It is also essential to distinguish between naturally infected mosquitoes and those infected under experimental conditions, as the infection context can significantly influence observed viral interactions. In experimental systems, infection methods

Table 2 Coinfection studies and outcomes between insect-specific orthoflaviviruses (ISOFs) and mosquito-borne orthoflaviviruses (MBOFs).

ISOF	MBOFs	Coinfection outcome	Experimental system	Effect	References
ARPV	WNV	Antagonistic	CD-1 mice	Limits WNV-induced disease through induction of a type I interferon response	Auguste et al. (2021)
CFAV	DENV-1	Antagonistic	C6/36 (*Ae. albopictus*)	Reduces viral titers	Baidaliuk et al. (2019)
CxFV	WNV	Antagonistic	C6/36 (*Ae. albopictus*)	WNV replication was suppressed (Reduction of viral titer)	Goenaga et al. (2020)
CxFV Izabal	WNV	Synergistic	*Cx. quinquefasciatus* Honduras	Increases the proportion of mosquitoes transmitting WNV. Increases in WNV transmission	Kent et al. (2010)
LAMV	WNV	Neutral	U4.4 (*Ae. albopictus*)	No detectable effect	Öhlund et al. (2022)
NHUV	WNV, JEV, and SLEV	Antagonistic	C6/36 (*Ae. albopictus*)	Reduces viral titers	Kenney et al. (2014)
NHUV	WNV	Antagonistic	C6/36 (*Ae. albopictus*) and C7-10 (*Ae. albopictus*)	Reduces viral titers	Goenaga et al. (2015)

NHUV	WNV	Neutral	*Cx. pipiens*	No detectable effect	Goenaga et al. (2015)
NHUV	WNV	Antagonistic	*Cx. quinquefasciatus* Sebring	Produces a greater reduction in WNV transmissibility at later time points following NHUV infection	Goenaga et al. (2015)
NHUV	ZIKV	Antagonistic	*Ae. aegypti*	Reduces ZIKV transmission without affecting infection rates or viral titers	Romo et al. (2018)

Abbreviations: ARPV, Aripo virus; CFAV, cell fusing agent virus; CxFV, Culex flavivirus; DENV, dengue virus; JEV, Japanese encephalitis virus; LAMV, Lammi virus; NHUV, Nhumirim virus; SLEV, St. Louis encephalitis virus; WNV, West Nile virus; ZIKV, Zika virus.

such as oral feeding and intrathoracic injection differ markedly in biological relevance. Oral infection involves key physiological barriers, including midgut infection and escape processes, while intrathoracic injection circumvents these barriers, potentially affecting viral replication, dissemination, and coinfection outcomes. Although laboratory models of simultaneous coinfection and intrathoracic injection remain valuable for elucidating virus–virus interactions and potential outcomes, they may not fully represent the complexity of natural infection dynamics or transmission processes (Ciota, 2019).

Of particular relevance for transmission control is superinfection exclusion (SIE), also known as homologous or heterologous interference, which occurs when a biological system infected by a primary virus becomes resistant or less permissive to subsequent infection by a second virus (Roundy et al., 2017). In homologous interference, the secondary virus is closely related to the primary virus, whereas in heterologous interference, it is genetically distinct but separate from the primary virus (Laureti et al., 2020). This phenomenon has been documented across a wide range of viral taxa and occurs in both vertebrate and invertebrate hosts (Auguste et al., 2021; Colmant et al., 2018; Öhlund et al., 2022).

4.1.2 Superinfection exclusion (SIE)

From an evolutionary perspective, superinfection exclusion could be considered a selective advantage mediated by virus-encoded mechanisms that reduce competition for limited cellular resources and means to maintain the stability of viral sequences in segmented viruses (Huang et al., 2008). SIE occurs by preventing the entry or replication of additional viruses within the same cell (Folimonova, 2012).

Several studies have demonstrated the biological relevance and complexity of SIE in mosquito–virus systems. By altering replication success and competitive outcomes among viruses, superinfection can modify the selective pressures acting on viral populations. SIE capacity has been evaluated for primary infections with AEFV, ARPV, BinJV, CFAV, CxFV, CxFV Izabal, LAMV, NHUV, PaRV, PCV, and QBV (Table 3), suggesting that this phenotype occurs broadly across phylogenetically distinct ISOF lineages.

Remarkably, homologous superinfection exclusion is more prevalent and generally more robust than heterologous SIE. As shown in Table 3, only a limited number of studies have examined primary ISOFs infection followed by secondary alphavirus infection, and among these,

Table 3 Superinfection studies and outcomes between insect-specific *Orthoflaviviruses* (ISOFs) and mosquito-borne orthoflaviviruses (MBOFs).

Primary infecting virus	MBOFs	Superinfection outcome	Experimental system	Effect	References
AEFV	ZIKV and DENV-2	Superinfection exclusion	*Ae. albopictus*	Detrimental to viral replication, associated with activation of the Janus kinase/signal transducer and activator of transcription (JAK/STAT), Toll, or RNA interference (RNAi) pathways	Fu et al. (2025)
ARPV	WNV	Superinfection exclusion	CD-1 mice	Limits WNV-induced disease through induction of a type I interferon response	Auguste et al. (2021)
BinJV	CHIKV, WNV and ZIKV	Superinfection exclusion	C6/36 (*Ae. albopictus*)	Reduces WNV, ZIKV, and CHIKV viral titers	Willemsen et al. (2025)
BinJV (persistently infected)	CHIKV, WNV and ZIKV	Superinfection exclusion (WNV and ZIKV) and Neutral (CHIKV)	C6/36 (*Ae. albopictus*)	Reduces WNV and ZIKV titers, although inhibition is weaker than during acute infection	Willemsen et al. (2025)
BinJV	CHIKV, WNV and ZIKV	Superinfection exclusion (CHIKV and ZIKV) and Neutral (WNV)	Aag2 (*Ae. aegypti*)	Reduces ZIKV and CHIKV titers, with stronger inhibition of CHIKV than observed in C6/36 cells	Willemsen et al. (2025)

(continued)

Table 3 Superinfection studies and outcomes between insect-specific *Orthoflaviviruses* (ISOFs) and mosquito-borne orthoflaviviruses (MBOFs). (*cont'd*)

Primary infecting virus	MBOFs	Superinfection outcome	Experimental system	Effect	References
BinJV (persistently infected)	CHIKV, WNV and ZIKV	Superinfection exclusion (CHIKV and ZIKV) and Neutral (WNV)	Aag2 (*Ae. aegypti*)	Reduces ZIKV and CHIKV titers, with stronger inhibition of CHIKV than observed in C6/36 cells	Willemsen et al. (2025)
BinJV	ZIKV	Superinfection exclusion	Aag2 Ago2-deficient (*Ae. aegypti*)	Reduces viral titers, primarily associated with cell death and loss of RNAi activity	Willemsen et al. (2025)
BinJVzsGreen	ZIKVmCherry	Superinfection exclusion	C6/36 (*Ae. albopictus*)	Inhibits ZIKVmCherry infection, preventing productive infection	Torres et al. (2022)
CFAV	DENV-1 and ZIKV	Superinfection exclusion	C6/36 (*Ae. albopictus*) and *Ae. aegypti*	Decreases the replicative fitness of ZIKV and DENV-1 in C6/36 cells and reduces systemic dissemination of infectious DENV-1 particles by disrupting the positive correlation between body RNA load and dissemination titer	Baidaliuk et al. (2019)
CFAV	DENV-2	Synergistic	Aa20 (*Ae. aegypti*)	Increases expression of ribonuclease kappa (RNASEK), a factor known to facilitate infection of viruses that depend on endocytosis and pH-dependent entry	Zhang et al. (2017)

CxFV	WNV	Superinfection exclusion	C6/36 (*Ae. albopictus*) and *Culex pipiens*	Reduces RNA copy number, suggesting early suppression of WNV replication in *Cx. pipiens* and C6/36 superinfection models	Bolling et al. (2012)
CxFV	JEV and DENV-2	Synergistic	CTR (*Culex tritaeniorhynchus* Giles)	Enhances viral release from superinfected cells	Kuwata et al. (2015)
CxFV Izabal	WNV	Neutral	C6/36 (Ae. albopictus) and *Cx. quinquefasciatus* Sebring	No detectable effect	Kent et al. (2010)
LamV	WNV	Superinfection exclusion	U4.4 (*Ae. albopictus*)	Reduces WNV RNA copy number and induces upregulation of signal peptidase complex (SPC) genes associated with endoplasmic reticulum functions and proper cleavage of *Orthoflavivirus* structural proteins, as well as heat-shock proteins (HSPs) involved in protein folding and stress responses	Öhlund et al. (2022)
NHUV	WNV, JEV, and SLEV	Superinfection exclusion	C6/36 (*Ae. albopictus*)	Reduces viral titers	Kenney et al. (2014)

(continued)

Table 3 Superinfection studies and outcomes between insect-specific *Orthoflaviviruses* (ISOFs) and mosquito-borne orthoflaviviruses (MBOFs). (*cont'd*)

Primary infecting virus	MBOFs	Superinfection outcome	Experimental system	Effect	References
NHUV	WNV	Superinfection exclusion	C6/36 (*Ae. albopictus*) and C7-10 (*Ae. albopictus*)	Reduces viral titers	Goenaga et al. (2015)
NHUV	CHIKV, DENV-2 and ZIKV	Superinfection exclusion (ZIKV and DENV-2) and Neutral (CHIKV)	C6/36 (*Ae. albopictus*)	Reduces DENV-2 titers and decreases intracellular ZIKV RNA replication and viral titers	Hall et al. (2025)
NHUV	ZIKV	Superinfection exclusion	*Ae. aegypti*	Reduces ZIKV infection rates in mosquitoes previously infected with NHUV, indicating inhibition primarily at infection establishment rather than post-infection spread	Romo et al. (2018)
PaRV	DENV-3, RRV and WNV	Superinfection exclusion (WNV and DENV-3) and Neutral (RRV)	C6/36 (*Ae. albopictus*)	Reduces WNV and DENV-3 titers and suppresses viral replication	McLean et al. (2021)
PCV	RRV, MVEV and WNV	Superinfection exclusion (MVEV and WNV) and Neutral (RRV)	C6/36 (*Ae. albopictus*)	Reduces infectious titers of MVEV and WNV and decreases cellular permissiveness	Hobson-Peters et al. (2013)

PCV	WNV	Superinfection exclusion	*Cx. annulirostris*	Reduces susceptibility to WNV infection and transmission	Hall-Mendelin et al. (2016)
PCV	CHIKV and ZIKV	Neutral	*Ae. aegypti* and *Ae. albopictus*	No detectable effect	Koh et al. (2021)
QBV	DENV-2 and WNV	Superinfection exclusion	C6/36 (*Ae. albopictus*)	Reduces viral titers	Ang et al. (2025)

Abbreviations: Family Flaviviridae (genus *Orthoflavivirus*): AEFV, Aedes flavivirus; ARPV, Aripo virus; BinJV, Binjari virus; CFAV, cell fusing agent virus; CxFV, Culex flavivirus; DENV, dengue virus; JEV, Japanese encephalitis virus; LamV, Lammi virus; MVEV, Murray Valley encephalitis virus; NHUV, Nhumirim virus; PaRV, Parramatta River virus; PCV, Palm Creek virus; QBV, Quang Binh virus; SLEV, St. Louis encephalitis virus; WNV, West Nile virus; ZIKV, Zika virus. **Family *Togaviridae* (genus *Alphavirus*):** CHIKV, chikungunya virus; RRV, Ross River virus.

approximately half reported neutral outcomes (Colmant et al., 2018; Hobson-Peters et al., 2013; Koh et al., 2021; Romo et al., 2018; Willemsen et al., 2025). This predominance is primarily due to the high genetic and functional similarity between primary and secondary viruses in homologous contexts, which facilitates direct competition for intracellular resources, replication machinery, and antiviral responses, reinforcing the view that functional similarity determines interference efficiency.

At the cellular level, following primary infection, orthoflaviviruses remodel the host endoplasmic reticulum to form specialized replication organelles that coordinate RNA synthesis, genome amplification, and virion assembly (Apte-Sengupta et al., 2014; McMahon and Gallop, 2005). Viral budding occurs adjacent to replication vesicle pores, allowing newly synthesized genomes to be efficiently captured by capsid proteins, likely ensuring packaging specificity through spatial proximity rather than a defined packaging signal (Junjhon et al., 2014; Welsch et al., 2009). Because intracellular resources such as membranes, lipids, and host cofactors are limited, the virus that establishes infection first can monopolize these components, creating a replication environment that disadvantages subsequently infecting viruses (Nasar et al., 2015; Torres et al., 2022; Willemsen et al., 2025).

Experimental evidence from studies of ZIKV and WNV superinfection with NHUV and LAMV indicates that orthoflavivirus SIE primarily operates at the level of viral RNA synthesis, rather than at virion attachment, entry, or translation (Romo et al., 2018; Zou et al., 2009). In sequential infection models, secondary viral genomes can enter cells and initiate early infection steps, but do not efficiently synthesize new RNA. This suggests that interference with the replication complex function primarily mediates exclusion, rather than receptor competition or entry blockade (Baidaliuk et al., 2019; Bolling et al., 2012; Daims and Wagner, 2018; Goenaga et al., 2015; Kenney et al., 2014; McLean et al., 2021; Romo et al., 2018). Romo et al. 2018 demonstrated a specific reduction in intracellular RNA replication, while Öhlund et al. 2022 reported upregulation of signal peptidase complex proteins (SPC), genes associated with endoplasmic reticulum (ER) functions, and increased expression of various heat-shock proteins (Öhlund et al., 2022; Romo et al., 2018).

The interval between primary and secondary infections also may be a critical factor in SIE. Pepin et al. 2008 demonstrated in *Ae. albopictus* cell lines, with longer delays between superinfections with DENV serotypes (DENV-2 and DENV-4) influenced the degree of suppression, with earlier-infecting serotypes having an advantage and later-infecting serotypes experiencing greater suppression (Pepin et al., 2008). Similarly, Baidaliuk et al. 2019 found that infection timing significantly shapes SIE dynamics; prior infection with CFAV markedly suppressed DENV-1 replication, thereby reducing titers. The negative effect was stronger when CFAV infection was established for a longer time period before arbovirus inoculation (Baidaliuk et al., 2019).

The time interval between infections allows the primary virus to establish intracellular dominance by expanding replication complexes, occupying membrane platforms, accumulating viral proteins, and consuming host resources. Therefore, longer delays between sequential infections generally result in stronger SIE, as the first virus secures both spatial and biochemical control over the intracellular replication environment (Baidaliuk et al., 2019; Öhlund et al., 2022; Romo et al., 2018). Reinforcing this view, studies of superinfection indicate that exclusion of secondary infection typically does not occur until at least one hour after initial virus infection (Eaton, 1979; Johnston et al., 1974). This transition suggests that late-stage infected cells become structurally and metabolically resistant to additional viral genomes, providing a mechanistic explanation for the temporal restriction observed in orthoflavivirus superinfection.

Beyond direct competition, mosquitoes can develop a primed antiviral state following prior pathogen exposure. This phenomenon, known as immune priming, refers to enhanced resistance or accelerated immune responses during subsequent infections (Gutiérrez-Millán et al., 2025; Ratcliffe et al., 2024). Immune priming functionally resembles vertebrate immunological memory, despite the absence of adaptive immune cells or lymphoid tissues in insects (Gutiérrez-Millán et al., 2025). In mosquito–virus systems, immune priming is associated with reduced viral replication and increased expression of antiviral pathways, particularly RNAi-related components such as Dicer-2 and Ago2 (Vargas et al., 2020). The interaction between immune priming and canonical antiviral pathways may influence mosquito antiviral defenses, including RNAi, which can restrict viral replication and affect viral diversity (Olson and Blair, 2015). Further studies have shown that mosquitoes generate small RNAs from orthoflavivirus-specific double-stranded RNA (dsRNA), which triggers the RNAi response shortly after ingestion of a blood meal and may establish immune priming for subsequent MBOFs infection (Keene et al., 2004; Sanchez-Vargas et al., 2004).

RNAi is a small RNA–mediated nucleic acid degradation system comprising distinct pathways defined by the type of small RNAs produced. While the microRNA (miRNA) pathway regulates endogenous gene expression and the Piwi-interacting RNA (piRNA) and endogenous small interfering RNA (siRNA) pathways primarily control transposons, the exogenous siRNA pathway functions as the main antiviral mechanism in mosquitoes (Fu et al., 2025; Lee et al., 2019; Sanchez-Vargas et al., 2004; Willemsen et al., 2025). During orthoflavivirus replication, viral dsRNA intermediates are recognized by pattern recognition receptors (PRRs), triggering innate immune responses (Blair, 2011; Lee et al., 2019). The Dicer-2/R2D2 complex processes viral dsRNA into ~21-nt siRNAs (Tikhe and Dimopoulos, 2021), which are loaded into the RNA-induced silencing complex (RISC). The guide strand then directs Argonaute-2 (Ago2) to degrade complementary viral RNA in a sequence-specific manner (Fig. 7) (Agboli et al., 2019; Lee et al., 2019). Importantly, RNAi restricts viral replication without inducing cytopathic effects (Blair, 2011). In this context, the detection of virus-derived small RNAs through small RNA sequencing has become an important approach for inferring active viral replication in mosquito hosts, providing stronger evidence of true host association than viral RNA detection alone, particularly in studies based on pooled mosquito samples and in investigations of ISOFs.

Primary infection may enhance antiviral RNAi, generating virus-derived siRNAs that cross-target related secondary viruses. Fu et al., 2025 reported increased Dicer-2 transcription during AEFV infection and following superinfection with ZIKV or DENV-2 compared with single infections (Fu et al., 2025). Since RNAi targeting efficiency may correlate with sequence homology, these results indicate preferential RNAi activation between phylogenetically related viruses. Supporting this, Willemsen et al., 2025 identified siRNAs mapping to the ZIKV prM–E region during infection with a BinJ–ZIKV chimera, which coincided with reduced ZIKV replication. In contrast, no similar inhibition was observed during WNV superinfection, consistent with insufficient sequence homology for cross-targeting (Willemsen et al., 2025).

The Janus kinase signal transducer and activator of transcription (JAK/STAT) pathway, along with the Toll pathway, constitutes a critical component of mosquito innate antiviral immunity. In the absence of an adaptive immune system in insects, these pathways serve as the primary defense against viral infections and may also contribute to SIE (Agaisse and Perrimon, 2004). Functionally, the JAK/STAT pathway is analogous to

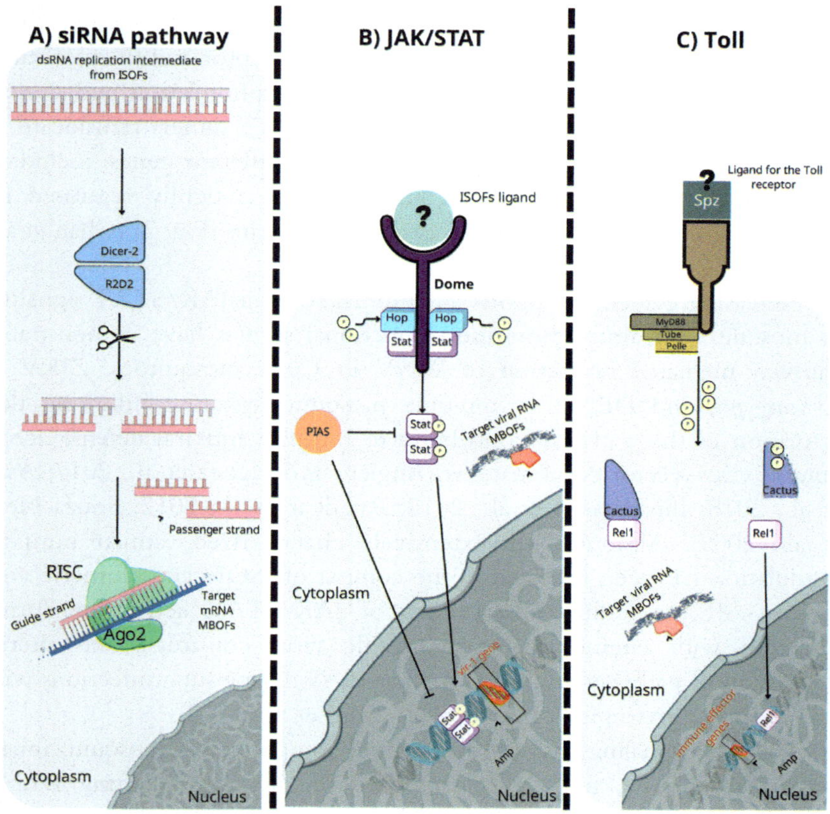

Fig. 7 Antiviral immune pathways potentially modulated during superinfection assays involving ISOFs and MBOFs. Schematic overview of the principal mosquito antiviral signaling routes that may influence SIE. (A) siRNA pathway: ISOF dsRNA generated during replication is recognized and cleaved by Dicer-2 in association with R2D2, producing small siRNAs. These are loaded into the RISC containing Ago2, where the guide strand directs sequence-specific degradation of complementary viral RNA, limiting replication. (B) *JAK/STAT pathway:* Putative ISOF ligands trigger activation of the Dome receptor, promoting phosphorylation of Hop and Stat proteins, Stat dimerization, and nuclear translocation to induce antiviral effector gene expression (e.g., *vir-1*). Negative regulation may occur through PIAS-mediated inhibition. (C) *Toll pathway:* Binding of the ligand Spätzle (Spz) to the Toll receptor activates the MyD88–Tube–Pelle signaling cascade, leading to phosphorylation and degradation of the inhibitor Cactus and release of the NF-κB–like transcription factor Rel1, which translocates to the nucleus and induces antimicrobial and antiviral genes. Dashed vertical lines separate pathways to emphasize that these systems can act independently or coordinately. Question marks indicate steps where virus-specific sensing mechanisms remain incompletely defined in mosquitoes.

the vertebrate interferon system. JAK/STAT signaling is initiated when ligand binding activates the Domeless (Dome) receptor and the associated kinase Hopscotch (Hop), leading to STAT phosphorylation and dimerization (Souza-Neto et al., 2009). Activated STAT dimers translocate to the nucleus and induce transcription of antiviral effector genes, including vir-1 (Souza-Neto et al., 2009). This pathway is tightly regulated by negative modulators such as SOCS and PIAS proteins (Fig. 7) (Chan et al., 2026; Tikhe and Dimopoulos, 2021).

Increasing evidence supports an antiviral role for JAK/STAT signaling in mosquitoes against arboviruses. Functional studies have demonstrated pathway-mediated restriction of WNV in *Culex* mosquitoes, ZIKV in *Ae. aegypti*, and DENV in multiple mosquito systems, indicating that activation of this pathway contributes to intrinsic antiviral defense across diverse virus–vector combinations (Angleró-Rodríguez et al., 2017; Asad et al., 2018; Jupatanakul et al., 2017; Paradkar et al., 2012; Souza-Neto et al., 2009). Although less extensively characterized, similar immune modulation has been reported in the context of ISOFs and chimeric viral systems SIE, including observations of JAK/STAT activation during infection with engineered insect-specific virus constructs and altered expression of pathway regulators such as PIAS during superinfections with ISOFs and arboviruses (Fu et al., 2025; Tan et al., 2023).

The Toll signaling pathway is a central component of mosquito innate antiviral immunity, initially characterized in *Ae. aegypti* during DENV infection (Ramirez and Dimopoulos, 2010; Xi et al., 2008). In mosquitoes, Toll activation regulates antiviral transcription primarily through the NF-κB transcription factor Rel1, which is negatively controlled by Cactus (Lau et al., 2003; Shin et al., 2005). Infection-associated upregulation of Spätzle (Spz) may promote receptor activation and trigger a conserved cascade involving MyD88, Tube, and Pelle (Fig. 7) (Cherry and Silverman, 2006; Ramirez and Dimopoulos, 2010; Shin et al., 2006). Downstream, Cactus degradation allows Rel1 nuclear translocation and induction of immune effector genes (Tikhe and Dimopoulos, 2021). Although most mechanistic evidence derives from arbovirus models (Angleró-Rodríguez et al., 2017; Jia et al., 2024; McFarlane et al., 2014; Paradkar et al., 2012; Xi et al., 2008), emerging data suggest that Toll signaling also participates in broader insect–virus interactions, including those involving ISOFs. Experimental studies show that modulation or silencing of Toll components, such as Cactus, alters viral replication (Fu et al., 2025; Nanfack-Minkeu et al., 2019), supporting an antiviral role. Additionally, evidence from *Wolbachia*–virus interactions indicates that Rel1-mediated Toll regulation can influence viral replication dynamics (Pan et al., 2012), suggesting that Toll signaling may contribute to virus–virus interactions and potentially shape ISOF maintenance or superinfection outcomes in mosquito populations.

The antiviral effects of siRNA, JAK/STAT, and Toll may act synergistically with resource competition (Fu et al., 2025; Willemsen et al., 2025). However, the limited number of studies directly assessing RNAi responses, JAK/STAT, and Toll expression during superinfection precludes definitive conclusions. Many investigations have been conducted, either in mosquito cells, particularly C6/36 cells, which are deficient in antiviral RNAi, JAK/STAT, and Toll, or in vivo, without quantifying small RNAs and transcript levels of antiviral proteins, thereby restricting mechanistic interpretation. While the occupation of replication complexes may limit physical access to replication sites, antiviral immune pathways may simultaneously diminish the availability of functional viral RNA templates. Although variation in the efficiency of the antiviral immune pathways across mosquito species and virus lineages may influence the evolutionary dynamics of ISOFs and MBOFs, potentially affecting their infectivity or virulence with different outcomes (Dodson et al., 2024, 2014; Lambrechts and Saleh, 2019; Zélé et al., 2014).

4.1.3 Superinfection enhancement

Under certain cellular conditions, interactions between superinfecting orthoflaviviruses may shift from competitive interference to functional facilitation. Experimental evidence suggests that replication enhancement can result from indirect host-mediated mechanisms rather than direct viral cooperation. For instance, Zhang et al. (2017) reported that infection with CFAV can transcriptionally upregulate cellular factors that favor entry or early infection steps of a secondary virus. In mosquito cells, prior CFAV infection increases ribonuclease kappa (RNASEK) expression, a host protein involved in endocytic uptake and endosomal acidification, thereby promoting DENV infection. Conversely, silencing RNASEK reduces DENV genomic RNA levels and infectious particle production (Zhang et al., 2017).

Additional enhancement mechanisms may function at later stages of infection. For example, in persistently infected *Cx. tritaeniorhynchus* cultured cells, CxFV infection enhances the release of DENV-2 and JEV from superinfected cells (Kuwata et al., 2015). This increase is associated with cellular stress or cytopathic effects induced by superinfection, which may accelerate the release of mature virions from intracellular compartments (Kuwata et al., 2015). Variations in viral budding kinetics may further influence these outcomes, as viruses that assemble or exit cells earlier can benefit more from superinfection-induced cellular changes. Collectively, these findings indicate that replication enhancement in orthoflavivirus-superinfected cells does not require cooperative genome packaging or mixed particle formation but can result from virus-driven reprogramming of host physiology that transiently increases cellular permissiveness to additional arboviruses.

4.2 Applications of ISOFs in vaccine and antiviral development

Increasing urbanization, globalization, human mobility, and climate change have facilitated the geographic expansion of mosquito vectors, reinforcing the urgent need for effective vaccines to control arboviral diseases, particularly those caused by viruses of the genus *Orthoflavivirus*, such as DENV (Naddaf, 2024; Wallau et al., 2023). Conventional vaccine strategies for these pathogens face significant challenges, including safety concerns, the risk of antibody-dependent enhancement, and the difficulty of inducing balanced, durable immunity (Kim, 2019). In this context, ISOFs have attracted growing interest as innovative vaccine platforms (Hall et al., 2025).

The development of vaccines based on ISOFs relies on the construction of chimeric viruses in which the structural genes of the ISOFs are replaced by those of distinct pathogenic orthoflaviviruses (Hall et al., 2025; Hobson-Peters et al., 2019; Piyasena et al., 2017). As a result, the engineered virus displays the structural proteins of the target arbovirus while retaining the nonstructural proteins and replication machinery derived from the ISOFs backbone, thereby restricting host tropism to invertebrate cells (Piyasena et al., 2019). Consequently, these chimeric viruses replicate efficiently in mosquito cell lines, enabling scalable production, but are incapable of productive replication in vertebrate cells. Importantly, the particles preserve the native antigenic conformation of viral envelope proteins and their conformational epitopes, allowing the induction of neutralizing antibodies and cellular immune responses without the risk of viral dissemination (Piyasena et al., 2019; Unali and Douam, 2025; Zhang et al., 2024). This intrinsic biological attenuation confers a favorable safety profile while maintaining immunogenicity, supporting the use of ISOFs-based chimeras as a versatile vaccine platform for arboviruses of medical and veterinary relevance and enabling rapid antigen exchange for emerging orthoflaviviruses (Hall et al., 2016; Kim, 2019).

Several chimeric vaccines based on ISOFs, such as BinJV (BinJV/DENV (Choo et al., 2021), BinJV/ZIKV (Hazlewood et al., 2020), BinJV/YFV (Yan et al., 2020), BinJV/JEV (Harrison et al., 2024), and BinJV/WNV (Vet et al., 2020)), ARPV (ARPV/ZIKV) (Porier et al., 2021), YN15-283-02 (YN15-283-02/WNV) (Zhang et al., 2024), and CHAOV (CHAOV/ ZIKV) (Dong et al., 2024), have been described. Fig. 8 illustrates the chimeric construction of ISOF-based vaccines, including the pathogenic virus used to construct the chimera, the host, and the animal model tested. Across the different platforms and animal models evaluated, a consistent phenotype is observed, characterized by high immunogenicity, induction of neutralizing antibodies, protection following viral challenge, and absence of disease, effectively combining the structural authenticity of whole-virus vaccines with the safety profile of replication-deficient platforms in vertebrate hosts (Hall et al., 2025; Unali and Douam, 2025).

Due to their phylogenetic relatedness to pathogenic orthoflaviviruses, while remaining restricted to insect cells, ISOFs provide comparative systems for investigating molecular determinants of host tropism, viral replication, and virus-host interactions (Goenaga et al., 2015; Wilkman et al., 2025). Analyses of genomic and functional differences between ISOFs and dual-host flaviviruses may help identify conserved viral proteins or replication steps that can serve as targets for antiviral intervention (Goenaga et al., 2015; Hall et al., 2016). Moreover, chimeric systems derived from ISOF backbones can be employed in controlled experimental settings to evaluate candidate antiviral compounds, enabling assessment of inhibitory effects on the structural proteins of medically relevant arboviruses while maintaining a restricted replication phenotype in vertebrate cells (Hazlewood et al., 2022; Zhang et al., 2024). Together, these approaches support the use of ISOFs as experimental platforms to advance the identification and characterization of novel antiviral strategies.

Despite their advantages, the use of ISOFs in vaccine and antiviral development presents important limitations (Hall et al., 2025; Kim, 2019). One major constraint is the restricted replication of ISOF-based chimeras in insect cells, which, although conferring safety, may limit antigen expression dynamics and immunological stimulation compared to live-attenuated vaccines that replicate in vertebrate hosts. In addition, large-scale manufacturing requires well-established mosquito cell culture systems, and regulatory familiarity with insect-cell-derived viral vaccines remains limited in some settings. Genetic stability of chimeric constructs must also be carefully monitored, particularly regarding the potential for recombination or

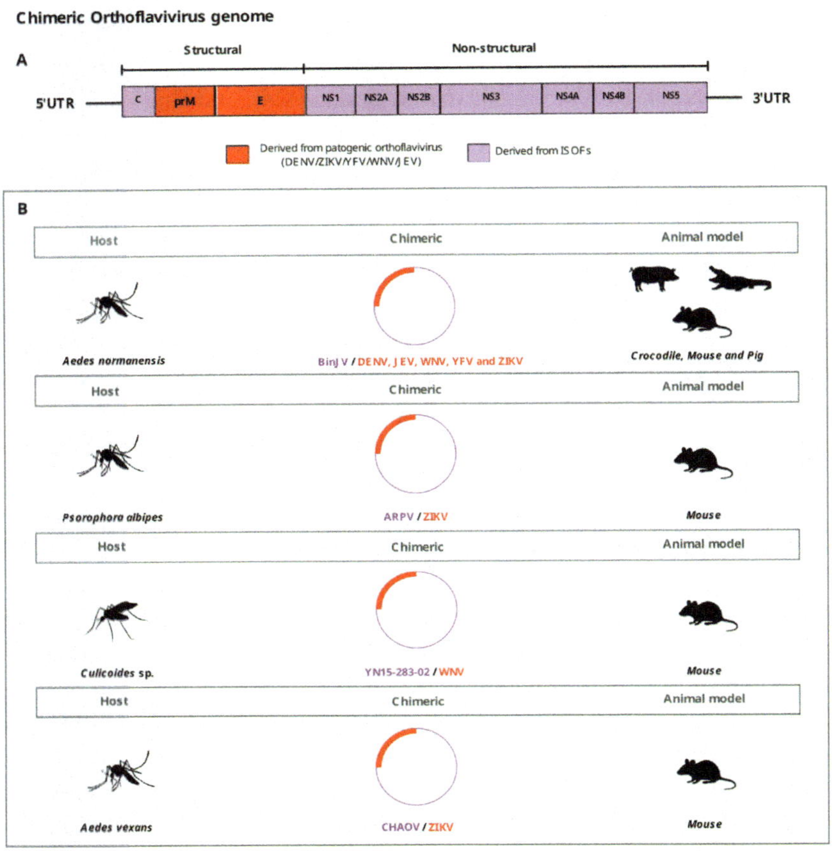

Fig. 8 Strategy for the construction of chimeric orthoflaviviruses using insect-specific orthoflaviviruses (ISOFs) as backbone platforms. (A) The schematic representation at the top illustrates the genome organization of a chimeric orthoflavivirus, highlighting structural (C, prM, E) and non-structural (NS1-NS5) regions. Red segments indicate genomic regions derived from pathogenic orthoflaviviruses (DENV, ZIKV, WNV, JEV, and YFV), whereas purple segments represent sequences derived from ISOFs. (B) Below, representative examples of chimeric viruses BinJV (BinJV/DENV, BinJV/ZIKV, BinJV/YFV, BinJV/JEV, and BinJV/WNV), ARPV (ARPV/ZIKV), YN15–283-02 (YN15–283-02/WNV), and CHAOV (CHAOV/ZIKV) are shown according to their insect host of origin and corresponding vertebrate animal models used for experimental evaluation. The circular configuration is a schematic representation for illustrative purposes.

unintended mutations during production (Choo et al., 2021; Harrison et al., 2024; Unali and Douam, 2025; Zhang et al., 2024). In antiviral research, ISOFs serve as indirect models, and differences in host factors between insect and vertebrate cells may limit the extrapolation of findings to human infection. Furthermore, while theoretical, concerns related to host-range evolution or the acquisition of dual-host tropism require continued evaluation to ensure long-term biosafety (Hall et al., 2016; Kim, 2019; Vasilakis and Tesh, 2015).

4.3 Biological control of arboviruses using ISOFs

ISOFs have emerged as promising tools for the biological control of arboviruses due to their natural ability to infect and replicate exclusively in mosquitoes, without posing a risk of infection to vertebrates (Blitvich and Firth, 2015; Chen et al., 2023; Wilkman et al., 2025). These viruses can interfere with vector competence through different mechanisms, including competitive superinfection, modulation of host antiviral pathways, and occupation of cellular niches required for the replication of pathogenic arboviruses such as WNV and JEV (Chan et al., 2026; Chen et al., 2023; Patterson et al., 2020; Wilkman et al., 2025).

One proposed strategy for biological control of arboviruses using ISOFs involves deliberately introducing ISOF-infected mosquitoes into natural vector populations. Once established, the virus can be maintained through vertical and horizontal transmission among insects, allowing ecological persistence without the need for continuous applications (Patterson et al., 2020; Wilkman et al., 2025). The presence of the ISOFs within the mosquito occupies cellular niches and activates basal antiviral mechanisms that limit the replication of pathogenic arboviruses (Bolling et al., 2015b, 2015a; Vasilakis and Tesh, 2015). Consequently, vector competence is reduced, and the probability of transmission to humans decreases, representing a self-sustaining, potentially more specific approach than conventional insecticide-based methods (Bolling et al., 2015a).

Another way in which ISOFs may act is through interactions with the endosymbiotic bacteria *Wolbachia*. This bacterium has been extensively studied for its ability to induce a nonspecific antiviral state in mosquitoes, particularly in species of the genus *Aedes*, thereby reducing the replication and transmission of several arboviruses by activating basal immune pathways and competing for cellular resources (Ant et al., 2023; Iturbe-Ormaetxe et al., 2011; Nazni et al., 2019). However, this blocking effect can be modulated by the presence of ISOFs, which also interfere with the host antiviral response. Consequently, interactions between *Wolbachia* and the mosquito virome may lead to either synergistic or antagonistic effects on vector competence, ultimately influencing the effectiveness and predictability of biological control strategies (Blitvich and Firth, 2015; Vasilakis and Tesh, 2015).

Despite the promising potential of ISOFs as tools for biological control, several limitations and challenges must be considered (Unali and Douam, 2025). The viral-blocking effect is not universal and may vary depending on the combination of mosquito species, ISOF, and target arbovirus, potentially leading to neutral or unpredictable outcomes under certain biological condition (Chan et al., 2026; Folimonova, 2012; Goenaga et al., 2020; Zou et al., 2009). The stability of ISOFs infections in natural mosquito populations depends on the efficiency of vertical transmission and on its impact on mosquito fitness, as biological costs may compromise long-term ecological persistence (Folimonova, 2012; Goenaga et al., 2015; Kuwata et al., 2015). Evolutionary pressures may also select for viral variants that escape ISOF-mediated interference, reducing ISOF efficacy over time. Furthermore, we cannot exclude the idea that ISOFs viruses could acquire dual-host tropism, potentially expanding their host range to include vertebrates, including humans (Elrefaey et al., 2020). In addition, in specific biological contexts, coinfection with an ISOF could produce the opposite of the intended effect, enhancing rather than suppressing arbovirus replication (Guggemos et al., 2024). For these reasons, ecological, evolutionary, and regulatory considerations pose important challenges for implementing ISOF-based strategies for arbovirus control to ensure that the deliberate release of infected mosquitoes does not result in unintended environmental consequences.

5. Conclusion and future perspectives

ISOFs remain an important subject for future investigation. Further studies are required to determine whether ISOFs, alone or in coinfection with other microorganisms such as the endosymbiotic bacteria *Wolbachia*, influence mosquito longevity, fitness, and vector competence, and to assess the ecological and epidemiological implications of these interactions. From an evolutionary perspective, it remains unclear whether ISOFs lost the ability to infect vertebrate cells or whether arboviruses acquired this capacity over time; more comparative analyses are needed to identify the molecular determinants underlying host restriction. Systematic evaluation of the theoretical risk of dual-host tropism is also necessary to support biosafety assessments for biological control strategies. In addition, the mechanisms that prevent ISOFs replication in vertebrate cells, as well as those governing infection, persistence, and interaction within mosquito hosts, remain to be fully characterized. Continued research is also warranted to explore the application of ISOFs as platforms for vaccine development and as tools for identifying novel antiviral targets.

6. Material and methods

6.1 Phylogenetic analyses

The majority of *Orthoflavivirus* sequences utilized in the phylogenetic analyses were sourced from ICTV reports and downloaded directly from NCBI. For viruses with accession numbers not corresponding to complete genomes, or those absent from ICTV reports (e.g., GUAPV, ARPV, KRBV, McPV, HaCV, PSFV, MAFV, Ochlerotatus scapularis flavivirus, BinJV, and HVV), sequences were identified through manual searches in NCBI using relevant search terms. The number of accessions for each sequence is indicated at the tips of the respective phylogenetic trees.

Nucleotide and polyprotein sequences for each dataset were downloaded and aligned using MAFFT v7.505 (Katoh and Standley, 2013) with the L-INS-i algorithm (–localpair –maxiterate 1000), which iteratively refines local pairwise alignments. Ambiguous and highly gapped regions were removed using trimAl v1.5. Rev1 with the automated option (Capella-Gutiérrez et al., 2009), following internal criteria. Alignments were visually inspected in AliView v1.30 (Larsson, 2014). The best-fit amino acid substitution model was determined using ModelFinder (Kalyaanamoorthy et al., 2017), implemented in IQ-TREE v1.6.9 (Nguyen et al., 2015), based on the Bayesian Information Criterion (BIC). All datasets were assigned the same amino acid substitution model. Bayesian phylogenetic reconstruction was conducted using the Markov chain Monte Carlo (MCMC) method in BEAST v1.10.4 (Suchard et al., 2018), employing the LG+ G4 amino acid substitution model, a strict molecular clock, and the Yule speciation process as the tree prior. For the dataset shown in Fig. 5, which includes several sequences from the same viral species, the Yule speciation process was replaced with a constant-size tree prior. MCMC chains were run for 10 million generations, with parameters sampled every 10,000 steps. Convergence and effective sample sizes (ESS) were assessed using Tracer v1.7.2 (Rambaut et al., 2018). Maximum clade credibility (MCC) trees were summarized using TreeAnnotator v1.10.4 (Suchard et al., 2018) after discarding the first 10 % of sampled trees as burn-in. The resulting tree was visualized and annotated in RStudio v4.3.3 (R Core Team, 2021) using the ggtree package (Xu et al., 2022).

A distance matrix was generated based on the whole-coding nucleotide sequences of ISOFs, and MBOFs were aligned using MAFFT v7.505 (Katoh and Standley, 2013) with the L-INS-i algorithm (–localpair –maxiterate 1000), which applies iterative refinement incorporating local pairwise alignment information. Ambiguous and highly gapped regions were removed using trimAl v1.5. with the automated option (Capella-Gutiérrez et al., 2009), resulting in a final alignment of 8149 nucleotide positions. The alignment was visually inspected in AliView v1.30 (Larsson, 2014) and imported into R for downstream analyses. Pairwise genetic distances were calculated using the dist.DNA function in the Phangorn package (Schliep, 2011), with pairwise deletion of missing data. The resulting identity matrix was visualized as a heatmap using the pheatmap package (https://github.com/raivokolde/pheatmap). All graphs and maps were generated using Rstudio v. 4.5.2.

Figure-credit note

All images were created using Canva Pro graphic features. Some visual elements were adapted from resources provided by the NIAID Visual & Medical Arts (2024). Links to the original elements and files are available:

Fig. 2: https://www.canva.com/design/DAHCKxIb8fo/CF_Ja013_Ox0c7kgJRSGyg/view?utm_content=DAHCKxIb8fo&utm_campaign=designshare&utm_medium=link2&utm_source=uniquelinks&utlId=h901dee31f6.

Fig. 3: https://www.canva.com/design/DAHCK6zkNFw/ui2TgGocHdeJ9k6Yyp6woA/view?utm_content=DAHCK6zkNFw&utm_campaign=designshare&utm_medium=link2&utm_source=uniquelinks&utlId=h1ea65704d7.

Fig. 7: https://www.canva.com/design/DAG5z-KkMyw/N8oOHCrA_BU-eSUJg7VqvA/view?utm_content=DAG5z-KkMyw&utm_campaign=designshare&utm_medium=link2&utm_source=uniquelinks&utlId=h4789f872af.

Fig. 8: https://www.canva.com/design/DAHCKgt1flU/GM6MCztduceSziR_im_Fjg/view?utm_content=DAHCKgt1flU&utm_campaign=designshare&utm_medium=link2&utm_source=uniquelinks&utlId=hd1c8171556.

Acknowledgements

This work was supported by the Fundação de Amparo à Pesquisa do Estado de São Paulo (FAPESP; grants 2022/03645–1 and 25/00398–1 to MLN, 2022/1467–0 to CAB, 2024/20412–6 to VB, and 2024/22180–5; 2025/17203–9 to IST). This study was partly funded by the INCT Viral Genomic Surveillance and One Health (grant 405786/2022–0) and the INCT Dengue Program (grant 465425/2014–3), and partly by the Centers for Research in Emerging Infectious Diseases (CREID) through the "Coordinating Research on Emerging Arboviral Threats Encompassing the Neotropics (CREATE-NEO)" grant U01AI151807 to NV from the National Institutes of Health (NIH, USA). MLN is a CNPq Research Fellow. The funders had no role in the study design, data collection, data analysis, interpretation, manuscript preparation, or the decision to publish the results. Grammarly AI was used only for language editing, grammar, spelling, and punctuation. The tool did not generate, modify, or interpret any scientific content. All analyses, interpretations, and conclusions are the authors' own.

References

Agaisse, H., Perrimon, N., 2004. The roles of JAK/STAT signaling in Drosophila immune responses. Immunol. Rev. 198, 72–82. https://doi.org/10.1111/j.0105-2896.2004.0133.x.

Agboli, E., Leggewie, M., Altinli, M., Schnettler, E., 2019. Mosquito-specific viruses—transmission and interaction. Viruses 11, 873. https://doi.org/10.3390/v11090873.

Alkan, C., Zapata, S., Bichaud, L., Moureau, G., Lemey, P., Firth, A.E., et al., 2015. Ecuador paraiso escondido virus, a new flavivirus isolated from New world sand flies in ecuador, is the first representative of a novel clade in the genus flavivirus. J. Virol. 89, 11773–11785. https://doi.org/10.1128/JVI.01543-15.

Anderson, J.F., Main, A.J., Ferrandino, F.J., 2020. Horizontal and vertical transmission of West Nile virus by Aedes vexans (Diptera: Culicidae). J. Med. Entomol. 57, 1614–1618. https://doi.org/10.1093/jme/tjaa049.

Ang, C.C.W., Ang, A.T.Y., Yam, J., Lee, R.X., Lim, M.J., Loh, Z.Y., et al., 2025. Spatiotemporal prevalence and characterization of the lineage I insect-specific flavivirus, Quang Binh virus, isolated from Culex gelidus mosquitoes in Singapore. J. Gen. Virol. 106. https://doi.org/10.1099/jgv.0.002105.

Angleró-Rodríguez, Y.I., MacLeod, H.J., Kang, S., Carlson, J.S., Jupatanakul, N., Dimopoulos, G., 2017. Aedes aegypti molecular responses to Zika virus: modulation of infection by the Toll and Jak/Stat immune pathways and virus host factors. Front. Microbiol. 8. https://doi.org/10.3389/fmicb.2017.02050.

Ant, T.H., Mancini, M.V., McNamara, C.J., Rainey, S.M., Sinkins, S.P., 2023. Wolbachia -virus interactions and arbovirus control through population replacement in mosquitoes. Pathog. Glob. Health 117, 245–258. https://doi.org/10.1080/20477724.2022.2117939.

Apte-Sengupta, S., Sirohi, D., Kuhn, R.J., 2014. Coupling of replication and assembly in flaviviruses. Curr. Opin. Virol. 9, 134–142. https://doi.org/10.1016/j.coviro.2014.09.020.

Asad, S., Parry, R., Asgari, S., 2018. Upregulation of Aedes aegypti Vago1 by Wolbachia and its effect on dengue virus replication. Insect Biochem. Mol. Biol. 92, 45–52. https://doi.org/10.1016/j.ibmb.2017.11.008.

Auguste, A.J., Langsjoen, R.M., Porier, D.L., Erasmus, J.H., Bergren, N.A., Bolling, B.G., et al., 2021. Isolation of a novel insect-specific flavivirus with immunomodulatory effects in vertebrate systems. Virology 562, 50–62. https://doi.org/10.1016/j.virol.2021.07.004.

Baidaliuk, A., Miot, E.F., Lequime, S., Moltini-Conclois, I., Delaigue, F., Dabo, S., et al., 2019. Cell-fusing agent virus reduces arbovirus dissemination in aedes aegypti mosquitoes *In Vivo*. J. Virol. 93. https://doi.org/10.1128/JVI.00705-19.

Batson, J., Dudas, G., Haas-Stapleton, E., Kistler, A.L., Li, L.M., Logan, P., et al., 2021. Single mosquito metatranscriptomics identifies vectors, emerging pathogens and reservoirs in one assay. Elife 10. https://doi.org/10.7554/eLife.68353.

Blackard, J.T., 2012. HCV superinfection and reinfection. Antivir. Ther 17, 1443–1448. https://doi.org/10.3851/IMP2460.

Blair, C.D., 2011. Mosquito RNAi is the major innate immune pathway controlling arbovirus infection and transmission. Future Microbiol. 6, 265–277. https://doi.org/10.2217/fmb.11.11.

Blitvich, B., Firth, A., 2017. A review of flaviviruses that have no known arthropod vector. Viruses 9, 154. https://doi.org/10.3390/v9060154.

Blitvich, B., Firth, A., 2015. Insect-specific flaviviruses: a systematic review of their discovery, host range, mode of transmission, superinfection exclusion potential and genomic organization. Viruses 7, 1927–1959. https://doi.org/10.3390/v7041927.

Bolling, B.G., Eisen, L., Moore, C.G., Blair, C.D., 2011. Insect-specific flaviviruses from culex mosquitoes in Colorado, with evidence of vertical transmission. Am. Soc. Tropical Med. Hyg. 85, 169–177. https://doi.org/10.4269/ajtmh.2011.10-0474.

Bolling, B.G., Olea-Popelka, F.J., Eisen, L., Moore, C.G., Blair, C.D., 2012. Transmission dynamics of an insect-specific flavivirus in a naturally infected Culex pipiens laboratory colony and effects of co-infection on vector competence for West Nile virus. Virology 427, 90–97. https://doi.org/10.1016/j.virol.2012.02.016.

Bolling, B.G., Vasilakis, N., Guzman, H., Widen, S.G., Wood, T.G., Popov, V.L., et al., 2015a. Insect-specific viruses detected in laboratory mosquito colonies and their potential implications for experiments evaluating arbovirus vector competence. Am. Soc. Tropical Med. Hyg. 92, 422–428. https://doi.org/10.4269/ajtmh.14-0330.

Bolling, B.G., Weaver, S.C., Tesh, R.B., Vasilakis, N., 2015b. Insect-specific virus discovery: significance for the arbovirus community. Viruses 7, 4911–4928. https://doi.org/10.3390/v7092851.

Brinton, M.A., 2002. The molecular biology of West Nile virus: a new invader of the western hemisphere. Annu. Rev. Microbiol. 56, 371–402. https://doi.org/10.1146/annurev.micro.56.012302.160654.

Brustolin, M., Pujhari, S., Terradas, G., Werling, K., Asad, S., Metz, H.C., et al., 2023. Vitro and in vivo coinfection and superinfection dynamics of mayaro and zika viruses in mosquito and vertebrate backgrounds. J. Virol. 97. https://doi.org/10.1128/jvi.01778-22.

Calisher, C.H., Higgs, S., 2018. The discovery of arthropod-specific viruses in hematophagous arthropods: an open door to understanding the mechanisms of arbovirus and arthropod evolution? Annu. Rev. Entomol. 63, 87–103. https://doi.org/10.1146/annurev-ento-020117-043033.

Calisher, C.H., Karabatsos, N., Dalrymple, J.M., Shope, R.E., Porterfield, J.S., Westaway, E.G., et al., 1989. Antigenic relationships between flaviviruses as determined by cross-neutralization tests with polyclonal antisera. J. Gen. Virology 70, 37–43. https://doi.org/10.1099/0022-1317-70-1-37.

Calzolari, M., Zé-Zé, L., Růžek, D., Vázquez, A., Jeffries, C., Defilippo, F., et al., 2012. Detection of mosquito-only flaviviruses in Europe. J. Gen. Virol. 93, 1215–1225. https://doi.org/10.1099/vir.0.040485-0.

Calzolari, M., Zé-Zé, L., Vázquez, A., Sánchez Seco, M.P., Amaro, F., Dottori, M., 2016. Insect-specific flaviviruses, a worldwide widespread group of viruses only detected in insects. Infection, Genet. Evolution 40, 381–388. https://doi.org/10.1016/j.meegid.2015.07.032.

Cammisa-Parks, H., Cisar, L.A., Kane, A., Stollar, V., 1992. The complete nucleotide sequence of cell fusing agent (CFA): Homology between the nonstructural proteins encoded by CFA and the nonstructural proteins encoded by arthropod-borne flaviviruses. Virology 189, 511–524. https://doi.org/10.1016/0042-6822(92)90575-A.

Capella-Gutiérrez, S., Silla-Martínez, J.M., Gabaldón, T., 2009. trimAl: a tool for automated alignment trimming in large-scale phylogenetic analyses. Bioinformatics 25, 1972–1973. https://doi.org/10.1093/bioinformatics/btp348.

Carrera, J.-P., Guzman, H., Beltrán, D., Díaz, Y., López-Vergès, S., Torres-Cosme, R., et al., 2015. Mercadeo virus: a novel mosquito-specific flavivirus from panama. Am. Soc. Tropical Med. Hyg. 93, 1014–1019. https://doi.org/10.4269/ajtmh.15-0117.

Carvalho, V.L., Long, M.T., 2021. Insect-specific viruses: an overview and their relationship to arboviruses of concern to humans and animals. Virology 557, 34–43. https://doi.org/10.1016/j.virol.2021.01.007.

Casals, J., 1971. Arboviruses: incorporation in a general system of virus classification. Comparative Virology. Elsevier, pp. 307–333. https://doi.org/10.1016/B978-0-12-470260-8.50014-2.

Chan, J.J.X., Zhao, Z., Vieira, C.J.S.P., Goh, J.Z.H., Slonchak, A., 2026. Current insights into superinfection exclusion in insect-specific orthoflaviviruses. Viruses 18, 115. https://doi.org/10.3390/v18010115.

Chancey, C., Grinev, A., Volkova, E., Rios, M., 2015. The global ecology and epidemiology of West Nile virus. Biomed. Res. Int. 2015, 1–20. https://doi.org/10.1155/2015/376230.

Charles, J., Tangudu, C.S., Firth, A.E., Blitvich, B.J., 2017. Complete genome sequences of two insect-specific flaviviruses. Arch. Virol. 162, 3913–3917. https://doi.org/10.1007/s00705-017-3552-5.

Charles, J., Tangudu, C.S., Hurt, S.L., Tumescheit, C., Firth, A.E., Garcia-Rejon, J.E., et al., 2018. Detection of novel and recognized RNA viruses in mosquitoes from the Yucatan Peninsula of Mexico using metagenomics and characterization of their in vitro host ranges. J. Gen. Virol. 99, 1729–1738. https://doi.org/10.1099/jgv.0.001165.

Chen, J., Deng, S., Peng, H., 2023. Insect-specific viruses used in biocontrol of mosquito-borne diseases. Interdiscip. Med. 1. https://doi.org/10.1002/INMD.20220001.

Cherry, S., Silverman, N., 2006. Host-pathogen interactions in drosophila: new tricks from an old friend. Nat. Immunol. 7, 911–917. https://doi.org/10.1038/ni1388.

Cholleti, H., Hayer, J., Abilio, A.P., Mulandane, F.C., Verner-Carlsson, J., Falk, K.I., et al., 2016. Discovery of novel viruses in mosquitoes from the zambezi valley of mozambique. PLoS One 11, e0162751. https://doi.org/10.1371/journal.pone.0162751.

Choo, J.J.Y., Vet, L.J., McMillan, C.L.D., Harrison, J.J., Scott, C.A.P., Depelsenaire, A.C.I., et al., 2021. A chimeric dengue virus vaccine candidate delivered by high density microarray patches protects against infection in mice. NPJ Vaccines 6, 66. https://doi.org/10.1038/s41541-021-00328-1.

Ciota, A.T., 2019. The role of co-infection and swarm dynamics in arbovirus transmission. Virus Res. 265, 88–93. https://doi.org/10.1016/j.virusres.2019.03.010.

Colmant, A.M.G., Hall-Mendelin, S., Ritchie, S.A., Bielefeldt-Ohmann, H., Harrison, J.J., Newton, N.D., et al., 2018. The recently identified flavivirus Bamaga virus is transmitted horizontally by Culex mosquitoes and interferes with West Nile virus replication in vitro and transmission in vivo. PLoS Negl. Trop. Dis. 12, e0006886. https://doi.org/10.1371/journal.pntd.0006886.

Colmant, A.M.G., Hobson-Peters, J., Bielefeldt-Ohmann, H., van den Hurk, A.F., Hall-Mendelin, S., Chow, W.K., et al., 2017. A new clade of insect-specific flaviviruses from Australian anopheles mosquitoes displays species-specific host restriction. mSphere 2. https://doi.org/10.1128/mSphere.00262-17.

Colmant, A.M.G., Hobson-Peters, J., Slijkerman, T.A.P., Harrison, J.J., Pijlman, G.P., van Oers, M.M., et al., 2021. Insect-specific flavivirus replication in mammalian cells is inhibited by physiological temperature and the Zinc-Finger antiviral protein. Viruses 13, 573. https://doi.org/10.3390/v13040573.

Cook, S., Bennett, S.N., Holmes, E.C., De Chesse, R., Moureau, G., de Lamballerie, X., 2006. Isolation of a new strain of the flavivirus cell fusing agent virus in a natural mosquito population from Puerto Rico. J. Gen. Virol. 87, 735–748. https://doi.org/10.1099/vir.0.81475-0.

Cook, S., Chung, B.Y.-W., Bass, D., Moureau, G., Tang, S., McAlister, E., et al., 2013. Novel virus discovery and genome reconstruction from field RNA samples reveals highly divergent viruses in dipteran hosts. PLoS One 8, e80720. https://doi.org/10.1371/journal.pone.0080720.

Cook, S., Moureau, G., Harbach, R.E., Mukwaya, L., Goodger, K., Ssenfuka, F., et al., 2009. Isolation of a novel species of flavivirus and a new strain of Culex flavivirus (Flaviviridae) from a natural mosquito population in Uganda. J. Gen. Virol. 90, 2669–2678. https://doi.org/10.1099/vir.0.014183-0.

Cook, S., Moureau, G., Kitchen, A., Gould, E.A., de Lamballerie, X., Holmes, E.C., et al., 2012. Molecular evolution of the insect-specific flaviviruses. J. Gen. Virol. 93, 223–234. https://doi.org/10.1099/vir.0.036525-0.
Costa, L.B., Barreto, F.K. de A., Barreto, M.C.A., Santos, T.H.P. dos, Andrade, M. de M.O. de, Farias, L.A.B.G., et al., 2023. Epidemiology and economic burden of Chikungunya: A systematic literature review. Trop. Med. Infect. Dis. 8, 301. https://doi.org/10.3390/tropicalmed8060301.
Crabtree, M.B., Nga, P.T., Miller, B.R., 2009. Isolation and characterization of a new mosquito flavivirus, Quang Binh virus, from Vietnam. Arch. Virol. 154, 857–860. https://doi.org/10.1007/s00705-009-0373-1.
Crabtree, M.B., Sang, R.C., Stollar, V., Dunster, L.M., Miller, B.R., 2003. Genetic and phenotypic characterization of the newly described insect flavivirus, Kamiti River virus. Arch. Virol. 148, 1095–1118. https://doi.org/10.1007/s00705-003-0019-7.
Cui, J.-Z., Xiong, X.-H., Wang, Q.-Y., Dong, H.-L., Liu, G., Chen, H.-P., 2025. Insect-specific flaviviruses have potential applications as a scaffold for pathogenic flavivirus vaccines. Vaccines (Basel) 13, 769. https://doi.org/10.3390/vaccines13070769.
da Silva Ferreira, R., de Toni Aquino da Cruz, L.C., de Souza, V.J., da Silva Neves, N.A., de Souza, V.C., Filho, L.C.F., et al., 2020. Insect-specific viruses and arboviruses in adult male culicids from Midwestern Brazil. Infection, Genet. Evolution 85, 104561. https://doi.org/10.1016/j.meegid.2020.104561.
Daims, H., Wagner, M., 2018. Nitrospira. Trends Microbiol. 26, 462–463. https://doi.org/10.1016/j.tim.2018.02.001.
de Oliveira Ribeiro, G., da Costa, A.C., Gill, D.E., Ribeiro, E.S.D., Rego, M.O. da S., Monteiro, F.J.C., et al., 2021. Guapiaçu virus, a new insect-specific flavivirus isolated from two species of Aedes mosquitoes from Brazil. Sci. Rep. 11, 4674. https://doi.org/10.1038/s41598-021-83879-6.
Dodson, B.L., Hughes, G.L., Paul, O., Matacchiero, A.C., Kramer, L.D., Rasgon, J.L., 2014. Wolbachia enhances West Nile virus (WNV) infection in the mosquito culex tarsalis. PLoS Negl. Trop. Dis. 8, e2965. https://doi.org/10.1371/journal.pntd.0002965.
Dodson, B.L., Pujhari, S., Brustolin, M., Metz, H.C., Rasgon, J.L., 2024. Variable effects of transient Wolbachia infections on alphaviruses in Aedes aegypti. PLoS Negl. Trop. Dis. 18, e0012633. https://doi.org/10.1371/journal.pntd.0012633.
Dong, H.-L., Chen, Z.-L., He, M.-J., Cui, J.-Z., Cheng, H., Wang, Q.-Y., et al., 2024. The chimeric chaoyang-zika vaccine candidate is safe and protective in mice. Vaccines (Basel) 12, 215. https://doi.org/10.3390/vaccines12020215.
Dudas, G., Obbard, D.J., 2015. Are arthropods at the heart of virus evolution? Elife 4. https://doi.org/10.7554/eLife.06837.
Eaton, B.T., 1979. Heterologous interference in Aedes albopictus cells infected with alphaviruses. J. Virol. 30, 45–55. https://doi.org/10.1128/jvi.30.1.45-55.1979.
Elrefaey, A.M., Abdelnabi, R., Rosales Rosas, A.L., Wang, L., Basu, S., Delang, L., 2020. Understanding the mechanisms underlying host restriction of insect-specific viruses. Viruses 12, 964. https://doi.org/10.3390/v12090964.
Erasmus, J.H., Weaver, S.C., 2017. Biotechnological applications of an insect-specific alphavirus. DNA Cell Biol. 36, 1045–1049. https://doi.org/10.1089/dna.2017.4019.
Evangelina, M., Victoria, M.M., José, G.J., 2013. Culex pipiens affected by joint infection of a mosquito iridescent virus and Strelkovimermis spiculatus. J. Invertebr. Pathol. 114, 295–297. https://doi.org/10.1016/j.jip.2013.08.010.
Fan, H., Zhao, Q., Guo, X., Sun, Q., Zuo, S., Wu, C., et al., 2016. Complete genome sequence of Xishuangbanna flavivirus, a novel mosquito-specific flavivirus from China. Arch. Virol. 161, 1723–1727. https://doi.org/10.1007/s00705-016-2827-6.

Fang, Y., Hang, T., Xue, J., Li, Y., Li, L., Wei, Z., et al., 2021a. Diversity, geography, and host range of emerging mosquito-associated viruses — China, 2010–2020. China CDC Wkly 3, 746–750. https://doi.org/10.46234/ccdcw2021.184.

Fang, Y., Tambo, E., Xue, J.-B., Zhang, Y., Zhou, X.-N., Khater, E.I.M., 2021b. Detection of DENV-2 and insect-specific flaviviruses in mosquitoes collected from Jeddah, Saudi Arabia. Front. Cell. Infect. Microbiol. 11. https://doi.org/10.3389/fcimb.2021.626368.

Fang, Y., Zhang, Y., Zhou, Z.-B., Shi, W.-Q., Xia, S., Li, Y.-Y., et al., 2018. Co-circulation of Aedes flavivirus, Culex flavivirus, and Quang Binh virus in Shanghai, China. Infect. Dis. Poverty 7, 75. https://doi.org/10.1186/s40249-018-0457-9.

Farfan-Ale, J.A., Loroño-Pino, M.A., Garcia-Rejon, J.E., Hovav, E., Powers, A.M., Lin, M., et al., 2009. Detection of RNA from a novel West Nile-like virus and high prevalence of an insect-specific flavivirus in mosquitoes in the Yucatan Peninsula of Mexico. Am. J. Trop. Med. Hyg. 80, 85–95.

Farfan-Ale, J.A., Loroño-Pino, M.A., Garcia-Rejon, J.E., Soto, V., Lin, M., Staley, M., et al., 2010. Detection of flaviviruses and orthobunyaviruses in mosquitoes in the Yucatan Peninsula of Mexico in 2008. Vector-Borne Zoonotic Dis. 10, 777–783. https://doi.org/10.1089/vbz.2009.0196.

Fauver, J.R., Grubaugh, N.D., Krajacich, B.J., Weger-Lucarelli, J., Lakin, S.M., Fakoli, L.S., et al., 2016. West African anopheles gambiae mosquitoes harbor a taxonomically diverse virome including new insect-specific flaviviruses, mononegaviruses, and totiviruses. Virology 498, 288–299. https://doi.org/10.1016/j.virol.2016.07.031.

Ferreira, D.D., Cook, S., Lopes, Â., de Matos, A.P., Esteves, A., Abecasis, A., et al., 2013. Characterization of an insect-specific flavivirus (OCFVPT) co-isolated from Ochlerotatus caspius collected in southern Portugal along with a putative new Negev-like virus. Virus Genes 47, 532–545. https://doi.org/10.1007/s11262-013-0960-9.

Folimonova, S.Y., 2012. Superinfection exclusion is an active virus-controlled function that requires a specific viral protein. J. Virol. 86, 5554–5561. https://doi.org/10.1128/JVI.00310-12.

Fu, Y., Zhao, W., Wu, S., Li, J., Liu, Q., Jiang, F., et al., 2025. Risk assessment of an Aedes flavivirus and its effect on pathogenic flavivirus replication in mosquitoes. Parasit. Vectors 18, 88. https://doi.org/10.1186/s13071-025-06711-4.

Gao, X., Liu, H., Li, X., Fu, S., Cao, L., Shao, N., et al., 2019. Changing geographic distribution of Japanese encephalitis virus genotypes, 1935–2017. Vector-Borne Zoonotic Dis. 19, 35–44. https://doi.org/10.1089/vbz.2018.2291.

Goenaga, S., Goenaga, J., Boaglio, E.R., Enria, D.A., Levis, S. del C., 2020. Superinfection exclusion studies using West Nile virus and Culex flavivirus strains from Argentina. Mem. Inst. Oswaldo Cruz 115. https://doi.org/10.1590/0074-02760200012.

Goenaga, S., Kenney, J., Duggal, N., Delorey, M., Ebel, G., Zhang, B., et al., 2015. Potential for co-infection of a mosquito-specific flavivirus, nhumirim virus, to block west nile virus transmission in mosquitoes. Viruses 7, 5801–5812. https://doi.org/10.3390/v7112911.

Gould, E., Solomon, T., 2008. Pathogenic flaviviruses. Lancet 371, 500–509. https://doi.org/10.1016/S0140-6736(08)60238-X.

Grard, G., Lemasson, J.-J., Sylla, M., Dubot, A., Cook, S., Molez, J.-F., et al., 2006. Ngoye virus: a novel evolutionary lineage within the genus Flavivirus. J. Gen. Virology 87, 3273–3277. https://doi.org/10.1099/vir.0.82071-0.

Grard, G., Moureau, G., Charrel, R.N., Holmes, E.C., Gould, E.A., de Lamballerie, X., 2010. Genomics and evolution of Aedes-borne flaviviruses. J. Gen. Virol. 91, 87–94. https://doi.org/10.1099/vir.0.014506-0.

Gritsun, T.S., Gould, E.A., 2006a. The 3′ untranslated regions of Kamiti River virus and cell fusing agent virus originated by self-duplication. J. Gen. Virol. 87, 2615–2619. https://doi.org/10.1099/vir.0.81950-0.

Gritsun, T.S., Gould, E.A., 2006. Origin and evolution of 3′utr of flaviviruses: long direct repeats as a basis for the formation of secondary structures and their significance for virus transmission. pp. 203–248. https://doi.org/10.1016/S0065–3527(06)69005–2.

Gritsun, T.S., Gould, E.A., 2006b. Direct repeats in the 3′ untranslated regions of mosquito-borne flaviviruses: possible implications for virus transmission. J. Gen. Virology 87, 3297–3305. https://doi.org/10.1099/vir.0.82235-0.

Gritsun, T.S., Lashkevich, V.A., Gould, E.A., 2003. Tick-borne encephalitis. Antivir. Res. 57, 129–146. https://doi.org/10.1016/S0166-3542(02)00206-1.

Guggemos, H.D., Kopp, A., Voigt, K., Fendt, M., Graff, S.L., Mfune, J.K.E., et al., 2024. Eilat virus isolated from Culex univittatus mosquitoes from the Namibian Zambezi Region influences in vitro superinfection with alpha- and flaviviruses in a virus-species-dependent manner. PLoS One 19, e0312182. https://doi.org/10.1371/journal.pone.0312182.

Gutiérrez-Millán, E., Rodríguez-Aguilar, E.D., Rodríguez, M.H., 2025. Molecular antiviral responses, immune priming and inheritance in insects. Virology 605, 110468. https://doi.org/10.1016/j.virol.2025.110468.

Guzman, H., Contreras-Gutierrez, M.A., Travassos da Rosa, A.P.A., Nunes, M.R.T., Cardoso, J.F., Popov, V.L., et al., 2018. Characterization of three new insect-specific flaviviruses: their relationship to the mosquito-borne flavivirus pathogens. Am. J. Trop. Med. Hyg. 98, 410–419. https://doi.org/10.4269/ajtmh.17-0350.

Haddow, A.D., Guzman, H., Popov, V.L., Wood, T.G., Widen, S.G., Haddow, A.D., et al., 2013. First isolation of Aedes flavivirus in the Western Hemisphere and evidence of vertical transmission in the mosquito Aedes (Stegomyia) albopictus (Diptera: Culicidae). Virology 440, 134–139. https://doi.org/10.1016/j.virol.2012.12.008.

Halbach, R., Junglen, S., van Rij, R.P., 2017. Mosquito-specific and mosquito-borne viruses: evolution, infection, and host defense. Curr. Opin. Insect Sci. 22, 16–27. https://doi.org/10.1016/j.cois.2017.05.004.

Hall, R.A., Bielefeldt-Ohmann, H., McLean, B.J., O'Brien, C.A., Colmant, A.M.G., Piyasena, T.B.H., et al., 2016. Commensal viruses of mosquitoes: host restriction, transmission, and interaction with arboviral pathogens. Evolut. Bioinforma. 12s2. https://doi.org/10.4137/EBO.S40740.

Hall, R.A., Nguyen, W., Khromykh, A.A., Suhrbier, A., 2025. Insect-specific virus platforms for arbovirus vaccine development. Front. Immunol. 16. https://doi.org/10.3389/fimmu.2025.1521104.

Hall-Mendelin, S., McLean, B.J., Bielefeldt-Ohmann, H., Hobson-Peters, J., Hall, R.A., van den Hurk, A.F., 2016. The insect-specific Palm Creek virus modulates West Nile virus infection in and transmission by Australian mosquitoes. Parasit. Vectors 9, 414. https://doi.org/10.1186/s13071-016-1683-2.

Harrison, J.J., Hobson-Peters, J., Colmant, A.M.G., Koh, J., Newton, N.D., Warrilow, D., et al., 2020. Antigenic characterization of new lineage II insect-specific flaviviruses in australian mosquitoes and identification of host restriction factors. mSphere 5. https://doi.org/10.1128/mSphere.00095-20.

Harrison, J.J., Nguyen, W., Morgan, M.S., Tang, B., Habarugira, G., de Malmanche, H., et al., 2024. A chimeric vaccine derived from Australian genotype IV Japanese encephalitis virus protects mice from lethal challenge. NPJ Vaccines 9, 134. https://doi.org/10.1038/s41541-024-00903-2.

Hazlewood, J.E., Rawle, D.J., Tang, B., Yan, K., Vet, L.J., Nakayama, E., et al., 2020. A Zika vaccine generated using the chimeric insect-specific Binjari virus platform protects against fetal brain infection in pregnant mice. Vaccines (Basel) 8, 496. https://doi.org/10.3390/vaccines8030496.

Hazlewood, J.E., Tang, B., Yan, K., Rawle, D.J., Harrison, J.J., Hall, R.A., et al., 2022. The chimeric Binjari-Zika vaccine provides long-term protection against ZIKA virus challenge. Vaccines (Basel) 10, 85. https://doi.org/10.3390/vaccines10010085.

Herreros-Moya, E., Sinka, M., Harris, A.F., Entwistle, J., Martin, A.C., Willis, K.J., 2025. The food of life: which nectar do mosquitoes feed on?—An evidence-based meta-analysis. Environ. Entomol. 54, 352–366. https://doi.org/10.1093/ee/nvaf009.

Hobson-Peters, J., Harrison, J.J., Watterson, D., Hazlewood, J.E., Vet, L.J., Newton, N.D., et al., 2019. A recombinant platform for flavivirus vaccines and diagnostics using chimeras of a new insect-specific virus. Sci. Transl. Med. 11. https://doi.org/10.1126/scitranslmed.aax7888.

Hobson-Peters, J., Yam, A.W.Y., Lu, J.W.F., Setoh, Y.X., May, F.J., Kurucz, N., et al., 2013. A new insect-specific flavivirus from northern Australia suppresses replication of West Nile virus and murray valley encephalitis virus in co-infected mosquito cells. PLoS One 8, e56534. https://doi.org/10.1371/journal.pone.0056534.

Holbrook, M., 2017. Historical perspectives on flavivirus research. Viruses 9, 97. https://doi.org/10.3390/v9050097.

Hoshino, K., Isawa, H., Tsuda, Y., Sawabe, K., Kobayashi, M., 2009. Isolation and characterization of a new insect flavivirus from Aedes albopictus and Aedes flavopictus mosquitoes in Japan. Virology 391, 119–129. https://doi.org/10.1016/j.virol.2009.06.025.

Hoshino, K., Isawa, H., Tsuda, Y., Yano, K., Sasaki, T., Yuda, M., et al., 2007. Genetic characterization of a new insect flavivirus isolated from Culex pipiens mosquito in Japan. Virology 359, 405–414. https://doi.org/10.1016/j.virol.2006.09.039.

Hoshino, K., Takahashi-Nakaguchi, A., Isawa, H., Sasaki, T., Higa, Y., Kasai, S., et al., 2012. Entomological surveillance for flaviviruses at migratory bird stopover sites in Hokkaido, Japan, and a New insect flavivirus detected in *Aedes galloisi* (Diptera: Culicidae). J. Med. Entomol. 49, 175–182. https://doi.org/10.1603/ME11123.

Huang, I.-C., Li, W., Sui, J., Marasco, W., Choe, H., Farzan, M., 2008. Influenza A virus neuraminidase limits viral superinfection. J. Virol. 82, 4834–4843. https://doi.org/10.1128/JVI.00079-08.

Huang, Y., Wang, S., Liu, H., Atoni, E., Wang, F., Chen, W., et al., 2023. A global dataset of sequence, diversity and biosafety recommendation of arbovirus and arthropod-specific virus. Sci. Data 10. https://doi.org/10.1038/s41597-023-02226-8.

Huhtamo, E., Cook, S., Moureau, G., Uzcátegui, N.Y., Sironen, T., Kuivanen, S., et al., 2014. Novel flaviviruses from mosquitoes: mosquito-specific evolutionary lineages within the phylogenetic group of mosquito-borne flaviviruses. Virology 464–465 320–329. https://doi.org/10.1016/j.virol.2014.07.015.

Huhtamo, E., Moureau, G., Cook, S., Julkunen, O., Putkuri, N., Kurkela, S., et al., 2012. Novel insect-specific flavivirus isolated from northern Europe. Virology 433, 471–478. https://doi.org/10.1016/j.virol.2012.08.038.

Huhtamo, E., Putkuri, N., Kurkela, S., Manni, T., Vaheri, A., Vapalahti, O., et al., 2009. Characterization of a novel flavivirus from mosquitoes in northern Europe that is related to mosquito-borne flaviviruses of the tropics. J. Virol. 83, 9532–9540. https://doi.org/10.1128/JVI.00529-09.

Im, J., Baek, J.-H., Durey, A., Kwon, H., Chung, M.-H., Lee, J.-S., 2020. Geographic distribution of tick-borne encephalitis virus complex. J. Vector Borne Dis. 57, 14. https://doi.org/10.4103/0972-9062.308794.

International Committee on Taxonomy of Viruses (ICTV), 2023. Family Flaviviridae: Genus Orthoflavivirus [WWW Document]. URL https://ictv.global/report/chapter/flaviviridae/flaviviridae/orthoflavivirus (Accessed 2.24.26).

Iturbe-Ormaetxe, I., Walker, T., O' Neill, S.L., 2011. Wolbachia and the biological control of mosquito-borne disease. EMBO Rep. 12, 508–518. https://doi.org/10.1038/embor.2011.84.

Jia, D., Luo, G., Guan, H., Yu, T., Sun, X., Du, Y., et al., 2024. Arboviruses antagonize insect Toll antiviral immune signaling to facilitate the coexistence of viruses with their vectors. PLoS Pathog. 20, e1012318. https://doi.org/10.1371/journal.ppat.1012318.

Johnston, R.E., Wan, K., Bose, H.R., 1974. Homologous interference induced by sindbis virus. J. Virol. 14, 1076–1082. https://doi.org/10.1128/jvi.14.5.1076-1082.1974.

Junglen, S., Kopp, A., Kurth, A., Pauli, G., Ellerbrok, H., Leendertz, F.H., 2009. A new flavivirus and a new vector: characterization of a novel flavivirus isolated from uranotaenia mosquitoes from a tropical rain forest. J. Virol. 83, 4462–4468. https://doi.org/10.1128/JVI.00014-09.

Junglen, S., Korries, M., Grasse, W., Wieseler, J., Kopp, A., Hermanns, K., et al., 2017. Host range restriction of insect-specific flaviviruses occurs at several levels of the viral life cycle. mSphere 2. https://doi.org/10.1128/mSphere.00375-16.

Junjhon, J., Pennington, J.G., Edwards, T.J., Perera, R., Lanman, J., Kuhn, R.J., 2014. Ultrastructural characterization and three-dimensional architecture of replication sites in dengue virus-infected mosquito cells. J. Virol. 88, 4687–4697. https://doi.org/10.1128/JVI.00118-14.

Jupatanakul, N., Sim, S., Angleró-Rodríguez, Y.I., Souza-Neto, J., Das, S., Poti, K.E., et al., 2017. Engineered Aedes aegypti JAK/STAT pathway-mediated immunity to dengue virus. PLoS Negl. Trop. Dis. 11, e0005187. https://doi.org/10.1371/journal.pntd.0005187.

Kalyaanamoorthy, S., Minh, B.Q., Wong, T.K.F., von Haeseler, A., Jermiin, L.S., 2017. ModelFinder: fast model selection for accurate phylogenetic estimates. Nat. Methods 14, 587–589. https://doi.org/10.1038/nmeth.4285.

Katoh, K., Standley, D.M., 2013. MAFFT multiple sequence alignment software version 7: improvements in performance and usability. Mol. Biol. Evol. 30, 772–780. https://doi.org/10.1093/molbev/mst010.

Keene, K.M., Foy, B.D., Sanchez-Vargas, I., Beaty, B.J., Blair, C.D., Olson, K.E., 2004. RNA interference acts as a natural antiviral response to O'nyong-nyong virus (Alphavirus; Togaviridae) infection of Anopheles gambiae. Proc. Natl Acad. Sci. 101, 17240–17245. https://doi.org/10.1073/pnas.0406983101.

Kenney, J.L., Solberg, O.D., Langevin, S.A., Brault, A.C., 2014. Characterization of a novel insect-specific flavivirus from Brazil: potential for inhibition of infection of arthropod cells with medically important flaviviruses. J. Gen. Virology 95, 2796–2808. https://doi.org/10.1099/vir.0.068031-0.

Kent, R.J., Crabtree, M.B., Miller, B.R., 2010. Transmission of West Nile virus by culex quinquefasciatus say infected with culex flavivirus Izabal. PLoS Negl. Trop. Dis. 4, e671. https://doi.org/10.1371/journal.pntd.0000671.

Kholodilov, I.S., Aibulatov, S.V., Khalin, A.V., Polienko, A.E., Klimentov, A.S., Belova, O.A., et al., 2024. Orthoflavivirus lammi in Russia: possible transovarial transmission and trans-stadial survival in aedes cinereus (Diptera, Culicidae). Viruses 16, 527. https://doi.org/10.3390/v16040527.

Kim, D.Y., Guzman, H., Bueno, R., Dennett, J.A., Auguste, A.J., Carrington, C.V.F., et al., 2009. Characterization of culex flavivirus (flaviviridae) strains isolated from mosquitoes in the United States and Trinidad. Virology 386, 154–159. https://doi.org/10.1016/j.virol.2008.12.034.

Kim, K., 2019. Current challenges in the development of vaccines and drugs against emerging vector-borne diseases. Curr. Med. Chem. 26, 2974–2986. https://doi.org/10.2174/0929867325666181105121146.

Kobayashi, D., Watanabe, M., Faizah, A.N., Amoa-Bosompem, M., Higa, Y., Tsuda, Y., et al., 2021. Discovery of a novel flavivirus (flaviviridae) from the horse fly, tabanus rufidens (Diptera: Tabanidae): the possible coevolutionary relationships between the classical insect-specific flaviviruses and host dipteran insects. J. Med. Entomol. 58, 880–890. https://doi.org/10.1093/jme/tjaa193.

Koh, C., Henrion-Lacritick, A., Frangeul, L., Saleh, M.-C., 2021. Interactions of the insect-specific palm creek Virus with Zika and Chikungunya viruses in aedes mosquitoes. Microorganisms 9, 1652. https://doi.org/10.3390/microorganisms9081652.

Kolodziejek, J., Pachler, K., Bin, H., Mendelson, E., Shulman, L., Orshan, L., et al., 2013. Barkedji virus, a novel mosquito-borne flavivirus identified in Culex perexiguus mosquitoes, Israel, 2011. J. Gen. Virol. 94, 2449–2457. https://doi.org/10.1099/vir.0.056200-0.

Kraemer, M.U., Sinka, M.E., Duda, K.A., Mylne, A.Q., Shearer, F.M., Barker, C.M., et al., 2015. The global distribution of the arbovirus vectors Aedes aegypti and Ae. albopictus. Elife 4. https://doi.org/10.7554/eLife.08347.

Kramer, L.D., Styer, L.M., Ebel, G.D., 2008. A global perspective on the epidemiology of West Nile virus. Annu. Rev. Entomol. 53, 61–81. https://doi.org/10.1146/annurev.ento.53.103106.093258.

Kuno, G., 2007. Host range specificity of flaviviruses: correlation with in vitro replication. J. Med. Entomol. 44, 93–101. https://doi.org/10.1093/jmedent/41.5.93.

Kuno, G., Chang, G.-J.J., 2005. Biological transmission of arboviruses: reexamination of and new insights into components, mechanisms, and unique traits as well as their evolutionary trends. Clin. Microbiol. Rev. 18, 608–637. https://doi.org/10.1128/CMR.18.4.608-637.2005.

Kuno, G., Chang, G.-J.J., Tsuchiya, K.R., Karabatsos, N., Cropp, C.B., 1998. Phylogeny of the genus flavivirus. J. Virol. 72, 73–83. https://doi.org/10.1128/JVI.72.1.73-83.1998.

Kuwata, R., Isawa, H., Hoshino, K., Sasaki, T., Kobayashi, M., Maeda, K., et al., 2015. Analysis of mosquito-borne flavivirus superinfection in culex tritaeniorhynchus (Diptera: Culicidae) cells persistently infected with culex flavivirus (flaviviridae). J. Med. Entomol. 52, 222–229. https://doi.org/10.1093/jme/tju059.

Lambrechts, L., Saleh, M.-C., 2019. Manipulating mosquito tolerance for arbovirus control. Cell Host Microbe 26, 309–313. https://doi.org/10.1016/j.chom.2019.08.005.

Larsson, A., 2014. AliView: a fast and lightweight alignment viewer and editor for large datasets. Bioinformatics 30, 3276–3278. https://doi.org/10.1093/bioinformatics/btu531.

Lau, G.W., Goumnerov, B.C., Walendziewicz, C.L., Hewitson, J., Xiao, W., Mahajan-Miklos, S., et al., 2003. The Drosophila melanogaster Toll pathway participates in resistance to infection by the gram-negative human pathogen pseudomonas aeruginosa. Infect. Immun. 71, 4059–4066. https://doi.org/10.1128/IAI.71.7.4059-4066.2003.

Laureti, M., Paradkar, P.N., Fazakerley, J.K., Rodriguez-Andres, J., 2020. Superinfection exclusion in mosquitoes and its potential as an arbovirus control strategy. Viruses 12, 1259. https://doi.org/10.3390/v12111259.

Lee, J.S., Grubaugh, N.D., Kondig, J.P., Turell, M.J., Kim, H.-C., Klein, T.A., et al., 2013. Isolation and genomic characterization of Chaoyang virus strain ROK144 from Aedes vexans nipponii from the Republic of Korea. Virology 435, 220–224. https://doi.org/10.1016/j.virol.2012.10.020.

Lee, W.-S., Webster, J.A., Madzokere, E.T., Stephenson, E.B., Herrero, L.J., 2019. Mosquito antiviral defense mechanisms: a delicate balance between innate immunity and persistent viral infection. Parasit. Vectors 12, 165. https://doi.org/10.1186/s13071-019-3433-8.

Lim, A., Shearer, F.M., Sewalk, K., Pigott, D.M., Clarke, J., Ghouse, A., et al., 2025. The overlapping global distribution of dengue, chikungunya, Zika and yellow fever. Nat. Commun. 16, 3418. https://doi.org/10.1038/s41467-025-58609-5.

Lindenbach, B.D., Murray, C.L., Thiel, H.-J., Rice, C., 2013. Flaviviridae. In: Knipe, D., Howley, P. (Eds.), Fields Virology. Lippincott Williams & Wilkins, Wolters Kluwer, Philadelphia, pp. 712–746.

Liu, L., Dong, H., Chen, H., Zhang, J., Ling, H., Li, Z., et al., 2010. Flavivirus RNA cap methyltransferase: structure, function, and inhibition. Front. Biol. (Beijing). 5, 286–303. https://doi.org/10.1007/s11515-010-0660-y.

Lobo, F.P., Mota, B.E.F., Pena, S.D.J., Azevedo, V., Macedo, A.M., Tauch, A., et al., 2009. Virus-host coevolution: common patterns of nucleotide motif usage in flaviviridae and their hosts. PLoS One 4, e6282. https://doi.org/10.1371/journal.pone.0006282.
Logan, R.A.E., Quek, S., Muthoni, J.N., von Eicken, A., Brettell, L.E., Anderson, E.R., et al., 2022. Vertical and horizontal transmission of cell fusing agent virus in aedes aegypti. Appl. Environ. Microbiol. 88. https://doi.org/10.1128/aem.01062-22.
Lutomiah, J.J.L., Mwandawiro, C., Magambo, J., Sang, R.C., 2007. Infection and vertical transmission of kamiti river virus in laboratory bred Aedes aegypti mosquitoes. J. Insect Sci. 7, 1–7. https://doi.org/10.1673/031.007.5501.
Markoff, L., 2003. 5′- and 3′-noncoding regions in flavivirus RNA. pp. 177–228. https://doi.org/10.1016/S0065-3527(03)59006-6.
McFarlane, M., Arias-Goeta, C., Martin, E., O'Hara, Z., Lulla, A., Mousson, L., et al., 2014. Characterization of Aedes aegypti innate-immune pathways that limit Chikungunya virus replication. PLoS Negl. Trop. Dis. 8, e2994. https://doi.org/10.1371/journal.pntd.0002994.
McLean, B.J., Hall-Mendelin, S., Webb, C.E., Bielefeldt-Ohmann, H., Ritchie, S.A., Hobson-Peters, J., et al., 2021. The insect-specific parramatta river virus is vertically transmitted by aedes vigilax mosquitoes and suppresses replication of pathogenic flaviviruses in vitro. Vector-Borne Zoonotic Dis. 21, 208–215. https://doi.org/10.1089/vbz.2020.2692.
McLean, B.J., Hobson-Peters, J., Webb, C.E., Watterson, D., Prow, N.A., Nguyen, H.D., et al., 2015. A novel insect-specific flavivirus replicates only in Aedes-derived cells and persists at high prevalence in wild Aedes vigilax populations in Sydney, Australia. Virology 486, 272–283. https://doi.org/10.1016/j.virol.2015.07.021.
McMahon, H.T., Gallop, J.L., 2005. Membrane curvature and mechanisms of dynamic cell membrane remodelling. Nature 438, 590–596. https://doi.org/10.1038/nature04396.
Misencik, M.J., Grubaugh, N.D., Andreadis, T.G., Ebel, G.D., Armstrong, P.M., 2016. Isolation of a novel insect-specific flavivirus from culiseta melanura in the northeastern United States. Vector-Borne Zoonotic Dis. 16, 181–190. https://doi.org/10.1089/vbz.2015.1889.
Moureau, G., Ninove, L., Izri, A., Cook, S., Lamballerie, X., De, et al., 2010. Flavivirus RNA in phlebotomine sandflies. Vector-Borne Zoonotic Dis. 10, 195–197. https://doi.org/10.1089/vbz.2008.0216.
Muttis, E., Miele, S.A.B., Belaich, M.N., Micieli, M.V., Becnel, J.J., Ghiringhelli, P.D., et al., 2012. First record of a mosquito iridescent virus in Culex pipiens L. (Diptera: Culicidae). Arch. Virol. 157, 1569–1571. https://doi.org/10.1007/s00705-012-1302-2.
Naddaf, M., 2024. Mosquito-borne diseases are surging in Europe — how worried are scientists? Nature 633, 749. https://doi.org/10.1038/d41586-024-03031-y.
Nanfack-Minkeu, F., Mitri, C., Bischoff, E., Belda, E., Casademont, I., Vernick, K.D., 2019. Interaction of RNA viruses of the natural virome with the African malaria vector, Anopheles coluzzii. Sci. Rep. 9, 6319. https://doi.org/10.1038/s41598-019-42825-3.
Nasar, F., Erasmus, J.H., Haddow, A.D., Tesh, R.B., Weaver, S.C., 2015. Eilat virus induces both homologous and heterologous interference. Virology 484, 51–58. https://doi.org/10.1016/j.virol.2015.05.009.
Nazni, W.A., Hoffmann, A.A., NoorAfizah, A., Cheong, Y.L., Mancini, M.V., Golding, N., et al., 2019. Establishment of Wolbachia strain wAlbB in Malaysian populations of aedes aegypti for dengue control. Curr. Biol. 29, 4241–4248.e5. https://doi.org/10.1016/j.cub.2019.11.007.
Nguyen, L.-T., Schmidt, H.A., von Haeseler, A., Minh, B.Q., 2015. IQ-TREE: A fast and effective stochastic algorithm for estimating maximum-likelihood phylogenies. Mol. Biol. Evol. 32, 268–274. https://doi.org/10.1093/molbev/msu300.

Öhlund, P., Delhomme, N., Hayer, J., Hesson, J.C., Blomström, A.-L., 2022. Transcriptome analysis of an Aedes albopictus cell line single- and dual-infected with lammi virus and WNV. Int. J. Mol. Sci. 23, 875. https://doi.org/10.3390/ijms23020875.

Öhlund, P., Lundén, H., Blomström, A.-L., 2019. Insect-specific virus evolution and potential effects on vector competence. Virus Genes 55, 127–137. https://doi.org/10.1007/s11262-018-01629-9.

Olmo, R.P., Todjro, Y.M.H., Aguiar, E.R.G.R., de Almeida, J.P.P., Ferreira, F.V., Armache, J.N., et al., 2023. Mosquito vector competence for dengue is modulated by insect-specific viruses. Nat. Microbiol. 8, 135–149. https://doi.org/10.1038/s41564-022-01289-4.

Olson, K.E., Blair, C.D., 2015. Arbovirus–mosquito interactions: RNAi pathway. Curr. Opin. Virol. 15, 119–126. https://doi.org/10.1016/j.coviro.2015.10.001.

Orba, Y., Matsuno, K., Nakao, R., Kryukov, K., Saito, Y., Kawamori, F., et al., 2021. Diverse mosquito-specific flaviviruses in the Bolivian Amazon basin. J. Gen. Virol. 102. https://doi.org/10.1099/jgv.0.001518.

Ortiz-Baez, A.S., Holmes, E.C., Charon, J., Pettersson, J.H.-O., Hesson, J.C., 2022. Meta-transcriptomics reveals potential virus transfer between Aedes communis mosquitoes and their parasitic water mites. Virus Evol. 8. https://doi.org/10.1093/ve/veac090.

Pan, X., Zhou, G., Wu, J., Bian, G., Lu, P., Raikhel, A.S., et al., 2012. Wolbachia induces reactive oxygen species (ROS)-dependent activation of the Toll pathway to control dengue virus in the mosquito Aedes aegypti. Proc. Natl. Acad. Sci. 109. https://doi.org/10.1073/pnas.1116932108.

Papa, A., Papadopoulou, E., Kalaitzopoulou, S., Tsioka, K., Mourelatos, S., 2014. Detection of West Nile virus and insect-specific flavivirus RNA in Culex mosquitoes, central Macedonia, Greece. Trans. R. Soc. Trop. Med. Hyg. 108, 555–559. https://doi.org/10.1093/trstmh/tru100.

Paradkar, P.N., Trinidad, L., Voysey, R., Duchemin, J.-B., Walker, P.J., 2012. Secreted Vago restricts West Nile virus infection in Culex mosquito cells by activating the Jak-STAT pathway. Proc. Natl. Acad. Sci. 109, 18915–18920. https://doi.org/10.1073/pnas.1205231109.

Patterson, E.I., Villinger, J., Muthoni, J.N., Dobel-Ober, L., Hughes, G.L., 2020. Exploiting insect-specific viruses as a novel strategy to control vector-borne disease. Curr. Opin. Insect Sci. 39, 50–56. https://doi.org/10.1016/j.cois.2020.02.005.

Pauvolid-Corrêa, A., Solberg, O., Couto-Lima, D., Kenney, J., Serra-Freire, N., Brault, A., et al., 2015. Nhumirim virus, a novel flavivirus isolated from mosquitoes from the Pantanal, Brazil. Arch. Virol. 160, 21–27. https://doi.org/10.1007/s00705-014-2219-8.

Pepin, K.M., Lambeth, K., Hanley, K.A., 2008. Asymmetric competitive suppression between strains of dengue virus. BMC Microbiol. 8, 28. https://doi.org/10.1186/1471-2180-8-28.

Piyasena, T.B.H., Newton, N.D., Hobson-Peters, J., Vet, L.J., Setoh, Y.X., Bielefeldt-Ohmann, H., et al., 2019. Chimeric viruses of the insect-specific flavivirus Palm Creek with structural proteins of vertebrate-infecting flaviviruses identify barriers to replication of insect-specific flaviviruses in vertebrate cells. J. Gen. Virol. 100, 1580–1586. https://doi.org/10.1099/jgv.0.001326.

Piyasena, T.B.H., Setoh, Y.X., Hobson-Peters, J., Newton, N.D., Bielefeldt-Ohmann, H., McLean, B.J., et al., 2017. Infectious DNAs derived from insect-specific flavivirus genomes enable identification of pre- and post-entry host restrictions in vertebrate cells. Sci. Rep. 7, 2940. https://doi.org/10.1038/s41598-017-03120-1.

Pontremoli, C., Forni, D., Clerici, M., Cagliani, R., Sironi, M., 2021. Alternation between taxonomically divergent hosts is not the major determinant of flavivirus evolution. Virus Evol. 7. https://doi.org/10.1093/ve/veab040.

Porier, D.L., Wilson, S.N., Auguste, D.I., Leber, A., Coutermarsh-Ott, S., Allen, I.C., et al., 2021. Enemy of my enemy: a novel insect-specific flavivirus offers a promising platform for a zika virus vaccine. Vaccines (Basel) 9, 1142. https://doi.org/10.3390/vaccines9101142.

R Core Team, 2021. R: A language and environment for statistical computing. R foundation for statistical computing, Vienna. [WWW Document]. URL https://www.R-project.org (Accessed 11.24.23).

Rambaut, A., Drummond, A.J., Xie, D., Baele, G., Suchard, M.A., 2018. Posterior summarization in bayesian phylogenetics using tracer 1.7. Syst. Biol. 67, 901–904. https://doi.org/10.1093/sysbio/syy032.

Ramirez, J.L., Dimopoulos, G., 2010. The Toll immune signaling pathway control conserved anti-dengue defenses across diverse Ae. aegypti strains and against multiple dengue virus serotypes. Dev. Comp. Immunol. 34, 625–629. https://doi.org/10.1016/j.dci.2010.01.006.

Ratcliffe, N.A., Mello, C.B., Castro, H.C., Dyson, P., Figueiredo, M., 2024. Immune reactions of vector insects to parasites and pathogens. Microorganisms 12, 568. https://doi.org/10.3390/microorganisms12030568.

Rathore, A.P.S., St. John, A.L., 2020. Cross-reactive immunity among flaviviruses. Front. Immunol. 11. https://doi.org/10.3389/fimmu.2020.00334.

Reed, W., Carroll, J., Agramonte, A., Lazear, J.W., 2001. The etiology of yellow fever — A preliminary note. Mil. Med. 166, 29–36. https://doi.org/10.1093/milmed/166.suppl_1.29.

Romo, H., Kenney, J.L., Blitvich, B.J., Brault, A.C., 2018. Restriction of Zika virus infection and transmission in Aedes aegypti mediated by an insect-specific flavivirus. Emerg. Microbes Infect. 7, 181. https://doi.org/10.1038/s41426-018-0180-4.

Roundy, C.M., Azar, S.R., Rossi, S.L., Weaver, S.C., Vasilakis, N., 2017. Insect-Specific Viruses 119 146 https://doi.org/10.1016/bs.aivir.2016.10.001.

Saiyasombat, R., Bolling, B.G., Brault, A.C., Bartholomay, L.C., Blitvich, B.J., 2011. Evidence of efficient transovarial transmission of culex flavivirus by culex pipiens (Diptera: Culicidae). J. Med. Entomol. 48, 1031–1038. https://doi.org/10.1603/ME11043.

Sanchez-Vargas, I., Travanty, E.A., Keene, K.M., Franz, A.W.E., Beaty, B.J., Blair, C.D., et al., 2004. RNA interference, arthropod-borne viruses, and mosquitoes. Virus Res. 102, 65–74. https://doi.org/10.1016/j.virusres.2004.01.017.

Sang, R.C., Gichogo, A., Gachoya, J., Dunster, M.D., Ofula, V., Hunt, A.R., et al., 2003. Isolation of a new flavivirus related to cell fusing agent virus (CFAV) from field-collected flood-water Aedes mosquitoes sampled from a dambo in central Kenya. Arch. Virol. 148, 1085–1093. https://doi.org/10.1007/s00705-003-0018-8.

Shepard, D.S., Undurraga, E.A., Halasa, Y.A., Stanaway, J.D., 2016. The global economic burden of dengue: a systematic analysis. Lancet Infect. Dis. 16, 935–941. https://doi.org/10.1016/S1473-3099(16)00146-8.

Shin, S.W., Bian, G., Raikhel, A.S., 2006. A toll receptor and a cytokine, Toll5A and Spz1C, are involved in toll antifungal immune signaling in the mosquito Aedes aegypti. J. Biol. Chem. 281, 39388–39395. https://doi.org/10.1074/jbc.M608912200.

Shin, S.W., Kokoza, V., Bian, G., Cheon, H.-M., Kim, Y.J., Raikhel, A.S., 2005. REL1, a homologue of drosophila dorsal, regulates toll antifungal immune pathway in the female mosquito aedes aegypti. J. Biol. Chem. 280, 16499–16507. https://doi.org/10.1074/jbc.M500711200.

Shroyer, D.A., 1990. Venereal transmission of st. louis encephalitis virus by culex quinquefasciatus males (Diptera: Culicidae). J. Med. Entomol. 27, 334–337. https://doi.org/10.1093/jmedent/27.3.334.

Simmonds, P., Becher, P., Bukh, J., Gould, E.A., Meyers, G., Monath, T., et al., 2017. ICTV virus taxonomy profile: flaviviridae. J. Gen. Virology. 98, 2–3. https://doi.org/10.1099/jgv.0.000672.
Simmonds, P., Butković, A., Grove, J., Mayne, R., Mifsud, J.C.O., Beer, M., et al., 2025. Taxonomic expansion and reorganization of Flaviviridae. Nat. Microbiol. 10, 3026–3037. https://doi.org/10.1038/s41564-025-02134-0.
Slonchak, A., Parry, R., Pullinger, B., Sng, J.D.J., Wang, X., Buck, T.F., et al., 2022. Structural analysis of 3'UTRs in insect flaviviruses reveals novel determinants of sfRNA biogenesis and provides new insights into flavivirus evolution. Nat. Commun. 13, 1279. https://doi.org/10.1038/s41467-022-28977-3.
Souza-Neto, J.A., Sim, S., Dimopoulos, G., 2009. An evolutionary conserved function of the JAK-STAT pathway in anti-dengue defense. Proc. Natl. Acad. Sci. 106, 17841–17846. https://doi.org/10.1073/pnas.0905006106.
Stokes, A., Bauer, J.H., Hudson, N.P., 1928. The transmission of yellow fever to macacus rhesus. J. Am. Med. Assoc. 90, 253. https://doi.org/10.1001/jama.1928.02690310005002.
Stollar, V., Thomas, V.L., 1975. An agent in the Aedes aegypti cell line (Peleg) which causes fusion of Aedes albopictus cells. Virology 64, 367–377. https://doi.org/10.1016/0042-6822(75)90113-0.
Suchard, M.A., Lemey, P., Baele, G., Ayres, D.L., Drummond, A.J., Rambaut, A., 2018. Bayesian phylogenetic and phylodynamic data integration using BEAST 1.10. Virus Evol. 4. https://doi.org/10.1093/ve/vey016.
Schliep, K.P., 2011. Phangorn: phylogenetic analysis in R. Bioinformatics 27 (4), 592–593.
Tan, L., Zhang, Y., Kim, D.Y., Li, R., 2023. Insect-specific chimeric viruses potentiated antiviral responses and inhibited pathogenic alphavirus growth in mosquito cells. Microbiol. Spectr. 11. https://doi.org/10.1128/spectrum.03613-22.
Tangudu, C.S., Charles, J., Nunez-Avellaneda, D., Hargett, A.M., Brault, A.C., Blitvich, B.J., 2021. Chimeric Zika viruses containing structural protein genes of insect-specific flaviviruses cannot replicate in vertebrate cells due to entry and post-translational restrictions. Virology 559, 30–39. https://doi.org/10.1016/j.virol.2021.03.014.
Tikhe, C.V., Dimopoulos, G., 2021. Mosquito antiviral immune pathways. Dev. Comp. Immunol. 116, 103964. https://doi.org/10.1016/j.dci.2020.103964.
Torres, F.J., Parry, R., Hugo, L.E., Slonchak, A., Newton, N.D., Vet, L.J., et al., 2022. Reporter flaviviruses as tools to demonstrate homologous and heterologous super-infection exclusion. Viruses 14, 1501. https://doi.org/10.3390/v14071501.
Tree, M.O., McKellar, D.R., Kieft, K.J., Watson, A.M., Ryman, K.D., Conway, M.J., 2016. Insect-specific flavivirus infection is restricted by innate immunity in the vertebrate host. Virology 497, 81–91. https://doi.org/10.1016/j.virol.2016.07.005.
Tyler, S., Bolling, B.G., Blair, C.D., Brault, A.C., Pabbaraju, K., Armijos, M.V., et al., 2011. Distribution and phylogenetic comparisons of a novel mosquito flavivirus sequence present in culex tarsalis mosquitoes from western canada with viruses isolated in California and Colorado. Am. Soc. Tropical Med. Hyg. 85, 162–168. https://doi.org/10.4269/ajtmh.2011.10-0469.
Unali, G., Douam, F., 2025. Orthoflavivirus vaccine platforms: current strategies and challenges. Vaccines (Basel) 13, 1015. https://doi.org/10.3390/vaccines13101015.
Vargas, V., Cime-Castillo, J., Lanz-Mendoza, H., 2020. Immune priming with inactive dengue virus during the larval stage of Aedes aegypti protects against the infection in adult mosquitoes. Sci. Rep. 10, 6723. https://doi.org/10.1038/s41598-020-63402-z.
Vasilakis, N., Tesh, R.B., 2015. Insect-specific viruses and their potential impact on arbovirus transmission. Curr. Opin. Virol. 15, 69–74. https://doi.org/10.1016/j.coviro.2015.08.007.
Vázquez, A., Sánchez-Seco, M.-P., Palacios, G., Molero, F., Reyes, N., Ruiz, S., et al., 2012. Novel flaviviruses detected in different species of mosquitoes in spain. Vector-Borne Zoonotic Dis. 12, 223–229. https://doi.org/10.1089/vbz.2011.0687.

Vet, L.J., Setoh, Y.X., Amarilla, A.A., Habarugira, G., Suen, W.W., Newton, N.D., et al., 2020. Protective efficacy of a chimeric insect-specific flavivirus vaccine against West Nile virus. Vaccines (Basel) 8, 258. https://doi.org/10.3390/vaccines8020258.

Villordo, S.M., Carballeda, J.M., Filomatori, C.V., Gamarnik, A.V., 2016. RNA structure duplications and flavivirus host adaptation. Trends Microbiol. 24, 270–283. https://doi.org/10.1016/j.tim.2016.01.002.

Villordo, S.M., Gamarnik, A.V., 2009. Genome cyclization as strategy for flavivirus RNA replication. Virus Res. 139, 230–239. https://doi.org/10.1016/j.virusres.2008.07.016.

Vogels, C.B.F., Rückert, C., Cavany, S.M., Perkins, T.A., Ebel, G.D., Grubaugh, N.D., 2019. Arbovirus coinfection and co-transmission: a neglected public health concern? PLoS Biol. 17, e3000130. https://doi.org/10.1371/journal.pbio.3000130.

Wallau, G.L., Abanda, N.N., Abbud, A., Abdella, S., Abera, A., Ahuka-Mundeke, S., et al., 2023. Arbovirus researchers unite: expanding genomic surveillance for an urgent global need. Lancet Glob. Health 11, e1501–e1502. https://doi.org/10.1016/S2214-109X(23)00325-X.

Wastika, C.E., Harima, H., Sasaki, M., Hang'ombe, B.M., Eshita, Y., Qiu, Y., et al., 2020. Discoveries of exoribonuclease-resistant structures of insect-specific flaviviruses isolated in zambia. Viruses 12, 1017. https://doi.org/10.3390/v12091017.

Weaver, S.C., Barrett, A.D.T., 2004. Transmission cycles, host range, evolution and emergence of arboviral disease. Nat. Rev. Microbiol. https://doi.org/10.1038/nrmicro1006.

Welsch, S., Miller, S., Romero-Brey, I., Merz, A., Bleck, C.K.E., Walther, P., et al., 2009. Composition and three-dimensional architecture of the dengue virus replication and assembly sites. Cell Host Microbe 5, 365–375. https://doi.org/10.1016/j.chom.2009.03.007.

Westaway, E.G., Brinton, M.A., Gaidamovich, Y.A., Horzinek, M.C., Igarashi, A., Käriäinen, L., et al., 1985. Flaviviridae. Intervirology 24, 183–192. https://doi.org/10.1159/000149642.

Wilkman, L., Kaarle, E., Luande, V.N., Lantto, R., Evander, M., Lwande, O.W., 2025. Insect-specific viruses: transmission dynamics and biological control strategies against arboviruses. Front. Microbiol. 16. https://doi.org/10.3389/fmicb.2025.1624662.

Willemsen, W., Helmes, N., Overheul, G.J., Henkens, M., Spruijt, R., van Rij, R.P., et al., 2025. Differential effect of acute versus persistent insect-specific flavivirus infection on superinfection exclusion of West Nile, Zika and chikungunya viruses in RNAi-competent and -deficient mosquito cells. One Health 20, 100960. https://doi.org/10.1016/j.onehlt.2024.100960.

Xi, Z., Ramirez, J.L., Dimopoulos, G., 2008. The aedes aegypti toll pathway controls dengue virus infection. PLoS Pathog. 4, e1000098. https://doi.org/10.1371/journal.ppat.1000098.

Xu, S., Li, L., Luo, X., Chen, M., Tang, W., Zhan, L., et al., 2022. Ggtree: a serialized data object for visualization of a phylogenetic tree and annotation data. iMeta 1. https://doi.org/10.1002/imt2.56.

Yan, K., Vet, L.J., Tang, B., Hobson-Peters, J., Rawle, D.J., Le, T.T., et al., 2020. A yellow fever virus 17D infection and disease mouse model used to evaluate a chimeric binjari-yellow fever virus vaccine. Vaccines (Basel) 8, 368. https://doi.org/10.3390/vaccines8030368.

Young, P.R., 2018. Arboviruses: a family on the move. pp. 1–10. https://doi.org/10.1007/978-981-10-8727-1_1.

Zélé, F., Nicot, A., Berthomieu, A., Weill, M., Duron, O., Rivero, A., 2014. Wolbachia increases susceptibility to plasmodium infection in a natural system. Proc. R. Soc. B: Biol. Sci. 281, 20132837. https://doi.org/10.1098/rspb.2013.2837.

Zhang, G., Asad, S., Khromykh, A.A., Asgari, S., 2017. Cell fusing agent virus and dengue virus mutually interact in Aedes aegypti cell lines. Sci. Rep. 7, 6935. https://doi.org/10.1038/s41598-017-07279-5.
Zhang, H.-Q., Li, N., Zhang, Z.-R., Deng, C.-L., Xia, H., Ye, H.-Q., et al., 2024. A chimeric classical insect-specific flavivirus provides complete protection against West Nile virus lethal challenge in mice. J. Infect. Dis. 229, 43–53. https://doi.org/10.1093/infdis/jiad238.
Zhang, Q.-Y., Li, X.-F., Niu, X., Li, N., Wang, H.-J., Deng, C.-L., et al., 2020. Short direct repeats in the 3′ untranslated region are involved in subgenomic flaviviral RNA production. J. Virol. 94. https://doi.org/10.1128/JVI.01175-19.
Zhang, X., Guo, X., Fan, H., Zhao, Q., Zuo, S., Sun, Q., et al., 2017. Complete genome sequence of Menghai flavivirus, a novel insect-specific flavivirus from China. Arch. Virol. 162, 1435–1439. https://doi.org/10.1007/s00705-017-3232-5.
Zhang, Y., Liang, D., Yuan, F., Yan, Y., Wang, Z., Liu, P., et al., 2022. Replication is the key barrier during the dual-host adaptation of mosquito-borne flaviviruses. Proc. Natl. Acad. Sci. 119. https://doi.org/10.1073/pnas.2110491119.
Zou, G., Zhang, B., Lim, P.-Y., Yuan, Z., Bernard, K.A., Shi, P.-Y., 2009. Exclusion of West Nile virus superinfection through RNA replication. J. Virol. 83, 11765–11776. https://doi.org/10.1128/JVI.01205-09.

CHAPTER FOUR

Molecular arms race: Viral adaptations to mosquito immune defenses

Gladys Gutierrez-Bugallo, Shengzhang Dong, and George Dimopoulos[*,1]
W. Harry Feinstone Department of Molecular Microbiology and Immunology, Bloomberg School of Public Health, Johns Hopkins University, Baltimore, United States
*Corresponding author. e-mail address: gdimopo1@jhu.edu

Contents

[1] Senior contributor

Advances in Insect Physiology, Volume 70
ISSN 0065-2806, https://doi.org/10.1016/bs.aiip.2026.05.003

Abstract

Mosquitoes (Diptera: Culicidae) are considered the deadliest animals on the planet because of the public health impact caused by the pathogens they transmit. They are present on every continent except Antarctica, with more than 3726 species identified. Mosquitoes harbor two distinct viral categories: mosquito-borne viruses and mosquito-specific viruses. While the first group comprises disease-causing viruses transmitted to vertebrate hosts during blood feeding by female mosquitoes, the second group of viruses are mainly maintained within insect populations through vertical transmission and do not infect vertebrates. Viruses establish persistent infections in mosquitoes through a delicate balance between insect immune responses and viral replication dynamics and immune evasion strategies. Mosquitoes mount potent antiviral defenses primarily via the RNA interference (RNAi) pathway, although other canonical immune pathways, including Toll, JAK/STAT, and IMD, also contribute to host survival and fitness. These immune mechanisms impose strong selective pressures on viruses, which have consequently evolved diverse strategies to evade mosquito defenses and achieve replication levels sufficient for transmission and persistence in nature. The interaction between mosquitoes and viruses occurs within a constant "molecular arms race" that shapes their co-evolution. In the case of mosquito-borne viruses, this evolutionary dynamic is further influenced by an additional selective layer imposed by infection of vertebrate hosts. Although viral immune evasion mechanisms in vertebrates have been extensively studied, the strategies employed by viruses to counteract antiviral defenses in mosquitoes remain poorly characterized. A deeper understanding of these reciprocal interactions is essential for the rational design of effective antiviral and vector-based intervention strategies.

1. Introduction

Mosquitoes (Culicidae) are a major group of arthropod vectors, comprising 3726 species across two subfamilies and more than 217 genera, distributed worldwide except Antarctica (Harbach, 2024). They host two principal virus groups: mosquito-borne viruses (MBVs), a major subset of arboviruses with ~200 species across seven families and ~20 genera (Huang et al., 2023), and mosquito-specific viruses (MSVs), also known as insect-specific viruses (ISVs), which replicate in mosquitoes but are not transmitted to vertebrates (Huang et al., 2023).

Only a small fraction of mosquito species are competent vectors of viruses (Wouters et al., 2024; Panda et al., 2024). Key genera—*Aedes, Culex, Haemagogus*, and *Sabethes*—transmit major human and veterinary pathogens, including dengue viruses (DENVs), yellow fever virus (YFV), Japanese encephalitis virus (JEV), chikungunya virus (CHIKV), Zika virus (ZIKV), Rift Valley fever virus (RVFV), and West Nile virus (WNV), among others;

making mosquitoes the deadliest animals to humans (Velde et al., 2024; WHO, 2024). MBVs are maintained through sylvatic cycles involving non-human hosts and urban cycles between humans and anthropophilic mosquitoes (Bray and Monath, 2013). Emergence and re-emergence of MBVs are driven by land-use change, human mobility, trade-associated transportation, urbanization, population growth, and climate change, posing increasing global public health threats (Gubler, 2002; Mayer et al., 2017; Stephens et al., 2021). In humans, arbovirus infection commonly causes acute febrile illness but can progress to severe manifestations, including neurological syndromes, hemorrhagic fever, chronic arthropathy, pregnancy complications, and congenital anomalies (Weaver and Barrett, 2004; de Souza and Weaver, 2024; de Souza et al., 2025). Arboviruses also cause significant disease in livestock, including neurological signs, abortion, and congenital malformations (Hubálek et al., 2014).

While MBVs pose the greatest concern due to their human and economic impact, MSVs provide a valuable model to study viral evolution and vector–virus interactions. MSVs replicate exclusively in mosquitoes and are thought to circulate among mosquito populations primarily via vertical transmission (Blitvich and Firth, 2015; Huang et al., 2023). Some ISVs are phylogenetically nested within arbovirus families (Vasilakis et al., 2019). Phylogenetic analyses suggest that ISVs may represent ancestral viral lineages or be interwoven within arbovirus clades (Calisher and Higgs, 2018; Li et al., 2015).

Both MBVs and MSVs must overcome physical, physiological, and immunological barriers to establish infection in mosquitoes. Mosquitoes mount potent antiviral defenses that limit viral replication and reduce fitness costs (Tikhe and Dimopoulos, 2021). Host–virus interactions drive reciprocal genetic shaping: hosts evolve sensors and effectors to detect and restrict infection, while viruses develop evasion strategies that influence persistence, transmission efficiency, and emergence (Daugherty and Malik, 2012; Demogines et al., 2013; Woolhouse et al., 2001). This dynamic arms race underscores how studying both MBVs and MSVs can illuminate the molecular mechanisms governing viral adaptation in mosquito vectors.

Because antivirals or vaccines are lacking for most arboviruses, control relies heavily on vector management (Monath, 2021). Conventional measures—such as insecticide use and breeding site reduction—have limited success, motivating innovative approaches including *Wolbachia*- and transgenesis-based mosquito suppression and transmission-blocking interventions (Caragata et al., 2020). The development of transmission-blocking strategies depends on a deep understanding of mosquito antiviral immunity and viral adaptation mechanisms, a field that remains largely unexplored. This chapter synthesizes current knowledge, identifies key gaps, and outlines future research directions.

2. Mosquito-associated viruses

2.1 Mosquito-borne viruses

Mosquito-borne viruses group a subset of arboviruses. This term does not correspond to a formal taxonomic rank but rather describes an ecological and transmission-based grouping of viruses that are maintained in nature through arthropod vectors. Also known as dual-host viruses, encompass approximately 530 entities (Gubler, 2001) distributed across nine families (Table 1). Notably, several of these families also include viruses that are not transmitted by arthropods. Arboviruses are predominantly RNA viruses; the sole known DNA arbovirus is African swine fever virus, a member of the family *Asfarviridae* (ICTV, 2023).

The genomic organization and structural features of arboviruses are highly diverse and include single-stranded RNA genomes of positive (+ssRNA) or negative (−ssRNA) polarity, as well as double-stranded RNA (dsRNA) genomes. Viruses belonging to most of the families are enveloped, whereas members of the *Reoviridae* family possess a non-enveloped capsid structure (Strauss and Strauss, 1994; Lescar et al., 2001; King et al., 2012). Genome segmentation is a common feature among arboviruses but varies substantially between families. Some of them are characterized by tripartite (−) ssRNA genomes, while others contain ten dsRNA genome segments (King et al., 2012). In contrast, *Flaviviridae* and *Togaviridae* have non-segmented (+) ssRNA genomes, encoding one and two major open reading frames (ORFs), respectively (Strauss and Strauss, 1994; Lescar et al., 2001).

Despite this diversity, arboviral genomes are generally translated into polyproteins that are subsequently cleaved into mature structural (capsid and envelope) and non-structural proteins by a combination of viral and host proteases. From an epidemiological standpoint, the most medically important MBVs are concentrated within a limited number of genera. These include *Orthoflavivirus*, *Alphavirus*, and *Orthobunyavirus* (Jupille et al., 2014) (Table 1).

Table 1 Viral families, key genera, representative viruses, and mosquito species serving as vectors or insect hosts for mosquito-borne and mosquito-specific viruses.

	Viral families	Relevant viral genera	Relevant viruses	Mosquito vector	References
Mosquito-borne viruses	Flaviviridae Togaviridae Peribunyaviridae Phenuiviridae Nairoviridae Reoviridae Rhabdoviridae Orthomyxoviridae Asfarviridae	*Orthoflavivirus* (Flaviviridae)	Dengue viruses (DENV1–4)	**Aedes aegypti* and *Ae. albopictus*	Gubler (1998)
			Yellow fever virus (YFV)	**Ae. aegypti*	Dauer and Carrera (1937)
			West Nile Virus (WNV)	**Culex pipiens* and *Cx quinquefasciatus*	Hurlbut (1956)
			Zika Virus (ZIKV)	**Ae. aegypti*	Marchete et al. (1969)
			Japanese encephalitis virus (JEV)	*Cx. tritaeniorhynchus*	Self et al. (1973)
		Alphavirus (Togaviridae)	Chikungunya virus (CHIKV)	**Ae. aegypti* and *Ae. albopictus*	Lumsden (1955), Tsetsarkin et al. (2007)
			O'nyong-nyong virus (ONNV)	*Anopheles funestus* and *An. gambiae*	Williams et al. (1965)
			Mayaro virus (MAYV)	*Haemagogus* spp	Hoch et al. (1981)

(continued)

Table 1 Viral families, key genera, representative viruses, and mosquito species serving as vectors or insect hosts for mosquito-borne and mosquito-specific viruses. (*cont'd*)

	Viral families	Relevant viral genera	Relevant viruses	Mosquito vector	References
		Orthobunyavirus (Peribunyaviridae)	La Crosse virus (LACV)	*Ae. triseriatus*	Watts et al. (1972)
			California encephalitis virus (CEV)	*Ae. dorsalis* and *Ae. melanimon*	Turell et al. (1982), Reeves and Hammon (1952)
Mosquito-specific viruses	Flaviviridae Togaviridae Peribunyaviridae Phenuiviridae Reoviridae Rhabdoviridae Orthomyxoviridae Mesoniviridae Tymoviridae Birnaviridae Nodaviridae Parvoviridae Iridoviridae Permutotetraviridae Iflaviridae Totiviridae *Negevirus*[a]	*Orthoflavivirus*, (Flaviviridae)	**Representative viruses**	**First detected in**	
			Cell-fusion agent virus (CFAV)	*Ae. aegypti* (Peleg) cell line	Stollar and Thomas (1975)
			Kamiti River virus (KRV)	*Ae. macintoshi*	Sang et al. (2003)
			Culex flavivirus (CxFV)	*Cx. pipiens*	Hoshino et al. (2007)
		Alphavirus (Togaviridae)	Eilat virus (EILV)	*An. coustani*	Nasar et al. (2012)
			Tai Forest alphavirus (TALV)	*Cx. decens*	Hermanns et al. (2017)

		Agua Salud alphavirus (ASALV)	*Cx. declarator*	Hermanns et al. (2020)
	Unassigned genus; sister to *Orthobunyavirus* (Peribunyaviridae)	Herbert virus (HEBV)	*Culex* spp, *Anopheles* spp, *Uranotaenia mashonaensis*	Marklewitz et al. (2015)
		Taï virus (TAIV)	*Cx. nebulosus*	Marklewitz et al. (2015)
		Kibale virus (KIBV)	*Cx. simpliciforceps*	Marklewitz et al. (2015)

[a]Negevirus is a proposed taxon. It is not formally recognized as a family by the International Committee on Taxonomy of Viruses. Asterisk (*) indicates urban transmission vectors; other forest species may be involved in sylvatic transmission.

2.2 Mosquito-specific viruses

In contrast to the dual-host viruses, a growing number of viruses have been identified that appear to be restricted to insect hosts. Insect-specific viruses (ISVs) replicate exclusively in arthropods, where they typically establish persistent infections and are maintained primarily through vertical transmission. The identification of viral and host determinants that restrict ISVs to arthropods provides key insights into arbovirus evolution and vector–virus interactions, and may inform the development of novel antiviral and anti-vectorial strategies against arboviral diseases (Blitvich and Firth, 2015).

The first ISV identified was the Cell-fusing agent virus (CFAV), discovered in 1975 in an *Ae. aegypti* cell line (Stollar and Thomas, 1975). Since then, ISV discovery has increased markedly, largely driven by advances in next-generation sequencing and bioinformatic tools (reviewed in Roundy et al., 2017). ISVs are globally distributed and have been detected in numerous mosquito species, as well as in other dipterans (Blitvich and Firth, 2015; Laredo-Tiscareño et al., 2024), across all continents where these insects occur (Huhtamo et al., 2012; Moonen et al., 2023).

Mosquito-specific viruses (MSVs), a subset of ISVs, as well as MBVs, belong to three major viral families: *Flaviviridae*, *Togaviridae*, and *Peribunyaviridae* (Blitvich and Firth, 2015; Vasilakis and Tesh, 2015; de Almeida et al., 2021), though additional MSVs have been identified in a broad range of viral families (Table 1) (Carvalho and Long, 2021). The identification of the insect-specific *Negevirus* taxon further highlights the extensive genetic and evolutionary diversity of ISVs (Vasilakis et al., 2013). Collectively, ISVs display heterogeneous evolutionary origins, with some showing close phylogenetic relationships to plant or vertebrate viruses (Vasilakis et al., 2013; Nunes et al., 2017; Charles et al., 2019). Like their dual-host counterparts, ISVs within these families retain the characteristic genome organization and virion architecture of their respective taxa (Carvalho and Long, 2021).

Particularly within the family *Flaviviridae*, based on phylogenetic relationships, insect-specific flaviviruses (ISFs) are divided into classical ISFs (cISFs), which are highly divergent and basal within the *Orthoflavivirus* genus, and dual-host affiliated ISFs (dISFs), which cluster with mosquito-borne flaviviruses, including arboviruses (Blitvich and Firth, 2015). The basal placement of cISFs suggests they may represent ancestral flavivirus lineages (Blitvich and Firth, 2015).

Recent studies indicate that host restriction in ISVs may occur at multiple stages of the viral replication cycle and is influenced by complex interactions among viral, host, and environmental factors (Elrefaey et al., 2020; Junglen et al., 2017). For example, some insect-specific alphaviruses exhibit reduced replication at temperatures above 31°C (Hermanns et al., 2020). In flaviviruses, divergence or loss of conserved envelope motifs, such as an eight–amino acid sequence in domain III, may impair viral entry into vertebrate cells. This mechanism was demonstrated using chimeric YFV expressing envelope proteins from the cISF Nienokoue virus, which resulted in blocks in viral entry, replication, and assembly in vertebrate cells (Junglen et al., 2017).

Vertical transmission is considered the primary mechanism maintaining ISVs in nature (Bolling et al., 2011; Haddow et al., 2013; Saiyasombat et al., 2011), although venereal transmission has also been reported (Logan et al., 2022; Peinado et al., 2022). Additionally, Negevirus members may be transmitted orally via plant-derived substrates such as nectar or aquatic plant material, consistent with their phylogenetic proximity to plant viruses (Vasilakis et al., 2013; Vasilakis and Tesh, 2015). Some ISVs, including Eilat virus (EILV) and Negev virus, can also experimentally infect adult mosquitoes via high-titer infectious blood meals (Nasar et al., 2014). Comparative analyses of ISV and arbovirus infection routes may therefore provide valuable insights into mosquito barriers to infection and viral strategies to overcome vector defenses.

2.3 Genome organization and replication of major mosquito-associated viral genera

For a successful persistence in nature, mosquito-associated viruses rely on specialized genome architectures, regulatory RNA elements, and multifunctional viral proteins that coordinate entry, genome replication, assembly, and dissemination.

Much of our mechanistic understanding of viral RNA replication is derived from studies in vertebrate cells, particularly mammalian systems. While the core principles of genome transcription, replication, and particle assembly are likely conserved in mosquito cells, host-specific differences are expected and remain insufficiently characterized (Ahola et al., 2021). Consequently, extrapolation from vertebrate models should be approached cautiously, highlighting the need for direct studies in mosquito hosts. Below, we summarize the genome organization and replication strategies of three major mosquito-associated viral genera, illustrating both shared themes and distinct biological solutions.

2.3.1 *Orthoflavivirus*

Orthoflaviviruses are small (~50 nm), enveloped (+) ssRNA viruses with ~11 kb genome that contains a 5′ cap but lacks a poly(A) tail (Donaldson et al., 2025; Ng et al., 2017). Highly structured untranslated regions (UTRs) at both genome termini regulate translation, replication, and immune evasion (Lee et al., 2021). The genome encodes a single polyprotein that is processed into three structural proteins (C, prM, E) and seven nonstructural proteins (NS1–NS5) (Misra and Kalita, 2010; Brinton, 1986; Coffey et al., 2013; Vasilakis et al., 2011).

Virus entry occurs through receptor-mediated endocytosis, followed by low-pH–triggered membrane fusion mediated by the E glycoprotein (Mathew et al., 2011; Anwar et al., 2022). After uncoating, viral RNA is translated at endoplasmic reticulum (ER) membranes and processed by viral and host proteases (Smit et al., 2011; Sehrawat et al., 2021). Genome replication takes place in membrane invaginations of the ER known as replication organelles, where a (-) RNA intermediate serves as a template for the synthesis of new genomic RNA (Welsch et al., 2009; Overby et al., 2010; Alvarez et al., 2005). The viral helicase and polymerase coordinate RNA synthesis and capping. Assembly occurs at the ER, generating immature particles that mature during transport through the Golgi via furin-mediated cleavage of prM, yielding infectious virions (Stadler et al., 1997; Yu et al., 2008; Yu et al., 2009; Uchida et al., 2014).

2.3.2 *Alphavirus*

Alphaviruses are larger (~70 nm), enveloped viruses with a (+) ssRNA genome of ~11–12 kb that contains both a 5′ cap and a 3′ poly(A) tail, enabling immediate translation upon entry (Strauss and Strauss, 1994; Chen et al., 2018). The genome comprises two ORFs flanked by regulatory UTRs involved in replication, host range, and immune modulation (Hahn et al., 1985; Sokoloski et al., 2010; Trobaugh et al., 2014).

The 5′ ORF encodes nonstructural proteins (nsP1–nsP4) that assemble into replication complexes on host membranes, forming membrane spherules that compartmentalize RNA synthesis and protect replication intermediates (Ahola et al., 2021; Frolov and Frolova, 2022). These complexes generate both full-length genomic RNA and a subgenomic RNA, which drives high-level expression of structural proteins (C, E3, E2, 6 K/TF, and E1) from the downstream ORF (Strauss and Strauss, 1994; Chmielewski et al., 2022). Structural proteins assemble into envelope spikes that mediate receptor binding and membrane fusion. Unlike orthoflaviviruses, alphaviruses assemble and bud at the plasma membrane, acquiring their lipid envelope during particle release. Alphavirus replication is rapid, enabling high virus yields within hours of infection (Hyde et al., 2014; Madden et al., 2020). In addition to their replication functions, some viral proteins (i.e. NsP2 and C) interfere with host transcription and nuclear transport, contributing to cytopathic effects (Hyde et al., 2015; Reynaud et al., 2015).

2.3.3 *Orthobunyavirus*

Orthobunyaviruses differ fundamentally from the previous genera by possessing a segmented, (-) ssRNA genome composed of three segments (L, M, and S) (Elliott, 2014). Complementary terminal sequences form panhandle structures that act as promoters for transcription and replication (Barr et al., 2003; Kohl et al., 2004). Each segment encodes distinct proteins: the L segment encodes the RNA-dependent RNA polymerase (RdRp), the S segment encodes the nucleoprotein and an immune-modulating nonstructural protein, and the M segment encodes envelope glycoproteins and a protein involved in assembly (Ogg and Patterson, 2007; Osborne and Elliott, 2000).

Orthobunyaviruses initiate transcription using a cap-snatching mechanism, whereby short, capped fragments are cleaved from host mRNAs to prime viral mRNA synthesis (Dunn et al., 1995). Genome replication proceeds through a full-length antigenome intermediate, with encapsidation by nucleoprotein preventing recognition by host defenses. The segmented genome allows reassortment during co-infection, generating genetic diversity and potentially altering viral fitness and pathogenicity (Beaty et al., 1983; Iroegbu and Pringle, 1981; Urquidi and Bishop, 1992). Virion assembly occurs in the ER–Golgi network, where glycoproteins form surface spikes that mediate attachment and membrane fusion. Mature particles are transported to the plasma membrane for release (Obijeski et al., 1976; Bowden et al., 2013; Elliott, 2014).

2.3.4 Comparative perspectives

Together, these genera illustrate distinct evolutionary solutions to replication in mosquito and vertebrate hosts. Orthoflaviviruses and alphaviruses share (+) ssRNA genomes but differ in genome organization, site of assembly, and particle maturation pathways. Alphaviruses uniquely employ subgenomic RNA to amplify structural protein expression and bud at the plasma membrane, whereas orthoflaviviruses assemble and mature within the secretory pathway. Orthobunyaviruses, in contrast, possess segmented negative-sense genomes, rely on cap-snatching for transcription, and exhibit reassortment-driven diversity. Despite these differences, all three groups exploit membrane-associated replication compartments and encode multifunctional proteins that balance efficient replication with evasion of host defenses (Jupille et al., 2014). Understanding these shared and divergent strategies provides a foundation for dissecting how mosquito immune pressures shape viral evolution and transmission.

3. Evolution in mosquito-associated viruses

3.1 Viral source of evolution

Arboviruses are distinguished among RNA viruses by their obligatory alternation between vertebrate and invertebrate hosts, a life cycle historically thought to impose strong evolutionary constraints and favor a strategy of 'compromise fitness' (Coffey et al., 2008; Coffey and Vignuzzi, 2011; Novella et al., 1999; Weaver et al., 1999). However, this trade-off model has been questioned, as replication in distinct hosts may involve sufficiently similar cellular environments to avoid strong fitness conflicts, or one host may dominate as the primary driver of viral evolution (Weaver et al., 2021; Vasilakis et al., 2009). In line with these alternative views, comparative analyses show that flavivirus evolutionary rates do not strictly track host range: mosquito-borne flaviviruses do not consistently evolve more slowly than single-host viruses, including ISVs, whereas tick-borne flaviviruses display particularly low substitution rates (Pontremoli et al., 2021). Together, these observations indicate that host alternation alone does not fully explain flavivirus evolutionary dynamics.

Regardless of the extent to which host alternation constrains long-term evolution, mosquito-associated viruses are characterized by pronounced genetic heterogeneity at the population level (Lequime et al., 2016). Rather than existing as uniform genotypes, these viruses circulate as dynamic mutant swarms composed of closely related variants structured around a dominant or "master" sequence, consistent with quasispecies theory. This organization allows viral populations to balance genetic stability with adaptability (Lauring and Andino, 2010). Diversity within mosquito-associated viruses is continuously generated through multiple mechanisms, including (i) the intrinsically error-prone replication of viral RNA-dependent RNA polymerases (Forrester et al., 2014), as well as (ii) genetic exchange via recombination and, (iii) in segmented viruses, reassortment (Novella et al., 2011). Collectively, these processes sustain high levels of intra-population diversity that underpin viral persistence, host adaptation, and evolutionary potential across diverse ecological contexts.

At the core of this population-level heterogeneity lies the intrinsically error-prone nature of RNA virus replication. Mosquito-associated viruses encode their own RdRps, whose lack of 3–5′ exonuclease proofreading activity results in low replication fidelity and the accumulation of mutations during genome synthesis. Consequently, RNA virus mutation rates are typically on the order of $\sim 10^{-4}$ substitutions per nucleotide copied, several orders of magnitude higher than those observed in DNA viruses or cellular organisms. Given the large population sizes achieved during infection in both vertebrate and invertebrate hosts, this high error rate ensures the continual generation of genetic variants, thereby sustaining the mutant swarm structure described above and providing the raw material upon which selection can act (Lauring and Andino, 2010).

Experimental studies highlight the importance of replication fidelity for arbovirus fitness. CHIKV variants with increased RdRp fidelity exhibit reduced infection and dissemination in *Ae. aegypti* and produce lower viremias and tissue titers in newborn mice. These findings demonstrate that increased replication fidelity, and the accompanying reduction in genetic diversity, imposes a fitness cost in both invertebrate and vertebrate hosts, highlighting the adaptive value of error-prone replication in arboviruses (Coffey et al., 2011).

Beyond mutation, recombination represents an additional source of genetic diversity in some mosquito-associated RNA viruses and requires coinfection of a single cell by two or more parental viruses, producing progeny genomes derived from multiple templates (Novella et al., 2011). Although recombination has been reported across all RNA virus families, its frequency and outcomes vary with genome polarity. With few exceptions, (-) RNA viruses predominantly undergo nonhomologous recombination, often generating defective interfering particles rather than viable progeny (Holland, 1991). In contrast, (+) RNA viruses can undergo homologous recombination, yielding fully functional recombinant genomes (Novella et al., 2011). Among mosquito-associated viruses, recombination has been documented in alphaviruses; a classic example is western equine encephalitis virus, which emerged through genetic exchange between eastern equine encephalitis virus and a Sindbis-like virus approximately 1300–1900 years ago (Weaver et al., 1997). By contrast, recombination among orthoflaviviruses appears to be exceedingly rare, with only sporadic events detected in natural populations (Holmes et al., 1999; Chare et al., 2003; Taucher et al., 2010; Faye et al., 2014).

Reassortment is a major mechanism of genetic exchange restricted to viruses with segmented genomes and occurs when genome segments from different parental viruses are independently replicated and packaged during coinfection, generating reassortant progeny. In orthobunyaviruses, coinfection of two parental strains can theoretically produce up to six reassortant genotypes in addition to the parental viruses, although reassortment is constrained by segment compatibility and typically occurs among closely related viruses within the same serogroup (Elliott, 2014). Mosquito vectors may provide permissive conditions for reassortment because they can acquire multiple viruses through sequential blood meals (Beaty et al., 1983). Reassortment can have substantial phenotypic consequences, including alterations in virulence, host range, and vector competence. In orthobunyaviruses, the M segment—encoding the envelope glycoproteins Gn and Gc and the nonstructural protein NSm—plays a central role in vector specificity; thus, reassortment involving this segment may facilitate adaptation to new arthropod vectors and promote emergence in novel ecological settings (Beaty et al., 1981; Briese et al., 2013; Heitmann et al., 2021; Bowen et al., 2025).

A classic example of reassortment-driven phenotypic change is Ngari virus, which contains the L and S segments of Bunyamwera virus and the M segment of Batai virus and is associated with severe hemorrhagic disease, despite both parental viruses typically causing mild febrile illness (Briese et al., 2006). More recently, the emergence and spread of Oropouche virus in the Americas and the Caribbean have been linked to a novel reassortant lineage carrying the M segment from viruses circulating in the eastern Amazon (2009–2018) combined with the L and S segments from viruses detected in Peru, Colombia, and Ecuador (2008–2021) (Naveca et al., 2024). This reassortant has been associated with increased replication efficiency, enhanced transmission, reduced susceptibility to neutralization by pre-existing immunity (Scachetti et al., 2025), and increased pathogenicity, with recent reports linking Oropouche virus infection to severe adverse pregnancy outcomes (Schwartz, 2025), underscoring the public health significance of reassortment-driven viral evolution.

3.2 Natural forces driving evolution in mosquito-associated viruses

Beyond variation generated through mutation, recombination, and reassortment, mosquito-associated viruses are also strongly influenced by demographic processes during their transmission cycle. Even within a single mosquito host, viruses frequently experience population bottlenecks (sharp reductions in the number of viral particles, resulting in loss of genetic diversity) during midgut infection, dissemination, and transmission to the vertebrate host. These bottlenecks can reduce the efficiency of natural selection, and stochastic processes such as genetic drift (random changes in variant frequencies in small populations) may prevail, leading to the fixation or loss of mutations (Weaver et al., 2021). Repeated bottlenecks, combined with the intrinsically high mutation rates of RNA viruses, increase the risk of fitness decline through the irreversible accumulation of deleterious mutations, a process known as Müller's ratchet (Muller, 1964), which has been demonstrated experimentally for several arboviruses (Duarte et al., 1992; Weaver et al., 1999).

Despite these constraints, many arboviruses circulate efficiently in nature and often display remarkably stable consensus genomes (Weaver et al., 1992). This apparent stability likely reflects the rapid restoration of intrahost diversity following bottlenecks, driven by fast, error-prone replication. Population-level sequencing studies have shown that viral populations can quickly re-establish complex mutant swarms, including minority variants that facilitate dissemination and transmission within the mosquito (Forrester et al., 2014; Lequime et al., 2016; Patterson et al., 2018; Fitzmeyer et al., 2023; Banho et al., 2025). Together, these observations indicate that the ability to recover genetic diversity after repeated bottlenecks is a key contributor to the evolutionary persistence of mosquito-associated viruses.

In addition to intrinsic genetic and demographic processes, interactions with the mosquito host further shape viral evolution. The mosquito immune system exerts selective pressures that influence which viral variants can replicate and be transmitted. This ongoing host–virus interplay resembles a molecular "arms race," in which viruses evolve mechanisms to evade or suppress immune responses while mosquitoes adapt to detect and control infection (Olson and Blair, 2015; Samuel et al., 2018). These dynamics contribute to the maintenance of viral diversity and naturally lead to the following discussion on mosquito antiviral defenses.

4. The mosquito antiviral defense

Over evolutionary time, mosquito vectors have assembled a multifaceted antiviral defense that blends innate immune pathways, tissue-specific barriers, and physiological and behavioral adaptations. These defenses shape whether a given mosquito becomes infected, how broadly the infection disseminates through its body, and whether a virus is delivered in saliva to infect a vertebrate host (Kumar et al., 2018; Lee et al., 2019). Together, these components determine vector competence and influence specificity for different virus families such as flaviviruses, alphaviruses, and bunyaviruses (Kain et al., 2022; Kingsolver et al., 2013). Understanding the mosquito antiviral defense is essential for predicting transmission dynamics, designing biologically informed control strategies, and anticipating how environmental change will alter vector-virus interactions.

4.1 Vector competence and specificity for different viruses

Vector competence describes a mosquito's intrinsic ability to acquire, replicate, and transmit a virus, a trait governed by a complex interplay of genetic, environmental, and microbial factors. Specificity arises from co-evolutionary pressures that dictate receptor compatibility and immune evasion. For instance, *Ae. aegypti* is highly competent for DENV and ZIKV, exhibiting high transmission rates under optimal conditions. Conversely, *Culex* species (*Cx. pipiens, Cx. quinquefasciatus*) function as primary vectors for WNV and JEV but display near-zero competence for DENV (Kain et al., 2022).

Genetic variation drives these distinct susceptibility profiles. Refractory *Ae. aegypti* strains often carry polymorphisms in immune genes, such as Dicer-2, which modulate viral dsRNA sensing and subsequent dissemination (Gestuveo et al., 2022; Merkling et al., 2023; Samuel et al., 2023). In *Cx. quinquefasciatus*, refractoriness to DENV is attributed to a midgut infection barrier where the virus cannot enter epithelial cells, likely due to the absence of specific entry receptors found in permissive *Aedes* vectors (Fernandes-Santos and Azeredo, 2022) (Fig. 1A).

Environmental and physiological stressors further modulate competence. While warmer climates generally shorten the extrinsic incubation period (EIP) and accelerate DENV replication in *Ae. aegypti*, extreme heat induces physiological stress. This upregulates heat-shock proteins, potentially bolstering antiviral defenses or reducing vector longevity, which truncates the transmission window (Abbasi, 2025) (Fig. 1B). Additionally,

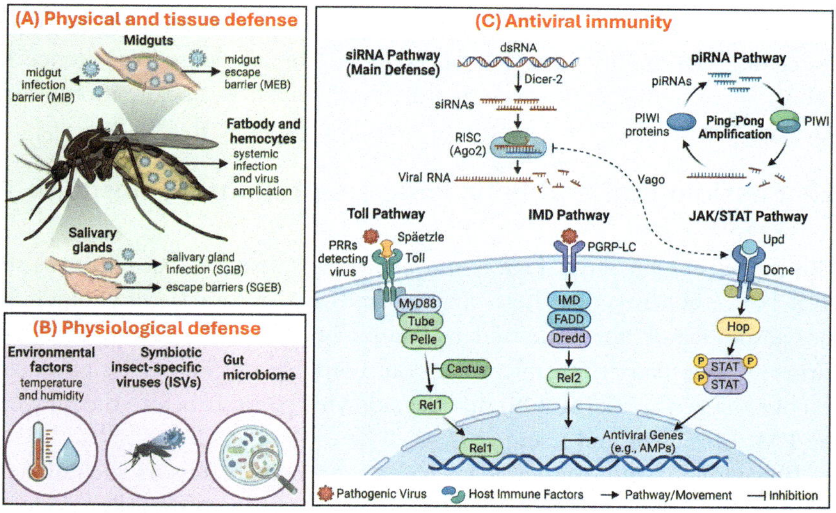

Fig. 1 Schematic representation of the multilayered antiviral defense systems in mosquitoes. The mosquito antiviral response comprises three interconnected layers: physical barriers, physiological modulators, and molecular immunity. (A) Physical and tissue defense restricts viral infection. The midgut serves as the initial site of infection, presenting the midgut infection barrier (MIB) and midgut escape barrier (MEB). Escaped virions establish systemic infection in the hemocytes and fat body before encountering the salivary gland infection (SGIB) and escape barriers (SGEB) required for transmission. (B) Physiological defense encompasses external and internal factors that modulate vector competence. Abiotic environmental stressors (e.g., temperature, humidity) and biotic factors, such as the gut microbiome and coinfection with insect-specific viruses (ISVs), influence the host's baseline susceptibility. (C) Antiviral immunity depicts the intracellular signaling cascades triggered by infection. The small interfering RNA (siRNA) pathway functions as the cornerstone of antiviral defense. This is complemented by the PIWI-interacting RNA (piRNA) pathway and canonical signaling pathways (Toll, IMD, and JAK/STAT), which collectively regulate the transcription of antiviral effectors to suppress viral replication.

insecticide resistance mechanisms, such as the upregulation of detoxification enzymes (e.g., Cytochrome P450s, GSTs), can exert pleiotropic effects on immunity. Resistant *Culex* mosquitoes, for example, may exhibit enhanced WNV dissemination due to oxidative stress tolerance or the reallocation of metabolic resources away from immune effectors (Marcombe et al., 2009).

Finally, the mosquito microbiome acts as a dynamic regulator of competence (Fig. 1B). MSVs like CFAV in *Ae. aegypti* can suppress arbovirus replication via superinfection exclusion (Elrefaey et al., 2020). Similarly, bacterial taxa, including *Serratia* and *Enterobacter*, prime IMD signaling and modify the peritrophic matrix (PM) (Carpenter and Clem, 2023). These interactions are highly variable; for example, *Ae. albopictus* competence for ZIKV varies significantly across populations, driven by local genotype-by-environment interactions (Azar et al., 2017; Obadia et al., 2022).

4.2 Physiological and behavioral adaptations in mosquito vectors

Mosquitoes exhibit physiological and behavioral plasticity that enhances survival but simultaneously alters viral transmission dynamics. Physiologically, mosquitoes rapidly form the midgut PM post-blood meal, creating a physical barrier to viral entry. Gut microbiota can reinforce this barrier by producing reactive oxygen species (ROS) that degrade viral particles before they traverse the PM (Carpenter and Clem, 2023).

Behaviorally, mosquitoes adapt to environmental stressors such as temperature fluctuations. While high temperatures may theoretically accelerate viral replication, mosquitoes actively seek cooler microhabitats to avoid desiccation. This thermoregulatory behavior can mitigate the temperature effects on viral kinetics, resulting in lower real-world transmission efficiency than predicted by laboratory models (Abbasi, 2025).

Conversely, arboviruses can manipulate vector behavior to enhance transmission. *Ae. aegypti* infected with DENV or ZIKV display increased locomotor activity and biting frequency. Mechanistically, this is driven by the upregulation of proteins such as Sialokinin and shifts in neurotransmitters, including dopamine and N-acetyl-L-tyrosine, which heighten the host-seeking drive (Lefteri et al., 2022). While some behavioral adaptations, such as altered oviposition choices, may dampen transmission potential, virus-induced hyper-activity can triple transmission capacity (Sim et al., 2012; Wei Xiang et al., 2022).

4.3 The mosquito immune arsenal

Mosquitoes lack the adaptive immune system characteristic of vertebrates, such as antibody production and T-cell memory. Instead, they rely on a sophisticated innate immune system capable of recognizing pathogen-associated molecular patterns (PAMPs) and mounting rapid, specific responses. The antiviral defense is orchestrated primarily by the RNA interference (RNAi) pathway, which functions as a genomic immune system, and a network of conserved signaling pathways (Toll, IMD, and JAK/STAT) that regulate the transcriptional landscape of the infected cell (Tikhe and Dimopoulos, 2021) (Fig. 1C).

4.3.1 *The RNA interference (RNAi) pathway*

RNA interference is universally recognized as the cornerstone of mosquito antiviral immunity. It is a sequence-specific silencing mechanism that targets viral RNAs for degradation, thereby limiting replication and preventing virus-induced mortality in the vector. The pathway is segmented into three distinct branches: the small interfering RNA (siRNA) pathway, the PIWI-interacting RNA (piRNA) pathway, and the microRNA (miRNA) pathway (Fig. 1C).

4.3.1.1 The exogenous siRNA pathway

The exogenous siRNA pathway acts as the immediate and most potent responder to active viral replication. Most arboviruses, including flaviviruses and alphaviruses, generate long dsRNA intermediates during their replicative cycles. This dsRNA acts as the primary PAMP for the siRNA pathway (Sánchez-Vargas et al., 2009; Blair, 2011; Feng et al., 2026).

The recognition mechanism is initiated by Dicer-2 (Dcr2), an RNase III enzyme that detects cytoplasmic dsRNA. In concert with the dsRNA-binding protein R2D2, Dcr2 cleaves the long dsRNA substrate into siRNAs, typically 21 nucleotides in length with characteristic 2-nucleotide 3′ overhangs (Wang et al., 2006). Following cleavage, siRNAs are loaded into the RNA-induced silencing complex (RISC), where the key effector protein, Argonaute-2 (Ago2), retains the guide strand. This guide strand directs RISC to complementary viral RNA, enabling Ago2's endonuclease activity to cleave the target, thereby dicing the viral genome or silencing viral gene expression (van Rij et al., 2006) (Fig 1C).

The biological significance of this pathway is profound. *Ae. aegypti* depleted of Dcr2 and Ago2 exhibit greater increases in viral RNA and infectious titers, transforming typically persistent, non-lethal infections into pathogenic ones. While this phenotype is consistently observed with alphaviruses such as CHIKV, SINV, and MAYV, DENV infections in Dcr2 mutants have yielded conflicting phenotypes across different studies (Sanchez-Vargas et al., 2009; Dong and Dimopoulos, 2023; Merkling et al., 2023; Samuel et al., 2023).

4.3.1.2 The piRNA pathway and endogenous viral elements

While historically characterized as a germline defense against transposable elements, the piRNA pathway has emerged as a critical somatic antiviral mechanism in *Aedes* mosquitoes (Blair, 2019). Unlike siRNAs, piRNAs are 24–30 nucleotides long and are processed in a Dicer-independent manner, loaded instead onto PIWI-clade Argonaute proteins (Ago3 and Piwi5–7 in *Aedes*) (Blair, 2019) (Fig 1C). The interplay between the piRNA pathway and viral infection is dynamic. Upon infection with Semliki Forest Virus (SFV), DENV, or CHIKV, mosquitoes produce virus-derived piRNAs (vpiRNAs) (Schnettler et al., 2013; Miesen et al., 2016; Wang et al., 2018; Fiorillo et al., 2022). In *Aedes*, the presence of viral DNA (vDNA) forms during infection suggests a mechanism where viral genomes are reverse-transcribed and potentially integrated or used as templates for piRNA biogenesis, reinforcing the host's long-term defense posture (Goic et al., 2016; Marconcini et al., 2021; Rodriguez-Andres et al., 2023).

A profound evolutionary insight into vector immunity is the discovery of Endogenous Viral Elements (EVEs), fragments of non-retroviral RNA virus genomes that have been reverse-transcribed and integrated into the mosquito genome (Palatini et al., 2017; Houé et al., 2019; Crava et al., 2021; Qu et al., 2023). In *Ae. aegypti* and *Ae. albopictus*, EVEs are frequently enriched within piRNA clusters and transcribed into antisense piRNAs that target cognate viral sequences. This provides a form of heritable, sequence-specific immunity that may pre-arm the mosquito against specific viral lineages (Tassetto et al., 2019).

Notably, *Culex* mosquitoes display distinct vpiRNA profiles and fewer EVEs despite an expansion of PIWI gene families, suggesting genus-specific divergence in deep-layer immune strategies. While the direct antiviral impact of the piRNA pathway is still a subject of debate (Samuel et al., 2018), correlations between PIWI protein expression and viral suppression suggest a functional role (Miesen et al., 2015). Furthermore, the observation that viral suppressors of RNAi (VSRs), such as FHV B2, can reduce vpiRNA accumulation suggests that viral countermeasures targeting dsRNA may broadly impair multiple small RNA-mediated immune mechanisms (Morazzani et al., 2012).

4.3.1.3 miRNA pathways

The antiviral roles of microRNAs (miRNAs) in insects remain less clearly defined than the siRNA or piRNA pathways (Bonning and Saleh, 2021). In mammals, miRNAs can establish antiviral states through transcriptional regulation (Cullen, 2011), whereas insects miRNAs often serve as a battleground for viral exploitation. Several mosquito-borne viruses have evolved to encode their own miRNAs, or manipulate host miRNAs, to promote viral persistence and evade the immune response (Hussain et al., 2008; Zhu et al., 2013; Zhu et al., 2016; Singh et al., 2014). A notable example is found in the Kunjin virus, where a viral miRNA derived from the 3' UTR (KUN-miR-1) targets the host transcription factor GATA4. By downregulating this factor, the virus indirectly enhances its own replication, illustrating how viruses can subvert host regulatory networks (Hussain et al., 2012). Therefore, the miRNA pathway in mosquitoes represents a complex landscape where the line between host defense and viral facilitation is frequently blurred.

4.3.2 Conserved signaling pathways: Toll, IMD, and JAK/STAT

Beyond nucleic acid recognition, mosquitoes employ broad-spectrum signaling pathways—Toll, IMD, and JAK/STAT—that were originally characterized for their roles in antibacterial and antifungal defense but have been co-opted for antiviral immunity (Kingsolver and Hardy, 2012).

4.3.2.1 The Toll pathway

The Toll pathway is homologous to the vertebrate Toll-like receptor (TLR) system and plays a dominant role in restricting flaviviruses, particularly DENV and ZIKV, in *Ae. aegypti* (Kumar et al., 2018; Xi et al., 2008).

Activation begins when the cytokine-like molecule Spätzle binds the transmembrane Toll receptor, triggering an intracellular cascade involving Myd88, Tube, and the kinase Pelle (Saucereau et al., 2022). Pelle phosphorylates Cactus (an IκB homolog), causing its degradation and releasing the transcription factor Rel1. Rel1 translocates to the nucleus to drive the expression of antimicrobial peptides (AMPs) such as defensins and cecropins, as well as other antiviral effectors (Tikhe and Dimopoulos, 2021) (Fig. 1C).

RNAi-mediated silencing of Cactus results in constitutive activation of the pathway and a significant reduction in DENV titers (Xi et al., 2008; Kumar et al., 2018). Conversely, silencing Rel1 renders mosquitoes hypersusceptible. The pathway's efficacy appears virus-specific: while potent against DENV, it shows limited impact on certain alphaviruses, such as CHIKV, suggesting distinct recognition mechanisms or viral evasion strategies (Tikhe and Dimopoulos, 2021).

4.3.2.2 The immune deficiency (IMD) pathway

The IMD pathway, parallel to mammalian TNF-α signaling, is activated by Peptidoglycan Recognition Proteins (PGRPs) like PGRP-LC and PGRP-LE, culminating in the activation of the transcription factor Rel2 (Russell et al., 2021; Tikhe and Dimopoulos, 2021) (Fig. 1C). While the direct antiviral role of IMD is less pronounced than that of Toll for DENV, it is essential for regulating the mosquito microbiota. Since the commensal microbiota significantly influences basal immune tone, the IMD pathway indirectly modulates vector competence. For example, *Wolbachia*-mediated viral blocking involves the upregulation of genes associated with the IMD pathway (e.g., PGRP-LE, Rel2) (Mushtaq et al., 2024).

4.3.2.3 The JAK/STAT pathway and the Vago-mediated cross-talk

The Janus Kinase/Signal Transducers and Activators of Transcription (JAK/STAT) pathway represents a critical systemic antiviral mechanism, particularly effective in the midgut tissue (Souza-Neto et al., 2009; Kingsolver and Hardy, 2012) (Fig. 1C). A unique feature of mosquito immunity is the molecular bridge between the RNAi and JAK/STAT pathways (Trammell et al., 2022). When Dicer-2 detects viral dsRNA, it not only processes siRNAs but also induces the secretion of a cytokine-like protein called Vago (Kingsolver and Hardy, 2012). Vago functions as a secreted signal that binds to JAK/STAT receptors on neighboring cells (Paradkar et al., 2012; Paradkar et al., 2014). Interestingly, studies indicate that Vago activation of JAK/STAT may occur independently of the canonical receptor Domeless (Dome) but remains dependent on the Janus Kinase Hopscotch (Hop). The activation of Hop leads to the phosphorylation and dimerization of STAT transcription factors, which enter the nucleus to upregulate virus-restriction factors. This pathway is vital for controlling DENV and WNV in *Aedes* and WNV in *Culex* (Souza-Neto et al., 2009; Anglero-Rodriguez et al., 2017). The ability of the pathway to propagate an antiviral state to uninfected cells makes it a key determinant of whether an infection remains localized or disseminates systemically. However, recent studies using Vago1 (VLG-1) mutants in Ae. aegypti found a modest proviral effect during DENV and ZIKV infections, suggesting a complex, dual role for Vago in viral pathogenesis (Couderc et al., 2025).

4.3.3 The interplay of host-virus genotypes in mosquito antiviral immunity

The functional efficacy of mosquito antiviral immunity is not a static trait of the species; rather, it is a dynamic outcome dictated by the specific genetic compatibility between the host and the viral strain. This phenomenon, categorized as genotype-by-genotype interaction, suggests that susceptibility and transmission potential are not fixed traits of a species, but vary significantly based on the unique pairing of host and pathogen genotypes. For instance, The African lineage of ZIKV exhibits inherently higher infectivity and transmission rates in *Ae aegypti* compared to the Asian lineage, a difference driven by accelerated intra-host infection kinetics (Aubry et al., 2021; Caldwell et al., 2024). This strain-specificity is further underscored by Dabo et al. (2024), whose work on African *Ae. aegypti* populations revealed that DENV susceptibility is highly heterogeneous and fundamentally governed by the specific pairing of the mosquito colony and the viral strain. Expanding on the molecular basis of these interactions, Torii et al. (2025) recently identified that the superior transmissibility of certain ZIKV lineages is a polygenic trait, where specific combinations of structural and non-structural viral genes interact with the mosquito's physiological barriers to determine the success of systemic dissemination. Collectively, these studies emphasize that even subtle genetic variations can pivot the mosquito's status from a refractory host to a highly competent vector, highlighting the complexity of antiviral defense in natural populations.

4.4 Tissue and physical defense

Viral infection in mosquitoes proceeds through distinct anatomical stages, each with its own barriers and defensive responses. At each step, the virus encounters distinct physical barriers and tissue-specific immune landscapes that determine vector competence (Franz et al., 2015) (Fig. 1A).

4.4.1 Physical barriers

The midgut is the first site of interaction between the virus and the vector. After ingestion, virus-infected blood is encased in the PM, which filters macromolecules and microbes. A well-formed PM can delay or reduce viral contact with the epithelium; its composition and thickness vary across species, feeding regimes, and age. The midgut epithelium has apical brush borders and tight junctions that limit paracellular passage, and below the epithelium lies the basal lamina (BL), a collagen-rich extracellular matrix that is particularly difficult for viruses to cross (Carpenter and Clem, 2023).

Together, these structures make up the midgut infection barrier (MIB) and midgut escape barrier (MEB). The MEB is particularly critical; the pore size of the unmodified basal lamina (~9–12 nm) is often smaller than the diameter of flaviviruses (~50 nm) and alphaviruses (~70 nm) (Kantor et al., 2018). Escape is thought to occur during the midgut distension associated with blood feeding, which, combined with the activity of Matrix Metalloproteinases (MMPs), transiently disrupts the laminin-collagen network, allowing virions to leak into the hemocoel (Dong et al., 2017) (Fig. 1A).

Once viruses traverse the MEB and enter the hemocoel, they must disseminate to distal tissues. The tracheal system, muscles, and hemolymph present additional physical and immunological hurdles. The salivary glands are encased in a basal lamina akin to the midgut; salivary gland infection (SGIB) and escape barriers (SGEB) regulate whether viruses colonize glandular epithelial cells and are subsequently secreted into saliva (Sanchez-Vargas et al., 2021) (Fig. 1A). Specific lobes of the salivary glands (e.g., medial vs. lateral) exhibit differential susceptibility to viruses like DENV and CHIKV, likely driven by the differential expression of basal lamina proteins and receptors (Bhattacharjee et al., 2023).

4.4.2 Tissue-specific defense

Beyond physical barriers, distinct tissues deploy specialized immune repertoires. The midgut epithelium deploys RNAi robustly upon viral contact, with Dicer-2 rapidly processing viral dsRNA and siRNA-RISC curtailing replication. Toll and IMD maintain a baseline antimicrobial tone tuned by the microbiome; shifts in bacterial composition alter susceptibility to viruses. Gut-specific lectins (e.g., C-type lectins) and receptor proteins mediate virus binding and entry; their expression patterns shape vector specificity.

The fat body, an analog of the vertebrate liver and adipose tissue, is a hub for systemic immune responses, nutrient storage, and detoxification. Viral infection can trigger JAK/STAT and NF-kB pathways in the fat body, leading to the production of antimicrobial peptides, stress proteins, and metabolic reprogramming that deprives viruses of lipid and nucleotide precursors (Cardoso-Jaime et al., 2025). Changes in cholesterol trafficking—particularly in *Wolbachia*-infected *Ae. aegypti*—have been implicated in viral inhibition, as viruses depend on cholesterol-rich membranes for replication complexes (Mushtaq et al., 2024).

Hemocytes orchestrate systemic surveillance (Fig. 1A). They produce cytokines and growth factors, phagocytose infected debris, and interact with the extracellular matrix to maintain tissue integrity. Their role in viral control is context dependent: hemocytes can clear virions, yet in some settings facilitate covert viral dissemination to secondary tissues such as the salivary glands (Hall et al., 2025; Leite et al., 2021). Hemocytes have a higher basal expression of antiviral genes compared to the fat body, suggesting they are a primary site of immune sensing in the hemocoel (Cardoso-Jaime et al., 2025).

Salivary glands show distinctive immune gene expression and small RNA responses, which can limit viral loads before saliva secretion (Dong et al., 2025). Salivary gland epithelial cells produce unique secretory profiles of proteins (e.g., Sialokinin, Apyrase) that modulate the vertebrate host environment; some of these proteins may influence viral egress, stability, or aggregation (Lefteri et al., 2022). Virus-induced changes in saliva composition can alter feeding behavior and bite duration, a feedback loop that influences transmission probability.

4.5 Ecology and evolution of antiviral defense

Mosquito antiviral defense is not static; it evolves under the pressures of pathogen load, host availability, climate, and control measures. Selection can favor mosquitoes with stronger barriers in environments with high virus prevalence, but trade-offs arise (Lambrechts, 2011; Contreras-Garduño et al., 2014). Enhanced immunity may come at the cost of reproductive success or longevity, reducing fitness in virus-scarce contexts (Gabrieli et al., 2021). Viral evolution counters mosquito defense, optimizing receptor usage and replication rates. For instance, a DENV-2 clade replacement event in Nicaragua has been linked to a replicative advantage of the replacing clade in the native mosquito vector, *Ae. aegypti* (Quiner et al., 2014).

As mosquitoes expand into new regions due to climate change and urbanization, they encounter novel microbiomes and temperature regimes, shifting the balance of defense and competence. The extrinsic incubation period (EIP), a key determinant of transmission, responds nonlinearly to temperature and immune activity (Abbasi, 2025). Larval nutritional stress can result in smaller adults with compromised midgut barriers, potentially increasing vector competence for viruses like ZIKV and WNV, despite reduced longevity (Herd et al., 2021).

5. Viral countermeasures to mosquito immunity

Mosquito innate immunity exerts strong selective pressure on viruses, shaping their evolution. To persist and transmit efficiently, arboviruses have evolved strategies to evade or modulate host defenses without completely suppressing them, maintaining a balance between replication and vector survival (Bonning and Saleh, 2021; Myles et al., 2008; Dong and Dimopoulos, 2023). Here, we focus on viral countermeasures, drawing also on insights from better-characterized systems (Fig. 1).

5.1 Anti-siRNA pathway mechanisms

RNAi represents a major antiviral barrier for RNA viruses. To overcome this defense, many viruses encode viral suppressors of RNAi (VSRs) that interfere with different steps of the siRNA pathway, thereby promoting viral replication and persistence (Fig. 2). VSRs appear to have emerged relatively recently in evolutionary terms and display striking diversity, with little or no detectable sequence or structural conservation despite occurring across viruses infecting a broad range of hosts. This diversity indicates that RNAi suppression has evolved independently multiple times through distinct molecular solutions and reflects the intense evolutionary arms race between viruses and their hosts (Bonning and Saleh, 2021).

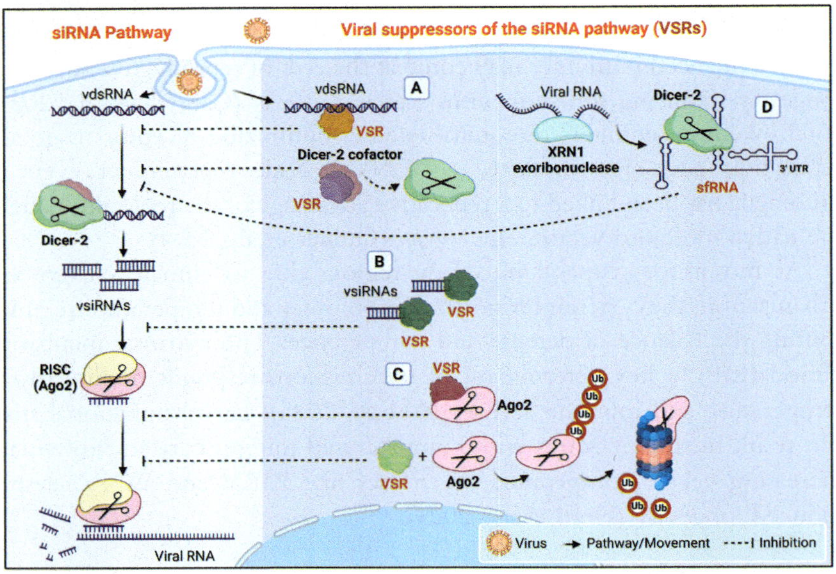

Fig. 2 Major mechanisms of action of viral suppressors of the siRNA pathway (VSRs). Viruses encode suppressor proteins that interfere with the host siRNA pathway at multiple stages. (A) VSRs can inhibit viral RNA recognition and the processing of viral double-stranded RNA (vdsRNA) intermediates. (B) VSRs may block RNA-induced silencing complex (RISC) assembly through sequestration of siRNAs. (C) VSRs can target Argonaute (Ago) proteins, thereby preventing RISC-mediated degradation of viral RNAs. (D) Some orthoflaviviruses produce noncoding sub-genomic flaviviral RNAs (sfRNAs) derived from incomplete degradation of genomic RNA by the cellular exoribonuclease XRN1. These structured elements from the 3′ untranslated region (UTR) act as decoy VSRs by engaging Dicer-2 and diverting it from processing viral dsRNA.

Since the first VSRs were identified in plant viruses (Anandalakshmi et al., 1998; Brigneti et al., 1998; Kasschau and Carrington, 1998), numerous RNAi-suppressive factors have been described. However, detailed mechanistic insight is available for only a subset of these proteins. This limitation largely arises from the multifunctional nature of many VSRs, which often fulfill additional essential roles in the viral life cycle, including genome replication, virion assembly, proteolytic processing, intracellular trafficking, transmission, and transcriptional regulation (Burgyán and Havelda, 2011). Consequently, genetic or functional disruption of VSRs frequently compromises viral viability, complicating efforts to dissect their specific contributions to RNAi antagonism.

Collectively, studies across plants, vertebrates, and invertebrates indicate that VSRs can target nearly every step of the silencing pathway, including viral RNA recognition, Dicer processing, RISC assembly, and Argonaute-mediated RNA targeting (Singh et al., 2009; Bivalkar-Mehla et al., 2011; Burgyán and Havelda, 2011; Song et al., 2011; Ali et al., 2015). However, given the focus on mosquito immunity, only the main mechanisms are summarized here (Fig. 2).

Inhibition of RNAi can occur at early stages by preventing recognition and processing of viral dsRNA by Dicer, for example through dsRNA binding or interference with host cofactors (Fig. 2A) (Deleris et al., 2006; Merai et al., 2006; Haas et al., 2008). Additional strategies include sequestration of siRNAs, thereby blocking their incorporation into the RISC (Fig. 2B) (Silhavy et al., 2002; Vargason et al., 2003; Ye et al., 2003; Chao et al., 2005; Lingel et al., 2005; Csorba et al., 2007; Ding and Voinnet, 2007; Aliyari et al., 2008; Chen et al., 2008), and direct targeting of Argonaute proteins, which disrupts slicing activity or promotes their degradation (Fig. 2C) (Pazhouhandeh et al., 2006; Zhang et al., 2006; Baumberger et al., 2007; Bortolamiol et al., 2007).

In insect systems, these mechanisms have been primarily characterized in viruses infecting *Drosophila melanogaster*. For instance, the 1A protein of Drosophila C virus (DCV) inhibits Dicer-2–mediated processing (van Rij et al., 2006), while the 1A protein of Cricket paralysis virus (CrPV) targets Ago2, reducing RNAi activity (Nayak et al., 2018; Watanabe et al., 2017). However, RNAi inhibition in arboviruses infecting mosquitoes remains less well characterized, with relatively few examples described.

Other mechanisms, such as modulation of host microRNAs (Varallyay et al., 2010), interference with epigenetic pathways (Wang et al., 2003; Bisaro, 2006), or the use of multiple VSRs (Bennasser et al., 2005;

Haasnoot et al., 2007; Karjee et al., 2010; Fabozzi et al., 2011; Aqil et al., 2013; Cui et al., 2015), have been reported mainly in plant and mammalian viruses and are not discussed here in detail.

5.2 VSRs described in mosquito-associated viruses

Comparatively little is known about viral strategies that suppress innate immune responses in mosquito-associated viruses. In many cases, mechanistic studies have been conducted primarily in vertebrate cell systems, leaving major gaps in our understanding of how these countermeasures operate in mosquito hosts. This represents a largely unexplored but promising area for future research, particularly given the biological differences between vertebrate and mosquito antiviral pathways.

5.2.1 Mosquito orthoflaviviruses' VSRs

Early studies in mammalian cells identified the DENV protein NS4B as a viral suppressor of RNAi in all four DENV serotypes (Kakumani et al., 2013). Using deletion mutants, the authors demonstrated that RNAi suppression is mediated by the middle and C-terminal regions of NS4B, and is independent of the well-established interferon antagonism of NS4B. Mechanistically, NS4B did not bind dsRNA directly but interfered with Dicer-mediated processing (Kakumani et al., 2013).

More recent work demonstrated that the DENV-2 NS2A protein also functions as a VSR across multiple experimental systems, including in vitro assays, mammalian cells, mice, and *Aedes* mosquito cells (Qiu et al., 2020). NS2A suppresses Dicer-dependent siRNA production, most likely by sequestering dsRNA and preventing its cleavage. Viruses engineered to lack NS2A-mediated RNAi suppression triggered robust production of virus-derived siRNAs in differentiated mammalian cells and exhibited severe replication defects in mosquito and mammalian cells as well as in vivo in mice. These defects were rescued in RNAi-deficient backgrounds, demonstrating that impaired replication resulted specifically from enhanced RNAi activity rather than unrelated cellular processes (Qiu et al., 2020).

NS2A proteins from multiple flaviviruses, including other DENV serotypes, JEV, WNV, and ZIKV, also suppress RNAi in vitro. JEV NS2A antagonizes RNAi during infection in both mosquito and human cells, suggesting that NS2A-mediated RNAi suppression represents a conserved strategy across medically relevant orthoflaviviruses (Qiu et al., 2020). Structurally and functionally, NS2A shares similarities with other characterized VSRs, such as the 3 A protein of Enterovirus A71 (Qiu et al., 2017).

Additional evidence for RNAi suppression in flaviviruses comes from studies of C proteins. Using recombinant Sindbis viruses expressing YFV capsid proteins in *Ae. aegypti* mosquitoes, Samuel et al. (2016) found that the YFV capsid protein exhibited robust RNA silencing suppressor activity and bound dsRNA, thereby inhibiting Dicer-mediated cleavage. Similar phenotypes were observed with C proteins from ZIKV, WNV, and DENV-2 in chimeric systems, suggesting that this activity may be conserved across orthoflaviviruses (Samuel et al., 2016). However, because these experiments relied on heterologous recombinant systems, their relevance to natural flavivirus infections in mosquitoes remains to be fully validated.

5.2.2 Mosquito alphaviruses' VSRs

Several alphaviruses may encode VSRs, with the C protein as a key candidate. In Semliki Forest virus (SFV), C suppresses RNAi in mammalian cells and in mosquito cells (Aag2), likely by blocking dsRNA processing or vsiRNA loading into RISC. Loss of C-mediated suppression reduces viral replication, which can be rescued by inhibiting RNAi, highlighting its functional importance. Similar activity has been observed in Sindbis virus, suggesting possible conservation across alphaviruses (Qian et al., 2020).

Additional studies propose that CHIKV nsP2 and nsP3 may function as VSRs. Using a GFP-based RNAi sensor cell line, expression of these proteins partially restored GFP expression, consistent with RNAi suppression (Mathur et al., 2016). These findings were supported in Aag2 and mammalian (HEK293T) cells and by EMSA assays. However, because these experiments did not involve authentic viral infection, their biological relevance remains uncertain and requires in vivo confirmation.

5.2.3 Mosquito orthobunyaviruses' VSRs

Evidence for VSR activity in orthobunyaviruses remains limited. In La Crosse virus (LACV), overexpression of the NSs protein reduced siRNA-mediated silencing in human 293 T cells, suggesting RNAi suppression (Soldan et al., 2005). Similarly, studies with Bunyamwera virus indicated that NSs is dispensable for replication in mosquito cell lines with impaired RNAi, indirectly supporting a potential VSR role (Szemiel et al., 2012). However, definitive evidence for RNAi suppression by NSs in either vertebrate hosts or arthropod vectors is lacking.

5.2.4 VSRs in other mosquito viral genera

Nodamura virus (NoV), a mosquito-transmitted alphanodavirus, encodes the B2 protein, a potent inhibitor of RNAi in mammalian cells

(Sullivan and Ganem, 2005). B2 binds both Dicer substrates and processed siRNAs, inhibiting RNAi at multiple stages. In vitro, B2 directly blocks Dicer cleavage of dsRNA without additional host factors. While this activity has not yet been demonstrated in mosquitoes for NoV, B2 proteins from related nodaviruses, particularly FHV, are well-established VSRs in insects (Aliyari et al., 2008; Chao et al., 2005; Lingel et al., 2005; Qi et al., 2011; Qi et al., 2012; Schuster et al., 2014).

Another example is Culex Y virus (CYV) (family *Birnaviridae*, genus *Entomobirnavirus*), a bisegmented dsRNA MSV isolated from *Culex pipiens* (Marklewitz et al., 2012). CYV infection suppresses RNAi triggered by both dsRNA and synthetic siRNAs in *D. melanogaster* S2 cells. Functional complementation assays demonstrated that the VP3 proteins of CYV and Drosophila X virus act as bona fide VSRs by restoring replication of a B2-deficient FHV replicon (van Cleef et al., 2014). These findings indicate that mosquito-specific viruses have evolved sophisticated RNAi antagonists that may influence vector competence and interactions with co-infecting pathogens.

5.3 Decoy RNAs as antiviral countermeasures

Orthoflaviviruses produce noncoding subgenomic flaviviral RNAs (sfRNAs) derived from the structured 3′ untranslated region (UTR), generated through incomplete degradation of genomic RNA by the cellular exoribonuclease XRN1 (Silva et al., 2010; Funk et al., 2010). Accumulating evidence from both vertebrate hosts and arthropod vectors indicates that sfRNA is a key viral strategy for evading host antiviral responses. In vertebrate systems, sfRNA broadly antagonizes innate immunity by interfering with interferon induction and downstream antiviral effector functions, largely through sequestration of host immune and RNA-binding proteins (Bidet et al., 2014; Manokaran et al., 2015; Soto-Acosta et al., 2018). Collectively, sfRNAs contribute to viral replication, transmission, and overall epidemiological fitness (Manokaran et al., 2015; Filomatori et al., 2017; Yeh et al., 2023).

In mosquitoes, sfRNAs suppress antiviral responses by sequestering components of the RNA interference (RNAi) pathway, consistent with a decoy-based VSR strategy (Fig. 2D) (Pijlman et al., 2008; Schnettler et al., 2012; Moon et al., 2015; Chen et al., 2024). sfRNAs from WNV and DENV suppress RNAi in both arthropod and mammalian cells. Their stable stem–loop structures resemble Dicer substrates, supporting a model in which sfRNA acts as a molecular decoy (Schnettler et al., 2012).

In vitro, sfRNA competitively inhibits Dicer-mediated cleavage of long dsRNA in a concentration-dependent manner, whereas RNAi triggered by transfected siRNAs remains largely unaffected, consistent with inhibition upstream of RISC (Schnettler et al., 2012). In mosquito cells, sfRNA enhances viral replicon replication, underscoring its functional relevance in the vector context (Schnettler et al., 2012; Schnettler et al., 2013). Consistently, DENV and Kunjin virus sfRNAs repress RNAi in both human cells and *Cx. quinquefasciatus*, with evidence of direct interaction with Dicer and Ago2 (Moon et al., 2015). More recently, ZIKV sfRNA was shown to antagonize RNAi in mammalian cells by binding components of the RNA-induced silencing complex and competing with siRNAs, indicating that sfRNA-mediated interference is conserved among flaviviruses and can target multiple steps of the pathway depending on the cellular context (Chen et al., 2024).

Comparable decoy-based mechanisms have been proposed for alphaviruses and RVFV, where specific genomic regions or defective interfering particles generate abundant siRNAs that are inefficiently loaded into Ago2, thereby diverting the RNAi machinery from productive antiviral targets (Siu et al., 2011; Sabin et al., 2013).

Although some authors have questioned the strength of evidence for sfRNA as a classical VSR in mosquitoes (Yeh and Pompon, 2018; Samuel et al., 2018), multiple studies indicate that sfRNA enhances mosquito transmission (Göertz et al., 2016; Pompon et al., 2017; Göertz et al., 2019; Slonchak et al., 2020; Yeh et al., 2023) and modulates immune signaling pathways (Slonchak et al., 2020), including suppression of Toll pathway components in salivary glands (Göertz et al., 2016; Pompon et al., 2017; Yeh and Pompon, 2018). Together, these findings support the idea that sfRNA contributes to viral fitness and transmission by dampening mosquito immune responses.

5.4 Balancing RNAi suppression, persistence, and alternative evasion strategies by arboviruses

Substantial evidence supports the existence of VSRs and RNAi antagonistic strategies across diverse RNA virus families and host taxa, yet definitive proof that arboviruses encode canonical VSR genes remains debated (Liu et al., 2019). In insect viruses, strong VSR activity correlates with increased virulence during natural infections (Nayak et al., 2010). However, potent RNAi suppression does not necessarily translate into severe host pathology or loss of persistence (O'Neal et al., 2014; Schnettler et al., 2012).

For example, DCV, CrPV, and FHV encode robust VSRs yet establish persistent infections in their insect hosts (Nayak et al., 2010; van Rij et al., 2006; Wang et al., 2006). Conversely, persistently infecting arboviruses may encode moderately potent VSRs that allow sufficient viral replication despite efficient mosquito RNAi machinery (Blair, 2011). Persistent infection likely reflects a finely tuned balance between viral replication kinetics, VSR activity, and vector antiviral defenses (O'Neal et al., 2014).

In addition to direct modulation of small RNA pathways, alternative evasion strategies exist. One involves physical shielding of viral replication intermediates within membrane-associated replication complexes (den Boon et al., 2010). (+) RNA viruses, including orthoflaviviruses and alphaviruses, induce extensive rearrangements of intracellular membranes, forming specialized replication organelles. These vesicular structures compartmentalize (-) RNA intermediates while facilitating efficient synthesis and export of positive-strand genomes. This compartmentalization likely limits dsRNA accessibility to cytoplasmic sensors and dsRNA-activated antiviral pathways, reducing the efficacy of RNAi and other innate immune responses (den Boon et al., 2010). Such spatial shielding may operate in parallel with, or partially substitute for, classical VSR activity, supporting persistent infections in mosquito vectors.

5.5 Evading and interfering with canonical immune pathways

While the siRNA pathway represents a robust and broadly conserved antiviral defense in mosquitoes, growing evidence supports contributory roles for other evolutionarily conserved immune pathways, including JAK–STAT, Imd, and Toll, in vector responses to viral infection (Kingsolver and Hardy, 2012). In contrast to RNAi antagonism, relatively little is known about viral countermeasures that target these pathways.

Interference with conserved immune signaling by arboviruses has been documented in various systems. For example, JEV inhibits JAK–STAT signaling by blocking STAT phosphorylation in *Ae. albopictus* C6/36 cells (Lin et al., 2004). More recently, it has also been reported to impair host DNA replication and to suppress Toll-like receptors as well as downstream components of the Toll pathway in the same system (Gao et al., 2025). In DENV models, viral infection broadly reshapes immune gene expression in mosquito cells. In the *Ae. aegypti* Aag2 cell line, DENV infection downregulates numerous immunity-related genes and impairs AMP induction following secondary bacterial challenge (Sim and Dimopoulos, 2010). Consistently, early suppression of several AMP genes has been reported in DENV-infected mosquitoes (Ramirez and Dimopoulos, 2010), suggesting that canonical immune signaling pathways are susceptible to viral manipulation.

Compelling evidence for viral suppression of canonical immune pathways comes from studies in *Ae. aegypti* infected with DENV-2. Pompon et al. (2017) examined the effects of the sfRNA on Toll, Imd, and JAK–STAT signaling in mosquito salivary glands and found that the sfRNA from a high epidemiological fitness isolate (PR6452) selectively inhibits the Toll pathway. This effect was mediated via transcriptional downregulation of immune components, notably Rel1a, a key transcriptional activator of Toll signaling. Rel1a expression was reduced 2.14-fold in salivary glands infected with PR6452 compared with a lower-fitness isolate. Downstream effector genes, including the antimicrobial peptide Cecropin G, were similarly suppressed, indicating broader attenuation of Toll-mediated immune activity. Although the precise molecular interactions between the viral sfRNA and host signaling components remain unresolved, these findings suggest that evolutionary variation in the sfRNA can modulate mosquito immune responses and enhance viral transmission by promoting immune suppression in salivary tissues (Pompon et al., 2017; Göertz et al., 2019; Slonchak et al., 2020; Yeh et al., 2023).

Evidence for viral antagonism of Toll-mediated immunity also comes from plant arboviruses in their insect vectors. In the brown planthopper *Sogatella furcifera*, a vector of Southern rice black-streaked dwarf virus (SRBSDV), a Toll-dependent response induces Defensin, which binds viral particles and restricts infection. SRBSDV counteracts this defense through the viral protein P7–1, which interacts with the E3 ligase SfREL to promote ubiquitination and degradation of the Toll adaptor MyD88, suppressing Defensin expression and enhancing viral replication (Jia et al., 2024). In the small brown planthopper *Laodelphax striatellus*, Toll signaling also restricts Rice stripe virus (RSV) infection through activation of LsToll, LsDorsal, and the E3 ligase LsPellino (He et al., 2020; He et al., 2024). However, RSV encodes antagonists that inhibit Toll signaling: NS4 blocks LsDorsal activation by interfering with LsMSK2, while NS3 promotes degradation of LsPellino via LsSOCS5, facilitating viral replication (Du et al., 2025). In the same vector, RBSDV activates the JAK–STAT pathway, inducing SOCS5 expression, which targets the anti-apoptotic protein BCL2 for proteasomal degradation and triggers apoptosis. Rather than limiting infection, this apoptosis appears to promote viral replication and persistence (Zhang et al., 2023).

5.6 Insect's strategy to counter viral suppression of antiviral immunity

In this continuous arms race between mosquito immune defenses and viruses, both partners are expected to evolve in parallel to achieve a balance between viral persistence and host survival, as predicted by the Red Queen theory (Van Valen, 1973). Consistent with this concept, a recent study in the *D. melanogaster* model identified a virus suppressor–interacting long non-coding RNA (VINR) that mediates a counter–counter-defense response to viral immune evasion (Zhang et al., 2020). VINR is accumulated upon recognition of the dsRNA-binding VSR encoded by DCV, the 1 A protein, and triggers a noncanonical innate immune signaling pathway. Mechanistically, VINR prevents the degradation of Cactin, leading to activation of a noncanonical antimicrobial response. VINR is required for the induction of antimicrobial peptide genes but is dispensable for antiviral RNAi. Accordingly, VINR-deficient flies exhibit increased susceptibility to both bacterial infection and DCV, highlighting an additional regulatory layer by which hosts can sense viral immune suppression and mount compensatory defenses (Zhang et al., 2020).

6. Implications for disease control and prevention

As described previously, mosquito–virus interactions occur under continuous reciprocal selective pressures that compel each component of the transmission system to adapt in order to maintain relative fitness (Dawkins and Krebs, 1979). In vector-borne diseases, this dynamic manifests as a persistent evolutionary arms race across a tripartite system encompassing the virus, the arthropod vector, and the vertebrate host, integrating hosts immune defenses, viral immune evasion strategies, and vector competence traits (Woolhouse et al., 2002; Lambrechts et al., 2010).

Selective pressures operating in any one component of this tripartite cycle can propagate across the transmission network, reshaping viral fitness landscapes, mosquito susceptibility, and vertebrate immunity. This multidimensional evolutionary environment contributes to the persistence of viral lineages, the emergence of novel variants, and temporal shifts in transmission dynamics (Weaver et al., 2021; Lewis et al., 2023). From a disease control perspective, intervention strategies therefore represent additional selective forces acting on pathogens, vectors, and vertebrates. Vaccination programs, antiviral therapies, insecticides, genetic control strategies, and environmental modifications may reduce disease burden in the short term but can also promote the selection of escape variants, resistance traits, or compensatory adaptations (Read et al., 2011; Little et al., 2012). Likewise, vector control measures may alter mosquito population structure, immune status, microbiota composition, or vector competence, with unintended consequences for pathogen transmission (Barillas-Mury et al., 2022; Andreazza et al., 2021). These evolutionary feedbacks highlight the limitations of static control approaches and reinforce the need for adaptive and integrated strategies.

Harnessing mosquito innate defenses represents a promising avenue for innovative interventions within this evolutionary framework. Field deployments of *Wolbachia* have achieved sustained reductions in dengue incidence in several urban settings (Mushtaq et al., 2024). Genetic approaches that enhance antiviral pathways—such as engineered activation of JAK/STAT signaling through overexpression of Dome or Hop, or upregulation of RNA interference components—have also been shown to reduce vector competence (Dong et al., 2022; Jupatanakul et al., 2017). Gene drive systems offer the potential to disseminate such refractory traits throughout mosquito populations, although their implementation requires careful design to minimize unintended ecological effects and the risk of viral evolutionary escape (Esvelt et al., 2014).

Anticipating evolutionary responses is therefore essential for sustainable disease control in vector-borne systems. Surveillance strategies integrating genomic monitoring of pathogens and vectors, detection of resistance markers, and ecological modeling can provide early warning of adaptive change. Interventions that impose diversified or rotating selective pressures, target conserved biological processes, or combine complementary approaches may reduce the likelihood of rapid evolutionary escape. Ultimately, durable disease prevention requires not only immediate efficacy but also long-term evolutionary resilience, recognizing that virus–vector–host systems are continuously co-evolving rather than approaching a stable equilibrium.

7. Knowledge gaps, challenges, and future directions

Despite advances in characterizing mosquito antiviral defenses, several critical knowledge gaps persist. The precise antiviral roles of virus-derived piRNAs (vpiRNAs) remain unresolved, particularly in *Culex* vectors (Abel et al., 2023; Rückert et al., 2019), where piRNA function and biogenesis are less well defined than in *Aedes*. Although viral immune suppressors have been proposed in mosquitoes, definitive functional evidence for viral antagonism remains limited, and it is unclear whether arboviruses encode specific inhibitors of major immune signaling pathways such as Toll, IMD, and JAK/STAT.

While RNAi is recognized as a central antiviral mechanism, the relative contribution of these signaling pathways varies by virus, tissue, and species, and their downstream effectors and cross-talk across environmental contexts remain incompletely characterized (Sim and Dimopoulos, 2010; Cheng et al., 2016). Beyond classical antimicrobial peptides, additional tissue-specific effectors—including metabolic enzymes, lipid transporters, and chaperones—warrant deeper exploration.

Our understanding of MSVs and their interactions with mosquito immune defenses is especially limited, including whether MSVs deploy immune evasion strategies analogous to arboviruses. Because vertical transmission is thought to be a major maintenance route for many MSVs and has been documented in numerous mosquito–virus systems, elucidating its contribution to viral persistence may provide insight into long-term host–virus co-adaptation (Peterson et al., 2024). Finally, integrating behavioral ecology, such as infection-induced changes in feeding and host-seeking, with molecular immune profiling will improve predictions of transmission under heterogeneous real-world conditions.

Beyond these core gaps, linking molecular immune mechanisms to functional outcomes at the organismal and field population levels remains a major challenge. Existing studies often use simplified laboratory models that do not capture the complexity of natural mosquito ecology, environmental variation, microbiota composition, and co-infections. Technical barriers remain for functional genetics in non-model mosquitoes and for the manipulation of ISVs, restricting mechanistic insight into natural virus–vector interactions. Comparative studies across mosquito genera and virus families will be essential to distinguish conserved versus lineage-specific immune strategies, as well as immune suppressors. Future efforts will therefore benefit from integrative approaches combining high-resolution omics, functional genetics, experimental evolution, and ecological field studies. Bridging these scales will be essential for translating mechanistic discoveries into robust, evolution-aware strategies for vector control and disease prevention, always under coordinated interdisciplinary efforts.

8. Conclusions

The relationship between the mosquito and the arbovirus is defined by a "delicate balance" between immune defense and tolerance (Lee et al., 2019). A mosquito immune system that is too effective results in the elimination of the virus, halting the transmission cycle. A system that is too permissive allows uncontrolled viral replication, killing the vector before transmission can occur. Evolution has thus selected for persistent infection—a state where the virus replicates at levels sufficient for transmission but sub-lethal to the mosquito, maintained by the regulatory breaks of the RNAi and other immune pathways. Vector competence and specificity emerge from the interplay of these defenses with viral strategies, microbiome composition, environmental conditions, and mosquito behavior. Leveraging these insights for control promises durable, biologically grounded reductions in arboviral disease, provided interventions anticipate evolutionary feedback and ecological complexity.

References

Abbasi, E., 2025. Climate change and vector-borne disease transmission: the role of insect behavioral and physiological adaptations. Integr. Org. Biol. 7, obaf011.

Abel, S.M., Hong, Z., Williams, D., et al., 2023. Small RNA sequencing of field Culex mosquitoes identifies patterns of viral infection and the mosquito immune response. Sci. Rep. 13, 10598. https://doi.org/10.1038/s41598-023-37571-6.

Ahola, T., McInerney, G., Merits, A., 2021. Alphavirus RNA replication in vertebrate cells. Adv. Virus Res. 111, 111–156. https://doi.org/10.1016/bs.aivir.2021.07.003.

Ali, P.S., John, J., Selvaraj, M., Kek, T.L., Salleh, M.Z., 2015. Nodamura virus B2 amino terminal domain sensitivity to small interfering RNA. Microbiol. Immunol. 59 (5), 299–304. https://doi.org/10.1111/1348-0421.12253.

Aliyari, R., Wu, Q., Li, H.W., Wang, X.H., Li, F., et al., 2008. Mechanism of induction and suppression of antiviral immunity directed by virus-derived small RNAs in Drosophila. Cell Host Microbe. 4, 387–397.

Alvarez, D.E., Lodeiro, M.F., Luduena, S.J., Pietrasanta, L.I., Gamarnik, A.V., 2005. Long-range RNA–RNA interactions circularize the dengue virus genome. J. Virol. 79, 6631–6643.

Anandalakshmi, R., et al., 1998. A viral suppressor of gene silencing in plants. Proc. Natl Acad. Sci. U S A 95, 13079–13084.

Andreazza, F., Oliveira, E.E., Martins, G.F., 2021. Implications of sublethal insecticide exposure and the development of resistance on mosquito physiology, behavior, and pathogen transmission. Insects 12, 917. https://doi.org/10.3390/insects12100917.

Anglero-Rodriguez, Y.I., MacLeod, H.J., Kang, S., Carlson, J.S., Jupatanakul, N., Dimopoulos, G., 2017. Aedes aegypti molecular responses to Zika virus: modulation of infection by the Toll and Jak/Stat immune pathways and virus host factors. Front. Microbiol. 8, 2050.

Anwar, M.N., Liu, H., Chen, W., Li, R., Luo, S., Liu, Q., 2022. The interactions of flaviviruses with cellular receptors: implications for virus attachment, internalization, and membrane fusion. Virology. https://doi.org/10.1016/j.virol.2022.02.001.

Aqil, M., Naqvi, A.R., Bano, A.S., Jameel, S., 2013. The HIV-1 Nef protein binds argonaute-2 and functions as a viral suppressor of RNA interference. PLoS One 8, e74472. https://doi.org/10.1371/journal.pone.0074472.

Aubry, F., Jacobs, S., Darmuzey, M., et al., 2021. Recent African strains of Zika virus display higher transmissibility and fetal pathogenicity than Asian strains. Nat. Commun. 12, 916. https://doi.org/10.1038/s41467-021-21199-z.

Azar, S.R., Roundy, C.M., Rossi, S.L., Huang, J.H., Leal, G., Yun, R., et al., 2017. Differential vector competency of Aedes albopictus populations from the Americas for Zika virus. Am. J. Trop. Med. Hyg. 97 (2), 330–339. https://doi.org/10.4269/ajtmh.16-0969.

Banho, C.A., de Carvalho Marques, B., Nascimento, O.B., Parra, M.C.P., Ferreira, M.V.M., Lemos, A.P., et al., 2025. Intrahost population dynamics of chikungunya virus in humans and naturally infected Aedes mosquitoes reveal transmission-driven diversity. Virus Evol. 12 (1), veaf102. https://doi.org/10.1093/ve/veaf102.

Barillas-Mury, C., et al., 2022. Understanding pathogen survival and transmission by arthropod vectors to prevent human disease. Science 377, eabc2757. https://doi.org/10.1126/science.abc2757.

Barr, J.N., Elliott, R.M., Dunn, E.F., Wertz, G.W., 2003. Segment-specific terminal sequences of Bunyamwera bunyavirus regulate genome replication. Virology 311, 326–338.

Baumberger, N., et al., 2007. The Polerovirus silencing suppressor P0 targets ARGONAUTE proteins for degradation. Curr. Biol. 17, 1609–1614.

Beaty, B.J., et al., 1981. Molecular basis of bunyavirus transmission by mosquitoes: role of the middle-sized RNA segment. Science 211, 1433–1435.

Beaty, B.J., Bishop, D.H., Gay, M., Fuller, F., 1983. Interference between bunyaviruses in Aedes triseriatus mosquitoes. Virology 127, 83–90.

Bennasser, Y., Le, S.Y., Benkirane, M., Jeang, K.T., 2005. Evidence that HIV-1 encodes an siRNA and a suppressor of RNA silencing. Immunity 22, 607–619. https://doi.org/10.1016/j.immuni.2005.03.010.

Bhattacharjee, S., Ghosh, D., Saha, R., Sarkar, R., Kumar, S., Khokhar, M., et al., 2023. Mechanism of immune evasion in mosquito-borne diseases. Pathogens 12.

Bidet, K., Dadlani, D., Garcia-Blanco, M.A., 2014. G3BP1, G3BP2 and CAPRIN1 are required for translation of interferon-stimulated mRNAs and are targeted by a dengue virus non-coding RNA. PLoS Pathog. 10, e1004242. https://doi.org/10.1371/journal.ppat.1004242.

Bisaro, D.M., 2006. Silencing suppression by geminivirus proteins. Virology 344, 158–168.

Bivalkar-Mehla, S., Vakharia, J., Mehla, R., Abreha, M., Kanwar, J.R., Tikoo, A., et al., 2011. Viral RNA silencing suppressors (RSS): novel strategy of viruses to ablate the host RNA interference (RNAi) defense system. Virus Res. 155, 1–9.

Blair, C.D., 2011. Mosquito RNAi is the major innate immune pathway controlling arbovirus infection and transmission. Future Microbiol. 6, 265–277.

Blair, C.D., 2019. Deducing the role of virus genome-derived PIWI-associated RNAs in the mosquito-arbovirus arms race. Front. Genet. 10, 1114.

Blitvich, B.J., Firth, A.E., 2015. Insect-specific flaviviruses: a systematic review of their discovery, host range, mode of transmission, superinfection exclusion potential and genomic organization. Viruses 7, 1927–1959. https://doi.org/10.3390/v7041927.

Bolling, B.G., Eisen, L., Moore, C.G., Blair, C.D., 2011. Insect-specific flaviviruses from Culex mosquitoes in Colorado, with evidence of vertical transmission. Am. J. Trop. Med. Hyg. 85 (1), 169–177. https://doi.org/10.4269/ajtmh.2011.10-0286.

Bonning, B.C., Saleh, M.C., 2021. The interplay between viruses and RNAi pathways in insects. Annu. Rev. Entomol. 66, 61–79. https://doi.org/10.1146/annurev-ento-033020-090410.

Bortolamiol, D., Pazhouhandeh, M., Marrocco, K., Genschik, P., Ziegler-Graff, V., 2007. The Polerovirus F box protein P0 targets ARGONAUTE1 to suppress RNA silencing. Curr. Biol. 17 (18), 1615–1621. https://doi.org/10.1016/j.cub.2007.07.061.

Bowden, T.A., Bitto, D., McLees, A., Yeromonahos, C., Elliott, R.M., Huiskonen, J.T., 2013. Orthobunyavirus ultrastructure and the curious tripodal glycoprotein spike. PLoS Pathog. 9, e1003374.

Bowen, J.M., Gunter, K., Lunel, A.M., Omoga, D.C.A., Jones, J.E., Giesel, H., et al., 2025. Probing orthobunyavirus reassortment using Bunyamwera and Batai viruses as models. PLoS Negl. Trop. Dis. 19 (5), e0013120. https://doi.org/10.1371/journal.pntd.0013120.

Bray, M., Monath, T.P., 2013. Arboviruses: epidemiology and ecology. In: Knipe, D.M., Howley, P.M. (Eds.), Fields Virology, 6th ed. Lippincott Williams & Wilkins, Philadelphia (PA), pp. 514–573. ISBN: 978-1-4511-0563-6.

Briese, T., Bird, B., Kapoor, V., Nichol, S.T., Lipkin, W.I., 2006. Batai and Ngari viruses: M segment reassortment and association with severe febrile disease outbreaks in East Africa. J. Virol. 80, 5627–5630.

Briese, T., Calisher, C.H., Higgs, S., 2013. Viruses of the family Bunyaviridae: are all available isolates reassortants? Virology 446, 207–216. https://doi.org/10.1016/j.virol.2013.07.030.

Brigneti, G., et al., 1998. Viral pathogenicity determinants are suppressors of transgene silencing in Nicotiana benthamiana. EMBO J. 17, 6739–6746.

Brinton, M.A., 1986. Replication of flaviviruses. In: Schlesinger, S., Schlesinger, M. (Eds.), The Togaviridae and Flaviviridae. Springer, New York, pp. 327–374.

Burgyán, J., Havelda, Z., 2011. Viral suppressors of RNA silencing. Trends Plant. Sci. 16, 265–272. https://doi.org/10.1016/j.tplants.2011.02.010. PMID:21439890.

Caldwell, J.M., Lambrechts, L., Rose, N.H., 2024. The role of vector population variation and climate in Zika virus transmission patterns in Africa: a modelling study. Lancet Planet. Health 8, e1020–e1029. https://doi.org/10.1016/S2542-5196(24)00276-6.

Calisher, C.H., Higgs, S., 2018. The discovery of arthropod-specific viruses in hematophagous arthropods: an open door to understanding the mechanisms of arbovirus and arthropod evolution? Annu. Rev. Entomol. 63, 87–103.

Caragata, E.P., Dong, S., Dong, Y., Simões, M.L., Tikhe, C.V., Dimopoulos, G., 2020. Prospects and pitfalls: next-generation tools to control mosquito-transmitted disease. Annu. Rev. Microbiol. 74, 455–475. https://doi.org/10.1146/annurev-micro-011320-025557.

Cardoso-Jaime, V., Tikhe, C.V., Tavadia, M., Dimopoulos, G., 2025. Aedes aegypti hemocytes mediate antiviral immunity. Int. J. Mol. Sci. 26.

Carpenter, A., Clem, R.J., 2023. Factors affecting arbovirus midgut escape in mosquitoes. Pathogens 12.

Carvalho, V.L., Long, M.T., 2021. Insect-specific viruses: an overview and their relationship to arboviruses of concern to humans and animals. Virology 557, 34–43. https://doi.org/10.1016/j.virol.2021.01.007.

Chao, J.A., Lee, J.H., Chapados, B.R., Debler, E.W., Schneemann, A., Williamson, J.R., 2005. Dual modes of RNA-silencing suppression by Flock House virus protein B2. Nat. Struct. Mol. Biol. 12, 952–957.

Chare, E.R., Gould, E.A., Holmes, E.C., 2003. Phylogenetic analysis reveals a low rate of homologous recombination in negative-sense RNA viruses. J. Gen. Virol 84, 2691–2703.

Charles, J., Hall, R.N., Kobayashi, D., et al., 2019. Phylogenetic analysis of insect-associated Tymoviridae and related taxa reveals insect-infecting lineages within plant virus families. Viruses 11, 853. https://doi.org/10.3390/v11090853.

Chen, H.Y., et al., 2008. Structural basis for RNA-silencing suppression by Tomato aspermy virus protein 2b. EMBO Rep. 9, 754–760.

Chen, X., Li, R.T., Chen, R.Y., Shi, P.D., Liu, Z.X., Lou, Y.N., et al., 2024. The subgenomic flaviviral RNA suppresses RNA interference through competing with siRNAs for binding RISC components. J. Virol. 98. https://doi.org/10.1128/jvi.01954-23. PMID:38289102. PMCID:PMC10878275.

Chen, R., Mukhopadhyay, S., Merits, A., Bolling, B., Nasar, F., et al., 2018. ICTV virus taxonomy profile: Togaviridae. J. Gen. Virol. 99, 761–762.

Cheng, G., Liu, Y., Wang, P., et al., 2016. Mosquito defense strategies against viral infection. Trends Parasitol. 32, 177–186.

Chmielewski, D., Schmid, M.F., Simmons, G., Jin, J., Chiu, W., 2022. Chikungunya virus assembly and budding visualized in situ using cryogenic electron tomography. Nat. Microbiol 7, 1270–1279.

Coffey, L.L., et al., 2008. Arbovirus evolution in vivo is constrained by host alternation. Proc. Natl Acad. Sci. USA 105, 6970–6975.

Coffey, L.L., Beeharry, Y., Bordería, A.V., Blanc, H., Vignuzzi, M., 2011. Arbovirus high fidelity variant loses fitness in mosquitoes and mice. Proc. Natl Acad. Sci. USA 108, 16038–16043. https://doi.org/10.1073/pnas.1111650108.

Coffey, L.L., Forrester, N., Tsetsarkin, K., Vasilakis, N., Weaver, S.C., 2013. Factors shaping the adaptive landscape for arboviruses: implications for the emergence of disease. Future Microbiol. 8, 155–176.

Coffey, L.L., Vignuzzi, M., 2011. Host alternation of chikungunya virus increases fitness while restricting population diversity and adaptability to novel selective pressures. J. Virol. 85, 1025–1035.

Contreras-Garduño, J., Rodríguez, M.C., Rodríguez, M.H., Alvarado-Delgado, A., Lanz-Mendoza, H., 2014. Cost of immune priming within generations: trade-off between infection and reproduction. Microbes Infect. 16, 261–267. https://doi.org/10.1016/j.micinf.2013.11.010.

Couderc, E., Crist, A.B., Daron, J., Varet, H., van Hout, F.A.H., Miesen, P., et al., 2025. Aedes aegypti VLG-1 challenges the assumed antiviral nature of Vago genes. BMC Biol. 23, 223. https://doi.org/10.1186/s12915-025-02325-5.

Crava, C.M., Varghese, F.S., Pischedda, E., Halbach, R., Palatini, U., Marconcini, M., et al., 2021. Population genomics in the arboviral vector Aedes aegypti reveals the genomic architecture and evolution of endogenous viral elements. Mol. Ecol. 30, 1594–1611. https://doi.org/10.1111/mec.15798.

Csorba, T., et al., 2007. The p122 subunit of Tobacco mosaic virus replicase is a potent silencing suppressor and compromises both siRNA- and microRNA-mediated pathways. J. Virol. 81, 11768–11780.

Cui, L., Wang, H., Ji, Y., Yang, J., Xu, S., Huang, X., et al., 2015. The nucleocapsid protein of coronaviruses acts as a viral suppressor of RNA silencing in mammalian cells. J. Virol. 89, 9029–9043.

Cullen, B.R., 2011. Viruses and microRNAs: RISCy interactions with serious consequences. Genes. Dev. 25, 1881–1894.

Dabo, S., Henrion-Lacritick, A., Lecuyer, A., et al., 2024. Extensive variation and strain-specificity in dengue virus susceptibility among African Aedes aegypti populations. PLoS Negl. Trop. Dis. 18, e0011862. https://doi.org/10.1371/journal.pntd.0011862.

Dauer, C.C., Carrera, G.M., 1937. Carlos Finlay's contribution to the epidemiology of yellow fever. Yale J Biol. Med. 9 (584), 1–588.

Daugherty, M.D., Malik, H.S., 2012. Rules of engagement: molecular insights from host-virus arms races. Annu. Rev. Genet. 46, 677–700. https://doi.org/10.1146/annurev-genet-110711-155522.

Dawkins, R., Krebs, J.R., 1979. Arms races between and within species. Proc. Biol. Sci. 205 (1161), 489–511. https://doi.org/10.1098/rspb.1979.0081.

de Almeida, J.P., Aguiar, E.R., Armache, J.N., Olmo, R.P., Marques, J.T., 2021. The virome of vector mosquitoes. Curr. Opin. Virol. 49, 7–12. https://doi.org/10.1016/j.coviro.2021.04.002. Epub 2021 May 12. PMID:33991759.

de Souza, W.M., Lecuit, M., Weaver, S.C., 2025. Chikungunya virus and other emerging arthritogenic alphaviruses. Nat. Rev. Microbiol. https://doi.org/10.1038/s41579-025-01177-8.

de Souza, W.M., Weaver, S.C., 2024. Effects of climate change and human activities on vector-borne diseases. Nat. Rev. Microbiol 22, 476–491. https://doi.org/10.1038/s41579-024-01026-0.

Deleris, A., Gallego-Bartolome, J., Bao, J., Kasschau, K.D., Carrington, J.C., Voinnet, O., 2006. Hierarchical action and inhibition of plant Dicer-like proteins in antiviral defense. Science 313 (5783), 68–71. https://doi.org/>10.1126/science.1128214. Epub 2006 Jun 1. Erratum in: Science. 2016 Jan 22;351(6271):aaf2336. https://doi.org/10.1126/science.aaf2336. PMID:16741077.

Demogines, A., Abraham, J., Choe, H., Farzan, M., Sawyer, S.L., 2013. Dual host-virus arms races shape an essential housekeeping protein. PLoS Biol. 11 (5), e1001571. https://doi.org/10.1371/journal.pbio.1001571.

den Boon, J.A., Diaz, A., Ahlquist, P., 2010. Cytoplasmic viral replication complexes. Cell Host Microbe. 8 (1), 77–85. https://doi.org/10.1016/j.chom.2010.06.010. PMID:20638644; PMCID:PMC2921950.

Ding, S.W., Voinnet, O., 2007. Antiviral immunity directed by small RNAs. Cell 130, 413–426.

Donaldson, M.K., Zanders, L.A., Jose, J., 2025. Functional roles and host interactions of orthoflavivirus non-structural proteins during replication. Pathogens 14, 184. https://doi.org/10.3390/pathogens14020184.

Dong, S., Balaraman, V., Kantor, A.M., Lin, J., Grant, D.G., Held, N.L., et al., 2017. Chikungunya virus dissemination from the midgut of Aedes aegypti is associated with temporal basal lamina degradation during bloodmeal digestion. PLoS Negl. Trop. Dis. 11, e0005976.

Dong, S., Ciomperlik-Patton, J., Zhao, Y., Dong, Y., Myles, K.M., Dimopoulos, G., 2025. Enhancing tissue-specific antiviral immunity to disrupt arbovirus transmission by mosquitoes. bioRxiv.

Dong, S., Dimopoulos, G., 2023. Aedes aegypti Argonaute 2 controls arbovirus infection and host mortality. Nat. Commun. 14, 5773.

Dong, Y., Dong, S., Dizaji, N.B., Rutkowski, N., Pohlenz, T., Myles, K., et al., 2022. The Aedes aegypti siRNA pathway mediates broad-spectrum defense against human pathogenic viruses and modulates antibacterial and antifungal defenses. PLoS Biol 20, e3001668.

Du, Y.X., Qi, Y.H., Lu, Y.H., Li, B.X., He, Y.J., Zhang, Y., et al., 2025. A plant virus attenuates the Toll immune pathway by degradation of Pellino to facilitate viral infection in insect vectors. J. Virol 99 (5), e0002125. https://doi.org/10.1128/jvi.00021-25. Epub 2025 Mar 31. PMID:40162791; PMCID:PMC12090757.

Duarte, E., Clarke, D., Moya, A., Domingo, E., Holland, J., 1992. Rapid fitness losses in mammalian RNA virus clones due to Muller's ratchet. Proc. Natl Acad. Sci. USA 89, 6015–6019.

Dunn, E.F., Pritlove, D.C., Jin, H., Elliott, R.M., 1995. Transcription of a recombinant bunyavirus RNA template by transiently expressed bunyavirus proteins. Virology 211, 133–143.

Elliott, R.M., 2014. Orthobunyaviruses: recent genetic and structural insights. Nat. Rev. Microbiol. 12, 673–685.

Elrefaey, A.M., Abdelnabi, R., Rosales Rosas, A.L., Wang, L., Basu, S., Delang, L., 2020. Understanding the mechanisms underlying host restriction of insect-specific viruses. Viruses 12.

Esvelt, K.M., Smidler, A.L., Catteruccia, F., Church, G.M., 2014. Concerning RNA-guided gene drives for the alteration of wild populations. eLife 3, e03401. https://doi.org/10.7554/eLife.03401.

Fabozzi, G., Nabel, C.S., Dolan, M.A., Sullivan, N.J., 2011. Ebolavirus proteins suppress the effects of small interfering RNA by direct interaction with the mammalian RNA interference pathway. J. Virol. 85 (6), 2512–2523. https://doi.org/10.1128/JVI.01160-10. Epub 2011 Jan 12. PMID:21228243; PMCID:PMC3067942.

Faye, O., Freire, C.C., Iamarino, A., et al., 2014. Molecular evolution of Zika virus during its emergence in the 20th century. PLoS Negl. Trop. Dis. 8 (1), e2636.

Feng, M., Liu, J., Wang, L., Swevers, L., 2026. dsRNA as pathogen-associated molecular pattern in innate immunity and multiple functions of the RNAi machinery complicate the use of RNAi in pest control. Front. Insect Sci. 5, 1749008. https://doi.org/10.3389/finsc.2025.1749008.

Fernandes-Santos, C., Azeredo, E.L., 2022. Innate immune response to dengue virus: Toll-like receptors and antiviral response. Viruses 14.

Filomatori, C.V., Carballeda, J.M., Villordo, S.M., Aguirre, S., Pallarés, H.M., Maestre, A.M., et al., 2017. Dengue virus genomic variation associated with mosquito adaptation defines the pattern of viral non-coding RNAs and fitness in human cells. PLoS Pathog. 13, e1006265. https://doi.org/10.1371/journal.ppat.1006265.

Fiorillo, C., Yen, P.S., Colantoni, A., Mariconti, M., Azevedo, N., Lombardo, F., et al., 2022. MicroRNAs and other small RNAs in Aedes aegypti saliva and salivary glands following chikungunya virus infection. Sci. Rep. 12 (1), 9536. https://doi.org/10.1038/s41598-022-13780-3.

Fitzmeyer, E.A., Gallichotte, E.N., Weger-Lucarelli, J., Kapuscinski, M.L., Abdo, Z., Pyron, K., et al., 2023. Loss of West Nile virus genetic diversity during mosquito infection due to species-dependent population bottlenecks. iScience 26 (10), 107711. https://doi.org/10.1016/j.isci.2023.107711.

Forrester, N.L., Coffey, L.L., Weaver, S.C., 2014. Arboviral bottlenecks and challenges to maintaining diversity and fitness during mosquito transmission. Viruses 6 (10), 3991–4004. https://doi.org/10.3390/v6103991.

Franz, A.W., Kantor, A.M., Passarelli, A.L., Clem, R.J., 2015. Tissue barriers to arbovirus infection in mosquitoes. Viruses 7, 3741–3767.

Frolov, I., Frolova, E.I., 2022. Molecular virology of chikungunya virus. In: Heise, M. (Ed.), Chikungunya Virus. Springer International Publishing, pp. 1–31.

Funk, A., et al., 2010. RNA structures required for production of subgenomic flavivirus RNA. J. Virol 84, 11407–11417.

Gabrieli, P., Caccia, S., Varotto-Boccazzi, I., Arnoldi, I., Barbieri, G., Comandatore, F., et al., 2021. Mosquito trilogy: microbiota, immunity and pathogens, and their implications for the control of disease transmission. Front. Microbiol. 12, 630438.

Gao, Y., Shi, X., Li, B., Shi, N., Zhao, G., Li, Z., et al., 2025. Signaling pathway of DNA replication and toll-like receptors was inhibited in C6/36 cells infected with Japanese encephalitis virus. Virol. J. 22 (1), 362. https://doi.org/10.1186/s12985-025-02956-8.

Gestuveo, R.J., Parry, R., Dickson, L.B., Lequime, S., Sreenu, V.B., Arnold, M.J., et al., 2022. Mutational analysis of Aedes aegypti Dicer 2 provides insights into the biogenesis of antiviral exogenous small interfering RNAs. PLoS Pathog. 18, e1010202.

Göertz, G.P., Fros, J.J., Miesen, P., Vogels, C.B.F., van der Bent, M.L., Geertsema, C., et al., 2016. Noncoding subgenomic flavivirus RNA is processed by the mosquito RNA interference machinery and determines West Nile virus transmission by Culex pipiens mosquitoes. J. Virol. 90 (22), 10145–10159. https://doi.org/10.1128/JVI.00930-16. PMID:27581979; PMCID:PMC5105652.

Göertz, G.P., van Bree, J.W.M., Hiralal, A., Fernhout, B.M., Steffens, C., Boeren, S., et al., 2019. Subgenomic flavivirus RNA binds the mosquito DEAD/H-box helicase ME31B and determines Zika virus transmission by Aedes aegypti. Proc. Natl Acad. Sci. USA 116, 19136–19144. https://doi.org/10.1073/pnas.1905146116.

Goic, B., Stapleford, K.A., Frangeul, L., Doucet, A.J., Gausson, V., Blanc, H., et al., 2016. Virus-derived DNA drives mosquito vector tolerance to arboviral infection. Nat. Commun. 7, 12410. https://doi.org/10.1038/ncomms12410.

Gubler, D.J., 1998. Dengue and dengue hemorrhagic fever. Clin. Microbiol. Rev 11, 480–496. https://doi.org/10.1128/CMR.11.3.480.

Gubler, D.J., 2001. Human arbovirus infections worldwide. Ann. N. Y. Acad. Sci. 951, 13–24.

Gubler, D.J., 2002. The global emergence/resurgence of arboviral diseases as public health problems. Arch. Med. Res. 33, 330–342.

Haas, G., et al., 2008. Nuclear import of CaMV P6 is required for infection and suppression of the RNA silencing factor DRB4. EMBO J. 27, 2102–2112.

Haasnoot, J., de Vries, W., Geutjes, E.J., Prins, M., de Haan, P., Berkhout, B., 2007. The Ebola virus VP35 protein is a suppressor of RNA silencing. PLoS Pathog. 3 (6), e86. https://doi.org/10.1371/journal.ppat.0030086.

Haddow, A.D., Guzman, H., Popov, V.L., Wood, T.G., Widen, S.G., Tesh, R.B., et al., 2013. First isolation of Aedes flavivirus in the Western Hemisphere and evidence of vertical transmission in Aedes albopictus (Diptera: Culicidae). Virology 440 (2), 134–139. https://doi.org/10.1016/j.virol.2012.12.008.

Hahn, C.S., Strauss, E.G., Strauss, J.H., 1985. Sequence analysis of three Sindbis virus mutants temperature-sensitive in the capsid protein autoprotease. Proc. Natl Acad. Sci. USA 82 (14), 4648–4652.

Hall, D.R., Johnson, R.M., Kwon, H., Ferdous, Z., Laredo-Tiscareno, S.V., Blitvich, B.J., et al., 2025. Mosquito immune cells enhance dengue and Zika virus infection in Aedes aegypti. Nat. Commun. 16, 5891.

Harbach, R.E., 2024. Composition and Nature of the Culicidae (Mosquitoes). CABI, Wallingford (UK) ISBN: 978-1-80062-799-4.

He, Y.J., Lu, G., Qi, Y.H., Zhang, Y., Zhang, X.D., Huang, H.J., et al., 2020. Activation of Toll immune pathway in an insect vector induced by a plant virus. Front. Immunol. 11, 613957. https://doi.org/10.3389/fimmu.2020.613957.

He, Y.J., Lu, G., Xu, B.J., Mao, Q.Z., Qi, Y.H., Jiao, G.Y., et al., 2024. Maintenance of persistent transmission of a plant arbovirus in its insect vector mediated by the Toll-Dorsal immune pathway. Proc. Natl Acad. Sci. USA 121, e2315982121. https://doi.org/10.1073/pnas.2315982121.

Heitmann, A., Gusmag, F., Rathjens, M.G., Maurer, M., Schicht, F.K., Jansen, S., et al., 2021. Mammals preferred: reassortment of Batai and Bunyamwera orthobunyavirus occurs in mammalian but not insect cells. Viruses 13 (9), 1702. https://doi.org/10.3390/v13091702.

Herd, C.S., Grant, D.G., Lin, J., Franz, A.W.E., 2021. Starvation at the larval stage increases the vector competence of Aedes aegypti females for Zika virus. PLoS Negl. Trop. Dis. 15, e0010003.

Hermanns, K., Marklewitz, M., Zirkel, F., Overheul, G.J., Page, R.A., Loaiza, J.R., et al., 2020. Agua Salud alphavirus defines a novel lineage of insect-specific alphaviruses discovered in the New World. J. Gen. Virol. 101 (1), 1887–1897. https://doi.org/10.1099/jgv.0.001344.

Hermanns, K., Zirkel, F., Kopp, A., Marklewitz, M., Rwego, I.B., Estrada, A., Gillespie, T.R., Drosten, C., Junglen, S., 2017. Discovery of a novel alphavirus related to Eilat virus. J. Gen. Virol. 98, 43–49. https://doi.org/10.1099/jgv.0.000694.

Hoch, A.L., Peterson, N.E., LeDuc, J.W., Pinheiro, F.P., 1981. An outbreak of Mayaro virus disease in Belterra, Brazil. III. Entomological and ecological studies. Am. J. Trop. Med. Hyg. 30, 689–698. https://doi.org/10.4269/ajtmh.1981.30.689.
Holland, J.J., 1991. Defective viral genomes. In: Knipe, D. (Ed.), Fundamental Virology. Raven Press, New York, pp. 151–165.
Holmes, E.C., Worobey, M., Rambaut, A., 1999. Phylogenetic evidence for recombination in Dengue virus. Mol. Biol. Evol. 16 (3), 405–409.
Hoshino, K., Isawa, H., Tsuda, Y., Yano, K., Sasaki, T., Yuda, M., Takasaki, T., Kobayashi, M., Sawabe, K., 2007. Genetic characterization of a new insect flavivirus isolated from *Culex pipiens* mosquito in Japan. Virology 359, 405–414. https://doi.org/10.1016/j.virol.2006.09.039.
Houé, V., Bonizzoni, M., Failloux, A.B., 2019. Endogenous non-retroviral elements in genomes of Aedes mosquitoes and vector competence. Emerg. Microbes Infect. 8 (1), 542–555. https://doi.org/10.1080/22221751.2019.1599302.
Huang, Y., Wang, S., Liu, H., et al., 2023. A global dataset of sequence, diversity and biosafety recommendation of arbovirus and arthropod-specific virus. Sci. Data 10, 305. https://doi.org/10.1038/s41597-023-02226-8.
Hubálek, Z., Rudolf, I., Nowotny, N., 2014. Arboviruses pathogenic for domestic and wild animals. Adv. Virus Res. 89, 201–275. https://doi.org/10.1016/B978-0-12-800172-1.00005-7.
Huhtamo, E., Moureau, G., Cook, S., Julkunen, O., Putkuri, N., Kurkela, S., et al., 2012. Novel insect-specific flavivirus isolated from northern Europe. Virology 433 (2), 471–478. https://doi.org/10.1016/j.virol.2012.08.038.
Hurlbut, H.S., 1956. West Nile virus infection in arthropods. Am. J. Trop. Med. Hyg. 5, 76–85. https://doi.org/10.4269/ajtmh.1956.5.76.
Hussain, M., Taft, R.J., Asgari, S., 2008. An insect virus-encoded microRNA regulates viral replication. J. Virol. 82, 9164–9170.
Hussain, M., Torres, S., Schnettler, E., Funk, A., Grundhoff, A., et al., 2012. West Nile virus encodes a microRNA-like small RNA in the 3′ untranslated region which upregulates GATA4 mRNA and facilitates virus replication in mosquito cells. Nucleic Acids Res. 40, 2210–2223.
Hyde, J.L., Chen, R., Trobaugh, D.W., Diamond, M.S., Weaver, S.C., et al., 2015. The 5′ and 3′ ends of alphavirus RNAs—Non-coding is not non-functional. Virus Res. 206, 99–107.
Hyde, J.L., Gardner, C.L., Kimura, T., White, J.P., Liu, G., et al., 2014. A viral RNA structural element alters host recognition of nonself RNA. Science 343 (6172), 783–787.
International Committee on Taxonomy of Viruses (ICTV), 2023. Virus taxonomy: 2023 release. Available at: https://ictv.global/taxonomy.
Iroegbu, C.U., Pringle, C.R., 1981. Genetic interactions among viruses of the Bunyamwera complex. J. Virol. 37, 383–394.
Jia, D., Luo, G., Guan, H., Yu, T., Sun, X., Du, Y., et al., 2024. Arboviruses antagonize insect Toll antiviral immune signaling to facilitate the coexistence of viruses with their vectors. PLoS Pathog. 20 (6), e1012318. https://doi.org/10.1371/journal.ppat.1012318.
Junglen, S., Korries, M., Grasse, W., Wieseler, J., Kopp, A., Hermanns, K., et al., 2017. Host range restriction of insect-specific flaviviruses occurs at several levels of the viral life cycle. mSphere 2 (1), e00375-16. https://doi.org/10.1128/mSphere.00375-16.
Jupatanakul, N., Sim, S., Anglero-Rodriguez, Y.I., Souza-Neto, J., Das, S., Poti, K.E., et al., 2017. Engineered Aedes aegypti JAK/STAT pathway-mediated immunity to dengue virus. PLoS Negl. Trop. Dis. 11, e0005187.
Jupille, H., Vega-Rua, A., Rougeon, F., Failloux, A.B., 2014. Arboviruses: variations on an ancient theme. Future Virol. 9 (8), 733–751. https://doi.org/10.2217/fvl.14.62.

Kain, M.P., Skinner, E.B., Athni, T.S., Ramirez, A.L., Mordecai, E.A., van den Hurk, A.F., 2022. Not all mosquitoes are created equal: a synthesis of vector competence experiments reinforces virus associations of Australian mosquitoes. PLoS Negl. Trop. Dis. 16, e0010768.

Kakumani, P.K., Ponia, S.S., Sood, S.R.K., Chinnappan, V., Banerjea, M., Medigeshi, A.C., et al., 2013. Role of RNA interference (RNAi) in dengue virus replication and identification of NS4B as an RNAi suppressor. J. Virol. 87 (16), 8870–8883. https://doi.org/10.1128/JVI.02774-12. PMID:23741001; PMCID:PMC3754049.

Kantor, A.M., Grant, D.G., Balaraman, V., White, T.A., Franz, A.W.E., 2018. Ultrastructural analysis of chikungunya virus dissemination from the midgut of the yellow fever mosquito, Aedes aegypti. Viruses 10.

Karjee, S., Minhas, A., Sood, V., Ponia, S.S., Banerjea, A.C., Chow, V.T., et al., 2010. The 7a accessory protein of severe acute respiratory syndrome coronavirus acts as an RNA silencing suppressor. J. Virol. 84, 10395–10401. https://doi.org/10.1128/JVI.00748-10.

Kasschau, K.D., Carrington, J.C., 1998. A counterdefensive strategy of plant viruses: suppression of posttranscriptional gene silencing. Cell 95, 461–470.

King, A.M.Q., Adams, M.J., Carstens, E.B., Lefkowitz, E.J., (Eds.), (2012). Virus taxonomy. Elsevier; 2012. Chapters: Family Bunyaviridae, p. 725–741; Family Orthomyxoviridae, p. 749–761; Family Reoviridae, p. 541–637; Family Rhabdoviridae, p. 686–713.

Kingsolver, M.B., Hardy, R.W., 2012. Making connections in insect innate immunity. Proc. Natl Acad. Sci. U S A 109, 18639–18640.

Kingsolver, M.B., Huang, Z., Hardy, R.W., 2013. Insect antiviral innate immunity: pathways, effectors, and connections. J. Mol. Biol. 425, 4921–4936.

Kohl, A., Dunn, E.F., Lowen, A.C., Elliott, R.M., 2004. Complementarity, sequence and structural elements within the 3' and 5' noncoding regions of the Bunyamwera orthobunyavirus S segment determine promoter strength. J. Gen. Virol. 85, 3269–3278.

Kumar, A., Srivastava, P., Sirisena, P., Dubey, S.K., Kumar, R., Shrinet, J., et al., 2018. Mosquito innate immunity. Insects 9.

Lambrechts, L., 2011. Quantitative genetics of Aedes aegypti vector competence for dengue viruses: towards a new paradigm? Trends Parasitol 27 (3), 111–114. https://doi.org/10.1016/j.pt.2010.12.001.

Lambrechts, L., Scott, T.W., Gubler, D.J., 2010. Consequences of the expanding global distribution of Aedes albopictus for dengue virus transmission. PLoS Negl. Trop. Dis. 4 (5), e646. https://doi.org/10.1371/journal.pntd.0000646. PMID:20520794; PMCID:PMC2876112.

Laredo-Tiscareño, S.V., Garza-Hernandez, J.A., Tangudu, C.S., Dankaona, W., Rodríguez-Alarcón, C.A., Adame-Gallegos, J.R., et al., 2024. Discovery of novel viruses in Culicoides biting midges in Chihuahua, Mexico. Viruses 16 (7), 1160. https://doi.org/10.3390/v16071160. PMID:39066322; PMCID:PMC11281482.

Lauring, A.S., Andino, R., 2010. Quasispecies theory and the behavior of RNA viruses. PLoS Pathog. 6 (7), e1001005. https://doi.org/10.1371/journal.ppat.1001005.

Lee, E., Bujalowski, P.J., Teramoto, T., Gottipati, K., Scott, S.D., Padmanabhan, R., et al., 2021. Structures of flavivirus RNA promoters suggest two binding modes with NS5 polymerase. Nat. Commun. 12, 2530.

Lee, W.S., Webster, J.A., Madzokere, E.T., Stephenson, E.B., Herrero, L.J., 2019. Mosquito antiviral defense mechanisms: a delicate balance between innate immunity and persistent viral infection. Parasit. Vectors 12, 165.

Lefteri, D.A., Bryden, S.R., Pingen, M., Terry, S., McCafferty, A., Beswick, E.F., et al., 2022. Mosquito saliva enhances virus infection through sialokinin-dependent vascular leakage. Proc. Natl Acad. Sci. U S A 119, e2114309119.

Leite, T., Ferreira, A.G.A., Imler, J.L., Marques, J.T., 2021. Distinct roles of hemocytes at different stages of infection by dengue and Zika viruses in Aedes aegypti mosquitoes. Front. Immunol. 12, 660873.

Lequime, S., Fontaine, A., Ar Gouilh, M., Moltini-Conclois, I., Lambrechts, L., 2016. Genetic drift, purifying selection and vector genotype shape dengue virus intra-host genetic diversity in mosquitoes. PLoS Genet. 12 (6), e1006111. https://doi.org/10.1371/journal.pgen.1006111.

Lescar, J., Roussel, A., Wien, M.W., et al., 2001. The fusion glycoprotein shell of Semliki Forest virus: an icosahedral assembly primed for fusogenic activation at endosomal pH. Cell 105 (1), 137–148.

Lewis, J., Gallichotte, E.N., Randall, J., Glass, A., Foy, B.D., Ebel, G.D., et al., 2023. Intrinsic factors driving mosquito vector competence and viral evolution: a review. Front. Cell Infect. Microbiol. 13, 1330600. https://doi.org/10.3389/fcimb.2023.1330600.

Li, C.X., et al., 2015. Unprecedented genomic diversity of RNA viruses in arthropods reveals the ancestry of negative-sense RNA viruses. eLife 4, e05378.

Lin, C.C., Chou, C.M., Hsu, Y.L., Lien, J.C., Wang, Y.M., Chen, S.T., et al., 2004. Characterization of two mosquito STATs, AaSTAT and CtSTAT. J. Biol. Chem. 279, 3308–3317.

Lingel, A., Simon, B., Izaurralde, E., Sattler, M., 2005. The structure of the flock house virus B2 protein, a viral suppressor of RNA interference, shows a novel mode of double-stranded RNA recognition. EMBO Rep. 6, 1149–1155.

Little, T.J., Allen, J.E., Babayan, S.A., Matthews, K.R., Colegrave, N., 2012. Harnessing evolutionary biology to combat infectious disease. Nat. Med. 18 (2), 217–220. https://doi.org/10.1038/nm.2572. PMID:22310693; PMCID:PMC3712261.

Liu, J., Swevers, L., Kolliopoulou, A., Smagghe, G., 2019. Arboviruses and the challenge to establish systemic and persistent infections in competent mosquito vectors: the interaction with the RNAi mechanism. Front. Physiol. 10, 890. https://doi.org/10.3389/fphys.2019.00890. PMID:31354527; PMCID:PMC6638189.

Logan, R.A.E., Quek, S., Muthoni, J.N., von Eicken, A., Brettell, L.E., Anderson, E.R., et al., 2022. Vertical and horizontal transmission of cell fusing agent virus in Aedes aegypti. Appl. Env. Microbiol. 88https://doi.org/10.1128/aem.01062-22. e01062-22.

Lumsden, W.H.R., 1955. An epidemic of virus disease in Southern Province, Tanganyika Territory, in 1952–53. II. General description and epidemiology. Trans. R. Soc. Trop. Med. Hyg. 49, 33–57. https://doi.org/10.1016/0035-9203(55)90081-X.

Madden, E.A., Plante, K.S., Morrison, C.R., Kutchko, K.M., Sanders, W., et al., 2020. Using SHAPE-MaP to model RNA secondary structure and identify 3′ UTR variation in chikungunya virus. J. Virol 94 (24).

Manokaran, G., Finol, E., Wang, C., Gunaratne, J., Bahl, J., Ong, E.Z., et al., 2015. Dengue subgenomic RNA binds TRIM25 to inhibit interferon expression for epidemiological fitness. Science 350, 217–221. https://doi.org/10.1126/science.aab3369.

Marchette, N.J., Garcia, R., Rudnick, A., 1969. Isolation of Zika virus from *Aedes aegypti* mosquitoes in Malaysia. Am. J. Trop. Med. Hyg. 18, 411–415. https://doi.org/10.4269/ajtmh.1969.18.411.

Marcombe, S., Poupardin, R., Darriet, F., Reynaud, S., Bonnet, J., Strode, C., et al., 2009. Exploring the molecular basis of insecticide resistance in the dengue vector Aedes aegypti: a case study in Martinique Island (French West Indies). BMC Genomics 10, 494.

Marconcini, M., Pischedda, E., Houé, V., Palatini, U., Lozada-Chávez, N., Sogliani, D., et al., 2021. Profile of small RNAs, vDNA forms and viral integrations in late chikungunya virus infection of Aedes albopictus mosquitoes. Viruses 13 (4), 553. https://doi.org/10.3390/v13040553.

Marklewitz, M., Gloza-Rausch, F., Kurth, A., Kümmerer, B.M., Drosten, C., Junglen, S., 2012. First isolation of an Entomobirnavirus from free-living insects. J. Gen. Virol. 93 (Pt 11), 2431–2435. https://doi.org/10.1099/vir.0.045435-0.

Marklewitz, M., Zirkel, F., Kurth, A., Drosten, C., Junglen, S., 2015. Evolutionary and phenotypic analysis of live virus isolates suggests arthropod origin of a pathogenic RNA virus family. Proc. Natl. Acad. Sci. U S A 112, pp. 7536–7541. https://doi.org/10.1073/pnas.1502036112.

Mathew, A., West, K., Kalayanarooj, S., et al., 2011. B-cell responses during primary and secondary dengue virus infections in humans. J. Infect. Dis. 204 (10), 1514–1522.

Mathur, K., Anand, A., Dubey, S.K., Sanan-Mishra, N., Bhatnagar, R.K., Sunil, S., 2016. Analysis of chikungunya virus proteins reveals that non-structural proteins nsP2 and nsP3 exhibit RNA interference (RNAi) suppressor activity. Sci. Rep. 6, 38065. https://doi.org/10.1038/srep38065.

Mayer, S.V., Tesh, R.B., Vasilakis, N., 2017. The emergence of arthropod-borne viral diseases: a global perspective on dengue, chikungunya and Zika fevers. Acta Trop. 166, 155–163.

Merai, Z., et al., 2006. Double-stranded RNA binding may be a general plant RNA viral strategy to suppress RNA silencing. J. Virol. 80, 5747–5756.

Merkling, S.H., Crist, A.B., Henrion-Lacritick, A., Frangeul, L., Couderc, E., Gausson, V., et al., 2023. Multifaceted contributions of Dicer2 to arbovirus transmission by Aedes aegypti. Cell Rep. 42 112977.

Miesen, P., Girardi, E., van Rij, R.P., 2015. Distinct sets of PIWI proteins produce arbovirus- and transposon-derived piRNAs in Aedes aegypti mosquito cells. Nucleic Acids Res. 43, 6545–6556.

Miesen, P., Ivens, A., Buck, A.H., van Rij, R.P., 2016. Small RNA profiling in Dengue virus 2-infected Aedes mosquito cells reveals viral piRNAs and novel host miRNAs. PLoS Negl. Trop. Dis. 10 (2), e0004452. https://doi.org/10.1371/journal.pntd.0004452.

Misra, U.K., Kalita, J., 2010. Overview: Japanese encephalitis. Prog. Neurobiol 91 (2), 108–120.

Monath, T.P., 2021. The Arboviruses: Epidemiology and Ecology. CRC Press, London (UK).

Moon, S.L., Dodd, B.J., Brackney, D.E., Wilusz, C.J., Ebel, G.D., Wilusz, J., 2015. Flavivirus sfRNA suppresses antiviral RNA interference in cultured cells and mosquitoes and directly interacts with the RNAi machinery. Virology 485, 322–329. https://doi.org/10.1016/j.virol.2015.08.009.

Moonen, J.P., Schinkel, M., van der Most, T., Miesen, P., van Rij, R.P., 2023. Composition and global distribution of the mosquito virome – a comprehensive database of insect-specific viruses. One Health 16, 100490. https://doi.org/10.1016/j.onehlt.2023.100490.

Morazzani, E.M., Wiley, M.R., Murreddu, M.G., Adelman, Z.N., Myles, K.M., 2012. Production of virus-derived ping-pong-dependent piRNA-like small RNAs in the mosquito soma. PLoS Pathog. 8 (1), e1002470. https://doi.org/10.1371/journal.ppat.1002470.

Muller, H.J., 1964. The relation of recombination to mutational advance. Mutat. Res. 1, 2–9.

Mushtaq, I., Sarwar, M.S., Munzoor, I., 2024. A comprehensive review of Wolbachia-mediated mechanisms to control dengue virus transmission in Aedes aegypti through innate immune pathways. Front. Immunol. 15, 1434003.

Myles, K.M., Wiley, M.R., Morazzani, E.M., Adelman, Z.N., 2008. Alphavirus-derived small RNAs modulate pathogenesis in disease vector mosquitoes. Proc. Natl. Acad. Sci. U S A 105 (50), 19938–19943.

Nasar, F., Haddow, A.D., Tesh, R.B., Weaver, S.C., 2014. Eilat virus displays a narrow mosquito vector range. Parasit. Vectors 7, 595.

Nasar, F., Palacios, G., Gorchakov, R.V., Guzman, H., Da Rosa, A.P., Savji, N., Popov, V.L., Sherman, M.B., Lipkin, W.I., Tesh, R.B., Weaver, S.C., 2012. Eilat virus, a unique alphavirus with host range restricted to insects by RNA replication. Proc. Natl. Acad. Sci. U S A 109, pp. 14622–14627. https://doi.org/10.1073/pnas.1204787109.

Naveca, F.G., Almeida, T.A.P., Souza, V., et al., 2024. Human outbreaks of a novel reassortant Oropouche virus in the Brazilian Amazon region. Nat. Med. 30, 3509–3521. https://doi.org/10.1038/s41591-024-03300-3.

Nayak, A., Berry, B., Tassetto, M., Kunitomi, M., Acevedo, A., Deng, C., et al., 2010. Cricket paralysis virus antagonizes Argonaute 2 to modulate antiviral defense in Drosophila. Nat. Struct. Mol. Biol. 17 (5), 547–554. https://doi.org/10.1038/nsmb.1810.

Nayak, A., Kim, D.Y., Trnka, M.J., Kerr, C.H., Lidsky, P.V., et al., 2018. A viral protein restricts Drosophila RNAi immunity by regulating Argonaute activity and stability. Cell Host Microbe 24, 542–557.e9.

Ng, W.C., Soto-Acosta, R., Bradrick, S.S., Garcia-Blanco, M.A., Ooi, E.E., 2017. The 5′ and 3′ untranslated regions of the flaviviral genome. Viruses 9, 137.

Novella, I.S., Hershey, C.L., Escarmis, C., Domingo, E., Holland, J.J., 1999. Lack of evolutionary stasis during alternating replication of an arbovirus in insect and mammalian cells. J. Mol. Biol. 287, 459–465.

Novella, I.S., Presloid, J.B., Smith, S.D., Wilke, C.O., 2011. Specific and nonspecific host adaptation during arboviral experimental evolution. J. Mol. Microbiol. Biotechnol. 21 (1–2), 71–81. https://doi.org/10.1159/000332752.

Nunes, M.R.T., Contreras-Gutiérrez, M.A., Guzman, H., Martins, L.C., Barbirato, M.F., Savit, C., et al., 2017. Genetic characterization, molecular epidemiology, and phylogenetic relationships of insect-specific viruses in the taxon Negevirus. Virology 504, 152–167. https://doi.org/10.1016/j.virol.2017.01.022.

O'Neal, S.T., Samuel, G.H., Adelman, Z.N., Myles, K.M., 2014. Mosquito-borne viruses and suppressors of invertebrate antiviral RNA silencing. Viruses 6 (11), 4314–4331. https://doi.org/10.3390/v6114314.

Obadia, T., Gutierrez-Bugallo, G., Duong, V., Nuñez, A.I., Fernandes, R.S., Kamgang, B., et al., 2022. Zika vector competence data reveals risks of outbreaks: the contribution of the European ZIKAlliance project. Nat. Commun. 13 (1), 4490. https://doi.org/10.1038/s41467-022-32234-y.

Obijeski, J.F., Bishop, D.H., Murphy, F.A., Palmer, E.L., 1976. Structural proteins of La Crosse virus. J. Virol. 19, 985–997.

Ogg, M.M., Patterson, J.L., 2007. RNA binding domain of Jamestown Canyon virus S segment RNAs. J. Virol. 81, 13754–13760.

Olson, K.E., Blair, C.D., 2015. Arbovirus-mosquito interactions: RNAi pathway. Curr. Opin. Virol. 15, 119–126. https://doi.org/10.1016/j.coviro.2015.10.001.

Osborne, J.C., Elliott, R.M., 2000. RNA binding properties of bunyamwera virus nucleocapsid protein and selective binding to an element in the 5' terminus of the negative-sense S segment. J. Virol. 74, 9946–9952.

Overby, A.K., Popov, V.L., Niedrig, M., Weber, F., 2010. Tick-borne encephalitis virus delays interferon induction and hides its double-stranded RNA in intracellular membrane vesicles. J. Virol 84, 8470–8483.

Palatini, U., Miesen, P., Carballar-Lejarazu, R., Ometto, L., Rizzo, E., Tu, Z., et al., 2017. Comparative genomics shows that viral integrations are abundant and express piRNAs in the arboviral vectors Aedes aegypti and Aedes albopictus. BMC Genomics 18, 512. https://doi.org/10.1186/s12864-017-3903-3.

Panda, D., Pandit, R.S., Sahu, B., Kamaraju, R., Barik, T.K., 2024. Understanding mosquito faunal diversity: an approach to assess the burden of vector-borne diseases in three representative topographies (rural, urban, and peri-urban) of Ganjam District in Odisha State, India. J. Trop. Med. 2024, 9701356.

Paradkar, P.N., Duchemin, J.B., Voysey, R., Walker, P.J., 2014. Dicer-2-dependent activation of Culex Vago occurs via the TRAF-Rel2 signaling pathway. PLoS Negl. Trop. Dis. 8, e2823.

Paradkar, P.N., Trinidad, L., Voysey, R., Duchemin, J.B., Walker, P.J., 2012. Secreted Vago restricts West Nile virus infection in Culex mosquito cells by activating the Jak-STAT pathway. Proc. Natl Acad. Sci. U S A 109, 18915–18920.

Patterson, E.I., et al., 2018. Mosquito bottlenecks alter viral mutant swarm in a tissue and time-dependent manner with contraction and expansion of variant positions and diversity. Virus Evol. 4, vey001.

Pazhouhandeh, M., et al., 2006. F-box-like domain in the polerovirus protein P0 is required for silencing suppressor function. Proc. Natl Acad. Sci. U S A 103, 1994–1999.

Peinado, S.A., Aliota, M.T., Blitvich, B.J., Bartholomay, L.C., 2022. Biology and transmission dynamics of Aedes flavivirus. J. Med. Entomol. 59 (2), 659–666. https://doi.org/10.1093/jme/tjab197.

Peterson, A.J., Hall, R.A., Harrison, J.J., Hobson-Peters, J., Hugo, L.E., 2024. Unleashing nature's allies: comparing the vertical transmission dynamics of insect-specific and vertebrate-infecting flaviviruses in mosquitoes. Viruses 16 (9), 1499. https://doi.org/10.3390/v16091499.

Pijlman, G.P., Funk, A., Kondratieva, N., Leung, J., Torres, S., et al., 2008. A highly structured, nuclease-resistant, noncoding RNA produced by flaviviruses is required for pathogenicity. Cell Host Microbe 4, 579–591.

Pompon, J., Manuel, M., Ng, G.K., Wong, B., Shan, C., et al., 2017. Dengue subgenomic flaviviral RNA disrupts immunity in mosquito salivary glands to increase virus transmission. PLoS Pathog. 13, e1006535.

Pontremoli, C., Forni, D., Clerici, M., Cagliani, R., Sironi, M., 2021. Alternation between taxonomically divergent hosts is not the major determinant of flavivirus evolution. Virus Evol. 7 (1), veab040. https://doi.org/10.1093/ve/veab040.

Qi, N., Cai, D., Qiu, Y., Xie, J., Wang, Z., et al., 2011. RNA binding by a novel helical fold of b2 protein from Wuhan nodavirus mediates suppression of RNA interference and promotes dimerization. J. Virol. 85, 9543–9554.

Qi, N., Zhang, L., Qiu, Y., Wang, Z., Si, J., et al., 2012. Targeting of dicer-2 and RNA by a viral RNA silencing suppressor in Drosophila cells. J. Virol. 86, 5763–5773.

Qian, Q., Zhou, H., Shu, T., Mu, J., Fang, Y., et al., 2020. The capsid protein of Semliki Forest virus antagonizes RNA interference in mammalian cells. J. Virol. 94 (3), e01233-19. https://doi.org/10.1128/JVI.01233-19.

Qiu, Y., Xu, Y.P., Wang, M., Miao, M., Zhou, H., et al., 2020. Flavivirus induces and antagonizes antiviral RNA interference in both mammals and mosquitoes. Sci. Adv. 6 (6), eaax7989. https://doi.org/10.1126/sciadv.aax7989.

Qiu, Y., Xu, Y., Zhang, Y., Zhou, H., Deng, Y.Q., et al., 2017. Human virus-derived small RNAs can confer antiviral immunity in mammals. Immunity 46 (6), 992–1004.e5. https://doi.org/10.1016/j.immuni.2017.05.006.

Qu, J., Betting, V., van Iterson, R., Kwaschik, F.M., van Rij, R.P., 2023. Chromatin profiling identifies transcriptional readthrough as a conserved mechanism for piRNA biogenesis in mosquitoes. Cell Rep. 42 (3), 112257. https://doi.org/10.1016/j.celrep.2023.112257.

Quiner, C.A., Parameswaran, P., Ciota, A.T., Ehrbar, D.J., Dodson, B.L., et al., 2014. Increased replicative fitness of a dengue virus 2 clade in native mosquitoes: potential contribution to a clade replacement event in Nicaragua. J. Virol. 88 (22), 13125–13134. https://doi.org/10.1128/JVI.01822-14.

Ramirez, J.L., Dimopoulos, G., 2010. The Toll immune signaling pathway controls conserved anti-dengue defenses across diverse Aedes aegypti strains and against multiple dengue virus serotypes. Dev. Comp. Immunol. 34, 625–629.

Read, A.F., Day, T., Huijben, S., 2011. The evolution of drug resistance and the curious orthodoxy of aggressive chemotherapy. Proc. Natl Acad. Sci. U S A 108 (2), 10871–10877. https://doi.org/10.1073/pnas.1100299108.

Reeves, W.C., Hammon, W.M.D., 1952. California encephalitis virus, a newly described agent: III. Mosquito infection and transmission. J. Immunol. 69, 511–514. https://doi.org/10.4049/jimmunol.69.5.511.

Reynaud, J.M., Kim, D.Y., Atasheva, S., et al., 2015. IFIT1 differentially interferes with translation and replication of alphavirus genomes and promotes type I interferon induction. PLoS Pathog. 11 (4), e1004863.

Rodriguez-Andres, J., Axford, J., Hoffmann, A., Fazakerley, J., 2023. Mosquito transgenerational antiviral immunity is mediated by vertical transfer of virus DNA sequences and RNAi. iScience 27 (1), 108598. https://doi.org/10.1016/j.isci.2023.108598.

Roundy, C.M., Azar, S.R., Rossi, S.L., Weaver, S.C., Vasilakis, N., 2017. Insect-specific viruses: a historical overview and recent developments. Adv. Virus Res. 98, 119–146. https://doi.org/10.1016/bs.aivir.2016.10.001. Epub 2016 Nov 17. PMID:28433051.

Rückert, C., Prasad, A.N., Garcia-Luna, S.M., Robison, A., Grubaugh, N.D., Weger-Lucarelli, J., et al., 2019. Small RNA responses of Culex mosquitoes and cell lines during acute and persistent virus infection. Insect Biochem. Mol. Biol. 109, 13–23. https://doi.org/10.1016/j.ibmb.2019.04.008. Epub 2019 Apr 5. PMID:30959110; PMCID:PMC6516063.

Russell, T.A., Ayaz, A., Davidson, A.D., Fernandez-Sesma, A., Maringer, K., 2021. Imd pathway-specific immune assays reveal NF-κB stimulation by viral RNA PAMPs in Aedes aegypti Aag2 cells. PLoS Negl. Trop. Dis. 15, e0008524.

Sabin, L.R., Zheng, Q., Thekkat, P., Yang, J., Hannon, G.J., Gregory, B.D., et al., 2013. Dicer-2 processes diverse viral RNA species. PLoS One 8, e55458. https://doi.org/10.1371/journal.pone.0055458. Epub 2013 Feb 12. PMID:23424633; PMCID:PMC3570552.

Saiyasombat, R., Bolling, B.G., Brault, A.C., Bartholomay, L.C., Blitvich, B.J., 2011. Evidence of efficient transovarial transmission of Culex flavivirus by Culex pipiens (Diptera: Culicidae). J. Med. Entomol. 48, 1031–1038. https://doi.org/10.1603/ME11043. PMID:21936322.

Samuel, G.H., Adelman, Z.N., Myles, K.M., 2018. Antiviral immunity and virus-mediated antagonism in disease vector mosquitoes. Trends Microbiol. 26, 447–461. https://doi.org/10.1016/j.tim.2017.12.005. Epub 2018 Jan 31. PMID:29395729; PMCID:PMC5910197.

Samuel, G.H., Pohlenz, T., Dong, Y., Coskun, N., Adelman, Z.N., Dimopoulos, G., et al., 2023. RNA interference is essential to modulating the pathogenesis of mosquito-borne viruses in Aedes aegypti. Proc. Natl Acad. Sci. U S A 120, e2213701120.

Samuel, G.H., Wiley, M.R., Badawi, A., Adelman, Z.N., Myles, K.M., 2016. Yellow fever virus capsid protein is a potent suppressor of RNA silencing that binds double-stranded RNA. Proc. Natl Acad. Sci. USA 113, 13863–13868. https://doi.org/10.1073/pnas.1600544113. Epub 2016 Nov 14. PMID:27849599; PMCID:PMC5137771.

Sanchez-Vargas, I., Olson, K.E., Black, W.C., 2021. The genetic basis for salivary gland barriers to arboviral transmission. Insects 12.

Sanchez-Vargas, I., Scott, J.C., Poole-Smith, B.K., Franz, A.W., Barbosa-Solomieu, V., Wilusz, J., et al., 2009. Dengue virus type 2 infections of Aedes aegypti are modulated by the mosquito's RNA interference pathway. PLoS Pathog. 5, e1000299.

Sánchez-Vargas, I., Scott, J.C., Poole-Smith, B.K., Franz, A.W., Barbosa-Solomieu, V., Wilusz, J., et al., 2009. Dengue virus type 2 infections of Aedes aegypti are modulated by the mosquito's RNA interference pathway. PLoS Pathog 5, e1000299. https://doi.org/10.1371/journal.ppat.1000299.

Sang, R.C., Gichogo, A., Gachoya, J., Dunster, M.D., Ofula, V., Hunt, A.R., Crabtree, M.B., Barry, R.M., Dunster, L.M., 2003. Isolation of a new flavivirus related to cell fusing agent virus (CFAV) from field-collected flood-water *Aedes* mosquitoes sampled from a dambo in central Kenya. Arch. Virol. 148, 1085–1093. https://doi.org/10.1007/s00705-003-0018-8.

Saucereau, Y., Wilson, T.H., Tang, M.C.K., Moncrieffe, M.C., Hardwick, S.W., Chirgadze, D.Y., et al., 2022. Structure and dynamics of Toll immunoreceptor activation in Aedes aegypti. Nat. Commun. 13, 5110.

Scachetti, G.C., Forato, J., Claro, I.M., Hua, X., Salgado, B.B., Vieira, A., et al., 2025. Re-emergence of Oropouche virus between 2023 and 2024 in Brazil: an observational epidemiological study. Lancet Infect. Dis. 25, 166–175.

Schnettler, E., Donald, C.L., Human, S., Watson, M., Siu, R.W.C., et al., 2013. Knockdown of piRNA pathway proteins results in enhanced Semliki Forest virus production in mosquito cells. J. Gen. Virol. 94, 1680–1689.

Schnettler, E., Sterken, M.G., Leung, J.Y., Metz, S.W., Geertsema, C., et al., 2012. Noncoding flavivirus RNA displays RNA interference suppressor activity in insect and mammalian cells. J. Virol. 86, 13486–13500.

Schuster, S., Zirkel, F., Kurth, A., van Cleef, K.W., Drosten, C., et al., 2014. A unique nodavirus with novel features: Mosinovfirus expresses two subgenomic RNAs, a capsid gene of unknown origin, and a suppressor of the antiviral RNA interference pathway. J. Virol. 88, 13447–13459.

Schwartz, D.A., 2025. Novel reassortants of Oropouche virus are causing maternal-fetal infection during pregnancy, stillbirth, congenital microcephaly and malformation syndromes. Genes. (Basel) 16, 87. https://doi.org/10.3390/genes16010087.

Sehrawat, S., Khasa, R., Deb, A., Prajapat, S.K., Mallick, S., Basu, A., et al., 2021. Valosin-containing protein/p97 plays critical roles in the Japanese encephalitis virus life cycle. J. Virol. 95, e02336-20.

Self, L.S., Shin, H.K., Kim, K.H., Lee, K.W., Chow, C.Y., Hong, H.K., 1973. Ecological studies on *Culex tritaeniorhynchus* as a vector of Japanese encephalitis. Bull. World Health Organ 49, 41–47.

Silhavy, D., et al., 2002. A viral protein suppresses RNA silencing and binds silencing-generated, 21–25 nucleotide double-stranded RNAs. EMBO J. 21, 3070–3080.

Silva, P.A., Pereira, C.F., Dalebout, T.J., Spaan, W.J., Bredenbeek, P.J., 2010. An RNA pseudoknot is required for production of yellow fever virus subgenomic RNA by the host nuclease XRN1. J. Virol. 84, 11395–11406.

Sim, S., Dimopoulos, G., 2010. Dengue virus inhibits immune responses in Aedes aegypti cells. PLoS One 5, e10678.

Sim, S., Ramirez, J.L., Dimopoulos, G., 2012. Dengue virus infection of Aedes aegypti salivary gland and chemosensory apparatus induces genes that modulate infection and blood-feeding behavior. PLoS Pathog. 8, e1002631. https://doi.org/10.1371/journal.ppat.1002631.

Singh, G., Popli, S., Hari, Y., Malhotra, P., Mukherjee, S., Bhatnagar, R.K., 2009. Suppression of RNA silencing by flock house virus B2 protein via PAZ domain interaction. FASEB J. 23, 1845–1857.

Singh, C.P., Singh, J., Nagaraju, J., 2014. bmnpv-miR-3 facilitates BmNPV infection by modulating viral P6.9 expression. Insect Biochem. Mol. Biol. 49, 59–69.

Siu, R.W., Fragkoudis, R., Simmonds, P., Donald, C.L., Chase-Topping, M.E., Barry, G., et al., 2011. Antiviral RNA interference responses induced by Semliki Forest virus infection of mosquito cells. J. Virol. 85, 2907–2917. https://doi.org/10.1128/JVI.02052-10. Epub 2010 Dec 29. PMID:21191029; PMCID:PMC3067965.

Slonchak, A., Hugo, L.E., Freney, M.E., Hall-Mendelin, S., Amarilla, A.A., Torres, F.J., et al., 2020. Zika virus noncoding RNA suppresses apoptosis and is required for virus transmission by mosquitoes. Nat. Commun. 11, 2205. https://doi.org/10.1038/s41467-020-16093-3.

Smit, J.M., Moesker, B., Rodenhuis-Zybert, I., Wilschut, J., 2011. Flavivirus cell entry and membrane fusion. Viruses 3, 160–171.

Sokoloski, K.J., Dickson, A.M., Chaskey, E.L., Garneau, N.L., Wilusz, C.J., Wilusz, J., 2010. Sindbis virus usurps the cellular HuR protein to stabilize its transcripts and promote productive infections in mammalian and mosquito cells. Cell Host Microbe 8 (2), 196–207.

Soldan, S.S., Plassmeyer, M.L., Matukonis, M.K., 2005. González-Scarano F. La Crosse virus nonstructural protein NSs counteracts the effects of short interfering RNA. J. Virol 79 (1), 234–244. https://doi.org/10.1128/JVI.79.1.234-244.2005. PMID:15596819; PMCID:PMC538693.

Song, L., Gao, S., Jiang, W., Chen, S., Liu, Y., Zhou, L., et al., 2011. Silencing suppressors: viral weapons for countering host cell defenses. Protein Cell 2, 273–281.

Soto-Acosta, R., Xie, X., Shan, C., Baker, C.K., Shi, P.Y., Rossi, S.L., et al., 2018. Fragile X mental retardation protein is a Zika virus restriction factor that is antagonized by subgenomic flaviviral RNA. eLife 7, e39023. https://doi.org/10.7554/eLife.39023.

Souza-Neto, J.A., Sim, S., Dimopoulos, G., 2009. An evolutionary conserved function of the JAK-STAT pathway in anti-dengue defense. Proc. Natl Acad. Sci. U S A 106, 17841–17846.

Stadler, K., Allison, S.L., Schalich, J., Heinz, F.X., 1997. Proteolytic activation of tick-borne encephalitis virus by furin. J. Virol. 71, 8475–8481.

Stephens, P.R., Gottdenker, N., Schatz, A.M., Schmidt, J.P., Drake, J.M., 2021. Characteristics of the 100 largest modern zoonotic disease outbreaks. Philos. Trans. R. Soc. Lond. B Biol. Sci. 376, 20200535. https://doi.org/10.1098/rstb.2020.0535.

Stollar, V., Thomas, V.L., 1975. An agent in the Aedes aegypti cell line (Peleg) which causes fusion of Aedes albopictus cells. Virology 64 (2), 367–377. https://doi.org/10.1016/0042-6822(75)90113-0.

Strauss, J.H., Strauss, E.G., 1994. The alphaviruses: gene expression, replication, and evolution. Microbiol. Rev. 58 (3), 491–562.

Sullivan, C.S., Ganem, D., 2005. A virus-encoded inhibitor that blocks RNA interference in mammalian cells. J. Virol. 79 (12), 7371–7379. https://doi.org/10.1128/JVI.79.12.7371-7379.2005. PMID:15919892; PMCID:PMC1143619.

Szemiel, A.M., Failloux, A.B., Elliott, R.M., 2012. Role of Bunyamwera orthobunyavirus NSs protein in infection of mosquito cells. PLoS Negl. Trop. Dis. 6 (9), e1823. https://doi.org/10.1371/journal.pntd.0001823. PMID:23029584; PMCID:PMC3459826.

Tassetto, M., Kunitomi, M., Whitfield, Z.J., Dolan, P.T., Sanchez-Vargas, I., Garcia-Knight, M., et al., 2019. Control of RNA viruses in mosquito cells through the acquisition of vDNA and endogenous viral elements. eLife 8, e42769.

Taucher, C., Berger, A., Mandl, C.W., 2010. A trans-complementing recombination trap demonstrates a low propensity of flaviviruses for intermolecular recombination. J. Virol. 84 (1), 599–611.

Tikhe, C.V., Dimopoulos, G., 2021. Mosquito antiviral immune pathways. Dev. Comp. Immunol. 116, 103964.

Torii, S., Lord, J.S., Lavina, M., et al., 2025. Polygenic viral factors enable efficient mosquito-borne transmission of African Zika virus. Nat. Commun. 16, 9594. https://doi.org/10.1038/s41467-025-64627-0.

Trammell, C.E., Ramirez, G., Sanchez-Vargas, I., St Clair, L.A., Ratnayake, O.C., Luckhart, S., et al., 2022. Coupled small molecules target RNA interference and JAK/STAT signaling to reduce Zika virus infection in Aedes aegypti. PLoS Pathog. 18, e1010411.

Trobaugh, D.W., Gardner, C.L., Sun, C., Haddow, A.D., Wang, E., et al., 2014. RNA viruses can hijack vertebrate microRNAs to suppress innate immunity. Nature 506 (7487), 245–248.

Tsetsarkin, K.A., Vanlandingham, D.L., McGee, C.E., Higgs, S., 2007. A single mutation in chikungunya virus affects vector specificity and epidemic potential. PLoS Pathog. 3, e201. https://doi.org/10.1371/journal.ppat.0030201.

Turell, M.J., Reeves, W.C., Hardy, J.L., 1982. Transovarial and trans-stadial transmission of California encephalitis virus in *Aedes dorsalis* and *Aedes melanimon*. Am. J. Trop. Med. Hyg. 31, 1021–1029. https://doi.org/10.4269/ajtmh.1982.31.1021.

Uchida, L., Espada-Murao, L.A., Takamatsu, Y., Okamoto, K., Hayasaka, D., Yu, F., et al., 2014. The dengue virus conceals double-stranded RNA in the intracellular membrane to escape from an interferon response. Sci. Rep. 4, 7395.

Urquidi, V., Bishop, D.H., 1992. Non-random reassortment between the tripartite RNA genomes of La Crosse and snowshoe hare viruses. J. Gen. Virol. 73, 2255–2265.

van Cleef, K.W., van Mierlo, J.T., Miesen, P., Overheul, G.J., Fros, J.J., Schuster, S., et al., 2014. Mosquito and Drosophila entomobirnaviruses suppress dsRNA- and siRNA-induced RNAi. Nucleic Acids Res. 42 (13), 8732–8744. https://doi.org/10.1093/nar/gku528. PMID:24939903; PMCID:PMC4117760.

van Rij, R.P., Saleh, M.C., Berry, B., Foo, C., Houk, A., Antoniewski, C., et al., 2006. The RNA silencing endonuclease Argonaute 2 mediates specific antiviral immunity in Drosophila melanogaster. Genes. Dev. 20, 2985–2995.

Van Valen, L., 1973. A new evolutionary law. Evolut. Theory 1, 1–30.

Varallyay, E., et al., 2010. Plant virus-mediated induction of miR168 is associated with repression of ARGONAUTE1 accumulation. EMBO J. 29, 3507–3519.

Vargason, J., et al., 2003. Size selective recognition of siRNA by an RNA silencing suppressor. Cell 115, 799–811.

Vasilakis, N., et al., 2009. Mosquitoes put the brake on arbovirus evolution: experimental evolution reveals slower mutation accumulation in mosquito than vertebrate cells. PLoS Pathog. 5, e1000467.

Vasilakis, N., et al., 2019. Exploiting the legacy of the arbovirus hunters. Viruses 11, 471. https://doi.org/10.3390/v11050471.

Vasilakis, N., Cardosa, J., Hanley, K.A., Holmes, E.C., Weaver, S.C., 2011. Fever from the forest: prospects for the continued emergence of sylvatic dengue virus and its impact on public health. Nat. Rev. Microbiol. 9 (7), 532–541.

Vasilakis, N., Forrester, N.L., Palacios, G., Nasar, F., Savji, N., Rossi, S.L., et al., 2013. Negevirus: a proposed new taxon of insect-specific viruses with wide geographic distribution. J. Virol. 87 (5), 2475–2488. https://doi.org/10.1128/JVI.00776-12.

Vasilakis, N., Tesh, R.B., 2015. Insect-specific viruses and their potential impact on arbovirus transmission. Curr. Opin. Virol. 15, 69–74. https://doi.org/10.1016/j.coviro.2015.08.007.

Velde, F.V., Overgaard, H.J., Bastien, S., 2024. An integrated human behavioral model for mosquito-borne disease control: a scoping review of behavior change theories used to identify key behavioral determinants. Heliyon 10, 4252.

Wang, H., et al., 2003. Adenosine kinase is inactivated by geminivirus AL2 and L2 proteins. Plant. Cell 15, 3020–3032.

Wang, X.H., Aliyari, R., Li, W.X., Li, H.W., Kim, K., Carthew, R., et al., 2006. RNA interference directs innate immunity against viruses in adult Drosophila. Science 312, 452–454.

Wang, Y., Jin, B., Liu, P., Li, J., Chen, X., Gu, J., 2018. piRNA profiling of Dengue virus type 2-infected Asian tiger mosquito and midgut tissues. Viruses 10 (4), 213. https://doi.org/10.3390/v10040213.

Watanabe, M., Iwakawa, H.O., Tadakuma, H., Tomari, Y., 2017. Biochemical and single-molecule analyses of the RNA silencing suppressing activity of CrPV-1A. Nucleic Acids Res. 45, 10837–10844.

Watts, D.M., Morris, C.D., Wright, R.E., DeFoliart, G.R., Hanson, R.P., 1972. Transmission of La Crosse virus (California encephalitis group) by the mosquito *Aedes triseriatus*. J. Med. Entomol 9, 125–127. https://doi.org/10.1093/jmedent/9.2.125.

Weaver, S.C., Barrett, A.D.T., 2004. Transmission cycles, host range, evolution and emergence of arboviral disease. Nat. Rev. Microbiol. 2, 789–801. https://doi.org/10.1038/nrmicro1006.

Weaver, S.C., Brault, A.C., Kang, W., Holland, J.J., 1999. Genetic and fitness changes accompanying adaptation of an arbovirus to vertebrate and invertebrate cells. J. Virol. 73, 4316–4326.

Weaver, S.C., Forrester, N.L., Liu, J., et al., 2021. Population bottlenecks and founder effects: implications for mosquito-borne arboviral emergence. Nat. Rev. Microbiol. 19, 184–195. https://doi.org/10.1038/s41579-020-00482-8.

Weaver, S.C., Kang, W., Shirako, Y., Rumenapf, T., Strauss, E.G., Strauss, J.H., 1997. Recombinational history and molecular evolution of western equine encephalomyelitis complex alphaviruses. J. Virol. 71 (1), 613–623.

Weaver, S.C., Rico-Hesse, R., Scott, T.W., 1992. Genetic diversity and slow rates of evolution in New World alphaviruses. Curr. Top. Microbiol. Immunol. 176, 99–117.

Wei Xiang, B.W., Saron, W.A.A., Stewart, J.C., Hain, A., Walvekar, V., Misse, D., et al., 2022. Dengue virus infection modifies mosquito blood-feeding behavior to increase transmission to the host. Proc. Natl Acad. Sci. USA 119, e2208629119.

Welsch, S., Miller, S., Romero-Brey, I., Merz, A., Bleck, C.K.E., Walther, P., et al., 2009. Composition and three-dimensional architecture of the dengue virus replication and assembly sites. Cell Host Microbe 5, 365–375.

Williams, M.C., Woodall, J.P., Corbet, P.S., Gillett, J.D., 1965. O'nyong-nyong fever: an epidemic virus disease in East Africa. 8. Virus isolations from *Anopheles* mosquitoes. Trans. R. Soc. Trop. Med. Hyg. 59, 300–306. https://doi.org/10.1016/0035-9203(65)90012-X.

Woolhouse, M.E.J., Taylor, L.H., Haydon, D.T., 2001. Population biology of multihost pathogens. Science 292, 1109–1112. https://doi.org/10.1126/science.1059026.

Woolhouse, M., Webster, J., Domingo, E., Charlesworth, B., Levin, B.R., 2002. Biological and biomedical implications of the co-evolution of pathogens and their hosts. Nat. Genet. 32, 569–577. https://doi.org/10.1038/ng1202-569.

World Health Organization (WHO), 2024. Vector-Borne DiseasesWHO, Geneva. https://www.who.int/news-room/fact-sheets/detail/vector-borne-diseases.

Wouters, R.M., et al., 2024. Local environmental factors drive distributions of ecologically contrasting mosquito species (Diptera: Culicidae). Sci. Rep. 14, 19315.

Xi, Z., Ramirez, J.L., Dimopoulos, G., 2008. The Aedes aegypti toll pathway controls dengue virus infection. PLoS Pathog. 4, e1000098.

Ye, K., et al., 2003. Recognition of small interfering RNA by a viral suppressor of RNA silencing. Nature 426, 874–878.

Yeh, S.C., Pompon, J., 2018. Flaviviruses produce a subgenomic flaviviral RNA that enhances mosquito transmission. DNA Cell Biol. 37 (3), 154–159. https://doi.org/10.1089/dna.2017.4059.

Yeh, S.C., Strilets, T., Tan, W.L., Castillo, D., Medkour, H., Rey-Cadilhac, F., et al., 2023. The anti-immune dengue subgenomic flaviviral RNA is present in vesicles in mosquito saliva and is associated with increased infectivity. PLoS Pathog. 19, e1011224. https://doi.org/10.1371/journal.ppat.1011224.

Yu, I.M., Holdaway, H.A., Chipman, P.R., Kuhn, R.J., Rossmann, M.G., Chen, J., 2009. Association of the pr peptides with dengue virus at acidic pH blocks membrane fusion. J. Virol. 83, 12101–12107.

Yu, I.M., Zhang, W., Holdaway, H.A., Li, L., Kostyuchenko, V.A., Chipman, P.R., et al., 2008. Structure of the immature dengue virus at low pH primes proteolytic maturation. Science 319, 1834–1837.

Zhang, X., et al., 2006. Cucumber mosaic virus-encoded 2b suppressor inhibits Arabidopsis Argonaute1 cleavage activity to counter plant defense. Genes. Dev. 20, 3255–3268.
Zhang, Y., Li, B.X., Mao, Q.Z., Zhuo, J.C., Huang, H.J., Lu, J.B., et al., 2023. The JAK-STAT pathway promotes persistent viral infection by activating apoptosis in insect vectors. PLoS Pathog. 19, e1011266.
Zhang, L., Xu, W., Gao, X., Li, W., Qi, S., Guo, D., et al., 2020. lncRNA sensing of a viral suppressor of RNAi activates non-canonical innate immune signaling in Drosophila. Cell Host Microbe 27 (1), 115–128.e8.
Zhu, M., Wang, J., Deng, R., Xiong, P., Liang, H., Wang, X., 2013. A microRNA encoded by Autographa californica nucleopolyhedrovirus regulates expression of viral gene ODV-E25. J. Virol. 87, 13029–13034.
Zhu, M., Wang, J., Deng, R., Wang, X., 2016. Functional regulation of an Autographa californica nucleopolyhedrovirus-encoded microRNA, AcMNPV-miR-1, in baculovirus replication. J. Virol. 90, 6526–6537.

CHAPTER FIVE

Vector competence: Evolutionary drivers and constraints

Emily N. Gallichotte and Gregory D. Ebel*
Department of Microbiology, Immunology and Pathology, College of Veterinary Medicine and Biomedical Sciences, Center for Vector-borne Infectious Diseases, Colorado State University, Fort Collins, CO, United States
*Corresponding author. e-mail address: gregory.ebel@colostate.edu

Contents

Abstract

The term vector competence (VC) refers to the efficiency with which an arthropod can become infected with, support replication of, and ultimately transmit a given pathogen. In the context of arthropod-borne virus (arbovirus) outbreaks, estimating arthropod species-specific VC through controlled laboratory studies is critical to assess the risk that implicated species pose to humans and animals, and to design appropriate interventions to limit this risk. An extensive literature clearly demonstrates significant variation in VC between mosquito genera, species, and populations of the same species. However, the factors that control VC under natural conditions are surprisingly poorly understood. This review discusses VC through an evolutionary lens, focusing on molecular and abiotic determinants of VC and how VC may be altered as mosquitoes adapt to new and/or changing environments. Barriers to advancing our understanding of VC are highlighted, as well as opportunities to circumvent these barriers.

1. Vector competence: definitions and central concepts

In public health, medical, and veterinary entomology, the term vector competence (VC) refers to the efficiency with which an arthropod vector can transmit a pathogen (Hardy et al., 1983; Kramer and Ebel, 2003). It is a

Advances in Insect Physiology, Volume 70
ISSN 0065-2806, https://doi.org/10.1016/bs.aiip.2026.05.005

fundamental property of the host-pathogen interaction that applies to most agents that are transmitted by arthropods. While VC and many of the concepts covered here are generally applicable to other arthropod-pathogen systems such as those that cause Lyme disease and malaria, we will limit our discussion to mosquito-borne arboviruses (i.e., arthropod-borne viruses) that infect vertebrates.

VC is partly controlled by mosquito genetics (Bosio et al., 2000; Graham et al., 2003; Bennett et al., 2005), and significant variation has been documented in VC for a given virus among different mosquito species (some are highly competent, whereas others are completely refractory (Ebel et al., 2005; Turell et al., 1982b; Turell et al., 2005)) and between populations of the same species (Hardy et al., 1976; Hardy et al., 1978). VC is a quantitative trait – it is rare for 100 % of individuals of a given species/population to be competent for a given virus. VC is also a dynamic trait. For instance, mosquitoes that do not transmit a viral species today may later become competent for its transmission. Evidence has been presented that viruses and mosquitoes coevolve such that local viruses are most efficiently transmitted by local mosquitoes (Dabo et al., 2024), suggesting genotype-by-genotype interactions in arbovirus-vector pairings. Understanding the evolutionary mechanisms that underpin virus transmission by mosquitoes is thus a central topic in the study of virus-host interactions.

Biological transmission of a virus by a mosquito vector can occur by several modes, including horizontal transmission involving cycling between vertebrates and mosquitoes, vertical transmission between infected female mosquitoes and their offspring, and venereal transmission between infected male mosquitoes and uninfected adult females during mating. Venereal transmission is thought to occur relatively infrequently, and thus will not be a major focus of this review, except to point out that it has been documented among particular virus-mosquito pairs (Sanchez-Vargas et al., 2018; Campos et al., 2017; Thompson, 1983).

Horizontal transmission of mosquito-borne arboviruses requires intimate interaction between the virus and mosquito that is initiated when an uninfected female mosquito feeds on a viremic vertebrate host. The steps required for the transmission of arboviruses by competent vectors have been reviewed extensively, as have the physiological 'barriers' that may limit the ability of some mosquito species to successfully transmit viruses (Hardy et al., 1983; Lewis et al., 2023). Briefly, the circulating virus is taken into the mosquito midgut along with blood during the feeding process.

Viruses then attach to and enter cells of the midgut epithelium, where they undergo initial replication and spread within the cells of the midgut. In many cases infection can spread to most of the midgut cells, inducing apoptotic and transcriptional changes that may dramatically affect midgut physiology (Bartholomay et al., 2010; Smartt et al., 2009; Shin et al., 2020). Ultimately, the virus egresses from the midgut and infects secondary tissues, including fat body, muscle, neurologic tissues, and, critically, the salivary glands (Lewis et al., 2023). Infection of the salivary gland is followed by the release of infectious virus into salivary secretions, where it may be expectorated along with immune-modulating salivary factors when a second bloodmeal is acquired. This entire process is enhanced when a mosquito feeds several times during its lifespan (Johnson et al., 2023). Several sequential steps are therefore required for an arbovirus to be horizontally transmitted by a competent vector, all of which must occur successfully for the virus to perpetuate (Fig. 1). Recent work has addressed the mechanism of dissemination of arboviruses within competent vectors, demonstrating that within *Aedes* and *Culex* mosquitoes, virus dissemination between tissues occurs most efficiently through virus 'packages' that may be either extracellular vesicles (Niu et al., 2026) or mosquito hemocytes (Hall et al., 2025). Many of these steps, including the ability to package viruses within cells or vesicles, may be under significant evolutionary pressure because they are central to the virus lifecycle, however how this impacts the mosquito remains unknown.

Vertical transmission from an infected female mosquito to her offspring is also critical to the perpetuation of some arboviruses, including several orthobunyaviruses (Turell et al., 1982a; Hartman et al., 2021; Turner and Christofferson, 2024; Tesh and Gubler, 1975). Vertical transmission may also occur less frequently among many other arboviruses, but the impact of this transmission mode on long-term virus perpetuation is less clear (Phumee et al., 2019; Sanchez-Vargas et al., 2018; Goddard et al., 2003; Watts et al., 1987b). The virus-vector interactions that produce vertical transmission are variable and dependent on the virus-vector system. For La Crosse virus, infection of the ovaries of female *Ae. triseriatus* occurs, ensuring that nearly all of the resulting offspring are infected (Miller et al., 1977). When mosquito ovaries are infected, vertical transmission is also termed transovarial transmission (TOT). Due to the high infection rate in transovarially infected offspring, the virus can be maintained for several generations without the need for the mosquito to acquire the virus from an infected vertebrate (Turell et al., 1982a). In contrast, most other arboviruses are transmitted less

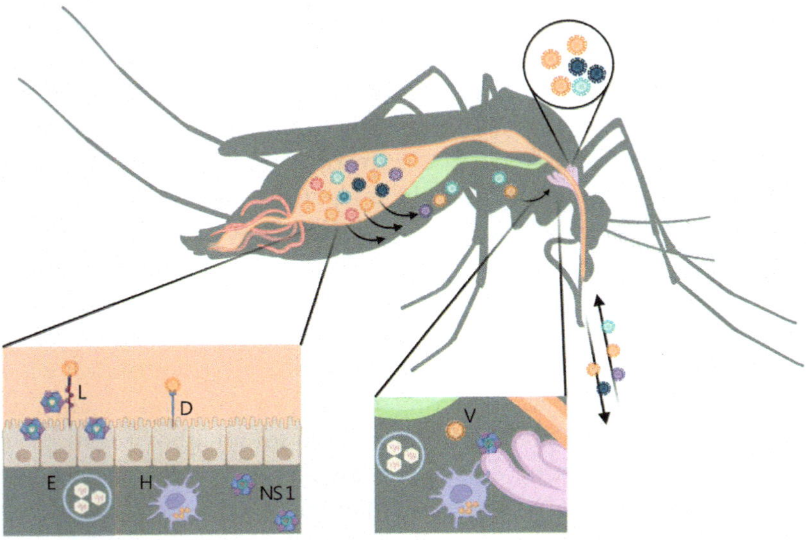

Fig. 1 Virus-mosquito adaptations and constraint during horizontal arbovirus transmission. Arboviruses have evolved mechanisms that permit their perpetuation in mosquitoes. These adaptations are intimately linked to several specific virus-cell interactions, any one of which may fail, rendering a mosquito incompetent. Viruses taken in during mosquito bloodfeeding attach to the midgut epithelium via C-type lectins such as DC-SIGN (**D**) or LDLR family receptors (**L**) expressed on cell surfaces. Hexameric NS1 (**NS1**) secreted by virus-infected vertebrate cells may be present in the host bloodmeal and facilitate virus infection of, and egress from, midgut tissue. Dissemination of arboviruses into secondary sites of replication, ultimately including the salivary glands, occurs via several mechanisms that are currently under investigation. In addition to infection by free viruses (**V**) within the hemolymph, it has recently been shown that viral genome-containing capsids are enclosed within endosomes (**E**), and/or present within mosquito hemocytes (**H**), either or both of which may facilitate spread within mosquito vectors. The degree to which hexameric secreted NS1 may also contribute to intra-mosquito spread remains to be determined. Ultimately, transmission occurs in competent vectors when virus is released into mosquito salivary secretions and is expectorated into the skin of a subsequent bloodmeal host. Similar adaptations and constraints could be imagined to control vertical and venereal transmission, but these are less well understood at present.

efficiently via the vertical route due to a lack of ovarian infection. While the mechanism for this type of vertical infection is not clear, it is thought that the virus contaminates eggs during oviposition (trans-egg vertical transmission, reviewed in (Lequime et al., 2016)), resulting in a lower infection prevalence compared to transovarial transmission (Nelms et al., 2013). Vertical transmission, including TOT, is thought to facilitate the overwintering of several arboviruses in infected eggs, larvae, or diapausing adults in regions where mosquito activity ceases for several months of the year. Moreover, vertical transmission of arboviruses is a central mode of perpetuation for a subset of arboviruses such as La Crosse virus in temperate North America, and in some circumstances, it is an important ancillary mode of perpetuation that facilitates virus persistence during periods of mosquito inactivity. Again, in many cases, the ability of a given virus to undergo vertical transmission may be under some degree of natural selection.

Regardless of transmission mode, there is a robust literature delineating the importance of VC, and how arboviruses adapt to maximize it. It seems obvious that arboviruses, like other organisms, would tend to maximize their replication rate to ensure perpetuation. One excellent example of this is chikungunya virus (CHIKV), which has emerged in recent decades as a significant global pathogen. During its initial expansion throughout the Indian Ocean basin, an outbreak occurred on La Réunion Island that resulted in over 266,000 human cases (Borgherini et al., 2007). Genomic and VC studies demonstrated that epidemic CHIKV in this setting was associated with a mutation in the E1 glycoprotein that resulted in the substitution of an alanine for a valine at position 226 (Tsetsarkin et al., 2007; Tsetsarkin et al., 2006; Schuffenecker et al., 2006). This mutation allowed CHIKV to be transmitted efficiently by *Ae. albopictus*, a species that had not previously been implicated in epidemic transmission and which was more abundant than the main vector, *Ae. aegypti*. This increase in the variety of competent vectors is thought to have significantly increased the number of human cases that occurred. A similar pattern was noted with West Nile virus (WNV) early during its expansion in the US following its introduction in 1999 (Hadfield et al., 2019). WNV strains collected during 2002 tended to have a seemingly conservative amino acid substitution at position 159 (valine to alanine) of the envelope glycoprotein (Moudy et al., 2007). WNV strains with this mutation had a shorter extrinsic incubation period (EIP, the number of days between initial infection and when transmission can occur) compared to the parental strains that did not have the V159A substitution (Ebel et al., 2004). In this case, although the overall VC of mosquitoes for this virus was unchanged, the speed at which the virus replicated and disseminated in the mosquito was increased significantly reducing EIP, meaning mosquitoes are infectious and capable of transmitting the virus sooner. These studies, along with several others e.g. (Liu et al., 2017), demonstrate that, as might be expected, viruses evolve rapidly to adapt to local vectors to increase their transmission.

VC is a complex phenotype that requires several distinct interactions to occur successfully. This poses several fascinating evolutionary questions. For example: what are the advantages for the viruses of a dual-host lifecycle that requires efficient horizontal transmission? It seems far simpler for a virus to specialize on a single host type. Several viruses that only circulate in either a vertebrate host or mosquito, but not both, are phylogenetically quite closely related to arboviruses. Entebbe bat virus (ENTV), for example, has been isolated only from bats, and it is an orthoflavivirus that is related to the yellow fever group of mosquito-borne viruses (Kading et al., 2015; Lumsden et al., 1961). While ENTV retains the ability to replicate in mosquito cell cultures and infect mosquito midguts, it fails to disseminate from the midgut into the hemocoel of tested mosquito colonies (Fujii et al., 2025) and has never been detected in mosquitoes in the field (Mayanja et al., 2021; Mossel et al., 2017). Similarly, a large group of insect-specific orthoflaviviruses (ISV) (e.g., Nhumirim virus (Kenney et al., 2014)) in addition to classic insect-specific orthoflaviviruses such as cell-fusing agent virus, are most closely related to mosquito-borne orthoflaviviruses (Blitvich and Firth, 2015; Ebel, 2017). Similar phylogenetic patterns exist in many groups of viruses that were formerly considered principally arboviruses (Nasar et al., 2015, Nasar et al., 2012). Thus, the ability to replicate, disseminate and be transmitted by vertebrate and invertebrate hosts appears to have been gained and/or lost frequently in the evolutionary history of viruses. Thus, the non-vector-borne viruses, which sit within clades that contain arboviruses provide a key opportunity to probe the evolutionary determinants of vector-vertebrate transmission, which have thus far largely remained unexplored although recent work has investigated these origins in bunyaviruses (Junglen, 2016).

2. Molecular control of VC

Due to the complexity and intimate nature of the virus-vector interaction, and the number of steps that must occur successfully for viruses to persist in nature, there are a wide array of viral evolutionary adaptations that underpin VC. These include adaptations both in the nucleic acids and encoded proteins of viruses.

Most arboviruses are RNA viruses, and viral RNA sequences have been used to assess the likelihood that a given virus may be arthropod-borne when viral isolates are not available for phenotypic characterization

(Fauver et al., 2016). Phylogenetic studies that determine the relatedness among viruses can, of course, place viruses into clades that are generally phenotypically similar in terms of host utilization (Gould et al., 2003). Dinucleotide and codon usage biases have also been used to assess host associations because they tend to match those of the hosts on which they mostly rely. A robust literature has clearly demonstrated strong codon bias in different host types, and that this bias has significant impacts on protein expression and stability within host cells (Peterson et al., 2004; Lobo et al., 2009; Hershberg and Petrov, 2008; Lawrie et al., 2013). Much of the bias in codon usage for viruses is attributable to the avoidance of CpG and UpA dinucleotides (Takata et al., 2017), which can lead to degradation of CpG-rich viral RNA sequences by several mechanisms. Flaviviruses transmitted by arthropods tend to have less pronounced codon bias than those flaviviruses that rely only on vertebrates (Sexton and Ebel, 2019). In addition, viruses that have been engineered to contain altered codon usage that results in increased CpG and UpA dinucleotides are highly attenuated in vertebrates but not in arthropods (Stauft et al., 2018; Li et al., 2018). In addition, viral RNA secondary structure is critical in facilitating virus replication and host immune antagonism (Moon et al., 2015). Thus, beyond the sequence of a given viral RNA molecule, the way the viral RNA codes for proteins, and how secondary structure functions to enhance virus replication, bear the evolutionary signature of longstanding host association.

The initial step in successful replication of a virus in any host is the ability to recognize host cells. This interaction is reliant on viral structural proteins that are present on the surface of mature virions. For viruses that rely on multiple hosts, this requires recognition of highly conserved receptor molecules or the ability to attach to abundantly expressed glycans that are expressed in most mosquito tissues. WNV and Japanese encephalitis viruse (JEV), for example, can utilize the C-type lectin DC-SIGN for attachment and infection of host vertebrate cells (Wang et al., 2016; Shimojima et al., 2014; Martina et al., 2008; Davis et al., 2006). Recently, several studies have addressed more specific vertebrate host receptors for arboviruses. In particular, members of the low-density lipoprotein family have been identified as vertebrate entry factors for multiple encephalitic alphaviruses (Ma et al., 2024), yellow fever virus (Chong et al., 2025), tick-borne encephalitis virus (Mittler et al., 2025), Jamestown Canyon virus (Frey et al., 2025) and likely many others. What is notably missing is the translation of these findings to arthropods. Given the lack of clarity and relative intractability of defining a single receptor molecule that controls arbovirus infection in arthropods, the extent to which any vertebrate-defined receptor is also relevant in mosquitoes remains to be determined. This constitutes a fundamental gap in the existing literature that should be addressed quickly using relevant *in vivo* systems (i.e., live mosquitoes and ticks), many of which have tractable systems for altering the expression of candidate 'receptor' molecules (Viswanatha et al., 2021; Rozen-Gagnon et al., 2021). Studies such as this would permit a far more objective assessment of the evolutionary adaptations and constraints that contribute to VC across arthropod taxa.

One notable recent exception to the general lack of consideration of the arthropod vector when addressing the molecular determinants of virus-host specificity is a recent publication that identifies the orthoflavivirus capsid protein of dengue virus (DENV), Zika virus (ZIKV) and JEV as a critical determinant of vector specificity via its interaction with the mosquito valosin-containing protein (VCP) (Niu et al., 2026). Viral genome-containing capsids bind to VCP, allowing their incorporation into extracellular vesicles, which promote spread within the mosquito by protecting them from the acidic environment of the *Aedes* and *Culex* hemolymph. Two amino acid residues in VCP determine its binding capacity for different viruses. This work is significant because the mechanisms that limit virus infection of different mosquito vectors are not well understood. Specifically, if receptor-ligand interactions drive vector specificity, it is hard to imagine how, for example, a C-type lectin would be so different that, for example, *Ae. aegpyti* could be highly susceptible to one virus but not another, or that different mosquito species display vastly different VC for the same virus, as has been noted throughout the history of VC studies. Moreover, it is becoming increasingly clear that post-entry determinants within diverse viruses have a significant impact on VC.

Arthropod-borne viral replicase complexes are similar to most other RNA viruses in that they lack error-checking and mismatch-repair mechanisms, leading to misincorporation rates much higher than those of eukaryotic polymerases (Holland et al., 1982). This, coupled with their short generation times and large population sizes (WNV, for example, can reach 10^9 plaque-forming units per mL in the blood of infected birds), can result in highly genetically diverse populations, which may, in turn, facilitate rapid evolution in changing environments (Clarke et al., 1994, Quesnel-Vallieres et al., 2008; Twiddy et al., 2003; Kerr and Best, 1998; Holland et al., 1982). This general phenomenon has been documented extensively for both arthropod-borne and directly transmitted viruses (Liu et al., 2017; Hernandez-Alonso et al., 2015; Quesnel-Vallieres et al., 2008; Laskus et al., 1998; Farci et al., 1997; Ebel et al., 2004; Tsetsarkin et al., 2007; Ciota et al., 2007a; Ciota et al., 2007b; Ciota et al., 2009). A second benefit of viral genomic diversity is that it may facilitate arboviral evasion from the mosquito RNAi-based antiviral response (Fig. 2). This is possible because mosquito RNAi, triggered by

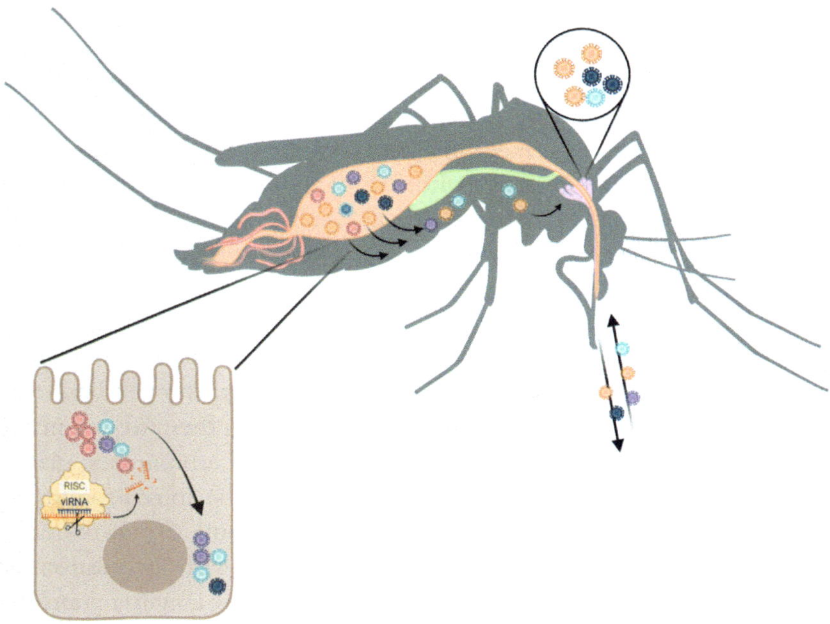

Fig. 2 Virus population dynamics within mosquito vectors during horizontal transmission. Mosquitoes have major evolutionary influences on the population dynamics of the viruses they transmit. Genetically diverse virus populations generated via error-prone replication experience strong bottlenecks upon infection of the mosquito midgut that remove most of the generated diversity at random, i.e., without respect to the overall fitness of the variant. Purifying selection in mosquitoes is weak, which permits the perpetuation of low fitness variants. Within individual mosquito cells, the dominant antiviral response, RNA interference (RNAi), mounts a virus-sequence-specific response to viral replicative intermediates that removes the most common viral genotypes, creating an intracellular milieu that favors rare viral genotypes and the emergence of new variants. This process of population reduction and expansion occurs at each stage of infection within mosquitoes. Thus, within many arbovirus systems, mosquitoes present a chaotic evolutionary environment, where bottlenecks blunt the power of natural selection, and diversifying selection imposed by RNAi serves as a source of new viral variants.

double-stranded RNA viral replicative intermediates (Campbell et al., 2008; Travanty et al., 2004), requires sequence specificity of guide strands loaded into the RNA-induced silencing complex (RISC) for efficient degradation of cognate strands (Fire et al., 1998; Martinez et al., 2002; Zamore et al., 2000) (imperfect guide-target matches tend to result in translational repression rather than degradation). A high degree of genetic diversity in viral RNA sequences within mosquitoes ensures that many of them escape RNAi because they poorly match common sequences loaded into the RISC (Brackney et al., 2009; Gitlin et al., 2005). An additional consequence of this is that rare variants have the chance to increase in population frequency within mosquitoes due to a mechanism akin to negative frequency-dependent selection. This has been noted in several publications, including several studies of arboviruses that demonstrate high viral diversity within arthropods and increased fitness of these diverse populations compared to populations that are less diverse (Brackney et al., 2009; Jerzak et al., 2007; Jerzak et al., 2005; Brackney et al., 2011), or where diversity is limited by engineering replicase complexes to more faithfully replicate input RNA sequences (Coffey et al., 2011; Beaucourt et al., 2011; Vignuzzi et al., 2006). Under this conceptual model, error-prone replication of arboviruses can be viewed as an evolutionary adaptation of arboviruses allowing them to circumvent mosquito immunity. Moreover, the error rate of arboviruses may be optimized, or 'tuned,' to facilitate successful infection of arthropods that rely on RNAi for antiviral defense (Wilke and Adami, 2003).

A major force exerted by mosquito vectors on the virus population they transmit is stochastic, rather than selective, reductions in virus population size. These reductions are termed 'bottlenecks,' and they have been extensively documented to occur at each successive step of mosquito infection (Smith et al., 2008; Ciota et al., 2012; Forrester et al., 2012; Weger-Lucarelli et al., 2018; Aliota et al., 2018; Fitzmeyer et al., 2023). For populations such as RNA viruses that undergo error-prone replication, this presents a significant challenge, as successive nonselective reductions in populations size can lead to fitness reductions through a Muller's ratchet-like mechanism, where deleterious mutations increase in frequency after stochastic population reductions, and therefore accumulate over time (Escarmis et al., 1996; Escarmis et al., 2009). This phenomenon has been demonstrated to occur in arbovirus populations following mosquito transmission (Grubaugh et al., 2016), raising important questions about how viruses that perpetuate using arthropod vectors are able to persist in nature.

Finally, many viruses have evolved mechanisms to modify the extracellular environment to maximize the likelihood of successful infection. An excellent example of this is the orthoflavivirus-encoded nonstructural protein 1 (NS1), which has been shown to dramatically modify the host environment. NS1 is a secreted protein that is frequently found in hexameric form in the blood of vertebrates (Watterson et al., 2016). While this protein has been shown to impact the integrity of the vertebrate host endothelium (Chen et al., 2016; Beatty et al., 2015), it also seems to alter the permeability of the arthropod midgut (Liu et al., 2016; Quezada-Ruiz et al., 2026). Injury to host endothelial tissues represents a unifying link between host types, where a function in one host (e.g., disruption of midgut endothelium to facilitate vector competence) contributes to disease and/or transmission in another (e.g., vascular leak in DENV (Beatty et al., 2015)). Whether the primary function of NS1, and its evolutionary origin, is in the mosquito, i.e., to maximize the likelihood of successful midgut (or salivary gland) infection, or in vertebrates, i.e., to improve dissemination and immune evasion, remains to be determined.

3. Influence of abiotic stress on VC via evolutionary mechanisms

Mosquitoes lack the ability to regulate their temperature and are thus subject to ambient environmental conditions. The temperatures to which they are exposed vary diurnally, seasonally, and according to local climatic conditions, which range from tropical to temperate. Therefore, understanding whether and how mosquitoes may adapt to varying thermal stresses, and how this may ultimately impact VC, is extremely complicated (reviewed in (Bellone and Failloux, 2020)). What is quite clear, however, is that temperature has a major impact on the VC of mosquitoes for most viruses (Hardy et al., 1983; Watts et al., 1987a; Kramer et al., 1983; Hardy et al., 1980; Lambrechts et al., 2011). Briefly, at higher temperatures, viruses replicate more efficiently, resulting in a shortened EIP. This shortened EIP in turn increases the likelihood of transmission, because a greater proportion of mosquitoes in the population will be capable of transmitting virus, which can have significant impacts on virus epidemiology, regardless of whether it is temperature-driven. Relatively few studies have examined the impact of temperature on the evolutionary dynamics of arboviruses in mosquitoes (Fay et al., 2021; Ware-Gilmore et al., 2025). A study of ZIKV in *Ae. aegypti* and *Ae. albopictus* held at different temperatures revealed that higher temperatures generally produce more viral genetic diversity (Murrieta et al., 2021). Whether this is due to temperature-driven increases in virus replication or shifts from resistance to tolerance in mosquitoes as has been seen with an ISV (cell-fusing agent virus) (Perdomo et al., 2025) requires further investigations. However, this study also documented that when mosquitoes were under diurnally varying temperatures, virus RNA sequences showed strong evidence of purifying selection (i.e., d_N/d_S ratios less than 1). This finding stands in contrast to much of the literature on the selective pressures on viruses in mosquitoes, which suggests that purifying selection within mosquitoes is relaxed (Jerzak et al., 2005, Jerzak et al., 2007; Jerzak et al., 2008). Additional studies have been conducted using field-collected virus strains to assess whether slowly changing climatic conditions may also impact virus populations. These studies are relatively rare, but tend to suggest that climate may impact virus evolution on observable timescales (Fay et al., 2024; Fay et al., 2025). The collective literature, therefore, contains some intriguing hints on how temperature may impact virus population biology within competent mosquitoes, but additional experimental and observational work is required to develop a clear understanding of how temperature impacts virus evolution in the context of VC.

3.1 Humidity

While dramatically less studied compared to temperature, humidity (and conversely dehydration stress) is an important variable that impacts mosquito physiology and many life history traits (Brown et al., 2023). These include longevity, blood-feeding propensity, and mosquito behavior, all of which can impact the virus-vector interaction, including both through impacts on vectorial capacity and some aspects of VC (Brown et al., 2023; Holmes et al., 2022; Hagan et al., 2018). Multiple studies have modeled how humidity impacts arbovirus transmission in nature (Whitehorn and Yacoub, 2019; Cox et al., 2025; Pascoe et al., 2022; Mousavi et al., 2024), however, only three studies have experimentally evaluated the impact of altered humidity on VC, limiting meta-analyses and our ability to define the role of humidity alone, independent of changes to temperature.

A study using ZIKV in *Ae. aegypti* found that infection rates increased as relative humidity decreased (Abu et al., 2024). However, disseminated infection rates and levels of disseminated infectious virus were highest at intermediate compared to low and high humidity. Two additional studies tested the impact of altered humidity on Mayaro virus (MAYV) VC in *Ae. aegypti.* The first study found that altering relative humidity for <24 h had no impact on any aspect of VC (infection, dissemination or transmission rates, or viral titers) regardless of whether the dehydration stress occurred prior to or after infection (Manzano-Alvarez et al., 2023). The second study found that long-term exposure (7–14 days) to dehydration stress did not impact infection, dissemination or transmission rates, however at low humidity conditions, there were significantly increased levels of disseminated virus, suggesting dehydration stress may impact viral dissemination (Wahaab et al., 2025).

While these studies suggest that dehydration stress and/or the mosquito's response to it might increase some aspects of infection in *Ae. aegypti,* additional studies across multiple mosquito species, viruses, and exposures (percent relative humidity and length) are needed to better define if and how relative humidity alters VC. Importantly, it is known that dehydration stress alters mosquito biology and physiology, which can have important implications for VC and viral evolution (Brown et al., 2023). Other than survival and blood-feeding, these studies did not evaluate other life history traits or demonstrate that the mosquitoes were biologically stressed in response to the altered humidity. Therefore, it's possible the length of dehydration exposure used was not enough to sufficiently stress the mosquitoes and alter their biology/physiology, which might have limited the impact on VC. Subsequent studies should investigate prolonged exposure to altered humidity, determine the extent of host response and adaptation, and evaluate VC and viral evolution.

4. Standards and data

Many of the most studied arboviruses in VC experiments are those that have caused major outbreaks or pose an ongoing threat to human or animal health (e.g., WNV, DENV, CHIKV, etc.). As a result, most mosquito-virus combinations are untested in experimental settings, even those with potential for emergence or spread (e.g., Everglades virus in *Ae. aegypti*) (Chen et al., 2023). The unevenness of research effort makes it challenging to compare mosquito and viral factors impacting VC due to the lack of experimental outcomes for most virus-vector pairings. This ultimately means that performing genome-scale association studies to predict future virus-vector compatibility and define key genetic determinants of VC is currently not possible to the degree that it should be.

Much work remains to be done experimentally and in vector genomics to assess functional and evolutionary relationships across virus-mosquito pairings.

Another major limitation to this type of meta-analyses, and identification of both intrinsic and extrinsic factors impacting VC, is the lack of consistent data reporting from experimental VC studies. There are many critical experimental variables within VC studies, all of which can impact outcomes (e.g., extrinsic incubation temperature, mosquito generation, virus passage history, etc.); however, many of these parameters go unreported in the literature. Additionally, while many studies report derived rates to quantify and compare VC (e.g., the proportion of mosquitoes with a disseminated infection of the total number of infected mosquitoes), by not also reporting the complete dataset, meta-analyses and comparisons across studies cannot be done without raw numbers.

Metadata standards are common (and sometimes required) in many scientific fields, ensuring that datasets are well-described and structured, and thus accessible to the scientific community. For example, the Minimum Information for Publication of Quantitative Real-Time PCR Experiments (MIQE) standards define the minimum information for publication of quantitative real-time PCR experiments (Bustin et al., 2009; Wu et al., 2022). To address this gap in the arbovirology field, a 'minimum data standard for vector competence experiments' was proposed, which aims at capturing the critical experimental details and outcomes (Wu et al., 2022). This data standard aligns with FAIR (Findability, Accessibility, Interoperability, and Reusability) guiding principles, which aim at improving the infrastructure supporting the reuse of published data (Wilkinson et al., 2016). However, unlike sequencing databases like Sequence Read Archive (SRA), Genbank, or GISAID, there are no centralized repositories for experimental VC data, so we rely on authors reporting results within their publications (Azman et al., 2024; Gallichotte et al., 2025).

An example of how this template (Wu et al., 2022) may be used is demonstrated by a recent publication assessing Oropouche virus experimental VC studies. This work conducted meta-analyses to quantitatively compare susceptibility and VC of multiple mosquito and midge species (Gallichotte et al., 2025). After recovering and standardizing data, it was clear that results across historical and current studies were surprisingly comparable: midges are more competent vectors compared to mosquitoes, regardless of experimental factors like mosquito origin or titer of virus in the bloodmeal. Ensuring data compatibility and consistent reporting should significantly improve the predictive capacity of models aiming to discern the factors that facilitate longstanding virus-vector interactions.

5. Summary

Hosts and parasites coevolve on evolutionary timescales. Due to their capacity for rapid evolution, viruses, including those transmitted by arthropods, have provided outstanding model systems through which we can better understand principles of evolutionary theory. Examination of the intimate nature of mosquito-virus interactions, i.e., the competence of mosquitoes to transmit viruses, has revealed key insights into the long-standing nature of these interactions and the points in these interactions where selection and constraint have shaped them. This review has covered several viral, host, and abiotic factors that impact VC. Nonetheless, several important questions remain to be answered. For example, while viruses are notoriously adaptable to different hosts, some barriers to host switching appear to be insurmountable (Gallichotte et al., 2023). Can the evolutionary mechanisms that arboviruses utilize overcome these barriers in any realistic scenario? As we gain more data on the molecular basis for virus-vector incompatibility answering this question becomes more tractable.

Experimental infections of mosquitoes are relatively straightforward, and often the best way to study aspects of VC. However, conducting these experiments under conditions (and for lengths of time) that allow evolution to occur is far more challenging (e.g., maintaining mosquitoes at increased temperatures for multiple generations, then assessing impacts on VC and viral evolution, as compared to only altering temperature for the duration of the infection). Additionally, both viruses and mosquitoes can evolve in ways that experiments do not measure (e.g., changes to the mosquito metabolome may occur, but only transcriptional responses were measured). Additionally, researchers almost exclusively use laboratory colonies of mosquitoes, which are genetically restricted due to lower population sizes compared to wild mosquitoes (Ross et al., 2019), and may poorly represent several virus-vector interactions. For example, it may be that ENTV is a true arbovirus, but we have not yet tested VC in its competent mosquito species because we do not have it in a colony, and colonization may be impossible (which is common among mosquitoes). Therefore, many of the conclusions from laboratory studies may not translate to field conditions.

Most studies of viruses in the mosquito *Cx. tarsalis* use a single colony that has been maintained in laboratories for over 20 years (Provost-Javier et al., 2010). Future work should include wild-caught or less laboratory-adapted strains. Expanding the literature beyond commonly tested virus-vector pairs could, moreover, provide key insights into the determinants of VC. Similarly, additional genomic resources for a broader array of vector species would prove invaluable in making rational predictions regarding likely virus emergence trajectories and charting the course of future virus evolution. As humanity faces the ongoing challenges of arboviruses emerging as global pathogens in the context of land use and environmental change, gathering this data and conducting these key experiments represents a pressing yet tractable challenge.

Acknowledgements

The authors thank Dr. Emily Fitzmeyer for assistance with figures.

Research in the Ebel laboratory concerning mosquito-virus interactions is supported by funds from the US National Institute of Allergy and Infectious Diseases, National Institutes of Health, under grants AI067380 and AI173206, and from the US National Science Foundation under DBI 2021909, 2213854, and 2515340.

References

Abu, A.E.I., Becker, M., Accoti, A., Sylla, M., Dickson, L.B., 2024. Low humidity enhances Zika virus infection and dissemination in Aedes aegypti mosquitoes. mSphere 9, e0040124.

Aliota, M.T., Dudley, D.M., Newman, C.M., Weger-Lucarelli, J., Stewart, L.M., Koenig, M.R., et al., 2018. Molecularly barcoded Zika virus libraries to probe in vivo evolutionary dynamics. PLoS Pathog. 14, e1006964.

Azman, I.K., Chan, Y.F., Chua, C.L., Abd Mutalib, Z.A., Dass, S.C., Gill, B.S., et al., 2024. A change in circulating chikungunya virus variant impacts Aedes aegypti vector competence and spatiotemporal distribution of disease in Malaysia. PLoS Negl. Trop. Dis. 18, e0012632.

Bartholomay, L.C., Waterhouse, R.M., Mayhew, G.F., Campbell, C.L., Michel, K., Zou, Z., et al., 2010. Pathogenomics of Culex quinquefasciatus and meta-analysis of infection responses to diverse pathogens. Science 330, 88–90.

Beatty, P.R., Puerta-Guardo, H., Killingbeck, S.S., Glasner, D.R., Hopkins, K., Harris, E., 2015. Dengue virus NS1 triggers endothelial permeability and vascular leak that is prevented by NS1 vaccination. Sci. Transl. Med. 7, 304ra141.

Beaucourt, S., Borderia, A.V., Coffey, L.L., Gnadig, N.F., Sanz-Ramos, M., Beeharry, Y., et al., 2011. Isolation of fidelity variants of RNA viruses and characterization of virus mutation frequency. J. Vis. Exp.

Bellone, R., Failloux, A.B., 2020. The role of temperature in shaping mosquito-borne viruses transmission. Front. Microbiol. 11, 584846.

Bennett, K.E., Flick, D., Fleming, K.H., Jochim, R., Beaty, B.J., Black, W.C.T., 2005. Quantitative trait loci that control dengue-2 virus dissemination in the mosquito Aedes aegypti. Genetics 170, 185–194.

Blitvich, B.J., Firth, A.E., 2015. Insect-specific flaviviruses: a systematic review of their discovery, host range, mode of transmission, superinfection exclusion potential and genomic organization. Viruses 7, 1927–1959.

Borgherini, G., Poubeau, P., Staikowsky, F., Lory, M., Le Moullec, N., Becquart, J.P., et al., 2007. Outbreak of chikungunya on reunion Island: early clinical and laboratory features in 157 adult patients. Clin. Infect. Dis. 44, 1401–1407.

Bosio, C.F., Fulton, R.E., Salasek, M.L., Beaty, B.J., Black, W.C.T., 2000. Quantitative trait loci that control vector competence for dengue-2 virus in the mosquito Aedes aegypti. Genetics 156, 687–698.

Brackney, D.E., Beane, J.E., Ebel, G.D., 2009. RNAi targeting of West Nile virus in mosquito midguts promotes virus diversification. PLoS Pathog. 5, e1000502.

Brackney, D.E., Pesko, K.N., Brown, I.K., Deardorff, E.R., Kawatachi, J., Ebel, G.D., 2011. West Nile virus genetic diversity is maintained during transmission by Culex pipiens quinquefasciatus mosquitoes. PLoS One 6, e24466.

Brown, J.J., Pascual, M., Wimberly, M.C., Johnson, L.R., Murdock, C.C., 2023. Humidity - The overlooked variable in the thermal biology of mosquito-borne disease. Ecol. Lett. 26, 1029–1049.

Bustin, S.A., Benes, V., Garson, J.A., Hellemans, J., Huggett, J., Kubista, M., et al., 2009. The MIQE guidelines: minimum information for publication of quantitative real-time PCR experiments. Clin. Chem. 55, 611–622.

Campbell, C.L., Black, W.C.T., Hess, A.M., Foy, B.D., 2008. Comparative genomics of small RNA regulatory pathway components in vector mosquitoes. BMC Genomics 9, 425.

Campos, S.S., Fernandes, R.S., Dos Santos, A.A.C., De Miranda, R.M., Telleria, E.L., Ferreira-De-Brito, A., et al., 2017. Zika virus can be venereally transmitted between Aedes aegypti mosquitoes. Parasit Vectors 10, 605.

Chen, B., Sweeny, A.R., Wu, V.Y., Christofferson, R.C., Ebel, G., Fagre, A.C., et al., 2023. Exploring the mosquito-arbovirus network: a survey of vector competence experiments. Am. J. Trop. Med. Hyg. 108, 987–994.

Chen, H.R., Chuang, Y.C., Lin, Y.S., Liu, H.S., Liu, C.C., Perng, G.C., et al., 2016. Dengue virus nonstructural protein 1 Induces vascular leakage through macrophage migration inhibitory factor and autophagy. PLoS Negl. Trop. Dis. 10, e0004828.

Chong, Z., Hui, S., Qiu, X., Palakurty, S., Sariol, A., Kaszuba, T., et al., 2025. Multiple LDLR family members act as entry receptors for yellow fever virus. Nature.

Ciota, A.T., Ehrbar, D.J., Van Slyke, G.A., Payne, A.F., Willsey, G.G., Viscio, R.E., et al., 2012. Quantification of intrahost bottlenecks of West Nile virus in Culex pipiens mosquitoes using an artificial mutant swarm. Infect. Genet. Evol.

Ciota, A.T., Jia, Y., Payne, A.F., Jerzak, G., Davis, L.J., Young, D.S., et al., 2009. Experimental passage of St. Louis encephalitis virus in vivo in mosquitoes and chickens reveals evolutionarily significant virus characteristics. PLoS One 4, e7876.

Ciota, A.T., Lovelace, A.O., Jones, S.A., Payne, A., Kramer, L.D., 2007a. Adaptation of two flaviviruses results in differences in genetic heterogeneity and virus adaptability. J. Gen. Virol. 88, 2398–2406.

Ciota, A.T., Ngo, K.A., Lovelace, A.O., Payne, A.F., Zhou, Y., Shi, P.Y., et al., 2007b. Role of the mutant spectrum in adaptation and replication of West Nile virus. J. Gen. Virol. 88, 865–874.

Clarke, D.K., Duarte, E.A., Elena, S.F., Moya, A., Domingo, E., Holland, J., 1994. The red queen reigns in the kingdom of RNA viruses. Proc. Natl. Acad. Sci. USA 91, 4821–4824.

Coffey, L.L., Beeharry, Y., Borderia, A.V., Blanc, H., Vignuzzi, M., 2011. Arbovirus high fidelity variant loses fitness in mosquitoes and mice. Proc. Natl. Acad. Sci. USA 108, 16038–16043.

Cox, V.M., De Melo Iani, F.C., Hinsley, W., Peixoto, P.S., Coelho, F.C., Prete, C.A., et al., 2025. Spatiotemporal relationships between extreme weather events and arbovirus transmission across Brazil. medRxiv 2025. 08.01.25332773.

Dabo, S., Henrion-Lacritick, A., Lecuyer, A., Jiolle, D., Paupy, C., Ayala, D., et al., 2024. Extensive variation and strain-specificity in dengue virus susceptibility among African Aedes aegypti populations. PLoS Negl. Trop. Dis. 18, e0011862.

Davis, C.W., Nguyen, H.Y., Hanna, S.L., Sanchez, M.D., Doms, R.W., Pierson, T.C., 2006. West Nile virus discriminates between DC-SIGN and DC-SIGNR for cellular attachment and infection. J. Virol. 80, 1290–1301.

Ebel, G.D., 2017. Promiscuous viruses-how do viruses survive multiple unrelated hosts? Curr. Opin. Virol. 23, 125–129.

Ebel, G.D., Carricaburu, J., Young, D., Bernard, K.A., Kramer, L.D., 2004. Genetic and phenotypic variation of West Nile virus in New York, 2000-2003. Am. J. Trop. Med. Hyg. 71, 493–500.

Ebel, G.D., Rochlin, I., Longacker, J., Kramer, L.D., 2005. Culex restuans (Diptera: Culicidae) relative abundance and vector competence for West Nile Virus. J. Med. Entomol. 42, 838–843.

Escarmis, C., Davila, M., Charpentier, N., Bracho, A., Moya, A., Domingo, E., 1996. Genetic lesions associated with Muller's ratchet in an RNA virus. J. Mol. Biol. 264, 255–267.

Escarmis, C., Perales, C., Domingo, E., 2009. Biological effect of Muller's Ratchet: distant capsid site can affect picornavirus protein processing. J. Virol. 83, 6748–6756.

Farci, P., Bukh, J., Purcell, R.H., 1997. The quasispecies of hepatitis C virus and the host immune response. Springer Semin. Immunopathol. 19, 5–26.

Fauver, J.R., Grubaugh, N.D., Krajacich, B.J., Weger-Lucarelli, J., Lakin, S.M., Fakoli, L.S., et al., 2016. West African Anopheles gambiae mosquitoes harbor a taxonomically diverse virome including new insect-specific flaviviruses, mononegaviruses, and totiviruses. Virology 498, 288–299.

Fay, R.L., Cruz-Loya, M., Keyel, A.C., Price, D.C., Zink, S.D., Mordecai, E.A., et al., 2024. Population-specific thermal responses contribute to regional variability in arbovirus transmission with changing climates. iScience 27, 109934.

Fay, R.L., Cruz-Loya, M., Maffei, J.G., Mordecai, E.A., Ciota, A.T., 2025. Rising temperatures contribute to West Nile virus diversification and increased transmission potential. Sci. Rep. 15, 25016.

Fay, R.L., Ngo, K.A., Kuo, L., Willsey, G.G., Kramer, L.D., Ciota, A.T., 2021. Experimental evolution of West Nile Virus at higher temperatures facilitates broad adaptation and increased genetic diversity. Viruses 13.

Fire, A., Xu, S., Montgomery, M.K., Kostas, S.A., Driver, S.E., Mello, C.C., 1998. Potent and specific genetic interference by double-stranded RNA in Caenorhabditis elegans. Nature 391, 806–811.

Fitzmeyer, E.A., Gallichotte, E.N., Weger-Lucarelli, J., Kapuscinski, M.L., Abdo, Z., Pyron, K., et al., 2023. Loss of West Nile virus genetic diversity during mosquito infection due to species-dependent population bottlenecks. iScience 26, 107711.

Forrester, N.L., Guerbois, M., Seymour, R.L., Spratt, H., Weaver, S.C., 2012. Vector-borne transmission imposes a severe bottleneck on an RNA virus population. PLoS Pathog. 8, e1002897.

Frey, Z.D., Price, D.A., Connors, K.A., Rush, R.E., Brown, G., Sterling, C.E., et al., 2025. LRP1 facilitates Jamestown Canyon virus infection of neurons. J. Virol. 99, e0184125.

Fujii, M., Gallichotte, E.N., Sanchez-Vargas, I., Enney, B.M., Malsick, L.E., Ebel, G.D., et al., 2025. Development of entebbe bat virus reverse genetics system and phenotypic evaluation of rescued virus reveals host-specific replication patterns in mosquitoes. Am. J. Trop. Med. Hyg.

Gallichotte, E.N., Ebel, G.D., Carlson, C.J., 2025. Vector competence for Oropouche virus: a systematic review of pre-2024 experiments. PLoS Negl. Trop. Dis. 19, e0013014.

Gallichotte, E.N., Samaras, D., Murrieta, R.A., Sexton, N.R., Robison, A., Young, M.C., et al., 2023. The incompetence of mosquitoes-Can Zika virus be adapted to infect culex tarsalis cells? mSphere 8, e0001523.

Gitlin, L., Stone, J.K., Andino, R., 2005. Poliovirus escape from RNA interference: short interfering RNA-target recognition and implications for therapeutic approaches. J. Virol. 79, 1027–1035.

Goddard, L.B., Roth, A.E., Reisen, W.K., Scott, T.W., 2003. Vertical transmission of West Nile virus by three California Culex (Diptera: Culicidae) species. J. Med. Entomol. 40 (6), 743.

Gould, E.A., De Lamballerie, X., Zanotto, P.M., Holmes, E.C., 2003. Origins, evolution, and vector/host coadaptations within the genus Flavivirus. Adv. Virus Res. 59, 277–314.

Graham, D.H., Holmes, J.L., Beaty, B.J., Black, W.C.T., 2003. Quantitative trait loci conditioning transovarial transmission of La Crosse virus in the eastern treehole mosquito, Ochlerotatus triseriatus. Insect Mol. Biol. 12, 307–318.

Grubaugh, N.D., Weger-Lucarelli, J., Murrieta, R.A., Fauver, J.R., Garcia-Luna, S.M., Prasad, A.N., et al., 2016. Genetic drift during systemic arbovirus infection of mosquito vectors leads to decreased relative fitness during host switching. Cell Host Microbe 19, 481–492.

Hadfield, J., Brito, A.F., Swetnam, D.M., Vogels, C.B.F., Tokarz, R.E., Andersen, K.G., et al., 2019. Twenty years of West Nile virus spread and evolution in the Americas visualized by Nextstrain. PLoS Pathog. 15, e1008042.

Hagan, R.W., Didion, E.M., Rosselot, A.E., Holmes, C.J., Siler, S.C., Rosendale, A.J., et al., 2018. Dehydration prompts increased activity and blood feeding by mosquitoes. Sci. Rep. 8, 6804.

Hall, D.R., Johnson, R.M., Kwon, H., Ferdous, Z., Laredo-Tiscareno, S.V., Blitvich, B.J., et al., 2025. Mosquito immune cells enhance dengue and Zika virus infection in Aedes aegypti. Nat. Commun. 16, 5891.

Hardy, J.L., Apperson, G., Asman, S.M., Reeves, W.C., 1978. Selection of a strain of Culex tarsalis highly resistant to infection following ingestion of western equine encephalomyelitis virus. Am. J. Trop. Med. Hyg. 27, 313–321.

Hardy, J.L., Houk, E.J., Kramer, L.D., Reeves, W.C., 1983. Intrinsic factors affecting vector competence of mosquitoes for arboviruses. Annu. Rev. Entomol. 28, 229–262.

Hardy, J.L., Reeves, W.C., Sjogren, R.D., 1976. Variations in the susceptiblility of field and laboratory populations of Culex tarsalis to experimental infection with western equine encephalomyelitis virus. Am. J. Epidemiol. 103, 498–505.

Hardy, J.L., Rosen, L., Kramer, L.D., Presser, S.B., Shroyer, D.A., Turell, M.J., 1980. Effect of rearing temperature on transovarial transmission of St. Louis encephalitis virus in mosquitoes. Am. J. Trop. Med. Hyg. 29, 963–968.

Hartman, D.A., Bergren, N.A., Kondash, T., Schlatmann, W., Webb, C.T., Kading, R.C., 2021. Susceptibility and barriers to infection of Colorado mosquitoes with Rift Valley fever virus. PLoS Negl. Trop. Dis. 15, e0009837.

Hernandez-Alonso, P., Garijo, R., Cuevas, J.M., Sanjuan, R., 2015. Experimental evolution of an RNA virus in cells with innate immunity defects. Virus Evol. 1, vev008.

Hershberg, R., Petrov, D.A., 2008. Selection on codon bias. Annu. Rev. Genet. 42, 287–299.

Holland, J., Spindler, K., Horodyski, F., Grabau, E., Nichol, S., Vandepol, S., 1982. Rapid evolution of RNA genomes. Science 215, 1577–1585.

Holmes, C.J., Brown, E.S., Sharma, D., Nguyen, Q., Spangler, A.A., Pathak, A., et al., 2022. Bloodmeal regulation in mosquitoes curtails dehydration-induced mortality, altering vectorial capacity. J. Insect. Physiol. 137, 104363.

Jerzak, G., Bernard, K.A., Kramer, L.D., Ebel, G.D., 2005. Genetic variation in West Nile virus from naturally infected mosquitoes and birds suggests quasispecies structure and strong purifying selection. J. Gen. Virol. 86, 2175–2183.

Jerzak, G.V., Bernard, K., Kramer, L.D., Shi, P.Y., Ebel, G.D., 2007. The West Nile virus mutant spectrum is host-dependant and a determinant of mortality in mice. Virology 360, 469–476.

Jerzak, G.V., Brown, I., Shi, P.Y., Kramer, L.D., Ebel, G.D., 2008. Genetic diversity and purifying selection in West Nile virus populations are maintained during host switching. Virology 374, 256–260.

Johnson, R.M., Cozens, D.W., Ferdous, Z., Armstrong, P.M., Brackney, D.E., 2023. Increased blood meal size and feeding frequency compromise Aedes aegypti midgut integrity and enhance dengue virus dissemination. PLoS. Negl. Trop. Dis. 17, e0011703.

Junglen, S., 2016. Evolutionary origin of pathogenic arthropod-borne viruses-a case study in the family Bunyaviridae. Curr. Opin. Insect Sci. 16, 81–86.

Kading, R.C., Kityo, R., Nakayiki, T., Ledermann, J., Crabtree, M.B., Lutwama, J., et al., 2015. Detection of entebbe Bat virus after 54 years. Am. J. Trop. Med. Hyg. 93, 475–477.

Kenney, J.L., Solberg, O.D., Langevin, S.A., Brault, A.C., 2014. Characterization of a novel insect-specific flavivirus from Brazil: potential for inhibition of infection of arthropod cells with medically important flaviviruses. J. Gen. Virol. 95, 2796–2808.

Kerr, P.J., Best, S.M., 1998. Myxoma virus in rabbits. Rev. Sci. Tech. 17, 256–268.

Kramer, L.D., Ebel, G.D., 2003. Dynamics of flavivirus infection in mosquitoes. Adv. Virus Res. 60, 187–232.

Kramer, L.D., Hardy, J.L., Presser, S.B., 1983. Effect of temperature of extrinsic incubation on the vector competence of Culex tarsalis for western equine encephalomyelitis virus. Am. J. Trop. Med. Hyg. 32, 1130–1139.

Lambrechts, L., Paaijmans, K.P., Fansiri, T., Carrington, L.B., Kramer, L.D., Thomas, M.B., et al., 2011. Impact of daily temperature fluctuations on dengue virus transmission by Aedes aegypti. Proc. Natl. Acad. Sci. USA 108, 7460–7465.

Laskus, T., Radkowski, M., Wang, L.F., Jang, S.J., Vargas, H., Rakela, J., 1998. Hepatitis C virus quasispecies in patients infected with HIV-1: correlation with extrahepatic viral replication. Virology 248, 164–171.

Lawrie, D.S., Messer, P.W., Hershberg, R., Petrov, D.A., 2013. Strong purifying selection at synonymous sites in D. melanogaster. PLoS Genet. 9, e1003527.

Lequime, S., Paul, R.E., Lambrechts, L., 2016. Determinants of arbovirus vertical transmission in mosquitoes. PLoS Pathog 12, e1005548.

Lewis, J., Gallichotte, E.N., Randall, J., Glass, A., Foy, B.D., Ebel, G.D., et al., 2023. Intrinsic factors driving mosquito vector competence and viral evolution: a review. Front Cell. Infect. Microbiol. 13, 1330600.

Li, P., Ke, X., Wang, T., Tan, Z., Luo, D., Miao, Y., et al., 2018. Zika virus attenuation by codon pair deoptimization induces sterilizing immunity in mouse models. J. Virol. 92.

Liu, J., Liu, Y., Nie, K., Du, S., Qiu, J., Pang, X., et al., 2016. Flavivirus NS1 protein in infected host sera enhances viral acquisition by mosquitoes. Nat. Microbiol. 1, 16087.

Liu, Y., Liu, J., Du, S., Shan, C., Nie, K., Zhang, R., et al., 2017. Evolutionary enhancement of Zika virus infectivity in Aedes aegypti mosquitoes. Nature 545, 482–486.

Lobo, F.P., Mota, B.E., Pena, S.D., Azevedo, V., Macedo, A.M., Tauch, A., et al., 2009. Virus-host coevolution: common patterns of nucleotide motif usage in flaviviridae and their hosts. PLoS One 4, e6282.

Lumsden, W.H., Williams, M.C., Mason, P.J., 1961. A virus from insectivorous bats in Uganda. Ann. Trop. Med. Parasitol. 55, 389–397.

Ma, H., Adams, L.J., Raju, S., Sariol, A., Kafai, N.M., Janova, H., et al., 2024. The low-density lipoprotein receptor promotes infection of multiple encephalitic alphaviruses. Nat. Commun. 15, 246.

Manzano-Alvarez, J., Terradas, G., Holmes, C.J., Benoit, J.B., Rasgon, J.L., 2023. Dehydration stress and Mayaro virus vector competence in Aedes aegypti. J. Virol. 97, e0069523.

Martina, B.E., Koraka, P., Van Den Doel, P., Rimmelzwaan, G.F., Haagmans, B.L., Osterhaus, A.D., 2008. DC-SIGN enhances infection of cells with glycosylated West Nile virus in vitro and virus replication in human dendritic cells induces production of IFN-alpha and TNF-alpha. Virus Res. 135, 64–71.

Martinez, J., Patkaniowska, A., Urlaub, H., Luhrmann, R., Tuschl, T., 2002. Single-stranded antisense siRNAs guide target RNA cleavage in RNAi. Cell 110, 563–574.

Mayanja, M.N., Mwiine, F.N., Lutwama, J.J., Ssekagiri, A., Egesa, M., Thomson, E.C., et al., 2021. Mosquito-borne arboviruses in Uganda: history, transmission and burden. J. Gen. Virol. 102.

Miller, B.R., Defoliart, G.R., Yuill, T.M., 1977. Vertical transmission of La Crosse virus (California encephalitis group): transovarial and filial infection rates in Aedes triseriatus (Diptera: Culicidae). J. Med. Entomol. 14, 437–440.

Mittler, E., Tse, A.L., Tran, P.T., Florez, C., Janer, J., Varnaite, R., et al., 2025. LRP8 is a receptor for tick-borne encephalitis virus. Nature 646, 945–952.

Moon, S.L., Dodd, B.J., Brackney, D.E., Wilusz, C.J., Ebel, G.D., Wilusz, J., 2015. Flavivirus sfRNA suppresses antiviral RNA interference in cultured cells and mosquitoes and directly interacts with the RNAi machinery. Virology 485, 322–329.

Mossel, E.C., Crabtree, M.B., Mutebi, J.P., Lutwama, J.J., Borland, E.M., Powers, A.M., et al., 2017. Arboviruses isolated from mosquitoes collected in Uganda, 2008-2012. J. Med. Entomol. 54, 1403–1409.

Moudy, R.M., Meola, M.A., Morin, L.L., Ebel, G.D., Kramer, L.D., 2007. A newly emergent genotype of West Nile virus is transmitted earlier and more efficiently by Culex mosquitoes. Am. J. Trop. Med. Hyg. 77, 365–370.

Mousavi, F.S., Shenagari, M., Reza Hashemnia, S.M., Zandi, M., 2024. Climate change and arboviruses_a growing threat to public health. N. Microbes N. Infect. 62, 101532.

Murrieta, R.A., Garcia-Luna, S.M., Murrieta, D.J., Halladay, G., Young, M.C., Fauver, J.R., et al., 2021. Impact of extrinsic incubation temperature on natural selection during Zika virus infection of Aedes aegypti and Aedes albopictus. PLoS Pathog. 17, e1009433.

Nasar, F., Gorchakov, R.V., Tesh, R.B., Weaver, S.C., 2015. Eilat virus host range restriction is present at multiple levels of the virus life cycle. J. Virol. 89, 1404–1418.

Nasar, F., Palacios, G., Gorchakov, R.V., Guzman, H., Da Rosa, A.P., Savji, N., et al., 2012. Eilat virus, a unique alphavirus with host range restricted to insects by RNA replication. Proc. Natl. Acad. Sci. USA 109, 14622–14627.

Nelms, B.M., Fechter-Leggett, E., Carroll, B.D., Macedo, P., Kluh, S., Reisen, W.K., 2013. Experimental and natural vertical transmission of West Nile virus by California Culex (Diptera: Culicidae) mosquitoes. J. Med. Entomol. 50, 371–378.

Niu, J., Ma, J., Zhu, Y., Wang, G., Xu, X., Wang, M., et al., 2026. Mosquito-capsid interactions contribute to flavivirus vector specificity. Nature 651, 1039–1050.

Pascoe, L., Clemen, T., Bradshaw, K., Nyambo, D., 2022. Review of importance of weather and environmental variables in agent-based arbovirus models. Int. J. Env. Res. Public Health 19.

Perdomo, H.D., Khorramnejad, A., Cham, N.M., Kropf, A., Sogliani, D., Bonizzoni, M., 2025. Prolonged exposure to heat enhances mosquito tolerance to viral infection. Commun. Biol. 8, 168.

Peterson, K.J., Lyons, J.B., Nowak, K.S., Takacs, C.M., Wargo, M.J., Mcpeek, M.A., 2004. Estimating metazoan divergence times with a molecular clock. Proc. Natl. Acad. Sci. USA 101, 6536–6541.

Phumee, A., Chompoosri, J., Intayot, P., Boonserm, R., Boonyasuppayakorn, S., Buathong, R., et al., 2019. Vertical transmission of Zika virus in Culex quinquefasciatus Say and Aedes aegypti (L.) mosquitoes. Sci. Rep. 9, 5257.

Provost-Javier, K.N., Chen, S., Rasgon, J.L., 2010. Vitellogenin gene expression in autogenous Culex tarsalis. Insect Mol. Biol. 19, 423–429.

Quesnel-Vallieres, M., Lemay, M., Lapointe, N., Martin, S.R., Soudeyns, H., 2008. HCV quasispecies evolution during treatment with interferon alfa-2b and ribavirin in two children coinfected with HCV and HIV-1. J. Clin. Virol. 43, 236–240.

Quezada-Ruiz, E., Silva-Olivares, A., Talamas-Lara, D., Diaz, G., Alanis-Munoz, S., Pacheco, C.A., et al., 2026. The dengue virus NS1 protein alters Aedes aegypti midgut permeability and favors virus dissemination. mBio 17, e0317325.

Ross, P.A., Endersby-Harshman, N.M., Hoffmann, A.A., 2019. A comprehensive assessment of inbreeding and laboratory adaptation in Aedes aegypti mosquitoes. Evol. Appl. 12, 572–586.

Rozen-Gagnon, K., Yi, S., Jacobson, E., Novack, S., Rice, C.M., 2021. A selectable, plasmid-based system to generate CRISPR/Cas9 gene edited and knock-in mosquito cell lines. Sci. Rep. 11, 736.

Sanchez-Vargas, I., Harrington, L.C., Doty, J.B., Black, W.C.T., Olson, K.E., 2018. Demonstration of efficient vertical and venereal transmission of dengue virus type-2 in a genetically diverse laboratory strain of Aedes aegypti. PLoS. Negl. Trop. Dis. 12, e0006754.

Schuffenecker, I., Iteman, I., Michault, A., Murri, S., Frangeul, L., Vaney, M.C., et al., 2006. Genome microevolution of chikungunya viruses causing the Indian Ocean outbreak. PLoS Med. 3, e263.

Sexton, N.R., Ebel, G.D., 2019. Effects of arbovirus multi-host life cycles on dinucleotide and codon usage patterns. Viruses 11.

Shimojima, M., Takenouchi, A., Shimoda, H., Kimura, N., Maeda, K., 2014. Distinct usage of three C-type lectins by Japanese encephalitis virus: DC-SIGN, DC-SIGNR, and LSECtin. Arch. Virol. 159, 2023–2031.

Shin, D., Kang, S., Smartt, C.T., 2020. Profiling transcripts of vector competence between two different aedes aegypti populations in Florida. Viruses 12.

Smartt, C.T., Richards, S.L., Anderson, S.L., Erickson, J.S., 2009. West Nile virus infection alters midgut gene expression in Culex pipiens quinquefasciatus Say (Diptera: Culicidae). Am. J. Trop. Med. Hyg. 81, 258–263.

Smith, D.R., Adams, A.P., Kenney, J.L., Wang, E., Weaver, S.C., 2008. Venezuelan equine encephalitis virus in the mosquito vector Aedes taeniorhynchus: infection initiated by a small number of susceptible epithelial cells and a population bottleneck. Virology 372, 176–186.

Stauft, C.B., Shen, S.H., Song, Y., Gorbatsevych, O., Asare, E., Futcher, B., et al., 2018. Extensive recoding of dengue virus type 2 specifically reduces replication in primate cells without gain-of-function in Aedes aegypti mosquitoes. PLoS One 13, e0198303.

Takata, M.A., Goncalves-Carneiro, D., Zang, T.M., Soll, S.J., York, A., Blanco-Melo, D., et al., 2017. CG dinucleotide suppression enables antiviral defence targeting non-self RNA. Nature 550, 124–127.

Tesh, R.B., Gubler, D.J., 1975. Laboratory studies of transovarial transmission of La Crosse and other arboviruses by Aedes albopictus and Culex fatigans. Am. J. Trop. Med. Hyg. 24, 876–880.

Thompson, W.H., 1983. Lower rates of oral transmission of La Crosse virus by Aedes triseriatus venereally exposed after engorgement on immune chipmunks. Am. J. Trop. Med. Hyg. 32, 1416–1421.

Travanty, E.A., Adelman, Z.N., Franz, A.W., Keene, K.M., Beaty, B.J., Blair, C.D., et al., 2004. Using RNA interference to develop dengue virus resistance in genetically modified Aedes aegypti. Insect Biochem. Mol. Biol. 34, 607–613.

Tsetsarkin, K., Higgs, S., Mcgee, C.E., De Lamballerie, X., Charrel, R.N., Vanlandingham, D.L., 2006. Infectious clones of Chikungunya virus (La Reunion isolate) for vector competence studies. Vector Borne Zoonotic Dis. 6, 325–337.

Tsetsarkin, K.A., Vanlandingham, D.L., Mcgee, C.E., Higgs, S., 2007. A single mutation in chikungunya virus affects vector specificity and epidemic potential. PLoS Pathog. 3, e201.
Turell, M.J., Dohm, D.J., Sardelis, M.R., Oguinn, M.L., Andreadis, T.G., Blow, J.A., 2005. An update on the potential of north American mosquitoes (Diptera: Culicidae) to transmit West Nile Virus. J. Med. Entomol. 42, 57–62.
Turell, M.J., Hardy, J.L., Reeves, W.C., 1982a. Stabilized infection of California encephalitis virus in Aedes dorsalis, and its implications for viral maintenance in nature. Am. J. Trop. Med. Hyg. 31, 1252–1259.
Turell, M.J., Reeves, W.C., Hardy, J.L., 1982b. Transovarial and trans-stadial transmission of California encephalitis virus in Aedes dorsalis and Aedes melanimon. Am. J. Trop. Med. Hyg. 31, 1021–1029.
Turner, E.A., Christofferson, R.C., 2024. Exploring the transmission modalities of Bunyamwera virus. Exp. Biol. Med. (Maywood) 249, 10114.
Twiddy, S.S., Holmes, E.C., Rambaut, A., 2003. Inferring the rate and time-scale of dengue virus evolution. Mol. Biol. Evol. 20, 122–129.
Vignuzzi, M., Stone, J.K., Arnold, J.J., Cameron, C.E., Andino, R., 2006. Quasispecies diversity determines pathogenesis through cooperative interactions in a viral population. Nature 439, 344–348.
Viswanatha, R., Mameli, E., Rodiger, J., Merckaert, P., Feitosa-Suntheimer, F., Colpitts, T.M., et al., 2021. Bioinformatic and cell-based tools for pooled CRISPR knockout screening in mosquitos. Nat. Commun. 12, 6825.
Wahaab, A., Alvarez, J.M., Minh, H.T.N., Benoit, J.B., Rasgon, J.L., 2025. Effect of prolonged dehydration stress on the vector competence of Aedes aegypti for Mayaro virus. bioRxiv.
Wang, P., Hu, K., Luo, S., Zhang, M., Deng, X., Li, C., et al., 2016. DC-SIGN as an attachment factor mediates Japanese encephalitis virus infection of human dendritic cells via interaction with a single high-mannose residue of viral E glycoprotein. Virology 488, 108–119.
Ware-Gilmore, F., Jones, M.J., Mejia, A.J., Dennington, N.L., Audsley, M.D., Hall, M.D., et al., 2025. Evolution and adaptation of dengue virus in response to high-temperature passaging in mosquito cells. Virus. Evol. 11, veaf016.
Watterson, D., Modhiran, N., Young, P.R., 2016. The many faces of the flavivirus NS1 protein offer a multitude of options for inhibitor design. Antivir Res. 130, 7–18.
Watts, D.M., Burke, D.S., Harrison, B.A., Whitmire, R.E., Nisalak, A., 1987a. Effect of temperature on the vector efficiency of Aedes aegypti for dengue 2 virus. Am. J. Trop. Med. Hyg. 36, 143–152.
Watts, D.M., Clark, G.G., Crabbs, C.L., Rossi, C.A., Olin, T.R., Bailey, C.L., 1987b. Ecological evidence against vertical transmission of eastern equine encephalitis virus by mosquitoes (Diptera: Culicidae) on the Delmarva Peninsula, USA. J. Med. Entomol. 24, 91–98.
Weger-Lucarelli, J., Garcia, S.M., Ruckert, C., Byas, A., O'connor, S.L., Aliota, M.T., et al., 2018. Using barcoded Zika virus to assess virus population structure in vitro and in Aedes aegypti mosquitoes. Virology 521, 138–148.
Whitehorn, J., Yacoub, S., 2019. Global warming and arboviral infections. Clin. Med. (Lond) 19, 149–152.
Wilke, C.O., Adami, C., 2003. Evolution of mutational robustness. Mutat Res 522, 3–11.
Wilkinson, M.D., Dumontier, M., Aalbersberg, I.J., Appleton, G., Axton, M., Baak, A., et al., 2016. The FAIR guiding principles for scientific data management and stewardship. Sci. Data 3, 160018.
Wu, V.Y., Chen, B., Christofferson, R., Ebel, G., Fagre, A.C., Gallichotte, E.N., et al., 2022. A minimum data standard for vector competence experiments. Sci. Data 9, 634.
Zamore, P.D., Tuschl, T., Sharp, P.A., Bartel, D.P., 2000. RNAi: double-stranded RNA directs the ATP-dependent cleavage of mRNA at 21–23 nucleotide intervals. Cell 101, 25–33.

CHAPTER SIX

Evolution of host preference in mosquitoes and its impact on virus transmission

Matthew DeGennaro[a,b] **and Takeshi Morita**[a,b,*]
[a]Department of Biological Sciences, Florida International University, Miami, FL, United States
[b]Biomolecular Sciences Institute, Florida International University, Miami, FL, United States
*Corresponding author. e-mail address: tmorita@fiu.edu

Contents

Advances in Insect Physiology, Volume 70
ISSN 0065-2806, https://doi.org/10.1016/bs.aiip.2026.06.002

Abstract

Mosquito host preference plays a major role in viral persistence, zoonotic spillover, and urban transmission. This chapter synthesizes evidence from behavioral studies, molecular genetics, genomics, ecology, and field surveillance to explain how mosquitoes acquire host preferences, how their sensory systems guide host choice, and how these behaviors shape the transmission of viruses such as dengue, Zika, yellow fever, and West Nile. Attention is given to the evolution of human-seeking behavior in *Aedes aegypti*, the flexible feeding patterns of *Culex* mosquitoes, the ways urban environments drive behavioral and genetic change, and the growing use of molecular tools to identify blood-meal sources. Together, this chapter highlights findings that show that host preference is a dynamic ecological trait that must be accounted for in any effort to model or manage mosquito-borne virus transmission.

1. Introduction

Mosquito-borne viruses continue to impose a major burden on global public health. Dengue, Zika, chikungunya, yellow fever, West Nile, Japanese encephalitis, and Rift Valley fever repeatedly cause outbreaks that are associated with severe illness, long-term health effects, and substantial economic disruption (Kilpatrick and Randolph, 2012; Paixão et al., 2018; Wei et al., 2025). Despite their ecological and taxonomic diversity, all of these pathogens depend on a critical event for transmission, the mosquito blood meal. Each blood meal creates an opportunity for pathogens to pass between mosquitoes and vertebrate hosts, linking wildlife reservoirs, domestic animals, and human populations within a common transmission network. This includes any vertical transmission from infected adults to offspring, a rare and poorly studied aspect of disease endemicity, as these potential pathogen reservoirs must be transmitted to the human host to cause morbidity and mortality (Boyles et al., 2020; Ferreira-de-Lima and Lima-Camara, 2018).

A mosquito population's potential to transmit pathogens is its vectorial capacity, which integrates key factors such as population density, biting frequency, survival, and vector competence (Smith et al., 2012). Vectorial capacity also depends on mosquito population genetics, virus strain, and environmental conditions, as well as mosquito blood-feeding behavior (Kramer and Ciota, 2015). For this reason, mosquito host choice is not a minor behavioral detail but a core determinant of transmission. Mosquitoes that feed mainly on birds participate in transmission systems that differ markedly from those of mosquitoes that feed mainly on humans, and mosquitoes that feed across wildlife, livestock, and humans can connect otherwise separate host communities and facilitate spillover. Although mosquito-borne disease dynamics depend on vector competence, abundance, survival, climate, host immunity, and human behavior, host preference remains a central mechanism by which mosquito biology shapes epidemiological contact (Abbasi, 2025; Coutinho-Abreu et al., 2022; Takken and Verhulst, 2013).

Host preference can be defined as the repeatable tendency of a mosquito species or population to seek out, approach, land on, and feed on particular host types more often than would be expected from host availability alone (McIver, 1968; Takken and Verhulst, 2013). This distinction is important because hosts may be bitten frequently either because mosquitoes actively prefer them or because they are abundant, accessible, or poorly defended. The motivation to bite a host is also likely dependent on their nutritional or mating status (Clifton and Noriega, 2012; League et al., 2021). A clear understanding of mosquito host choice, therefore, requires separating intrinsic attraction from realized host use in the field.

This distinction has important implications for disease ecology. A mosquito may have some innate attraction to humans, yet feed mostly on other hosts where humans are scarce or difficult to access. Conversely, a species with broad sensory capacity may appear highly anthropophilic in urban settings because local host availability and habitat structure channel feeding onto humans (Fikrig and Harrington, 2021; Takken and Verhulst, 2013). Host preference is therefore best understood as the interaction between evolved sensory and behavioral biases and the ecological settings in which mosquitoes encounter hosts (Gao et al., 2025; Stone and Gross, 2018; Yan et al., 2021). This interaction helps determine whether transmission remains largely within wildlife, spills over only occasionally, or becomes sustained within human populations.

The epidemiological importance of host preference is especially evident when contrasting major mosquito-borne virus systems. In West Nile virus ecology, transmission is maintained primarily between bird-feeding *Culex* mosquitoes and avian hosts, whereas human infections increase when *Culex* mosquitoes also feed on mammals, thereby functioning as bridge vectors (Kothera et al., 2020; Riccetti et al., 2022; Rizzoli et al., 2015; Wehmeyer et al., 2024). In contrast, dengue and Zika are transmitted mainly in human-dominated settings, where repeated human feeding by *Aedes aegypti* sustains urban transmission (McBride et al., 2014; Pruszynski et al., 2020; Tenywa et al., 2025). These contrasting systems show that host preference helps determine not only who is exposed but also whether pathogens remain enzootic, sporadically spill over, or circulate efficiently in human populations.

Mosquitoes exhibit host-use patterns that range from broad generalism to strong specialization. Generalist mosquitoes may distribute bites across several vertebrate groups, in some settings reducing human exposure and in others increasing spillover by linking host communities that would otherwise remain weakly connected (Kilpatrick and Randolph, 2012; Riccetti et al., 2022). Specialist mosquitoes, by contrast, concentrate bites within narrower transmission networks, which can either stabilize wildlife circulation or intensify transmission within human populations depending on the hosts involved (McBride et al., 2014; Rizzoli et al., 2015; Rose et al., 2020). The ecological and epidemiological consequences of host choice therefore depend not only on which hosts are preferred, but also on how strongly that preference is expressed under varying natural conditions (Fikrig and Harrington, 2021; Takken and Verhulst, 2013; Yan et al., 2021).

This chapter approaches host preference as an evolutionary, mechanistic, ecological, and epidemiological trait. The opening section examines the evolution of host preference, with emphasis on ecological opportunity, natural selection, genomic divergence, and host shifts across mosquito lineages. The next section considers the sensory and molecular basis of host seeking, including olfactory, visual, thermal, humidity, and contact-dependent cues. A subsequent section explores the role of host preference in virus amplification, spillover, and urban transmission, with particular emphasis on *Ae. aegypti* and *Culex* mosquitoes, as well as the contrasting ecologies of Zika and West Nile viruses. Another section addresses the effects of urbanization, land-use change, climate, and vector control on host availability and realized host use. The chapter then turns to blood-meal analysis, metabarcoding, and xenosurveillance as tools for improving surveillance of mosquito-borne zoonoses. Taken together, these perspectives support the argument that mosquito host preference is a dynamic trait shaped by both evolutionary history and ecological context, and that this host preference should be treated as a central component of any effort to understand, predict, or manage mosquito-borne virus transmission.

2. Evolutionary drivers of host preference in mosquitoes

2.1 Ecological and selective drivers of host choice

Mosquito host preference evolves through a combination of ecological opportunity and inherited genetic variation. For most medically important female mosquitoes, blood feeding is necessary for egg production, but hosts differ in abundance, accessibility, defensive behavior, nutritional value, and the environments in which they are encountered. Selection therefore acts not simply on whether mosquitoes obtain a blood meal, but on whether they obtain blood in ways that maximize survival and reproductive success (Lyimo and Ferguson, 2009; Stone and Gross, 2018; Swart et al., 2023; Takken and Verhulst, 2013).

One of the strongest ecological drivers of host choice is the predictability of host access across space and time. Humans often live in dense settlements, follow fairly predictable daily routines, and create reliable mosquito breeding sites through stored water and household containers. In *Ae. aegypti*, this combination appears to have been central to the evolution of human preference and domestication. A stronger preference for humans has been associated with drier climates and more urbanized environments, consistent with the idea that dependence on human water storage made human-associated habitats more reliable and reinforced selection for human preference (Facchinelli et al., 2023; Powell and Tabachnick, 2013; Rose et al., 2023, 2020). In this framework, host preference evolves not only because a host is attractive, but because feeding on that host is repeatedly profitable under local ecological conditions.

Host choice is also shaped by broader ecological context. Host community diversity, habitat structure, competition, and the local spatial arrangement of vertebrates all influence which hosts mosquitoes encounter and how often they do so. Recent DNA-based diet studies suggest that mosquito feeding patterns are often more flexible than previously assumed, with local ecological conditions playing a major role in determining which hosts are used (Lee et al., 2025). Landscape and meteorological variables have likewise been linked to *Ae. aegypti* and *Aedes albopictus* infestation in urban settings, illustrating how ecological context shapes the circumstances under which host encounters occur (Costa-da-Silva et al., 2024). These findings suggest that evolution acts on host-seeking within real ecological mosaics, where host availability, habitat predictability, and environmental stress jointly determine the fitness consequences of feeding decisions.

2.2 Genomic and sensory bases of host-linked divergence

Ecological conditions can favor certain feeding strategies, but lasting host preference can evolve only when heritable variation is available for selection to shape. Chemosensory receptor evolution in insects has been linked to host specialization, particularly through gene duplication (McBride and Arguello, 2007; Robertson, 2018). In mosquitoes, some of the clearest evidence for this variation comes from the sensory systems involved in host detection. In *Ae. aegypti*, preference for humans has been linked to divergence in odorant receptor pathways and broader differences in chemosensory gene expression, indicating that host specialization is associated with molecular changes in odor detection and processing rather than with human proximity alone (DeGennaro et al., 2013; McBride et al., 2014; Tallon et al., 2019; Wolff and Riffell, 2018; Zhao et al., 2022). Comparable evidence is now emerging in *Culex pipiens*, where differential expression of antennal chemosensory genes has been associated with host preference across biotypes, suggesting that sensory divergence may be a common route through which host use evolves across mosquito lineages (Menon et al., 2025).

Genomic studies further show that differences in host use can be detected across both recent and long evolutionary timescales. Early work using closely related species in the *Anopheles gambiae* complex suggested that chromosomal inversions were associated with differences in host preference (Coluzzi et al., 1977). Studies in *Anopheles* mosquitoes continue to work towards understanding the genetic basis of host preference and adaptation to host habitats (Ayala et al., 2017; Besansky et al., 2004; Main et al., 2016). Similarly, additional studies in *Ae. aegypti* suggests that chromosomal inversions and other features of genome structure may also contribute to host preference, helping maintain differences among populations even when they continue to exchange genes (Kent et al., 2025; Liang et al., 2025). In *Ae. aegypti*, genome-wide analyses have traced the spread of the human-preferring domestic form and identified adaptive genetic signatures linked to domestication, host seeking, and other ecologically important traits (Crawford et al., 2025; Lozada-Chávez et al., 2025). These findings indicate that host preference is not a purely plastic trait, but one that can leave detectable signatures in population history and genome structure.

Host specialization may also arise without relying entirely on new mutations. In *Ae. aegypti*, a recent genomic study suggests that domestication and preference for humans developed at least in part from genetic variation already present in ancestral populations, along with local adaptation and changes in the regulation of chemosensory, neuronal, and metabolic pathways (Lozada-Chávez et al., 2025). This helps explain why host preference can evolve repeatedly, sometimes quite rapidly, when ecological conditions favor a particular host. It also suggests that when mosquitoes invade new environments, already evolved host-seeking traits can spread into new regions and encounter new selective pressures and host communities. In this way, genomic divergence, sensory evolution, and ecological context are best understood as closely connected parts of host-preference evolution rather than as separate processes.

2.3 Comparative examples of host shifts across mosquito lineages

Ae. aegypti provides one of the clearest examples of a mosquito lineage that shifted toward humans. In this species, a strong preference for humans is closely linked to domestication, breeding in human-associated habitats, and sensory adaptation to human odor cues. Behavioral, ecological, and genomic evidence suggests that this specialization first evolved in Africa as mosquitoes became closely associated with human settlements, and that the human-adapted domestic form later spread worldwide (Crawford et al., 2025; Lozada-Chávez et al., 2025; McBride et al., 2014; Rose et al., 2023, 2020). The result is a mosquito lineage in which host preference, habitat use, and epidemiological importance have evolved together rather than independently.

Aedes albopictus provides a contrasting example of ecological flexibility. Unlike *Ae. aegypti*, this species often uses a broader range of hosts, and actual feeding patterns tend to vary with local host availability and habitat conditions. Field studies in suburban and invaded areas show that *Ae. albopictus* often feeds on humans, pets, and other vertebrates, depending on which hosts are locally available, rather than exhibiting strict host preference (Fikrig et al., 2022; Fikrig and Harrington, 2021; Richards et al., 2006). This flexibility may explain why invasion into a new region can quickly change the epidemiological importance of a population, even before longer-term evolutionary change occurs.

A third pattern is observed in the *Cx. pipiens* complex, where host use varies among biotypes, hybrids, and ecological settings. In general, feeding mainly on birds helps maintain West Nile virus and related flaviviruses in bird-mosquito cycles, whereas more frequent feeding on mammals raises the likelihood of spillover. Blood-meal and host-selection studies suggest that these differences are influenced by hybridization, urban habitats, and local host communities rather than by a fixed species-level preference (Apperson et al., 2004; Blom et al., 2024; Kothera et al., 2020; Molaei et al., 2006). In Europe, the underground form *molestus*, which is more associated with mammals, and the surface form *pipiens*, which is more associated with birds, illustrate how ecological divergence within a species complex can alter transmission potential (Farajollahi et al., 2011; Haba et al., 2025; Haba and McBride, 2022).

Mosquito host-preference evolution also extends beyond the familiar contrast between bird-feeding and mammal-feeding species. Some mosquitoes specialize on cold-blooded hosts such as amphibians, reptiles, or fish, demonstrating that host preference can evolve across diverse ecological contexts (Reinhold and Lahondère, 2024). Frog-biting mosquitoes are a good example. Some species, such as *Uranotaenia lowii*, feed mainly on amphibians, and mosquitoes may even specialize on different frog species within the same environment (Gould and Valdez, 2024; Silva et al., 2020). These cases show that specialization on humans is only one example of a broader evolutionary ability to diverge in host use.

These examples show that mosquito host preference can range from broad flexibility to strong specialization. Which pattern emerges depends on the combined effects of natural selection, sensory differences, genomic background, and ecological opportunity. Host preference is therefore best viewed as an adaptive trait shaped by both history and ecology, and one that can change with domestication, invasion, environmental change, and local host communities.

3. Mechanisms underlying host preference

Host preference in mosquitoes arises through a step-by-step sensory process in which mosquitoes detect, compare, and evaluate host cues at different distances. Host seeking does not depend on a single cue, but on the combined use of odors, visual signals, heat, humidity, and contact-dependent information gathered after landing. Because different cues matter more at different stages of approach, host preference reflects how mosquitoes combine and prioritize these signals in particular ecological settings rather than how they respond to any one cue alone (Coutinho-Abreu et al., 2022; Montell, 2025; Takken, 1990; Wolff and Riffell, 2018).

3.1 Olfactory cues and host odor discrimination

Olfaction provides the main basis for long-range host activation and discrimination. Carbon dioxide is one of the most important activators of mosquito host-seeking behavior (Dekker et al., 2005; McMeniman et al., 2014). Still, it does not by itself explain how mosquitoes distinguish among host species or among individual hosts. That discrimination depends on more complex blends of skin-, breath-, and microbiome-derived odorants. These odors arise from overlapping sources, including eccrine sweat, sebaceous secretions, and microbial metabolism on the skin surface, and together they provide the chemical landscape through which mosquitoes identify hosts (Coutinho-Abreu et al., 2022; Montell, 2025; Takken, 1990).

In *Ae. aegypti*, several lines of evidence show that human attraction depends strongly on acidic skin volatiles and carboxylic acids. Behavioral studies and receptor-based experiments show that acidic odor compounds are important for normal attraction to human scent, and odor-coding studies indicate that certain classes of skin-derived acids contribute especially strongly to human attraction (Chen et al., 2019; Raji et al., 2019; Ray et al., 2023). Lactic acid alters how mosquitoes respond to human odor, indicating that certain chemicals in human sweat actively guide host-seeking behavior (Raji et al., 2019; Steib et al., 2001). More broadly, these studies show that mosquitoes do not respond to human odor as a single, uniform signal, but to complex chemical blends whose composition differs among host species.

The skin microbiome further modifies this process. Microbial metabolism converts host secretions into additional volatile compounds, and these microbiome-derived products can alter mosquito landing and feeding behavior. Experimental manipulation of skin-associated microbes has shown that reducing the production of certain attractive compounds can substantially reduce mosquito attraction and feeding, providing direct evidence that host-associated microbial communities shape mosquito host choice (Coutinho-Abreu et al., 2024; Liu et al., 2024; Lucas-Barbosa et al., 2023). A recent survey further suggests that some of the most distinctive features of human body odor derive from sebaceous secretions and their volatile derivatives, helping to explain why humans differ chemically from many other vertebrates in ways relevant to mosquito attraction (Zung and McBride, 2025).

Olfactory discrimination also operates within host species. Repeated behavioral assays have shown that some humans are consistently much more attractive to *Ae. aegypti* than others, and that these rankings remain stable across time. Furthermore, highly attractive individuals tend to produce higher levels of skin-derived carboxylic acids (Obaldia et al., 2022). This pattern is consistent with earlier evidence indicating that attractiveness to mosquitoes is partly inherited, suggesting that differences among hosts are biologically meaningful rather than temporary or environmentally driven (Fernández-Grandon et al., 2015). Recent evidence also suggests that different mosquito species do not rank the same humans in the same way, implying that host preference is shaped not only by host chemistry but also by species-specific sensory interpretation of that chemistry (Marrero et al., 2026). Olfaction, therefore, contributes to both species-level host choice and fine-scale heterogeneity in host contact within human populations.

At the molecular level, olfactory-based behaviors depend on several classes of specialized receptors expressed in chemosensory appendages, primarily in the antenna, maxillary palp, and proboscis (Goldman et al., 2025; Herre et al., 2022; Matthews et al., 2016). In *Ae. aegypti*, a highly divergent family of chemosensory receptors called Odorant Receptors plays a critical role in detecting volatile odorants. Loss of the obligate co-receptor subunit, Orco, significantly reduces the strong preference for humans over guinea pigs, indicating that odorant receptor pathways play a major role in distinguishing hosts apart (DeGennaro et al., 2013). Acidic compounds associated with humans are detected in part through the Ionotropic Receptor pathways, helping to explain why carboxylic acids and related compounds are important for human attraction (Raji et al., 2019; Ray et al., 2023). Carbon dioxide is detected by gustatory receptor complexes, and disruption of the CO_2 receptor subunit Gr3 in *Ae. aegypti* greatly reduces attraction to host-related cues, showing that CO_2 detection is an important early step in multimodal host-seeking behavior (Kumar et al., 2020; McMeniman et al., 2014). Together, these studies show that mosquitoes distinguish hosts through the combined action of several receptor families rather than a single chemical pathway.

3.2 Visual cues and odor-gated host detection

Vision becomes especially important once odor has activated host seeking. Instead of acting independently, visual attraction is strongly shaped by olfactory context. In *Ae. aegypti*, wind-tunnel experiments showed that carbon dioxide exposure greatly increased attraction to certain wavelengths, especially cyan, orange, and red. That attraction to human skin tones disappeared when longer wavelengths were filtered out or when carbon dioxide detection or long-wavelength opsin function was disrupted (Alberto et al., 2022). These findings provide clear evidence that odor influences which visual cues become behaviorally important during host approach.

Visual responses are also flexible rather than fixed. When only carbon dioxide is present, *Ae. aegypti* shows a different pattern of color preference than when floral odors, egg-laying cues, or vertebrate host odors are also present, suggesting that visual evaluation depends on the wider sensory context in which a target is encountered (Blake and Riffell, 2025). Similar species-specific color preferences have been reported in *Ae. albopictus* and *Cx. pipiens*, indicating that visual host orientation differs among mosquito taxa in ways that match broader differences in host use (Jung et al., 2021). Vision, therefore, does more than identify a host-like object by helping mosquitoes distinguish among targets together with odor-driven activation and the weighting of other cues.

This visual component also has a clear molecular basis. In *Ae. aegypti*, CRISPR-based disruption of long-wavelength opsin genes strongly reduced vision-guided attraction to visual targets, showing that host-related visual behavior depends on specific photoreceptor pathways rather than on a simple response to contrast alone (Zhan et al., 2021). The mosquito genome contains several opsin genes, and the long-wavelength opsins appear to be especially important for detecting the orange and red wavelengths associated with human skin. Disruption of both long-wavelength opsins had a stronger effect than disruption of either one alone, suggesting that these genes work in partly overlapping ways during visual host seeking (Zhan et al., 2021). These results match behavioral evidence that attraction to orange and red wavelengths depends on both smell and long-wavelength visual processing (Alberto et al., 2022). Visual host orientation is therefore best viewed as one part of a broader, context-dependent host-seeking process.

3.3 Thermal, infrared, and humidity cues in close-range host localization

As mosquitoes approach a host, heat, infrared radiation, and humidity become more important for pinpointing the target. These short-range cues help mosquitoes move from general host seeking to precise localization by detecting the warmth and moisture associated with a living host (Montell, 2025; Takken, 1990).

Experiments under semi-natural conditions show that heat alone is not enough to drive strong host attraction. In both *Ae. aegypti* and *An. gambiae*, mosquitoes were much more likely to orient to and land on warmed targets approximating human skin temperature when those targets were also presented with carbon dioxide and human odor. By contrast, heated targets without the appropriate odor context were only weakly attractive or unattractive, indicating that warmth is interpreted as a host cue only when paired with other signals that identify the target as biologically relevant (Giraldo et al., 2023; McMeniman et al., 2014). In *An. gambiae*, semi-field experiments further showed that the most attractive human subjects had odor blends enriched in volatile carboxylic acids and certain microbiome-related compounds, reinforcing the close link between odor composition and thermotaxis during host selection (Giraldo et al., 2023). Recent work on *Ae. aegypti* has extended this picture by showing that thermal infrared radiation is part of the host-seeking repertoire and can serve as an effective mid-range directional cue when combined with carbon dioxide and other host-associated signals (Chandel et al., 2024). Together, these findings show that close-range host localization depends on the coordinated interpretation of odor, carbon dioxide, heat, humidity, and infrared radiation rather than on any single cue.

At the molecular level, thermal sensing in mosquitoes depends on specialized ion channels expressed in sensory appendages. One important component of this system is the ion channel TRPA1. In *Ae. aegypti*, female mosquitoes are attracted to targets with temperatures similar to those of vertebrate hosts. Still, they avoid targets that are too hot, showing that mosquitoes respond to a specific host-like temperature range rather than to heat in general. TRPA1 is required for this behavioral tuning because loss-of-function mosquitoes can no longer distinguish between harmless warmth and excessively hot surfaces (Corfas and Vosshall, 2015). More recent work also shows that TRPA1-dependent thermosensory neurons in the antennae contribute to infrared-guided approach to hosts, indicating that this pathway supports both temperature discrimination and host-directed orientation (Chandel et al., 2024; Corfas and Vosshall, 2015). Another class of ion channels shown to be important for mosquito thermotaxis is the ionotropic receptors. In *An. gambiae*, the ionotropic receptor subunit IR21a is required for heat-seeking and warmth-triggered blood-feeding behavior. This is especially notable because IR21a is related to a sensory pathway that functions as a cooling receptor in *Drosophila melanogaster*, suggesting that *An. gambiae* attraction to host heat may have evolved by modifying an evolutionarily more ancient temperature-sensing system for a new role in host-seeking behavior (Greppi et al., 2020; Laursen et al., 2023).

Humidity sensing provides an additional layer of close-range guidance in mosquitoes. In both *Ae. aegypti* and *An. gambiae*, IR93a-dependent sensory neurons are required for detecting nearby hosts and locating humid oviposition sites after blood feeding, thereby linking host-seeking and post-feeding behavior through shared hygrosensory pathways (Laursen et al., 2023). Additional work in *Ae. aegypti* has shown that dry-air-sensitive cells rely on IR40a, whereas humid-air-sensitive cells depend on IR68a, and that these pathways contribute differently to blood feeding and oviposition-site search

(Tang et al., 2024). Together, these findings indicate that mosquito humidity sensing is not a single uniform modality, but rather a set of molecularly and functionally distinct pathways that are differentially recruited across host-seeking and oviposition behaviors.

3.4 Contact chemosensation and post-landing host evaluation

Host choice does not end when a mosquito lands on a host. Once a mosquito reaches the skin surface, contact chemosensation helps determine whether probing continues, whether feeding begins, and whether the surface is accepted or rejected. This stage involves active sensory evaluation through structures such as the labellum, stylet, and legs rather than passive contact alone.

In *Ae. albopictus*, the labellum contains several classes of taste sensilla that respond to cues associated with humans, flowers, and oviposition sites. Experimental work shows that both excitatory and inhibitory responses occur at this stage and that inhibitory responses are especially informative, as they predict whether biting will continue (Baik et al., 2024). The same study found that some bitter compounds suppress both physiological and behavioral responses to sugar, whereas key tastants associated with human skin and sweat act together to promote biting. These findings suggest that post-landing decisions are controlled by a specialized gustatory system rather than by a simple continuation of earlier attraction.

Comparable evidence from *Ae. aegypti* shows that contact chemosensation is distributed across several peripheral structures. Functional studies have shown that the stylet and the proboscis can distinguish blood-like from nectar-like mixtures through combinatorial coding of contact cues. DEET can repel via olfactory detection or on contact by mosquito legs and suppress blood feeding before ingestion (DeGennaro et al., 2013; Dennis et al., 2019; Jové et al., 2020). At the molecular level, the TRP channel Painless1 has been implicated in gustatory avoidance of fatty acids on contact, extending the molecular basis of host evaluation beyond canonical gustatory receptors alone (Dhakal et al., 2026). Although similarly detailed work on other mosquito species remains limited, broader examination indicates that the labellum, stylet, and tarsi function as contact-sensing organs across mosquito genera, suggesting that post-landing evaluation is a general feature of mosquito feeding behavior (Coutinho-Abreu et al., 2022; Montell, 2025; Werner-Reiss et al., 1999). Host preference, therefore, does not end at the stage of long-range and mid-range attractive cues, but continues through contact-dependent acceptance or rejection at the host skin surface.

3.5 Neural coding, integration, internal state, and learning

The sensory systems involved in host-seeking behavior do not operate independently. Host preference emerges from the integration of multiple sensory inputs, and the outcome of that integration is shaped by physiological state, previous experience, and behavioral context. This helps explain why the same host cue can be evaluated differently across mosquito species, across individuals, and even across feeding states within the same individual.

Neural recording and imaging studies in *Ae. aegypti* show that the mosquito brain does not represent human odor through a simple labeled-line system, but through activity distributed across many neurons. Recordings from the antennal lobe indicate that a single odorant can activate multiple projection neurons in different glomeruli, and that both odor identity and behavioral relevance are encoded in these broader patterns of neural activity (Singh et al., 2023). Recent circuit and behavioral studies further suggest that *Ae. aegypti* processes human and animal odors differently, with dedicated olfactory glomeruli tuned to features that help distinguish humans from non-human hosts and are especially effective at driving host-seeking behavior (Zhao et al., 2022; Zhao and McBride, 2020). More recent work also shows that host-seeking behavior is not only multisensory but also flexible across sensory systems. In *Ae. aegypti*, impairment of one sensory pathway does not fully abolish attraction to humans, as mosquitoes can compensate by relying more heavily on other cues. The results indicate that host seeking is supported by cross-modal sensory compensation, with multiple sensory modalities working together to preserve attraction even when one pathway is disrupted (Morita et al., 2025). Together, these findings suggest that host preference depends not only on which odor molecules are detected but also on how the brain combines, compares, and prioritizes sensory information.

Internal state further shapes how mosquitoes respond to host cues. In *Ae. aegypti*, associative learning can change responses to those cues, and these effects depend on physiological condition and neuromodulatory signaling (Vinauger et al., 2018). Experimental work also shows that mating and sugar feeding alter blood-feeding physiology and behavior, indicating that the same host cues may be evaluated differently depending on energetic and reproductive state (League et al., 2021). Responses to humidity also vary with internal state, linking host-seeking to hydration status and to the later search for egg-laying sites (Laursen et al., 2023; Tang et al., 2024). Host seeking is therefore not a fixed sensory reflex, but a flexible decision-making process shaped by sensory input, neural processing, physiological state, and prior experience.

Mosquito host preference emerges from the integrated action of multiple sensory systems. Smell, vision, heat, humidity, and contact cues are combined into a context-dependent behavioral decision. This process involves a wide range of molecular and neural components, including odorant receptors, ionotropic receptors, gustatory receptors, TRP channels, peripheral sensory pathways, and central neural circuits. Host preference is therefore not the result of a single cue or receptor, but the product of a flexible sensory system that mosquitoes use to identify and prioritize hosts.

4. Host preference and mosquito-borne virus transmission

4.1 Host specificity, amplification, and spillover

Host specificity shapes mosquito-borne virus transmission by determining which vertebrate hosts mosquitoes bite often enough to sustain pathogen circulation. In wildlife-associated systems, repeated feeding on competent reservoir or amplification hosts helps maintain enzootic transmission, whereas feeding that also connects those hosts to humans or livestock creates opportunities for bridge transmission and spillover. The epidemiological importance of a mosquito, therefore, depends not only on vector competence, abundance, or survival, but also on whether feeding remains concentrated within reservoir hosts or extends broadly enough to connect reservoir, amplification, and incidental hosts (Gao et al., 2025; Wu et al., 2025).

West Nile virus illustrates this pattern clearly, with transmission maintained mainly between bird-feeding *Culex* mosquitoes and birds, which serve as the main amplifying hosts. In contrast, humans and horses are usually incidental dead-end hosts that contribute little to onward transmission (Farajollahi et al., 2011; Riccetti et al., 2022; Rizzoli et al., 2015). In this system, the key epidemiological shift occurs when mosquito feeding crosses the bird-mammal boundary often enough to expose humans or other mammals to what is otherwise an avian transmission cycle. A broader survey of host-selection studies emphasizes that this shift depends on actual host use in local ecological settings rather than on a fixed species trait alone (Riccetti et al., 2022).

Japanese encephalitis follows a similar but more complex pattern. Ardeid wading birds, such as egrets and herons, help maintain the virus in nature, whereas pigs and livestock are especially important for viral amplification. When *Culex* mosquitoes feed frequently on birds and pigs and opportunistically on humans, these feeding links can connect enzootic and domestic transmission cycles, increasing spillover risk (Lindahl et al., 2012, 2013; Mulvey et al., 2021; Wu et al., 2025). In this type of system, transmission depends less on any single host class than on how mosquito feeding links several vertebrate compartments.

The same principle applies to other arboviruses. In Eastern Equine Encephalitis virus, strong feeding on birds by key vectors supports wildlife transmission cycles, reinforcing the broader point that bird-oriented feeding can stabilize enzootic amplification (Hassan et al., 2003; Simpson et al., 2009). In Rift Valley fever, mosquito feeding can focus transmission on livestock rather than on birds. Studies in Kenya found that mosquitoes frequently fed on cattle, sheep, and goats, indicating that domestic livestock can play a central role in shaping host contact and transmission. As a result, transmission may become organized mainly around farm animals rather than wildlife reservoirs (Tchouassi et al., 2016). Together, these examples show that host specificity helps determine which vertebrate carries most of the transmission burden and whether mosquitoes maintain enzootic cycles, generate bridge transmission, or support sustained amplification.

By contrast, in urban systems dominated by strongly human-preferring mosquitoes, repeated feeding on humans can maintain transmission largely within human populations. Dengue is the clearest example, and urban chikungunya, Zika, and yellow fever can follow the same pattern once those viruses enter densely populated settings where *Ae. aegypti* or other mosquitoes that frequently bite humans repeatedly encounter susceptible people (Weaver, 2005; Weaver and Forrester, 2015). In these systems, host preference influences not only who is exposed but also determines whether humans remain incidental hosts or become the primary amplifying population.

4.2 Case studies: Aedes aegypti - Zika and Dengue, Culex spp. - West Nile virus

Aedes aegypti illustrates why human-specialized mosquitoes are so effective in urban transmission. In many affected regions, this species served as the main urban vector because it combines a strong preference for humans, domestic breeding, a tendency to remain in and around human dwellings, and frequent human contact. Blood-meal analysis in the Florida Keys showed that most identified meals in *Ae. aegypti* were human, consistent with the broader literature indicating that human blood indices are often very high in urban and peri-urban populations (Pruszynski et al., 2020). Similar patterns were reported in nonincorporated urban communities along the U.S.-Mexico border, where blood-meal analysis again showed strong feeding on humans despite lower-density urban conditions (Vera et al., 2023). Earlier field studies from Latin America, Southeast Asia, and Africa likewise found that *Ae. aegypti* commonly feeds predominantly on humans in domestic settings, with human feeding approaching near-exclusivity in some densely inhabited neighborhoods (Liebman et al., 2014; Ponlawat and Harrington, 2005; Scott et al., 1993). Together, these studies help explain why *Ae. aegypti* is quite effective at sustaining urban transmission once a human-amplified virus is introduced.

Zika provides a particularly clear example of this process. Because *Ae. aegypti* feeds repeatedly on humans, frequent human contact can facilitate rapid urban spread when mosquito populations are also highly susceptible to infection. Experimental infections showed that globally invasive *Ae. aegypti* populations were more susceptible to Zika virus than several noninvasive populations, suggesting that invasion history can further increase epidemic potential when repeated human feeding is already present (Aubry et al., 2020). The Cape Verde Zika epidemic provides another example in which epidemic intensity was argued to depend on enhanced vectorial capacity rather than on host contact alone (Rose et al., 2022).

A similar interaction between host preference and competence has also been reported more broadly in *Ae. aegypti* arbovirus systems. More recent surveillance in Dar es Salaam detected DENV-2 circulation in highly human-preferential, outdoor-biting *Ae. aegypti*, indicating that strong human feeding can sustain intense arbovirus circulation even when most host contact occurs outdoors rather than indoors (Tenywa et al., 2025). Similar patterns have been reported for dengue and Zika with substantial population-level and strain-specific differences in susceptibility (Gonçalves et al., 2014; Souza-Neto et al., 2019). Taken together, these studies show that *Ae. aegypti* populations become especially dangerous when frequent human contact coincides with high viral susceptibility.

West Nile virus provides a contrasting model because transmission is maintained mainly in a bird-mosquito cycle. In both North America and Europe, bird-feeding *Culex* mosquitoes play a central role in maintenance. At the same time, birds serve as the primary amplifying hosts, and humans and horses are usually incidental dead-end hosts that contribute little to onward transmission (Farajollahi et al., 2011; Riccetti et al., 2022; Rizzoli et al., 2015). In this system, the key epidemiological question is not simply whether mosquitoes can bite humans, but how often feeding crosses from birds to mammals. Blood-meal studies show that *Culex* feeding often remains strongly focused on birds during enzootic amplification, but that host use can vary substantially across ecological settings. Classic studies from North America found that *Culex* vectors frequently fed predominantly on birds, while the degree of avian specialization varied by locality, season, host availability, and mosquito taxon (Apperson et al., 2004; Kilpatrick et al., 2006; Molaei et al., 2006). A broader survey reached the same general conclusion, emphasizing that realized host use depends heavily on local ecological conditions rather than on a single fixed species trait (Riccetti et al., 2022).

Human risk increases when feeding becomes more mixed and crosses the bird-mammal boundary often enough to create bridge transmission. Within the *Cx. pipiens* complex, blood-meal sources, and host-selection patterns have been associated with biotype identity and genetic admixture, suggesting that relatively small shifts in host use can alter the likelihood of spillover to humans (Kothera et al., 2020). More recent work in the Eastern Balkans and Camargue likewise showed that local *Cx. pipiens* host-feeding patterns have important implications for avian pathogen circulation and flavivirus risk, again underscoring that bridge transmission emerges from ecological variation in host feeding (Bobeva et al., 2025; Rodríguez-Valencia et al., 2025). Other field studies reached similar conclusions, indicating that spillover risk is shaped less by a fixed species-level tendency than by the timing and location of mosquito shifts from birds to mammals (Julian et al., 2010; Oduola and Awe, 2006). Taken together, these studies show that West Nile risk increases when bird-focused feeding broadens enough to connect avian amplification hosts to mammals.

4.3 Vectorial capacity, repeated feeding, and heterogeneity in host contact

Traditional studies of vectorial capacity emphasize mosquito density, biting rate, survival, competence, and how often humans are bitten (Smith et al., 2012). A mosquito may be abundant and competent yet contribute little to human transmission if most of its bites occur on non-human hosts. Conversely, a less abundant but strongly anthropophilic mosquito can have a disproportionate effect on human disease because its bites are concentrated on epidemiologically relevant hosts. Understanding how host preference can vary across vector species can improve the accuracy of models of vectorial capacity.

Repeated feeding behavior can amplify these effects by increasing both host contact and survival under stressful conditions. Modeling and experimental work showed that multiple blood-feeding bouts prolonged mosquito survival during dry periods and were predicted to increase viral transmission, indicating that host preference has stronger epidemiological consequences when feeding is frequent and repetitive rather than limited to one blood meal per gonotrophic cycle (Holmes et al., 2025). Earlier experimental work likewise showed that *Ae. aegypti* commonly take multiple blood meals within a single gonotrophic cycle, increasing opportunities for both acquisition and onward transmission (Ponlawat and Harrington, 2005; Scott et al., 1993). In highly anthropophilic mosquitoes, repeated or interrupted feeding can therefore place high-contact behavior directly into the human transmission network.

Environmental conditions further modify this relationship. Experimental studies found that higher temperatures reduced the vectorial capacity of *Aedes* mosquitoes for Zika virus, indicating that environmental stress can alter the relationships among host contact, survival, and competence (Onyango et al., 2020). A broader survey of *Ae. aegypti* vector competence similarly concluded that competence data must be interpreted alongside biting behavior, host contact, and ecology rather than in isolation (Souza-Neto et al., 2019). These findings reinforce the point that host preference is not an isolated behavioral parameter but rather one component of a larger transmission system in which environmental conditions influence how feeding behavior translates into epidemic potential.

Transmission consequences may also depend on how the infection alters mosquito behavior after exposure. Experimental work has shown that viral infection can interfere with mosquito host-seeking behavior, raising the possibility that pathogens may modify vector contact patterns in ways that alter onward transmission (Vogels et al., 2017). Although the magnitude and direction of such effects likely vary among systems, they reinforce the broader point that transmission depends not only on baseline host preference but also on the interaction between infection state and host-seeking behavior.

Taken together, these studies show that host preference affects far more than simple host identity. It shapes how vector competence is deployed across host populations, how often bridges form between host species, how repetitive feeding behavior amplifies contact, and how unevenly exposure is distributed within communities. Host preference is therefore one of the main behavioral pathways by which mosquito biology translates into transmission intensity.

5. Environmental and anthropogenic influences

Mosquito host preference is continually reshaped by urbanization, land-use change, climatic variation, and human interventions that alter host availability, habitat structure, and the costs and benefits of feeding on different hosts. For this reason, host preference is best understood not as a fixed species attribute, but as a dynamic ecological trait embedded in changing human landscapes (Takken and Verhulst, 2013; Vinauger and Chandrasegaran, 2024; Yan et al., 2021).

5.1 Urbanization, land-use change, and climate effects on host availability

Urbanization is one of the strongest present-day drivers of mosquito host preference because it alters both the host community and the physical environment in which host-seeking occurs. Cities concentrate humans, reduce many large vertebrate populations, elevate local temperatures, and create abundant artificial larval habitat through water storage, containers, drainage systems, and other infrastructure. These conditions often favor mosquitoes that are well adapted to locating and exploiting humans, especially the anthropophilic mosquito species such as *Ae. aegypti*. Across African populations of *Ae. aegypti*, stronger preference for humans has been associated with hotter, drier climates and more urbanized environments, supporting the view that human specialization is linked to human-dominated habitats rather than occurring equally across the species range (Rose et al., 2020). More broadly, ecological studies of urban mosquitoes have concluded that rapid urbanization can promote both the spread of *Aedes* mosquitoes and the persistence of the diseases they transmit by creating predictable access to hosts and persistent larval habitats (Kolimenakis et al., 2021).

Urban adaptation involves more than host choice alone. Urban adaptation also includes the physiological and life-history traits that allow mosquitoes to persist in human-modified habitats. Recent work on *Ae. aegypti* has shown that egg thermal tolerance is linked to urban adaptation and to traits that support persistence near humans, suggesting that cities may select not only for a stronger preference for humans but also for the ability to survive in hot, dry, and thermally variable environments (Chakraborty et al., 2025). This interpretation is consistent with the broader view arguing that thermotolerance is becoming increasingly important in blood-feeding arthropods as climate change, urban heat islands, and global movement reshape vector distributions and seasonal persistence (Bonizzoni and Lahondère, 2026). Climatic adaptation may therefore reinforce host specialization by allowing mosquitoes to remain active and abundant where humans are most concentrated.

Urban areas contain many different habitat types, and those differences influence where mosquitoes thrive. Studies from coastal cities have shown that *Aedes* infestation is not evenly distributed across urban space, but instead varies with combinations of vegetation, built cover, moisture, and weather conditions (Costa-da-Silva et al., 2024; Sallam et al., 2017). Work in northern Spain documented medically important mosquitoes in urban green spaces and cemeteries, showing that important urban vector habitats extend beyond houses and household containers (Goiri et al., 2024). Comparable studies have identified storm drains, catch basins, vacant lots, construction sites, and refuse around homes as productive habitats for urban mosquitoes, including *Ae. aegypti*, *Ae. albopictus*, and *Culex* species (Paploski et al., 2016; Unlu et al., 2013; Wilke et al., 2019). Together, these findings show that urban transmission risk arises from a patchwork of habitat types in which human-dominated and wildlife-associated host communities may overlap.

Human-built environments can also support less obvious mosquito ecologies with distinct evolutionary histories. An ancient urban underground mosquito lineage has been linked to subterranean human-made habitats, showing that adaptation to built environments can have deep historical roots rather than being only a recent response to modern cities (Haba et al., 2025). This interpretation is consistent with earlier work on the *Cx. pipiens* complex, which has long documented associations of the *molestus* form with underground and enclosed urban habitats and has shown that belowground and aboveground populations can differ in ecology and host use (Byrne and Nichols, 1999; Epstein et al., 2021; Farajollahi et al., 2011). More broadly, sewers, flooded basements, stormwater systems, urban water bodies, cemeteries, and peri-urban wetlands can support mosquito populations exposed to different host communities, thermal conditions, and control pressures (Crocker et al., 2017; Goiri et al., 2024; Unlu et al., 2013). These findings complicate any simple distinction between "domestic" and "wild" mosquitoes and suggest that host preference may evolve across a wider range of human-made habitats than is often assumed.

Land-use change outside cities can be just as important, as it reshapes the host communities and habitats that arboviral vectors such as *Aedes* and *Culex* encounter. Deforestation, irrigation, agriculture, livestock farming, and wetland modification often bring wildlife, domestic animals, and humans into closer contact while also creating or stabilizing larval habitat. Studies of *Aedes* and *Culex* in transformed landscapes show that host use often shifts with local vertebrate community structure, water management, and the arrangement of houses, animals, and breeding sites rather than remaining fixed at the species level (Cebrián-Camisón et al., 2020; Froxán-Grabalosa et al., 2025). More recent work in Kenya likewise found that land-use change shaped mosquito assemblages, host-feeding patterns, and arbovirus infection in landscapes where *Cx. pipiens sensu lato* and *Ae. aegypti* were both present, reinforcing the view that host contact in arboviral systems depends strongly on how land use reorganizes hosts, water, and opportunities for encounter (Getugi et al., 2026). Together, these studies show that land-use change can alter arbovirus risk not only by changing vector abundance but also by changing which hosts *Aedes* and *Culex* mosquitoes encounter and connect with.

Agricultural and livestock systems can be especially important because these systems can reorganize transmission around domestic amplifying hosts. In Kenya, host-choice experiments and blood-meal analysis showed substantial feeding on cattle, sheep, and goats by *Aedes*, *Culex*, and *Anopheles* species, demonstrating that domestic ungulates can strongly shape mosquito host choice and arbovirus ecology (Tchouassi et al., 2016). Experimental transmission work in a lamb-mosquito-lamb system further showed that successful mosquito feeding directly led to measurable onward transmission, linking host contact and feeding success to transmission efficiency (Bron et al., 2023). A similar logic applies to Japanese encephalitis and other multi-host systems in which pigs, birds, and humans occur close together, because land use can alter spillover risk by changing which host groups mosquitoes encounter most often and how frequently mosquitoes move between the hosts (Bouyer et al., 2025; Lindahl et al., 2012, 2013; Mulvey et al., 2021). Together, these examples show that farming systems can shape transmission not only by changing mosquito abundance but also by changing which domestic and wild hosts mosquitoes connect with.

These patterns are further complicated by climate change, which interacts with urbanization and land-use change and is predicted to alter the distribution of *Aedes* mosquitoes (Ryan et al., 2019). Temperature can affect mosquito development, survival, activity, gonotrophic timing, and biting frequency, whereas rainfall and humidity influence breeding-site persistence and the spatial distribution of hosts. Heat stress and altered rainfall can therefore shift both mosquito abundance and the relative advantages of feeding on different hosts. Experimental work has shown that temperature can alter feeding-related components of transmission potential in *Aedes* mosquitoes, including survival and vectorial capacity for arboviruses (Onyango et al., 2020). More broadly, climate change, urban warming, altered precipitation, and infrastructure-mediated microclimates are likely to reshape the biology of blood-feeding arthropods, affecting both behavior and distribution (Bonizzoni and Lahondère, 2026). These findings suggest that host preference is embedded within broader environmental change, because warming and climatic stress not only alter mosquito abundance but also change which hosts mosquitoes encounter, when those hosts are available, and how favorable different feeding choices become.

5.2 Human interventions and unintended evolutionary or behavioral consequences

Human interventions can also reshape host preference, either directly or indirectly. Population control, water management, housing modification, screening, insecticide use, and other control measures can all change the costs and benefits of feeding on different hosts. Under these conditions, mosquitoes may shift the timing, location, or target of host-seeking; over longer timescales, these altered costs may impose selection on host-related behavior. Work on *Ae. aegypti* and other urban arboviral vectors has emphasized that control effectiveness depends heavily on mosquito behavior in the human environment, including where mosquitoes rest, where they bite, and how they exploit indoor and outdoor spaces (Achee et al., 2019; Facchinelli et al., 2023; Pates and Curtis, 2005). Evolutionary arguments have therefore suggested that mosquitoes with strong human preference may evolve away from human feeding if attacking humans becomes sufficiently costly due to insecticides, physical barriers, or other control measures (Stone and Gross, 2018). Housing improvements and integrated environmental interventions may likewise alter host contact by reducing mosquito access to humans or by changing the availability of larval habitat near homes (Alqassim, 2024; Bouzid et al., 2016). This possibility is important because control tools may have consequences that extend beyond immediate reductions in mosquito density, potentially altering how arboviral vectors encounter and prioritize hosts.

Behavioral evidence is consistent with this broader framework. Host use and biting behavior can vary substantially even in highly urban settings, indicating that preference for humans is not expressed in the same way across all populations or contexts (Oduola and Awe, 2006; Pruszynski et al., 2020; Vera et al., 2023). At the same time, work on *Ae. aegypti* in Florida and East Africa shows that strongly human-preferring populations can remain epidemiologically important even when substantial biting occurs outdoors or across heterogeneous environments around homes (Pruszynski et al., 2020; Tenywa et al., 2025; Vera et al., 2023). These findings suggest that *Aedes* mosquito interventions focused only on indoor exposure leave a substantial fraction of transmission intact. More broadly, studies of arboviral vectors indicate that biting time, resting site, feeding location, and host use can all shape how mosquito populations respond to sustained control pressure, especially in human-dominated environments where indoor and outdoor host access differ (Achee et al., 2019; Facchinelli et al., 2023; Pates and Curtis, 2005; Vinauger and Chandrasegaran, 2024).

Human-associated environmental change also affects host behavior, which then feeds back into mosquito feeding patterns. Sleep schedules, outdoor evening activity, proximity to livestock, housing design, window and door screening, fan use, and household structure can all influence when and where mosquitoes encounter hosts. Studies of arbovirus transmission in changing human environments emphasize that these behavioral and infrastructural factors are central to realized exposure, especially in urban and peri-urban settings where *Aedes* and *Culex* mosquitoes exploit fine-scale differences in host accessibility, water storage, housing quality, and movement patterns (Cofone et al., 2025; Eder et al., 2018; Kolimenakis et al., 2021; Renard et al., 2023). From this perspective, host preference is not simply a property of the mosquito but rather a property of the mosquito–host encounter system, shaped jointly by mosquito and host behavior, infrastructure, and environmental change.

6. Zoonoses and disease emergence

6.1 Why host-use surveillance matters for predicting spillover

Host-use surveillance is vital for predicting spillover because it reveals whether mosquitoes feed on wildlife, domestic animals, or humans, influencing disease emergence. Mosquito-borne zoonoses often emerge at ecological interfaces where vectors, wildlife, domestic animals, and humans overlap. At these interfaces, host use helps determine whether mosquitoes remain within wildlife transmission cycles or connect those cycles to humans. For this reason, understanding disease emergence requires more than documenting vector presence or pathogen prevalence. It also requires identifying which hosts mosquitoes are actually feeding on in the field to determine whether transmission remains concentrated within reservoir communities or spills into human populations (Kilpatrick and Randolph, 2012).

This point is especially important because many arboviruses circulate silently in wildlife for long periods before recognized human outbreaks occur. Spillover may follow environmental disruption, shifts in host communities, climatic anomalies, or increases in bridge-vector abundance. In these situations, surveillance based solely on mosquito abundance or virus detection can miss the central ecological question of whether mosquitoes are actually transmitting pathogens between host groups. Work in disturbed rainforest systems has shown that combining host-selection analysis with flavivirus screening can reveal multiple plausible transmission pathways operating simultaneously, including links among wildlife, domestic animals, and humans that would not be apparent from vector detection alone (Hoyos et al., 2021). Integrated molecular studies of wildlife-associated mosquito communities have similarly shown that identifying both blood-meal origin and pathogen presence can clarify host-vector-pathogen interactions far better than taxonomic surveillance alone (Hernández-Triana et al., 2020). These findings suggest that host-use data are essential for understanding not only where mosquitoes occur but also the epidemiological role they play.

Host-use surveillance is particularly valuable in systems where environmental change and behavioral plasticity occur together. In a disturbed tropical forest, joint analysis of host selection and flavivirus occurrence revealed several potential transmission pathways operating simultaneously, supporting the view that ecological disruption can increase the number of host links available to mosquitoes rather than simply increasing mosquito abundance (Hoyos et al., 2021). Similar logic likely applies in suburban landscapes, irrigated agricultural systems, fragmented forests, and expanding cities, where mosquitoes may move among wildlife, livestock, and humans in ways that increase spillover probability even without dramatic changes in total vector density (Rizzoli et al., 2015; Rose et al., 2020; Tchouassi et al., 2016). From a surveillance perspective, the crucial question is therefore not only how many mosquitoes are present, but whether local ecological change is increasing the frequency of bridge feeding across host groups.

6.2 Blood-meal analysis, metabarcoding, and xenosurveillance

Blood-meal analysis remains one of the most direct ways to reconstruct actual mosquito-host contact. Host use in the field often differs from host preference measured under laboratory or semi-field conditions, so direct evidence of feeding is essential. Recent studies from dengue-endemic and arbovirus-prone settings have documented substantial geographic variation in the realized host use of *Ae. aegypti* and *Ae. albopictus*, including strong human feeding in some urban settings and more mixed feeding where domestic animals or other vertebrates are readily available (Dias et al., 2025; Gonçalves et al., 2023; Jeyaprakasam et al., 2024). Earlier field studies reached similar conclusions, showing that blood-feeding patterns in *Aedes* and *Culex* are highly context-dependent and can shift with host density, habitat type, season, and local vertebrate community structure (Apperson et al., 2004; Liebman et al., 2014; Molaei et al., 2006). Blood-meal analysis, therefore, provides a direct view of actual transmission pathways rather than inferred ones.

Recent methodological advances have expanded this approach beyond simple host identification. Mosquito blood meals can now be used to assess vertebrate diversity, infer pathogen exposure, and reconstruct aspects of local host-contact networks in both humans and wildlife (Vieira et al., 2024). Recent studies further support the feasibility of using mosquito blood meals for wildlife pathogen surveillance. However, those studies also emphasize important constraints related to blood-meal digestion, sample quality, and the probabilistic nature of pathogen detection (Wisely et al., 2025). These approaches are especially valuable where direct sampling of wildlife or humans is logistically difficult, expensive, or ethically constrained.

Xenosurveillance extends this logic by using blood-feeding arthropods to recover information about pathogens circulating in host populations. Foundational work has shown that mosquito blood meals can be used to detect human bacteria, parasites, and viruses, providing proof of principle for mosquito-based surveillance of the human pathogen landscape (Fauver et al., 2017; Grubaugh et al., 2015). Related studies of viral metagenomics and molecular xenomonitoring have shown that blood-feeding arthropods can provide access to host and pathogen information that might otherwise be missed by conventional sampling (Brinkmann et al., 2016; Cameron and Ramesh, 2021). Although xenosurveillance does not replace direct epidemiological or veterinary surveillance, it offers a practical, complementary strategy for detecting pathogen circulation and potential spillover signals in complex ecological settings.

Metabarcoding approaches have further increased the information yield of these methods. Instead of identifying only the mosquito or only the host, integrated workflows can now recover information on the mosquito species, the vertebrate host, and associated pathogens from the same blood-fed individual (Johnson et al., 2025). This is especially important for emergence studies because it allows surveillance to move beyond parallel data streams and toward more direct reconstruction of host–vector–pathogen linkages. In rapidly changing environments, where assumptions about host use may quickly become outdated, such integrated approaches are likely to become increasingly valuable.

6.3 Toward network-based surveillance of host-vector-pathogen systems

Traditional vector surveillance often asks whether a known vector species is present or whether a pathogen has been detected in mosquitoes. A more informative approach asks which hosts local mosquitoes use, how those contacts are distributed across the landscape, and whether pathogens are detectable in the same blood-fed individuals (Hernández-Triana et al., 2020; Johnson et al., 2025; Vieira et al., 2024). In this framework, emergence risk depends not only on vector presence but also on the structure of the interaction network connecting vectors, hosts, and pathogens.

This network perspective also has implications for spatial scale. Standard mosquito surveillance often treats the human population as a relatively uniform host category, but growing evidence suggests that this assumption is too coarse. Experimental work in *Ae. aegypti* showed stable and repeatable differences in human attractiveness over time and linked high attractiveness to elevated levels of skin-derived carboxylic acids (Obaldia et al., 2022). Broader studies likewise conclude that mosquito attraction to humans is highly heterogeneous and that this heterogeneity can have important consequences for transmission by concentrating bites on particular individuals or groups (Martinez et al., 2021). Additional work suggests that mosquito species do not necessarily rank hosts in the same way, opening the possibility that host-contact structure may vary at fine spatial scales even within the same household or neighborhood environment (Marrero et al., 2026). These patterns suggest that emergence risk may cluster much more locally than standard trap counts reveal. A more informative surveillance framework would therefore combine blood-meal analysis with host-odor ecology, fine-scale vector mapping, and environmental data on host accessibility.

Another reason to integrate host-use data into surveillance is that infection itself may alter mosquito behavior. Experimental work has shown that viral infection can interfere with mosquito host-seeking behavior, suggesting that pathogen detection and host-use data should ideally be interpreted together whenever possible (Vogels et al., 2017). If infected mosquitoes contact hosts differently from uninfected mosquitoes, then surveillance based solely on baseline feeding patterns may miss important heterogeneities in transmission. This reinforces the value of approaches that capture mosquitoes, hosts, and pathogens within a single analytical framework, rather than treating them as separate monitoring targets.

These approaches support a more informative form of surveillance in which hosts, vectors, and pathogens are monitored as parts of a single interaction network. Blood-meal analysis, metabarcoding, xenosurveillance, and molecular xenomonitoring now make this increasingly feasible and, in many settings, may offer the clearest way to identify which host contacts are sustaining local transmission and where the next spillover event is most likely to emerge.

7. Conclusion

Host preference is a central determinant of mosquito-borne virus transmission because host preference governs which hosts mosquitoes contact, how often those contacts occur, and whether transmission remains confined to animal reservoirs or extends into human populations. Across the evidence reviewed in this chapter, host preference emerges not as a minor behavioral characteristic but as a key factor linking mosquito evolution, sensory biology, ecology, and disease transmission.

Host preference is an evolving trait shaped by ecological opportunity, natural selection, and heritable variation rather than a fixed property of a species. Host preference is also expressed through a flexible sensory and behavioral system in which odor, vision, heat, humidity, contact cues, neural processing, and internal state are integrated into context-dependent host choice. Because host use helps determine whether pathogens remain in enzootic cycles, spill over only occasionally, or circulate efficiently in human populations, host preference has direct epidemiological consequences. Host preference is continually reshaped by urbanization, land-use change, climate, and intervention, which alter host availability, habitat structure, and the costs and benefits of feeding on different hosts. Because the hosts mosquitoes actually feed on in the wild often differ from the hosts they appear to prefer under laboratory conditions, direct evidence of feeding is essential for both surveillance and prediction.

To understand, predict, and manage mosquito-borne virus transmission, host preference must be treated as a central component of the system rather than as a background trait. Future progress will depend on integrating genomics, sensory biology, blood-meal ecology, experimental transmission, and fine-scale surveillance of host-vector contact. Host preference is not simply one trait among many in mosquito biology, but one of the clearest mechanisms by which ecological change translates into epidemic risk.

Acknowledgements

The development of this chapter was partially supported by the National Institute of Allergy and Infectious Diseases of the National Institutes of Health awards R01AI148551 and R01AI165560 (M.D.). We also thank the Centers for Disease Control and Prevention (CDC), Southeastern Center of Excellence in Vector-borne Disease (U01CK000662) for support to M.D. The funders did not have a role in the selection of studies for this review chapter, in the interpretation of literature, or in writing the manuscript.

References

Abbasi, E., 2025. *Aedes aegypti* and dengue: insights into transmission dynamics and viral lifecycle. Epidemiol. Infect. 153, e88. https://doi.org/10.1017/s0950268825100320.

Achee, N.L., Grieco, J.P., Vatandoost, H., Seixas, G., Pinto, J., Ching-NG, L., et al., 2019. Alternative strategies for mosquito-borne arbovirus control. PLoS Neglected Trop. Dis. 13, e0006822. https://doi.org/10.1371/journal.pntd.0006822.

Alberto, D.A.S., Rusch, C., Zhan, Y., Straw, A.D., Montell, C., Riffell, J.A., 2022. The olfactory gating of visual preferences to human skin and visible spectra in mosquitoes. Nat. Commun. 13, 555. https://doi.org/10.1038/s41467-022-28195-x.

Alqassim, A.Y., 2024. Social, behavioral and environmental determinants of vector-borne diseases: a narrative review of evidence and implications for integrated control approaches. J. Vector Borne Dis. 61, 525–535. https://doi.org/10.4103/jvbd.jvbd_34_24.

Apperson, C.S., Hassan, H.K., Harrison, B.A., Savage, H.M., Aspen, S.E., Farajollahi, A., et al., 2004. Host feeding patterns of established and potential mosquito vectors of west Nile Virus in the Eastern United States. Vector-Borne Zoonotic Dis. 4, 71–82. https://doi.org/10.1089/153036604773083013.

Aubry, F., Dabo, S., Manet, C., Filipović, I., Rose, N.H., Miot, E.F., et al., 2020. Enhanced Zika virus susceptibility of globally invasive *Aedes aegypti* populations. Science 370, 991–996. https://doi.org/10.1126/science.abd3663.

Ayala, D., Acevedo, P., Pombi, M., Dia, I., Boccolini, D., Costantini, C., et al., 2017. Chromosome inversions and ecological plasticity in the main African malaria mosquitoes. Evolution 71, 686–701. https://doi.org/10.1111/evo.13176.

Baik, L.S., Talross, G.J.S., Gray, S., Pattisam, H.S., Peterson, T.N., Nidetz, J.E., et al., 2024. Mosquito taste responses to human and floral cues guide biting and feeding. Nature 635, 639–646. https://doi.org/10.1038/s41586-024-08047-y.

Besansky, N.J., Hill, C.A., Costantini, C., 2004. No accounting for taste: host preference in malaria vectors. Trends Parasitol. 20, 249–251. https://doi.org/10.1016/j.pt.2004.03.007.

Blake, A.J., Riffell, J.A., 2025. Spectral preferences of mosquitos are altered by odors. J. Exp. Biol. 228, jeb250318. https://doi.org/10.1242/jeb.250318.

Blom, R., Krol, L., Langezaal, M., Schrama, M., Trimbos, K.B., Wassenaar, D., et al., 2024. Blood-feeding patterns of *Culex pipiens* biotype pipiens and pipiens/molestus hybrids in relation to avian community composition in urban habitats. Parasites Vectors 17, 95. https://doi.org/10.1186/s13071-024-06186-9.

Bobeva, A., Marinov, M.P., Klayn, S., Ilieva, M., Dimitrov, D., Zehtindjiev, P., 2025. Feeding preferences of mosquitoes (Culicidae) from the Eastern balkans and their role in transmission of avian Malaria. Integr. Zoöl. 21, 17–27. https://doi.org/10.1111/1749-4877.70018.

Bonizzoni, M., Lahondère, C., 2026. Blood-sucking arthropods in the Anthropocene: climate change thermotolerance, and global disease risks. Curr. Opin. Insect Sci. 75, 101506. https://doi.org/10.1016/j.cois.2026.101506.

Bouyer, J., Gil, D.A., Mora, I.P., Sorlí, V.D., Maiga, H., Mamai, W., et al., 2025. Suppression of *Aedes* mosquito populations with the boosted sterile insect technique in tropical and Mediterranean urban areas. Sci. Rep. 15, 17648. https://doi.org/10.1038/s41598-025-02795-1.

Bouzid, M., Brainard, J., Hooper, L., Hunter, P.R., 2016. Public health interventions for *Aedes* control in the time of Zikavirus– a meta-review on effectiveness of vector control strategies. PLoS Neglected Trop. Dis. 10, e0005176. https://doi.org/10.1371/journal.pntd.0005176.

Boyles, S.M., Mavian, C.N., Finol, E., Ukhanova, M., Stephenson, C.J., Hamerlinck, G., et al., 2020. Under-the-radar dengue virus infections in natural populations of *Aedes aegypti* mosquitoes. mSphere 5. https://doi.org/10.1128/msphere.00316-20.

Brinkmann, A., Nitsche, A., Kohl, C., 2016. Viral metagenomics on blood-feeding arthropods as a tool for human disease surveillance. Int. J. Mol. Sci. 17, 1743. https://doi.org/10.3390/ijms17101743.

Bron, G.M., Schreur, P.J.W., Jong, M.C.M. de, Keulen, L. van, Vloet, R.P.M., Koenraadt, C.J.M., et al., 2023. Quantifying Rift Valley fever virus transmission efficiency in a lamb-mosquito-lamb model. Front. Cell. Infect. Microbiol. 13, 1206089. https://doi.org/10.3389/fcimb.2023.1206089.

Byrne, K., Nichols, R.A., 1999. Culex pipiens in London underground tunnels: differentiation between surface and subterranean populations. Heredity 82, 7–15. https://doi.org/10.1046/j.1365-2540.1999.00412.x.

Cameron, M.M., Ramesh, A., 2021. The use of molecular xenomonitoring for surveillance of mosquito-borne diseases. Philos. Trans. R. Soc. B 376, 20190816. https://doi.org/10.1098/rstb.2019.0816.

Cebrián-Camisón, S., Puente, J.M. la, Figuerola, J., 2020. A literature review of host feeding patterns of invasive *Aedes* mosquitoes in Europe. Insects 11, 848. https://doi.org/10.3390/insects11120848.

Chakraborty, S., Zigmond, E., Shah, S., Dayal, D., Sylla, M., Akorli, J., et al., 2025. Thermal tolerance of *Aedes aegypti* mosquito eggs is associated with urban adaptation and human interactions. J. Therm. Biol. 131, 104167. https://doi.org/10.1016/j.jtherbio.2025.104167.

Chandel, A., DeBeaubien, N.A., Ganguly, A., Meyerhof, G.T., Krumholz, A.A., Liu, J., et al., 2024. Thermal infrared directs host-seeking behaviour in *Aedes aegypti* mosquitoes. Nature 633, 615–623. https://doi.org/10.1038/s41586-024-07848-5.

Chen, Z., Liu, F., Liu, N., 2019. Human odour coding in the yellow fever mosquito, *Aedes aegypti*. Sci. Rep., 13336. https://doi.org/10.1038/s41598-019-49753-2.

Clifton, M.E., Noriega, F.G., 2012. The fate of follicles after a blood meal is dependent on previtellogenic nutrition and juvenile hormone in *Aedes aegypti*. J. Insect Physiol. 58, 1007–1019. https://doi.org/10.1016/j.jinsphys.2012.05.005.

Cofone, L., Sabato, M., Paolo, C.D., Giovanni, S.D., Donato, M.A., Paglione, L., 2025. Urban, architectural, and socioeconomic factors contributing to the concentration of potential arbovirus vectors and arbovirosis in urban environments from a one health perspective: a systematic review. Sustainability 17, 4077. https://doi.org/10.3390/su17094077.

Coluzzi, M., Sabatini, A., Petrarca, Vi, Deco, M.A.D., 1977. Behavioural divergences between mosquitoes with different inversion karyotypes in polymorphic populations of the *Anopheles gambiae* complex. Nature 266, 832–833. https://doi.org/10.1038/266832a0.

Corfas, R.A., Vosshall, L.B., 2015. The cation channel TRPA1 tunes mosquito thermotaxis to host temperatures. Elife 4, e11750. https://doi.org/10.7554/elife.11750.

Costa-da-Silva, A.L., Dye-Braumuller, K.C., Wagner-Coello, H.U., Li, H., Johnson-Carson, D., Gunter, S.M., et al., 2024. Landscape and meteorological variables associated with *Aedes aegypti* and *Aedes albopictus* mosquito infestation in two southeastern U.S.A. coastal cities. J. Vector Ecol. 50, 28–38. https://doi.org/10.52707/1081-1710-50.1-28.

Coutinho-Abreu, I.V., Jamshidi, O., Raban, R., Atabakhsh, K., Merriman, J.A., Akbari, O.S., 2024. Identification of human skin microbiome odorants that manipulate mosquito landing behavior. Sci. Rep. 14, 1631. https://doi.org/10.1038/s41598-023-50182-5.

Coutinho-Abreu, I.V., Riffell, J.A., Akbari, O.S., 2022. Human attractive cues and mosquito host-seeking behavior. Trends Parasitol. 38, 246–264. https://doi.org/10.1016/j.pt.2021.09.012.

Crawford, J.E., Balcazar, D., Redmond, S., Rose, N.H., Youd, H.A., Lucas, E.R., et al., 2025. 1206 genomes reveal origin and movement of *Aedes aegypti* driving increased dengue risk. Science 389, eads3732. https://doi.org/10.1126/science.ads3732.

Crocker, W., Maute, K., Webb, C., French, K., 2017. Mosquito assemblages associated with urban water bodies; implications for pest and public health threats. Landsc. Urban. Plan 162, 115–125. https://doi.org/10.1016/j.landurbplan.2017.02.006.

DeGennaro, M., McBride, C.S., Seeholzer, L., Nakagawa, T., Dennis, E.J., Goldman, C., et al., 2013. *Orco* mutant mosquitoes lose strong preference for humans and are not repelled by volatile DEET. Nature 498, 487–491. https://doi.org/10.1038/nature12206.

Dekker, T., Geier, M., Cardé, R.T., 2005. Carbon dioxide instantly sensitizes female yellow fever mosquitoes to human skin odours. J. Exp. Biol. 208, 2963–2972. https://doi.org/10.1242/jeb.01736.

Dennis, E.J., Goldman, O.V., Vosshall, L.B., 2019. *Aedes aegypti* mosquitoes use their legs to sense DEET on contact. Curr. Biol. 29, 1551–1556.e5. https://doi.org/10.1016/j.cub.2019.04.004.

Dhakal, S., Bontempo, A.E., Singh, R., Dhavan, P., Montell, C., 2026. Gustatory avoidance of fatty acids by *Aedes aegypti* depends on an arthropod-specific TRP channel. Proc. Natl. Acad. Sci. 123, e2522818123. https://doi.org/10.1073/pnas.2522818123.

Dias, H.G., Garrido, I.O., Câmara, D.C.P., Stenn, T.M.S., Burkett-Cadena, N., Honório, N.A., et al., 2025. Blood-feeding patterns of mosquitoes (Diptera: Culicidae) collected in an area of high arbovirus transmission in West-Central Brazil. J. Méd. Ѐntomol. 62, 914–920. https://doi.org/10.1093/jme/tjaf041.

Eder, M., Cortes, F., Filha, N.T. de S., França, G.V.A. de, Degroote, S., Braga, C., et al., 2018. Scoping review on vector-borne diseases in urban areas: transmission dynamics, vectorial capacity and co-infection. Infect. Dis. Poverty 7, 90. https://doi.org/10.1186/s40249-018-0475-7.

Epstein, N.R., Saez, K., Polat, A., Davis, S.R., Aardema, M.L., 2021. The urban-adapted underground mosquito *Culex pipiens* form molestus maintains exogenously influenceable circadian rhythms. J. Exp. Biol. 224. https://doi.org/10.1242/jeb.242231.

Facchinelli, L., Badolo, A., McCall, P.J., 2023. Biology and behaviour of *Aedes aegypti* in the human environment: opportunities for vector control of arbovirus transmission. Viruses 15, 636. https://doi.org/10.3390/v15030636.

Farajollahi, A., Fonseca, D.M., Kramer, L.D., Kilpatrick, A.M., 2011. "Bird biting" mosquitoes and human disease: a review of the role of *Culex pipiens* complex mosquitoes in epidemiology. Infect., Genet. Evol. 11, 1577–1585. https://doi.org/10.1016/j.meegid.2011.08.013.

Fauver, J.R., Gendernalik, A., Weger-Lucarelli, J., Grubaugh, N.D., Brackney, D.E., Foy, B.D., et al., 2017. The use of xenosurveillance to detect human bacteria, parasites, and viruses in mosquito bloodmeals. Am. Soc. Trop. Med. Hyg. 97, 324–329. https://doi.org/10.4269/ajtmh.17-0063.

Fernández-Grandon, G.M., Gezan, S.A., Armour, J.A.L., Pickett, J.A., Logan, J.G., 2015. Heritability of attractiveness to mosquitoes. PLoS One 10, e0122716. https://doi.org/10.1371/journal.pone.0122716.

Ferreira-de-Lima, V.H., Lima-Camara, T.N., 2018. Natural vertical transmission of dengue virus in *Aedes aegypti* and *Aedes albopictus*: a systematic review. Parasites Vectors 11, 77. https://doi.org/10.1186/s13071-018-2643-9.

Fikrig, K., Harrington, L.C., 2021. Understanding and interpreting mosquito blood feeding studies: the case of *Aedes albopictus*. Trends Parasitol. 37, 959–975. https://doi.org/10.1016/j.pt.2021.07.013.

Fikrig, K., Martin, E., Dang, S., Fleur, K.S., Goldsmith, H., Qu, S., et al., 2022. The effects of host availability and fitness on *Aedes albopictus* blood feeding patterns in New York. Am. J. Trop. Med. Hyg. 106, 320–331. https://doi.org/10.4269/ajtmh.21-0157.

Froxán-Grabalosa, J., Mariani, S., Cerecedo-Iglesias, C., Richter-Boix, A., Torner, A.O., Pla, M., et al., 2025. Ecological drivers of arboviral disease risk: vector-host interfaces in a Mediterranean wetland of Northeastern Spain. PLOS Neglected Trop. Dis. 19, e0013447. https://doi.org/10.1371/journal.pntd.0013447.

Gao, H., Hu, W., Cui, C., Wang, Y., Zheng, Y., Jacobs-Lorena, M., et al., 2025. Emerging challenges for mosquito-borne disease control and the promise of symbiont-based transmission-blocking strategies. PLOS Pathog. 21, e1013431. https://doi.org/10.1371/journal.ppat.1013431.

Getugi, C., Kokwaro, E., Nyamache, A., Slothouwer, I., Rotich, G., Sang, R., et al., 2026. Land-use changes shape mosquito assemblages, host-feeding and arbovirus infection in Baringo County, Kenya. Infect., Genet. Evol. 139, 105913. https://doi.org/10.1016/j.meegid.2026.105913.

Giraldo, D., Rankin-Turner, S., Corver, A., Tauxe, G.M., Gao, A.L., Jackson, D.M., et al., 2023. Human scent guides mosquito thermotaxis and host selection under naturalistic conditions. Curr. Biol. 33, 2367–2382.e7. https://doi.org/10.1016/j.cub.2023.04.050.

Goiri, F., González, M.A., Cevidanes, A., Barandika, J.F., García-Peréz, A.L., 2024. Mosquitoes in urban green spaces and cemeteries in northern Spain. Parasites Vectors 17, 168. https://doi.org/10.1186/s13071-024-06263-z.

Goldman, O.V., DeFoe, A.E., Qi, Y., Jiao, Y., Weng, S.-C., Wick, B., et al., 2025. A single-nucleus transcriptomic atlas of the adult Aedes aegypti mosquito. Cell 188, 7267–7290.e26. https://doi.org/10.1016/j.cell.2025.10.008.

Gonçalves, A.A.L.M., Dias, A.H.C., Monteiro, D.D.S., Varela, I.B.F., Leal, S. da V., 2023. Blood meal survey reveals insights into mosquito-borne diseases on the island of Santiago, Cape Verde. Front. Trop. Dis. 4. https://doi.org/10.3389/fitd.2023.1070172.

Gonçalves, C.M., Melo, F.F., Bezerra, J.M., Chaves, B.A., Silva, B.M., Silva, L.D., et al., 2014. Distinct variation in vector competence among nine field populations of *Aedes aegypti* from a Brazilian dengue-endemic risk city. Parasites Vectors 7, 320. https://doi.org/10.1186/1756-3305-7-320.

Gould, J., Valdez, J.W., 2024. A little on the nose: a mosquito targets the nostrils of tree frogs for a blood meal. Ethology 130. https://doi.org/10.1111/eth.13424.

Greppi, C., Laursen, W.J., Budelli, G., Chang, E.C., Daniels, A.M., Giesen, L. van, et al., 2020. Mosquito heat seeking is driven by an ancestral cooling receptor. Science 367, 681–684. https://doi.org/10.1126/science.aay9847.

Grubaugh, N.D., Sharma, S., Krajacich, B.J., Fakoli III, L.S., Bolay, F.K., Diclaro II, J.W., et al., 2015. Xenosurveillance: a novel mosquito-based approach for examining the human-pathogen landscape. PLoS Neglected Trop. Dis. 9, e0003628. https://doi.org/10.1371/journal.pntd.0003628.

Haba, Y., Consortium‡, P., Korlević, P., McAlister, E., Lawniczak, M.K.N., Schumer, M., et al., 2025. Ancient origin of an urban underground mosquito. Science 390, eady4515. https://doi.org/10.1126/science.ady4515.

Haba, Y., McBride, L., 2022. Origin and status of *Culex pipiens* mosquito ecotypes. Curr. Biol. 32, R237–R246. https://doi.org/10.1016/j.cub.2022.01.062.

Hassan, H.K., Cupp, E.W., Hill, G.E., Katholi, C.R., Klingler, K., Unnasch, T.R., 2003. Avian host preference by vectors of eastern equine encephalomyelitis virus. Am. J. Trop. Med. Hyg. 69, 641–647. https://doi.org/10.4269/ajtmh.2003.69.641.

Hernández-Triana, L.M., Garza-Hernández, J.A., Morales, A.I.O., Prosser, S.W.J., Hebert, P.D.N., Nikolova, N.I., et al., 2020. An integrated molecular approach to untangling host-vector-pathogen interactions in mosquitoes (Diptera: Culicidae) from sylvan communities in Mexico. Front. Vet. Sci. 7, 564791. https://doi.org/10.3389/fvets.2020.564791.

Herre, M., Goldman, O.V., Lu, T.-C., Caballero-Vidal, G., Qi, Y., Gilbert, Z.N., et al., 2022. Non-canonical odor coding in the mosquito. Cell 185, 3104–3123.e28. https://doi.org/10.1016/j.cell.2022.07.024.

Holmes, C.J., Chakraborty, S., Ajayi, O.M., Uhran, M.R., Frigard, R., Stacey, C.L., et al., 2025. Multiple blood feeding bouts in mosquitoes allow for prolonged survival and are predicted to increase viral transmission during dry periods. iScience 28, 111760. https://doi.org/10.1016/j.isci.2025.111760.

Hoyos, J., Carrasquilla, M.C., León, C., Montgomery, J.M., Salyer, S.J., Komar, N., et al., 2021. Host selection pattern and flavivirus screening of mosquitoes in a disturbed Colombian rainforest. Sci. Rep. 11, 18656. https://doi.org/10.1038/s41598-021-98076-8.

Jeyaprakasam, N.K., Naidu, V.T.V., Vythilingam, I., Saeung, A., 2024. Blood meal analysis: unveiling the feeding preferences of *Aedes aegypti* and *Aedes albopictus* in a dengue-endemic area. Trop. Biomed. 41, 518–525. https://doi.org/10.47665/tb.41.4.013.

Johnson, B.J., Graham, M.C., Panahi, E., Vieira, C.J.S.P., Nath, N.S., Mason, P., et al., 2025. An all-in-one metabarcoding approach to mosquito and arbovirus xenosurveillance. Mol. Ecol. Resour. 25, e70022. https://doi.org/10.1111/1755-0998.70022.

Jové, V., Gong, Z., Hol, F.J.H., Zhao, Z., Sorrells, T.R., Carroll, T.S., et al., 2020. Sensory discrimination of blood and floral nectar by *Aedes aegypti* mosquitoes. Neuron. 108, 1163–1180.e12. https://doi.org/10.1016/j.neuron.2020.09.019.

Julian, E.G.-R., Bradley, J.B., Jose, A.F.-A., Maria, A.L.-P., Wilberth, A.C.C., Luis, F.F.-F., et al., 2010. Host-feeding preference of the mosquito, *Culex quinquefasciatus*, in Yucatan State, Mexico. J. Insect Sci. 10, 32. https://doi.org/10.1673/031.010.3201.

Jung, S.H., Kim, D., Jung, K.-S., Lee, D.-K., 2021. Color preference for host-seeking activity of *Aedes albopictus* and *Culex pipiens* (Diptera: Culicidae). J. Med. Entomol. 58, 2446–2452. https://doi.org/10.1093/jme/tjab100.

Kent, T.V., Schrider, D.R., Matute, D.R., 2025. Demographic history, genetic load, and the efficacy of selection in the globally invasive mosquito *Aedes aegypti*. Genome Biol. Evol. 17, evaf066. https://doi.org/10.1093/gbe/evaf066.

Kilpatrick, A.M., Kramer, L.D., Jones, M.J., Marra, P.P., Daszak, P., 2006. West Nile virus epidemics in north America are driven by shifts in mosquito feeding behavior. PLoS Biol. 4, e82. https://doi.org/10.1371/journal.pbio.0040082.

Kilpatrick, A.M., Randolph, S.E., 2012. Drivers, dynamics, and control of emerging vector-borne zoonotic diseases. Lancet 380, 1946–1955. https://doi.org/10.1016/s0140-6736(12)61151-9.

Kolimenakis, A., Heinz, S., Wilson, M.L., Winkler, V., Yakob, L., Michaelakis, A., et al., 2021. The role of urbanisation in the spread of *Aedes* mosquitoes and the diseases they transmit—A systematic review. PLoS Neglected Trop. Dis. 15, e0009631. https://doi.org/10.1371/journal.pntd.0009631.

Kothera, L., Mutebi, J.-P., Kenney, J.L., Saxton-Shaw, K., Ward, M.P., Savage, H.M., 2020. Bloodmeal, host selection, and genetic admixture analyses of *Culex pipiens* complex (Diptera: Culicidae) mosquitoes in Chicago, IL. J. Méd. Èntomol. 57, 78–87. https://doi.org/10.1093/jme/tjz158.

Kramer, L.D., Ciota, A.T., 2015. Dissecting vectorial capacity for mosquito-borne viruses. Curr. Opin. Virol 15, 112–118. https://doi.org/10.1016/j.coviro.2015.10.003.

Kumar, A., Tauxe, G.M., Perry, S., Scott, C.A., Dahanukar, A., Ray, A., 2020. Contributions of the conserved insect carbon dioxide receptor subunits to odor detection. Cell Rep. 31, 107510. https://doi.org/10.1016/j.celrep.2020.03.074.

Laursen, W.J., Budelli, G., Tang, R., Chang, E.C., Busby, R., Shankar, S., et al., 2023. Humidity sensors that alert mosquitoes to nearby hosts and egg-laying sites. Neuron 111, 874–887.e8. https://doi.org/10.1016/j.neuron.2022.12.025.

League, G.P., Degner, E.C., Pitcher, S.A., Hafezi, Y., Tennant, E., Cruz, P.C., et al., 2021. The impact of mating and sugar feeding on blood-feeding physiology and behavior in the arbovirus vector mosquito *Aedes aegypti*. PLoS Neglected Trop. Dis. 15, e0009815. https://doi.org/10.1371/journal.pntd.0009815.

Lee, M., O'Rorke, R., Clark, N.J., Webster, T.U., Wells, K., 2025. Diversity and plasticity in mosquito feeding patterns: a meta-analysis of 'Universal' DNA diet studies. Glob. Ecol. Biogeogr. 34. https://doi.org/10.1111/geb.70077.

Liang, J., Rose, N.H., Brusentsov, I.I., Lukyanchikova, V., Karagodin, D.A., Feng, Y., et al., 2025. Chromosomal inversions and their potential impact on the evolution of arboviral vector *Aedes aegypti*. Genome Biol. Evol. 17, evaf118. https://doi.org/10.1093/gbe/evaf118.

Liebman, K.A., Stoddard, S.T., Reiner, R.C., Perkins, T.A., Astete, H., Sihuincha, M., et al., 2014. Determinants of heterogeneous blood feeding patterns by *Aedes aegypti* in Iquitos, Peru. PLoS Neglected Trop. Dis. 8, e2702. https://doi.org/10.1371/journal.pntd.0002702.

Lindahl, J., Chirico, J., Boqvist, S., Thu, H.T.V., Magnusson, U., 2012. Occurrence of Japanese encephalitis virus mosquito vectors in relation to urban pig holdings. Am. J. Trop. Med. Hyg. 87, 1076–1082. https://doi.org/10.4269/ajtmh.2012.12-0315.

Lindahl, J.F., Ståhl, K., Chirico, J., Boqvist, S., Thu, H.T.V., Magnusson, U., 2013. Circulation of Japanese encephalitis virus in pigs and mosquito vectors within Can Tho City, Vietnam. PLoS Neglected Trop. Dis. 7, e2153. https://doi.org/10.1371/journal.pntd.0002153.

Liu, F., Coutinho-Abreu, I.V., Raban, R., Nguyen, T.T.D., Dimas, A.R., Merriman, J.A., et al., 2024. Engineered skin microbiome reduces mosquito attraction to mice. PNAS Nexus 3, pgae267. https://doi.org/10.1093/pnasnexus/pgae267.

Lozada-Chávez, A.N., Lozada-Chávez, I., Alfano, N., Palatini, U., Sogliani, D., Elfekih, S., et al., 2025. Adaptive genomic signatures of globally invasive populations of the yellow fever mosquito *Aedes aegypti*. Nat. Ecol. Evol. 9, 652–671. https://doi.org/10.1038/s41559-025-02643-5.

Lucas-Barbosa, D., Balvers, C., Bellantuono, A.J., Castillo, J.S., Costa-da-Silva, A.L., Moraes, C.M.D., et al., 2023. Competition matters: using in vitro community models to study the impact of human skin bacteria on mosquito attraction. Front. Ecol. Evol. 11, 1156311. https://doi.org/10.3389/fevo.2023.1156311.

Lyimo, I.N., Ferguson, H.M., 2009. Ecological and evolutionary determinants of host species choice in mosquito vectors. Trends Parasitol. 25, 189–196. https://doi.org/10.1016/j.pt.2009.01.005.

Main, B.J., Lee, Y., Ferguson, H.M., Kreppel, K.S., Kihonda, A., Govella, N.J., et al., 2016. The genetic basis of host preference and resting behavior in the major African Malaria vector, *Anopheles arabiensis*. PLoS Genet. 12, e1006303. https://doi.org/10.1371/journal.pgen.1006303.

Marrero, K.M., Castillo, J.S., Barbosa, D.L., Bellantuono, A.J., Marrero, M.A., Cid, D., et al., 2026. Individual humans attract different mosquito species. bioRxiv. https://doi.org/10.64898/2026.01.30.702931.

Martinez, J., Showering, A., Oke, C., Jones, R.T., Logan, J.G., 2021. Differential attraction in mosquito–human interactions and implications for disease control. Philos. Trans. R. Soc. B. 376, 20190811. https://doi.org/10.1098/rstb.2019.0811.

Matthews, B.J., McBride, C.S., DeGennaro, M., Despo, O., Vosshall, L.B., 2016. The neurotranscriptome of the *Aedes aegypti* mosquito. BMC Genomics 17, 32. https://doi.org/10.1186/s12864-015-2239-0.

McBride, C.S., Arguello, J.R., 2007. Five Drosophila genomes reveal nonneutral evolution and the signature of host specialization in the chemoreceptor superfamily. Genetics 177, 1395–1416. https://doi.org/10.1534/genetics.107.078683.

McBride, C.S., Baier, F., Omondi, A.B., Spitzer, S.A., Lutomiah, J., Sang, R., et al., 2014. Evolution of mosquito preference for humans linked to an odorant receptor. Nature 515, 222–227. https://doi.org/10.1038/nature13964.

McIver, S.B., 1968. Host preferences and discrimination by the mosquitoes *Aedes aegypti* and *Culex tarsali* (Diptera: Culicidae). J. Med. entomology 5, 422–428.

McMeniman, C.J., Corfas, R.A., Matthews, B.J., Ritchie, S.A., Vosshall, L.B., 2014. Multimodal integration of carbon dioxide and other sensory cues drives mosquito attraction to humans. Cell 156, 1060–1071. https://doi.org/10.1016/j.cell.2013.12.044.

Menon, R., Ignell, R., Hill, S.R., 2025. Differential expression of antennal chemosensory genes related to host preference of *Culex pipiens* biotypes. Parasites Vectors 18, 372. https://doi.org/10.1186/s13071-025-07028-y.

Molaei, G., Andreadis, T.G., Armstrong, P.M., Anderson, J.F., Vossbrinck, C.R., 2006. Host feeding patterns of culex mosquitoes and west Nile Virus transmission, north-eastern United States - volume 12, number 3—March 2006 - emerging infectious diseases Journal - CDC. Emerg. Infect. Dis. 12, 468–474. https://doi.org/10.3201/eid1203.051004.

Montell, C., 2025. The sensory arsenal mosquitoes use to find us. Trends Parasitol 41, 591–602. https://doi.org/10.1016/j.pt.2025.05.004.

Morita, T., Lyn, N.G., Heynitz, R.K. von, Goldman, O.V., Sorrells, T.R., DeGennaro, M., et al., 2025. Cross-modal sensory compensation increases mosquito attraction to humans. Sci. Adv. 11, eadn5758. https://doi.org/10.1126/sciadv.adn5758.

Mulvey, P., Duong, V., Boyer, S., Burgess, G., Williams, D.T., Dussart, P., et al., 2021. The ecology and evolution of Japanese encephalitis virus. Pathogens 10, 1534. https://doi.org/10.3390/pathogens10121534.

Obaldia, M.E.D., Morita, T., Dedmon, L.C., Boehmler, D.J., Jiang, C.S., Zeledon, E.V., et al., 2022. Differential mosquito attraction to humans is associated with skin-derived carboxylic acid levels. Cell 185, 4099–4116.e13. https://doi.org/10.1016/j.cell.2022.09.034.

Oduola, A.O., Awe, O.O., 2006. Behavioural biting preference of *Culex quinquefasciatus* in human host in Lagos metropolis Nigeria. J. Vector Borne Dis. 43, 16–20.

Onyango, M.G., Bialosuknia, S.M., Payne, A.F., Mathias, N., Kuo, L., Vigneron, A., et al., 2020. Increased temperatures reduce the vectorial capacity of *Aedes* mosquitoes for Zika virus. Emerg. Microbes & Infect. 9, 67–77. https://doi.org/10.1080/22221751.2019.1707125.

Paixão, E.S., Teixeira, M.G., Rodrigues, L.C., 2018. Zika, chikungunya and dengue: the causes and threats of new and re-emerging arboviral diseases. BMJ Glob. Heal. 3, e000530. https://doi.org/10.1136/bmjgh-2017-000530.

Paploski, I.A.D., Rodrigues, M.S., Mugabe, V.A., Kikuti, M., Tavares, A.S., Reis, M.G., et al., 2016. Storm drains as larval development and adult resting sites for *Aedes aegypti* and *Aedes albopictus* in Salvador, Brazil. Parasites Vectors 9, 419. https://doi.org/10.1186/s13071-016-1705-0.

Pates, H., Curtis, C., 2005. Mosquito behavior and vector control. Annu. Rev. Entomol. 50, 53–70. https://doi.org/10.1146/annurev.ento.50.071803.130439.

Ponlawat, A., Harrington, L.C., 2005. Blood feeding patterns of *Aedes aegypti* and *Aedes albopictus* in Thailand. J. Méd. Èntomol. 42, 844–849. https://doi.org/10.1093/jmedent/42.5.844.

Powell, J.R., Tabachnick, W.J., 2013. History of domestication and spread of *Aedes aegypti* - a review. Memórias Inst. Oswaldo Cruz 108, 11–17. https://doi.org/10.1590/0074-0276130395.

Pruszynski, C.A., Stenn, T., Acevedo, C., Leal, A.L., Burkett-Cadena, N.D., 2020. Human blood feeding by *Aedes aegypti* (Diptera: Culicidae) in the Florida Keys and a review of the literature. J. Med. Entomol. 57, 1640–1647. https://doi.org/10.1093/jme/tjaa083.

Raji, J.I., Melo, N., Castillo, J.S., Gonzalez, S., Saldana, V., Stensmyr, M.C., et al., 2019. *Aedes aegypti* mosquitoes detect acidic volatiles found in human odor using the IR8a pathway. Curr. Biol. 29, 1253–1262.e7. https://doi.org/10.1016/j.cub.2019.02.045.

Ray, G., Huff, R.M., Castillo, J.S., Bellantuono, A.J., DeGennaro, M., Pitts, R.J., 2023. Carboxylic acids that drive mosquito attraction to humans activate ionotropic receptors. PLOS Neglected Trop. Dis. 17, e0011402. https://doi.org/10.1371/journal.pntd.0011402.

Reinhold, J.M., Lahondère, C., 2024. Mosquitoes feeding on ectothermic hosts: from host seeking to pathogen transmission. Curr. Opin. Insect Sci. 66, 101273. https://doi.org/10.1016/j.cois.2024.101273.

Renard, A., Lombardini, F.P., Zapata, M.P., Porphyre, T., Bento, A., Suzán, G., et al., 2023. Interaction of human behavioral factors shapes the transmission of arboviruses by *Aedes* and *Culex* mosquitoes. Pathogens 12, 1421. https://doi.org/10.3390/pathogens12121421.

Riccetti, N., Fasano, A., Ferraccioli, F., Gomez-Ramirez, J., Stilianakis, N.I., 2022. Host selection and forage ratio in West Nile virus–transmitting *Culex* mosquitoes: challenges and knowledge gaps. PLOS Neglected Trop. Dis. 16, e0010819. https://doi.org/10.1371/journal.pntd.0010819.

Richards, S.L., Ponnusamy, L., Unnasch, T.R., Hassan, H.K., Apperson, C.S., 2006. Host-feeding patterns of *Aedes albopictus* (Diptera: Culicidae) in relation to availability of human and domestic animals in suburban landscapes of central North Carolina. J. Med. Entomol. 43, 543–551. https://doi.org/10.1093/jmedent/43.3.543.

Rizzoli, A., Bolzoni, L., Chadwick, E.A., Capelli, G., Montarsi, F., Grisenti, M., et al., 2015. Understanding west Nile virus ecology in Europe: *Culex pipiens* host feeding preference in a hotspot of virus emergence. Parasites Vectors 8, 213. https://doi.org/10.1186/s13071-015-0831-4.

Robertson, H.M., 2018. Molecular evolution of the major arthropod chemoreceptor gene families. Annu. Rev. Entomol. 64, 1–16. https://doi.org/10.1146/annurev-ento-020117-043322.

Rodríguez-Valencia, V., Olive, M., Goff, G.L., Faisse, M., Bourel, M., L'Ambert, G., et al., 2025. Host-feeding preferences of *Culex pipiens* and its potential significance for flavivirus transmission in the Camargue, France. Med. Vet. Entomol. 39, 614–625. https://doi.org/10.1111/mve.12802.

Rose, N.H., Badolo, A., Sylla, M., Akorli, J., Otoo, S., Gloria-Soria, A., et al., 2023. Dating the origin and spread of specialization on human hosts in *Aedes aegypti* mosquitoes. eLife 12, e83524. https://doi.org/10.7554/elife.83524.

Rose, N.H., Dabo, S., Leal, S. da V., Sylla, M., Diagne, C.T., Faye, Oumar, et al., 2022. Enhanced mosquito vectorial capacity underlies the Cape Verde Zika epidemic. PLOS Biol. 20, e3001864. https://doi.org/10.1371/journal.pbio.3001864.

Rose, N.H., Sylla, M., Badolo, A., Lutomiah, J., Ayala, D., Aribodor, O.B., et al., 2020. Climate and urbanization drive mosquito preference for humans. Curr. Biol. 30, 3570–3579.e6. https://doi.org/10.1016/j.cub.2020.06.092.

Ryan, S.J., Carlson, C.J., Mordecai, E.A., Johnson, L.R., 2019. Global expansion and redistribution of *Aedes*-borne virus transmission risk with climate change. PLoS Neglected Trop. Dis. 13, e0007213. https://doi.org/10.1371/journal.pntd.0007213.

Sallam, M.F., Fizer, C., Pilant, A.N., Whung, P.-Y., 2017. Systematic review: land cover, meteorological, and socioeconomic determinants of *Aedes* mosquito habitat for risk mapping. Int. J. Environ. Res. Public. Heal. 14, 1230. https://doi.org/10.3390/ijerph14101230.

Scott, T.W., Clark, G.G., Lorenz, L.H., Amerasinghe, P.H., Reiter, P., Edman, J.D., 1993. Detection of multiple blood feeding in *Aedes aegypti* (Diptera: Culicidae) during a single gonotrophic cycle using a histologic technique. J. Méd. Èntomol. 30, 94–99. https://doi.org/10.1093/jmedent/30.1.94.

Silva, W.A.P.P. de, Bernal, X.E., Chathuranga, W.G.D., Herath, B.P., Ekanayake, C., Abeysundara, H.T.K., et al., 2020. Feeding patterns revealed host partitioning in a community of frog-biting mosquitoes. Ecol. Entomol. 45, 988–996. https://doi.org/10.1111/een.12874.

Simpson, J.E., Folsom-O'Keefe, C.M., Childs, J.E., Simons, L.E., Andreadis, T.G., Diuk-Wasser, M.A., 2009. Avian host-selection by culex pipiens in experimental trials. PLoS ONE 4, e7861. https://doi.org/10.1371/journal.pone.0007861.

Singh, P., Goyal, S., Gupta, S., Garg, S., Tiwari, A., Rajput, V., et al., 2023. Combinatorial encoding of odors in the mosquito antennal lobe. Nat. Commun. 14, 3539. https://doi.org/10.1038/s41467-023-39303-w.

Smith, D.L., Battle, K.E., Hay, S.I., Barker, C.M., Scott, T.W., McKenzie, F.E., 2012. Ross, Macdonald, and a theory for the dynamics and control of mosquito-transmitted pathogens. PLoS Pathog. 8, e1002588. https://doi.org/10.1371/journal.ppat.1002588.

Souza-Neto, J.A., Powell, J.R., Bonizzoni, M., 2019. *Aedes aegypti* vector competence studies: a review. Infect., Genet. Evol. 67, 191–209. https://doi.org/10.1016/j.meegid.2018.11.009.

Steib, B.M., Geier, M., Boeckh, J., 2001. The effect of lactic acid on odour-related host preference of yellow fever mosquitoes. Chem. Senses 26, 523–528. https://doi.org/10.1093/chemse/26.5.523.

Stone, C., Gross, K., 2018. Evolution of host preference in anthropophilic mosquitoes. Malar. J. 17, 257. https://doi.org/10.1186/s12936-018-2407-1.

Swart, M.M. de, Balvers, C., Verhulst, N.O., Koenraadt, C.J.M., 2023. Effects of host blood on mosquito reproduction. Trends Parasitol. 39, 575–587. https://doi.org/10.1016/j.pt.2023.04.003.

Takken, W., 1990. The role of olfaction in host-seeking of mosquitoes: a review. Int. J. Trop. Insect Sci. 12, 287–295. https://doi.org/10.1017/s1742758400020816.

Takken, W., Verhulst, N.O., 2013. Host preferences of blood-feeding mosquitoes. Annu. Rev. Entomol. 58, 433–453. https://doi.org/10.1146/annurev-ento-120811-153618.

Tallon, A.K., Hill, S.R., Ignell, R., 2019. Sex and age modulate antennal chemosensory-related genes linked to the onset of host seeking in the yellow-fever mosquito, *Aedes aegypti*. Sci. Rep-uk 9, 43. https://doi.org/10.1038/s41598-018-36550-6.

Tang, R., Busby, R., Laursen, W.J., Keane, G.T., Garrity, P.A., 2024. Functional dissection of mosquito humidity sensing reveals distinct dry and moist cell contributions to blood feeding and oviposition. Proc. Natl. Acad. Sci. 121, e2407394121. https://doi.org/10.1073/pnas.2407394121.

Tchouassi, D.P., Okiro, R.O.K., Sang, R., Cohnstaedt, L.W., McVey, D.S., Torto, B., 2016. Mosquito host choices on livestock amplifiers of Rift Valley fever virus in Kenya. Parasites Vectors 9, 184. https://doi.org/10.1186/s13071-016-1473-x.

Tenywa, F.S.C., Hälg, S., Makame, H., Moore, J., Dogan, O., Mapipi, H.I., et al., 2025. DENV-2 circulation and host preference among highly anthropophilic, outdoor-biting *Aedes aegypti* in Dar es Salaam, Tanzania. Viruses 17, 818. https://doi.org/10.3390/v17060818.

Unlu, I., Farajollahi, A., Strickman, D., Fonseca, D.M., 2013. Crouching tiger, hidden trouble: urban sources of *Aedes albopictus* (Diptera: Culicidae) refractory to source-reduction. PLoS ONE 8, e77999. https://doi.org/10.1371/journal.pone.0077999.

Vera, A.J., Soliz, A.G., Mora-Covarrubias, A.D., la, Watts, D.M., 2023. Host feeding preference by Aedes aegypti at two nonincorporated urban communities along the U.S.-Mexico border. Southwest. Entomol. 48, 563–568. https://doi.org/10.3958/059.048.0307.

Vieira, C.J.S.P., Gyawali, N., Onn, M.B., Shivas, M.A., Shearman, D., Darbro, J.M., et al., 2024. Mosquito bloodmeals can be used to determine vertebrate diversity, host preference, and pathogen exposure in humans and wildlife. Sci. Rep. 14, 23203. https://doi.org/10.1038/s41598-024-73820-y.

Vinauger, C., Chandrasegaran, K., 2024. Context-specific variation in life history traits and behavior of *Aedes aegypti* mosquitoes. Front. Insect Sci. 4, 1426715. https://doi.org/10.3389/finsc.2024.1426715.

Vinauger, C., Lahondère, C., Wolff, G.H., Locke, L.T., Liaw, J.E., Parrish, J.Z., et al., 2018. Modulation of host learning in *Aedes aegypti* mosquitoes. Curr. Biol. 28, 333–344.e8. https://doi.org/10.1016/j.cub.2017.12.015.

Vogels, C.B.F., Fros, J.J., Pijlman, G.P., Loon, J.J.A. van, Gort, G., Koenraadt, C.J.M., 2017. Virus interferes with host-seeking behaviour of mosquito. J. Exp. Biol. 220, 3598–3603. https://doi.org/10.1242/jeb.164186.

Weaver, S.C., 2005. Infectious diseases from nature: mechanisms of viral emergence and persistence. Arch. Virol. Suppl. 33–44. https://doi.org/10.1007/3-211-29981-5_4.

Weaver, S.C., Forrester, N.L., 2015. Chikungunya: evolutionary history and recent epidemic spread. Antivir. Res. 120, 32–39. https://doi.org/10.1016/j.antiviral.2015.04.016.

Wehmeyer, M.L., Jaworski, L., Jöst, H., Şuleşco, T., Rauhöft, L., Afonso, S.M.M., et al., 2024. Host attraction and host feeding patterns indicate generalist feeding of *Culex pipiens s.s.* and *Cx. torrentium*. Parasites Vectors 17, 369. https://doi.org/10.1186/s13071-024-06439-7.

Wei, Y.-L., Wu, Z., Li, R.-L., Tang, F., 2025. Review of selected mosquito-borne diseases: arboviruses (dengue, chikungunya, Zika, West Nile, Japanese encephalitis, yellow fever) and parasitic diseases (malaria, lymphatic Filariasis). Front. Public. Heal. 13, 1712094. https://doi.org/10.3389/fpubh.2025.1712094.

Werner-Reiss, U., Galun, R., Crnjar, R., Liscia, A., 1999. Factors modulating the blood feeding behavior and the electrophysiological responses of labral apical chemoreceptors to adenine nucleotides in the mosquito *Aedes aegypti* (Culicidae). J. Insect Physiol. 45, 801–808. https://doi.org/10.1016/s0022-1910(98)00153-x.

Wilke, A.B.B., Chase, C., Vasquez, C., Carvajal, A., Medina, J., Petrie, W.D., et al., 2019. Urbanization creates diverse aquatic habitats for immature mosquitoes in urban areas. Sci. Rep. 9, 15335. https://doi.org/10.1038/s41598-019-51787-5.

Wisely, S.M., Torhorst, C.W., Botero-Cañola, S., Atsma, H., Burkett-Cadena, N.D., Reeves, L.E., 2025. Evaluation of mosquito blood meals as a tool for wildlife pathogen surveillance. Pathogens 14, 792. https://doi.org/10.3390/pathogens14080792.

Wolff, G.H., Riffell, J.A., 2018. Olfaction, experience and neural mechanisms underlying mosquito host preference. J. Exp. Biol. 221, jeb157131. https://doi.org/10.1242/jeb.157131.

Wu, Y., Wu, J., Wu, Q., Cai, H., Hu, J., Lin, Z., et al., 2025. The making of a dangerous vector: factors shaping the vector competence of *Aedes* mosquitoes. Front. Cell. Infect. Microbiol. 15, 1718737. https://doi.org/10.3389/fcimb.2025.1718737.

Yan, J., Gangoso, L., Ruiz, S., Soriguer, R., Figuerola, J., Puente, J.M. la, 2021. Understanding host utilization by mosquitoes: determinants, challenges and future directions. Biol. Rev. 96, 1367–1385. https://doi.org/10.1111/brv.12706.

Zhan, Y., Alberto, D.A.S., Rusch, C., Riffell, J.A., Montell, C., 2021. Elimination of vision-guided target attraction in *Aedes aegypti* using CRISPR. Curr. Biol. 31, 4180–4187.e6. https://doi.org/10.1016/j.cub.2021.07.003.

Zhao, Z., McBride, C.S., 2020. Evolution of olfactory circuits in insects. J. Comp. Physiol. A 206, 353–367. https://doi.org/10.1007/s00359-020-01399-6.

Zhao, Z., Zung, J.L., Hinze, A., Kriete, A.L., Iqbal, A., Younger, M.A., et al., 2022. Mosquito brains encode unique features of human odour to drive host seeking. Nature 605, 706–712. https://doi.org/10.1038/s41586-022-04675-4.

Zung, J.L., McBride, C.S., 2025. Sebaceous origins of human odor. Curr. Biol. 35, R303–R313. https://doi.org/10.1016/j.cub.2025.03.031.

CHAPTER SEVEN

The role of mosquito saliva in arbovirus transmission and vector competence

Renuka E. Joseph*, Rohit Chebbi, and Joel Vega-Rodriguez*
Laboratory of Malaria and Vector Research, National Institute of Allergy and Infectious Diseases, National Institutes of Health, Rockville, MD, United States
*Corresponding authors. e-mail address: renuka.elizabethjoseph@nih.gov; joel.vega-rodriguez@nih.gov

Contents

Advances in Insect Physiology, Volume 70
ISSN 0065-2806, https://doi.org/10.1016/bs.aiip.2026.06.001

Abstract

Mosquito saliva plays a central role in shaping vector–host–pathogen interactions and is a critical determinant of transmission. During blood feeding, mosquitoes inject a complex mixture of salivary proteins that facilitate vascular access by modulating hemostasis, inflammation, and immune responses at the bite site. These same processes influence pathogen establishment in the vertebrate host. Increasing evidence indicates that the effects of salivary proteins on arbovirus infection are multifaceted and highly context-dependent. While some salivary factors can directly inhibit viral entry or replication, others enhance infection by creating a local environment that promotes viral dissemination and the recruitment of permissive target cells. Beyond their role in the skin, salivary components can also shape the biochemical and immunological landscape of the ingested blood meal, influencing midgut physiology and microbial dynamics. These interactions can directly affect pathogen survival, replication, and transmission within the mosquito, highlighting saliva as a key regulator of vector competence across both host and vector environments. This dual role of mosquito saliva underscores the need for a more integrated framework that considers context, timing, and tissue-specific effects. Importantly, these insights open new opportunities for intervention. Targeting salivary components offers a promising vector-directed strategy to disrupt pathogen transmission, including the development of transmission-blocking vaccines that act at both the bite site and within the mosquito.

1. Introduction

Mosquito-borne arboviruses, including dengue virus (DENV; *Orthoflavivirus denguei*), Zika virus (ZIKV; *Orthoflavivirus zikaense*), chikungunya virus (CHIKV; *Alphavirus chikungunya*), and West Nile virus (WNV; *Orthoflavivirus nilense*), represent a major and expanding global health threat. These viruses are transmitted when mosquitoes take a blood meal, during which they inject saliva together with viral particles into the skin. Mosquito saliva is a complex mixture of bioactive proteins that enables efficient blood feeding by counteracting host hemostatic, inflammatory, and immune defenses (Ribeiro and Francischetti, 2003; Schneider and Higgs, 2008). As the vehicle for arbovirus and other pathogen transmission, mosquito saliva shapes the local skin environment in ways that directly influence infection success and disease severity.

Salivary proteins exert divergent effects on arbovirus transmission, with some enhancing transmission and others limiting it through distinct mechanisms, highlighting the context-dependent role of saliva in arbovirus pathogenesis (Conway et al., 2016; McCracken et al., 2020). These effects also reflect vector-specific differences in salivary composition among different mosquitoes. For example, *Aedes* mosquitoes encode the salivary peptide sialokinin, which induces rapid vascular leakage and is a key driver of saliva-mediated viral enhancement; this activity is absent in the saliva of *Anopheles* mosquitoes, which neither induces comparable endothelial permeability nor enhances arbovirus infection in vivo — consistent with the general inability of *Anopheles* species to transmit arboviruses efficiently (Lefteri et al., 2022).

While the effects of mosquito saliva at the bite site are well established, its role does not end in the skin. During blood feeding, mosquitoes ingest a substantial amount of their own saliva together with the blood meal, introducing salivary proteins into the midgut lumen where they mix with host-derived blood components that remain biologically active for several hours after ingestion (Pala et al., 2024; Ribeiro et al., 1984) (Fig. 1). Within this blood bolus, salivary proteins are positioned to interact with coagulation factors, complement proteins, and leukocytes, all of which can influence midgut physiology and pathogen survival. Salivary proteins likely retain their anti-hemostatic and immunomodulatory functions within the blood bolus, where they can influence pathogen infection in the mosquito midgut. Consistent with this, salivary apyrase from *Anopheles gambiae* mosquitoes has been shown to modulate hemostatic and inflammatory responses in the blood bolus and promote *Plasmodium* development within the mosquito midgut (Pala et al., 2024). However, how salivary proteins influence arbovirus infection within the mosquito midgut remains poorly defined, and this gap may in part reflect the widespread use of artificial membrane feeding systems that lack key host factors such as coagulation, immune, and inflammatory components, potentially obscuring these interactions.

In this chapter, we examine how mosquito salivary proteins shape arbovirus infection at the bite site, focusing on the molecular mechanisms that regulate early viral replication and disease severity. We also highlight emerging evidence that salivary proteins may influence the mosquito midgut environment following ingestion, an underexplored aspect of transmission biology. Finally, we discuss the broader implications of these interactions, including their potential as biomarkers of exposure and targets for vaccine and therapeutic development.

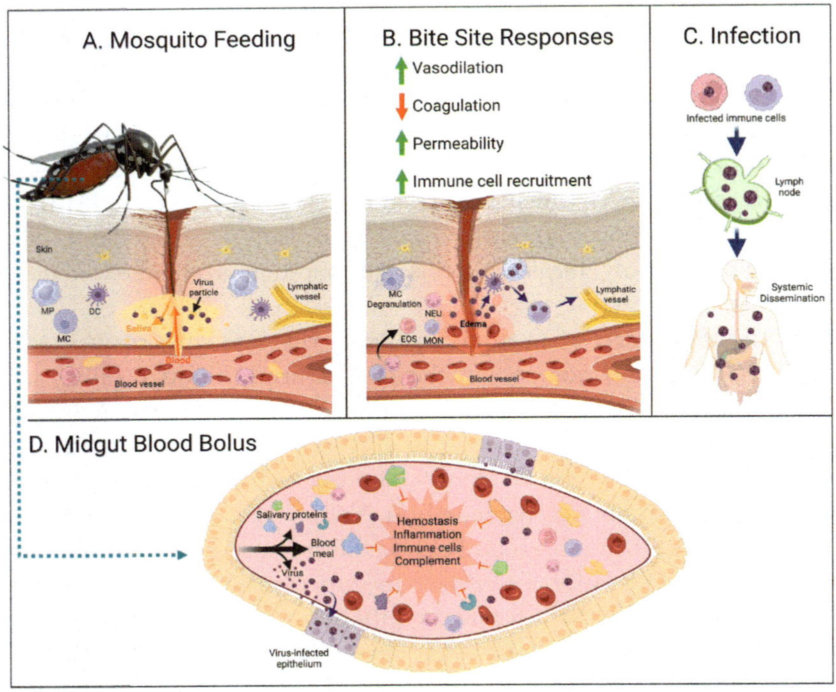

Fig. 1 Mosquito bite–induced modulation of host skin and mosquito midgut environments influencing pathogen transmission. A. During blood feeding, mosquitoes inject saliva together with virus particles into the skin. **B.** Salivary proteins modulate the local microenvironment by targeting vascular, hemostatic, and immune pathways, facilitating blood uptake. At the bite site, resident skin cells and innate immune cells, including mast cells (MC), macrophages (MP), and dendritic cells (DC), become activated, leading to mast cell degranulation, increased vascular permeability, and recruitment of circulating immune cells such as neutrophils (NEU), monocytes (MON), and eosinophils (EOS). These responses generate a dysregulated inflammatory environment characterized by enhanced cell recruitment and reduced antiviral defenses, promoting infection of susceptible cells. **C.** Infected cells can traffic to lymphatic vessels and draining lymph nodes, enabling viral amplification and subsequent systemic dissemination. **D.** Concurrently, mosquito saliva is ingested with the blood meal (A) and remains active within the blood bolus in the midgut, where it modulates hemostasis, inflammation, complement activity, and immune cell function. Together, these processes illustrate how mosquito saliva shapes host responses at the bite site and within the mosquito midgut, promoting pathogen transmission.

2. Mosquito feeding biology: Why saliva matters

Once the mosquito lands on a host it must find a blood vessel to acquire its blood meal. This process relies on specialized mouthparts and coordinated behavioral and biochemical mechanisms. The mosquito proboscis consists of an outer retractable sheath (labium), that encloses the fascicle, which contains the maxillae and mandibles for tissue penetration, the labrum for blood uptake, and the hypopharynx that houses the salivary duct (Gurera et al., 2018). During probing, these structures allow the mosquito to navigate through the skin and detect blood vessels.

While searching for a blood vessel, mosquitoes repeatedly insert and withdraw their proboscis while injecting saliva into the tissue (Fig. 1). Salivation begins prior blood ingestion (Griffiths and Gordon, 1952) and is essential for efficient vessel location, as mosquitoes with impaired salivation exhibit prolonged probing times and reduced feeding success (Ribeiro et al., 1984). Several salivary proteins, including *Aedes aegypti* sialokinin, aegyptin, D7L1 and D7L2, as well as *Anopheles* apyrase (Chagas et al., 2014; Martin-Martin et al., 2023, 2022; Ribeiro et al., 1985) contribute to this process by modulating vascular tone, hemostasis, and local inflammation, thereby facilitating vessel detection and access.

Importantly, saliva delivery during probing is tightly linked to pathogen transmission. Arboviruses and parasites are co-inoculated with saliva into the dermis, and the duration of probing is a key determinant of pathogen deposition. Prolonged probing may potentially alter transmission dynamics and influence the probability of successful infection (Rossignol et al., 1984). Once a vessel is located, mosquitoes can either cannulate it directly or lacerate it to create a blood pool from which they feed (Choumet et al., 2012). Continuous salivation during feeding is required to maintain blood flow by counteracting host hemostatic and inflammatory responses triggered by tissue injury. In the following section, we discuss the diverse functions of salivary proteins that modulate host physiology and shape the outcome of mosquito feeding and pathogen transmission.

3. Composition of mosquito saliva

3.1 Vasodilators

During blood feeding, the mosquito proboscis disrupts the skin, causing tissue injury and initiating host hemostatic pathways to prevent blood loss (Figs. 1 and 2). Among these, vasoconstriction — mediated by contraction of smooth muscle surrounding the blood vessel — is rapidly triggered at the bite site. This process is driven by vasoconstrictors and regulatory molecules, including serotonin, thromboxane A2, epinephrine, norepinephrine,

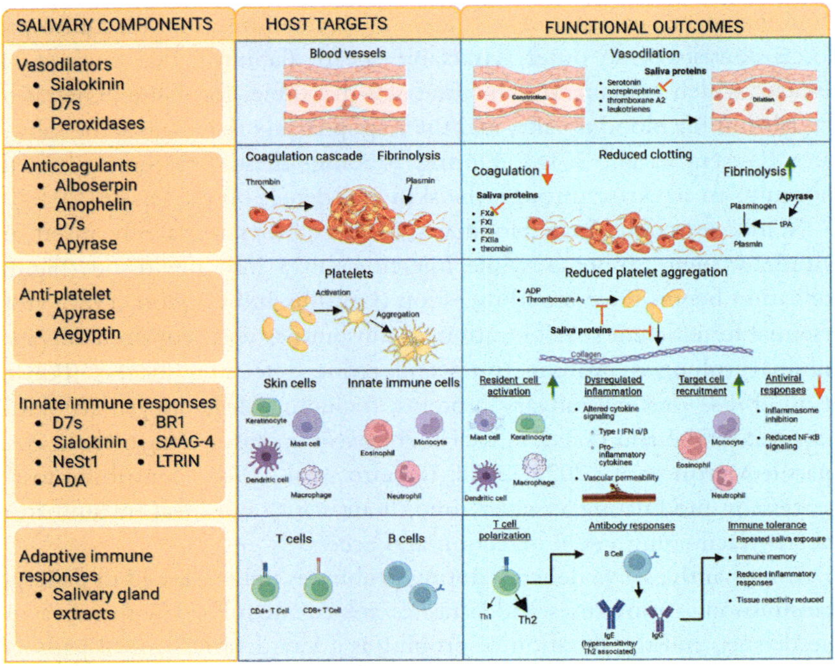

Fig. 2 Mosquito salivary components modulate host hemostasis and immune responses. Mosquito saliva contains bioactive molecules that facilitate blood feeding by targeting host vascular and hemostatic pathways. Vasodilators promote vessel dilation, while anticoagulants and anti-platelet factors inhibit coagulation and platelet aggregation, reducing clot formation. At the bite site, salivary proteins act on skin-resident cells and promote recruitment of circulating immune cells, leading to local activation, increased vascular permeability, and dysregulated inflammation. Concurrently, antiviral pathways, including type I interferon responses, are suppressed, creating a permissive environment for pathogen establishment. Saliva also shapes adaptive immunity through dendritic cell–mediated antigen presentation, driving T cell polarization, antibody responses, and immune memory or tolerance upon repeated exposure.

and prostaglandins. To successfully obtain a blood meal, mosquitoes must counteract these vasoconstrictive signals to maintain a steady blood flow, which is achieved by salivary molecules that inhibit vasoconstriction.

Sialokinin, a peptide in *Ae. aegypti* saliva, has been identified as a key regulator of vasoconstriction (Beerntsen et al., 1999; Champagne and Ribeiro, 1994; Martin-Martin et al., 2022). Similar to the mammalian neuropeptide substance P, sialokinin promotes vasodilation and inflammation by inducing nitric oxide (NO) production in endothelial cells through NK1R signaling. Mosquitoes lacking sialokinin exhibit prolonged probing times and reduced success in blood feeding (Martin-Martin et al., 2022). Additionally, sialokinin also promotes leukocyte recruitment to the bite site, which may influence downstream immune responses and viral transmission (Lefteri et al., 2022; Martin-Martin et al., 2022).

Another major group of salivary proteins involved in regulating vasoconstriction are the D7 family proteins, which are found in blood-feeding Nematocera including *Anopheles*, *Aedes*, and *Culex* mosquitoes, as well as in sand flies and black flies. These proteins are related to odorant-binding proteins and are widely present in hematophagous Dipterans. Members of the D7 family are classified into short (D7s or D7r) and long (D7L) forms, and are known to bind leukotrienes and a range of biogenic amines, including histamine, serotonin, norepinephrine, and dopamine (Alvarenga and Andersen, 2022). Serotonin and norepinephrine, which are rapidly released by activated platelets and neutrophils, act as potent vasoconstrictors by stimulating vascular smooth muscles contraction. Serotonin exerts this effect by binding to 5-HT2 serotonergic receptors, while norepinephrine acts through alpha-adrenergic receptors on smooth muscle cells. Various D7 proteins from different mosquito species bind either serotonin, norepinephrine, or both, dampening vasoconstrictor signaling (Alvarenga and Andersen, 2022). For example, *An. gambiae* D7L3 binds serotonin and inhibits serotonin-induced platelet aggregation and vasoconstriction (Smith et al., 2022).

Some D7 proteins also bind leukotrienes and thromboxane A2, both of which are strong vasoconstrictors. Leukotriene C4 can directly induce contraction of smooth muscle around blood vessels but can also indirectly promote vasoconstriction by activating the arachidonate cyclooxygenase pathway, which includes thromboxane A2 synthesis, and stimulating thromboxane A2 release (Fujimura et al., 1988; Whittle et al., 1981; Yonei and Guth, 1990). For example, D7L1 or DL2 from *Anopheles*, *Aedes,* and *Culex* mosquitoes bind leukotriene C4 and/or the thromboxane A2 analog U-46619, thereby inhibiting U-46619-induced platelet aggregation and vasoconstriction (Alvarenga and Andersen, 2022; Alvarenga et al., 2010; Calvo et al., 2009, 2006; Martin-Martin et al., 2021; Martin-Martin et al., 2020a; Smith et al., 2022).

An. albimanus and *An. gambiae* saliva also contain other vasodilators such as peroxidase enzyme with catechol oxidase activity (Ribeiro and Nussenzveig, 1993; Ribeiro et al., 1994; Ribeiro and Valenzuela, 1999). This enzyme catalyzes the oxidation of biogenic amines released during tissue injury, inactivating them and promoting vasodilation.

3.2 Anticoagulants and anti-platelet agents

Following activation of host hemostatic pathways at the bite site, blood coagulation takes place in two sequential stages: the primary stage involves the formation of an initial platelet plug at the injury site, while the secondary stage consists of the coagulation cascade. The coagulation cascade is initiated through two pathways: the extrinsic (tissue factor) pathway and the intrinsic (contact activation) pathway. Both pathways converge into a common pathway that leads to the activation of thrombin and the formation of a fibrin mesh, which stabilizes the platelet plug. The regulation of coagulation by mosquito salivary proteins is well characterized, with multiple proteins acting at different points in the coagulation cascade.

Mosquito salivary proteins inhibit the primary stage of coagulation through two main mechanisms that reduce platelet activation and aggregation (Fig. 2). The first mechanism involves the degradation of adenosine diphosphate (ADP), a potent activator of platelets, by salivary apyrase. Apyrase is a 5'-nucleotidase that hydrolyzes phosphate groups from adenosine triphosphate (ATP) and ADP, thereby preventing ADP-mediated platelet activation and aggregation via the G protein-coupled receptors P2Y1 and P2Y12 (Daniel et al., 1998). Apyrase is present in the saliva of several vectors including mosquitoes, sand flies, and triatomine bugs; and has been shown to strongly inhibit coagulation (Champagne et al., 1995; Francischetti, 2010; Pala et al., 2024).

The second mechanism disrupts collagen-mediated platelet activation, a process initiated upon collagen exposure after vascular damage, leading to platelet adhesion via glycoprotein VI (GPVI) and integrin receptors (Kuijpers et al., 2003). The *Ae. aegypti* salivary protein aegyptin, and its homologue in *Anopheles* mosquitoes, anopheline anti-platelet protein (AAPP), bind directly to collagen blocking access to platelet receptors—GPVI, integrin $\alpha 2\beta 1$, and von Willebrand factor (vWF)—and thereby preventing platelet adhesion. (Calvo et al., 2007; Chagas et al., 2014; Yoshida et al., 2008).

Mosquito salivary proteins also interfere with secondary hemostasis by targeting key enzymes and factors in the coagulation cascade and inhibiting clot formation (Fig. 2). One such protein is alboserpin, a serine protease inhibitor (serpin) identified in *Ae. albopictus*. Recombinant alboserpin exhibits tight, stoichiometric binding to activated factor X (FXa) in the picomolar range, inducing an irreversible inhibition (Calvo et al., 2011). FXa, a central enzyme in the common pathway of the coagulation cascade is part of the prothrombinase complex and catalyzes the conversion of prothrombin to thrombin. Inhibition of FXa by alboserpin prevents thrombin generation, thereby reducing fibrin formation and prolonging prothrombin time and activated partial thromboplastin time both in vitro and *ex vivo* (Calvo et al., 2011; Shrivastava et al., 2022). In addition, alboserpin displays high-affinity binding to heparin within the low nanomolar range, which may further enhance its anticoagulant properties (Calvo et al., 2011).

Another potent anticoagulant is anophelin, a 6.5-kDa peptide isolated from the salivary glands of *An. albimanus* (Francischetti et al., 1999). Anophelin inhibits blood clotting by directly binding to thrombin, acting as dual inhibitor of α-thrombin through interactions with both the anion binding exosite 1 and the catalytic site (Figueiredo et al., 2012; Francischetti et al., 1999). By targeting thrombin, anophelin prevents fibrin formation, ensuring uninterrupted blood flow during feeding.

Members of the D7 protein family, previously described for their role as vasodilators also play a significant role in modulating secondary hemostasis. *An. gambiae* D7L2, one of the three long-form D7 proteins, exerts anticoagulant activity by interacting with factors XII, XIIa, and XI in the intrinsic pathway of the coagulation cascade (Smith et al., 2022). Additionally, hamadarin (D7r1), found in the saliva of *An. stephensi*, suppresses activation of the intrinsic coagulation pathway by binding to factor XII (Isawa et al., 2002). Hamadarin also binds high-molecular-weight kininogen, which suppresses the subsequent release of bradykinin, a key mediator of inflammatory responses (Isawa et al., 2002).

Collectively, these salivary proteins target multiple components of the coagulation cascade, providing mosquitoes with multiple options to overcome host hemostatic defenses and facilitate efficient blood feeding.

3.3 Immunomodulatory proteins

Beyond their roles in counteracting hemostasis, mosquito salivary proteins also modulate host immune responses (Figs. 1 and 2). This immunomodulation facilitates prolonged feeding by reducing inflammation and pain at the bite site while creating a local environment that can enhance pathogen transmission. Here, we summarize key immunomodulatory components of mosquito saliva and their general effects on host immunity, while detailed innate and adaptive immune responses are discussed in Section 5.

In vitro studies have shown that *Ae. aegypti* salivary gland extracts (SGE) exert strong immunomodulatory effects on host immune cells. SGE suppress Th1 cytokines such as IL-2 and IFN-γ in antigen-stimulated mouse splenocytes, while Th2-associated cytokines such as IL-4 and IL-5 remain unaffected (Cross et al., 1994; Wanasen et al., 2004). SGE also inhibit lymphocyte proliferation and induce lymphocyte death at low concentrations (Bizzarro et al., 2013). However, the immune responses elicited by mosquito saliva appear highly dependent on experimental and biological context, cautioning against interpreting these findings as evidence of a uniformly Th2-biased response following natural mosquito biting.

In contrast, SGE from *Cx. quinquefasciatus* do not significantly affect murine splenocyte proliferation or cytokine production (Wanasen et al., 2004). *In vivo*, SGE further promote a Th2-biased environment and modulates local inflammatory responses (Schneider et al., 2004). *In vivo*, the picture is more complex. Controlled feeding of *Ae. aegypti* on humans elicits complex innate and adaptive immune responses beyond a simple Th2 shift, with a predominantly Th1/CD8-associated transcriptional signature evident at 48 h, including increased T-bet, CXCR3, CD4, CD8, IL-12RB and granzyme B, while canonical Th2 genes such as GATA3, IL-4 and IL-5 were not upregulated (Barningham et al., 2025; Guerrero et al., 2022). Collectively, these human studies point to a more complex immunological landscape, in which innate inflammation, endothelial activation, leukocyte recruitment, and adaptive T-cell responses are shaped by timing, prior exposure, and experimental conditions.

At the molecular level, individual salivary proteins contribute to these immunomodulatory effects through diverse and sometimes complementary mechanisms. Alboserpin, in addition to its previously described role in coagulation, exhibits anti-inflammatory properties. By inhibiting factor Xa (FXa), alboserpin suppresses FXa-mediated inflammatory signaling and endothelial activation (Shrivastava et al., 2022). Similarly, members of the D7 protein family contribute to immunomodulation by binding biogenic amines such as histamine and serotonin, which are key mediators of inflammation. By sequestering these molecules, D7 proteins dampen inflammatory responses and reduce host awareness at the bite site. In addition, *Ae. albopictus* D7L1 has been shown to reduce leukocyte recruitment in vivo, further highlighting its role in regulating local inflammation (Martin-Martin et al., 2020b; Smith et al., 2022).

In parallel, salivary proteins can directly modulate innate immune cell function. Neutrophil-stimulating factor 1 (NeSt1), a 34 kDa protein from *Ae. aegypti*, activates neutrophils and influences macrophage activity at the bite site (Hastings et al., 2019), while adenosine deaminase (ADA), abundantly secreted in *Ae. aegypti* saliva, modulates mast cell activation and promotes cytokine and chemokine production, thereby shaping the local inflammatory environment (Li et al., 2022; Mu et al., 2023; Ribeiro and Valenzuela, 2003).

Additional salivary proteins further highlight the breadth of these immunomodulatory strategies. AgBR1 induces neutrophil recruitment and pro-inflammatory cytokine expression, while salivary allergen/antigen 5-related gene 4 (SAAG-4), an interleukin-4–inducing protein, promotes IL-4 production and suppress IFN-γ in murine experimental systems. In contrast, the *An. gambiae* sporozoite-associated protein (AgSAP) suppresses local inflammatory responses in the skin (Arora et al., 2021; Boppana et al., 2009; Uraki et al., 2019b).

Together, these salivary proteins illustrate the diverse and coordinated strategies employed by mosquitoes to modulate host immunity, facilitating blood feeding and creating conditions that may enhance pathogen transmission.

4. Interactions of mosquito salivary proteins with the host immune system

Mosquito salivary proteins, while essential for overcoming host hemostatic defenses during blood feeding (Section 3), also directly engage host immune pathways (Fig. 2). These interactions initiate a cascade of immune responses at the bite site that evolves over time and with repeated exposure, involving coordinated interactions between innate and adaptive immunity (Fig. 1). As a result, mosquito saliva shapes local tissue environments, systemic immune responses, and susceptibility to pathogen infection. Here, we examine the cellular and molecular mechanisms through which mosquito salivary proteins modulate host immune systems.

4.1 Innate immune responses

The skin's innate immune system forms the body's first line of defense against physical injury and microbial invasion (Fig. 1). It includes structural barriers such as epidermis and dermis that contain keratinocytes and endothelial cells, resident immune cells like macrophages, dendritic cells, mast cells, and soluble factors such as complement proteins, cytokines, and antimicrobial peptides (Narayanan et al., 2025; Scurtu and Simionescu, 2021). When a mosquito pierces the skin, this system is rapidly activated by tissue damage and exposure to foreign salivary antigens. Damaged cells release alarmins and inflammatory mediators that stimulate vasodilation, recruit neutrophils and monocytes, and activate complement to neutralize potential pathogens (Depinay et al., 2006; Yang et al., 2017).

Mosquito saliva facilitates the suppression or evasion of some of the host's immediate immune defenses through multiple mechanisms that alter inflammation, complement activation, and innate immune signaling (Fig. 2). One of the best-characterized salivary immunomodulators is albicin, a 13.4-kDa SG7-family protein identified in *An. albimanus*. Albicin inhibits the alternative pathway of the complement system by binding to the C3 convertase enzyme complex and preventing activation of downstream components. This action suppresses the generation of inflammatory mediators such as C3a and C5a, thereby limiting local inflammation and leukocyte recruitment (Mendes-Sousa et al., 2016; Strayer et al., 2021). By dampening these early immune responses, albicin reduces pain, swelling, and vascular disruption, helping to maintain blood flow and prevent host detection, which allows the mosquito to probe and feed for longer periods with minimal interruption (Strayer et al., 2021). Similar complement-inhibitory activity has been observed in SGE from other *Anopheles* species, such as *An. aquasalis*, suggesting this mechanism is conserved across these mosquito species (Mendes-Sousa et al., 2018). However, studies show that *Ae. aegypti* saliva does not inhibit the complement system (Cavalcante et al., 2003; Pereira-Filho et al., 2020). In *Cx. quinquefasciatus*, female SGE have been shown to inhibit the classical and lectin complement pathways, while no inhibition was observed for the alternative pathway (de Melo Lara et al., 2021).

In addition to complement inhibition, mosquito saliva influences host innate signaling pathways. Studies using SGE show that mosquito saliva suppresses NF-κB–dependent inflammatory signaling in vitro. Exposure of murine macrophages and dendritic cells to *Ae. aegypti* SGE reduced production of the pro-inflammatory cytokines TNF-α and IL-12 following stimulation with lipopolysaccharide (LPS) and IFN-γ, indicating down-regulation of NF-κB–mediated transcriptional activity (Bizzarro et al., 2013; Schneider and Higgs, 2008; Wanasen et al., 2004). However, these in vitro observations should be distinguished from in vivo bite-site inflammation. Caution is warranted when extrapolating in vitro findings, since crude SGE preparations and serum-cultured immune cells only partially reproduce the composition of saliva and the tissue context encountered during natural mosquito biting. More recently, specific salivary proteins have been shown to exert the opposite effect. The *Ae. aegypti* neutrophil recruitment protein

(AaNRP) activates the TLR1/TLR4–MyD88–NF-κB pathway in skin macrophages, leading to increased expression of chemokines that recruit neutrophils and monocytes to the bite site (Wang et al., 2024). Together, these findings indicate that mosquito saliva can both suppress and activate NF-κB signaling depending on the experimental model, dose, cell type and mosquito species. Under natural biting conditions, *Aedes* saliva promotes rapid vascular and inflammatory changes that facilitate the recruitment of innate immune cells (Guerrero et al., 2022; Pingen et al., 2016).

While mast cells contribute to mosquito bite hypersensitivity and allergic inflammation, it is important to distinguish between several related but mechanistically distinct processes triggered by mosquito biting. Rapid vascular permeability and edema can arise from direct pharmacologic effects of salivary factors on endothelial function and vascular tone, whereas itch and neuroimmune signaling involve sensory-neuron activation and mediator release. Mast-cell degranulation is more closely associated with allergic and hypersensitivity responses, while longer-term adaptive responses involve shifts in Th1/Th2 polarization. Recent neuroimmune studies further emphasize the role of sensory neurons in shaping allergic itch and inflammatory responses (Flayer et al., 2024).

Mast cells, key mediators of cutaneous inflammation and allergic responses, are a major target of mosquito salivary modulation. Exposure to allergens or stimuli such as temperature changes or physical trauma from a mosquito bite can trigger mast-cell degranulation, releasing histamine, serotonin, proteases, and other inflammatory mediators. These mediators cause itching, swelling, and increased host awareness, potentially disrupting blood flow and reducing mosquito feeding efficiency. To counteract these effects, mosquito saliva contains D7 proteins that bind and sequester host inflammatory mediators. In *Ae. aegypti*, D7 proteins AeD7L1 and AeD7L2 bind biogenic amines—including histamine, serotonin, norepinephrine, and epinephrine—as well as lipid mediators such as cysteinyl leukotrienes in vitro (Calvo et al., 2006; Martin-Martin et al., 2021). Knockout of AeD7L1 or AeD7L2 significantly increased probing time on wild-type mice, an effect absent in leukotriene-deficient mice (B6.129S2-Alox5tm1Fun), demonstrating that these proteins counteract host inflammatory mediators to promote efficient blood feeding (Martin-Martin et al., 2023). Similarly, in *An. gambiae*, short-form D7 proteins (D7r1–D7r4) bind histamine and serotonin to neutralize mast cell–mediated inflammation, reducing itching, swelling, and host detection to facilitate uninterrupted feeding (Calvo et al., 2006; Mans et al., 2007).

Together, these mechanisms create a transiently immunomodulated environment at the bite site. By limiting inflammation through complement inhibition, cytokine modulation, and histamine sequestration, mosquitoes feed more efficiently while providing a temporary niche that favors infection by pathogens such as DENV, ZIKV, and the *Plasmodium* parasite (Pingen et al., 2016; Styer et al., 2011; Schneider and Higgs, 2008).

4.2 Adaptive immune responses and hypersensitivity

While innate immunity dominates the initial response to mosquito bites, repeated exposure activates adaptive immunity, involving both cellular and humoral components (Fontaine et al., 2011; Peng and Simons, 1998) (Fig. 2). In splenocyte cultures, murine splenic lymphocytes isolated 4–10 days after *Cx. pipiens* or *Ae. aegypti* feeding results in reduced IFN-γ and increased IL-4 and IL-10, consistent with suppressed Th1 and enhanced Th2 responses (Zeidner et al., 1999). However, recent human studies reveal a more nuanced pattern. Controlled *Ae. aegypti* feeding in humans induces mixed Th1/CD8 and Th2 signatures, with both responses varying by timing and prior exposure history (Barningham et al., 2025; Guerrero et al., 2022). In individuals sensitized to mosquito bites, IL-4– and IL-13–producing Th2 cells expand, promoting eosinophil recruitment, IgE class switching, and increased IL-5 production. This Th2-skewed response amplifies leukocyte infiltration—particularly eosinophils, mast cells, and $CD4^+$ T cells—at the bite site (Henrique et al., 2019). Mosquito saliva also regulates immune activity by reducing lymphocyte survival. Studies with *Ae. aegypti* and *An. stephensi* SGE show that T and B lymphocytes undergo caspase-dependent apoptosis after exposure, while memory T cells are less affected (Bizzarro et al., 2013). This selective cell death limits excessive immune activation but preserves memory responses.

With repeated bites, antibody production becomes a key feature of the adaptive response. Saliva-specific IgG and IgE antibodies form against mosquito salivary proteins. IgG can neutralize these proteins and reduce their pharmacologic effects, whereas IgE antibodies drive Type I hypersensitivity, the classic immediate allergic reaction. Upon re-exposure, IgE bound to mast-cell high-affinity IgE receptors (FcεRI) recognizes salivary allergens, triggering mast-cell degranulation and the release of histamine, prostaglandins, and leukotrienes. These mediators cause rapid vasodilation, itching, redness, and swelling, producing the characteristic *wheal-and-flare* reaction (Peng and Simons, 1998; Reunala et al., 1994a). Studies in human have identified D7 family proteins, apyrase (Aed a 1), and antigen-5–like proteins as major mosquito salivary allergens, with IgE reactivity confirmed across *Ae. aegypti*, *Ae. albopictus*, *Cx. quinquefasciatus*, and *An. dirus* (Calvo et al., 2006; Opasawatchai et al., 2020; Peng and Simons, 2004).

In some people, long-term exposure to mosquito bites leads to delayed-type hypersensitivity (DTH), marked by firm, inflamed lesions and lymphocyte infiltration at the bite site, reflecting a partial shift from Th2- to Th1-dominated immunity (Reunala et al., 1994b). In this study, Reunala et al. demonstrated that serum from a mosquito-allergic patient could passively transfer cutaneous hypersensitivity to non-reactive individuals, indicating that anti-saliva IgE antibodies are sufficient to mediate immediate bite reactions. These antibodies trigger histamine release upon exposure to mosquito salivary antigens, linking humoral responses to both immediate and late-phase cutaneous reactions. In rare cases, particularly in individuals with underlying mast-cell abnormalities such as systemic mastocytosis, mosquito bites can trigger severe allergic reactions including anaphylaxis (Reiter et al., 2013). Overall, mosquito saliva drives a dynamic balance between Th2-driven Type I allergy, lymphocyte regulation through apoptosis, and Th1-mediated DTH. The outcome depends on mosquito species, exposure frequency, host genetics, and concurrent infections (Vogt et al., 2018).

4.3 Immune tolerance and desensitization

Immune responses to mosquito bites evolve with repeated exposure, transitioning from sensitization and hypersensitivity to a state of tolerance or desensitization. This phenomenon exemplifies a process of immune adaptation, in which continuous, low-dose exposure to mosquito antigens leads to the downregulation of inflammatory responses and the establishment of immunological equilibrium (Oka et al., 2018; Oka and Ohtaki, 1989). Clinical and longitudinal studies in humans further demonstrate that this transition follows a defined trajectory, beginning with initial sensitization, followed by delayed-type T cell–mediated responses and the development of IgE-mediated hypersensitivity, and ultimately resulting in diminished cutaneous reactivity with continued exposure (Oka, 1989; Oka et al., 1989; Oka and Ohtaki, 1989; Peng et al., 1996; Peng and Simons, 1998).

Repeated exposure to mosquito salivary antigens induces the expansion of Tregs that secrete IL-10 and TGF-β—cytokines that suppress effector T-cell function and inhibit IgE-mediated mast-cell activation. These immunosuppressive cytokines also promote B-cell isotype switching toward IgG4, which competes with IgE for antigen binding (Depinay et al., 2006; Wanasen et al., 2004). The result is a reduction in the intensity of local allergic reactions and decreased recruitment of eosinophils and basophils, promoting immune tolerance (Depinay et al., 2006). Consistent with this, human studies have shown that individuals with strong immediate and delayed skin reactions exhibit elevated mosquito-specific IgE levels and lymphocyte proliferation responses, whereas desensitized individuals display reduced IgE responses and attenuated cellular reactivity despite continued exposure (Peng et al., 1996; Peng and Simons, 1998).

At the tissue level, tolerance involves reduced inflammatory cell infiltration and lower production of pro-inflammatory mediators. Mosquito bite–activated mast cells induce IL-10 production in draining lymph nodes, which suppresses antigen-specific T-cell responses and promotes a tolerogenic environment (Depinay et al., 2006). Over time, this regulatory network leads to desensitization, where repeated mosquito exposure triggers minimal clinical symptoms (Oka et al., 2018; Oka and Ohtaki, 1989). This attenuation is also reflected in age-dependent patterns, where children typically exhibit stronger hypersensitivity reactions that decline in adulthood with cumulative exposure, further supporting the development of naturally acquired tolerance (Peng and Simons, 1998).

Interestingly, while tolerance reduces host discomfort and allergic inflammation after a mosquito bite, it may inadvertently benefit the mosquito and its pathogens. A tolerant environment at the bite site ensures less immune-mediated interference, potentially enhancing the establishment and transmission efficiency of arboviruses and parasites (Schneider and Higgs, 2008; Vogt et al., 2018).

5. Mosquito saliva and pathogen transmission

Mosquito saliva plays a central role in shaping the outcome of pathogen transmission by modifying the host environment at the bite site (Fig. 1). During probing and feeding, saliva is introduced alongside pathogens such as arboviruses and *Plasmodium* parasites, where it influences early infection dynamics. These interactions create a localized microenvironment that facilitates pathogen survival, replication, and dissemination within the host. By altering immune cell recruitment and local tissue responses, mosquito saliva enhances infection efficiency and can increase disease severity (Schneider and Higgs, 2008; Styer et al., 2011; Vogt et al., 2018).

Consequently, the interplay between salivary components, host responses, and pathogens represents a key determinant of vector competence and transmission dynamics (Visser et al., 2023).

5.1 Modulation of arbovirus pathogenesis by mosquito saliva

Aedes and *Culex* mosquitoes are competent vectors of several medically important arboviruses, including DENV, ZIKV Semliki Forest virus (SFV; *Alphavirus semliki*), and WNV. Studies in both mice and non-human primates have shown that viral infectivity, viremia, and disease severity are greater when infection occurs via mosquito bite compared to needle injection (Cox et al., 2012; McCracken et al., 2020; Styer et al., 2011). This enhancement can also be reproduced when a needle-inoculated virus is introduced at the same saliva-containing bite site produced by an uninfected mosquito or co-injected with SGE (Cox et al., 2012; Schneider et al., 2004; Styer et al., 2011). In contrast, viruses injected at a site distant from an uninfected bite do not produce this increase in infection. These findings demonstrate that mosquito saliva, rather than tissue damage caused by the mosquito mouth parts, drives the local enhancement of viral infection (Schneider and Higgs, 2008; Styer et al., 2011).

However, this enhancement is not universal. The outcome of arboviral infection varies depending on factors such as mosquito species, salivary composition, specific salivary proteins, virus strain, and host model (Cox et al., 2012; McCracken et al., 2020; Schneider and Higgs, 2008). This section examines how mosquito salivary proteins influence virus infection and severity in the vertebrate host.

5.2 Saliva induces vascular leakage and recruitment of immune cells

Mosquito bites trigger an immediate inflammatory response, characterized by itching and localized edema (Fig. 1). *In vitro* and in vivo studies demonstrate that SGE disrupt endothelial barrier integrity and promotes vascular permeability in the skin (Schmid et al., 2016). A key initiator of this edema is sialokinin, a tachykinin-like peptide found in *Ae. aegypti* saliva. Sialokinin, as previously described, acts directly on endothelial cells to increase vascular permeability, resulting in plasma extravasation and local swelling (Lefteri et al., 2022; Martin-Martin et al., 2022). In parallel, mast cell degranulation induced by mosquito bite, especially in sensitized individuals, releases histamine, leukotrienes, and proteases that further induce vascular leakage (Henrique et al., 2019; Peng and Simons,1998; Reunala et al., 1994b). Together, these processes create a microenvironment that promotes arbovirus retention at the bite, which benefits early viral replication, as first demonstrated in a mouse model of SFV infection. Saliva-induced edema retained virus locally, delaying drainage into the lymph nodes and suppressing early INF-γ responses (Pingen et al., 2016). This prolonged local persistence promotes infection of cutaneous cells, enhancing viral replication in the skin and increasing the likelihood of systemic dissemination (Pingen et al., 2016). Similar edema-driven enhancement has been observed for ZIKV, DENV, WNV, and CHIKV, indicating that this mechanism is conserved across multiple arboviruses (Agarwal et al., 2016; Cox et al., 2012; Lefteri et al., 2022; Pingen et al., 2016; Schmid et al., 2016, Styer et al., 2011).

In contrast, *Anopheles* mosquitoes, which are poor vectors for most arboviruses, lack sialokinin in their saliva—a key factor responsible for inducing vascular permeability and edema at the bite site. As a result, bites from *Anopheles* mosquitoes do not promote significant local edema. Consistent with this, no enhancement of SFV infection was observed in a murine model when co-inoculated with *Anopheles* saliva, supporting that sialokinin-driven edema is a major determinant of saliva-mediated arbovirus enhancement (Lefteri et al., 2022). This distinction likely contributes, among other factors, to the ability of *Aedes* mosquitoes, which express sialokinin, to efficiently transmit multiple arboviruses, in contrast to *Anopheles* species. Whether similar vascular-modulating factors exist in *Culex* mosquitoes remains unknown, despite their importance as major arbovirus vectors, highlighting a critical gap in our understanding of saliva-mediated arbovirus transmission mechanisms.

5.3 Saliva-mediated recruitment of virus-permissive immune cells

The increased vascular permeability induced by mosquito saliva not only causes plasma extravasation but also promotes recruitment of immune cells to the bite site (Fig. 1). The resulting vascular leakage and plasma extravasation create a pro-inflammatory microenvironment, stimulating endothelial and tissue-resident cells to produce cytokines and chemokines, including IL-1β and CXCL2 (Lefteri et al., 2022; Pingen et al., 2016). These mediators drive the early recruitment of neutrophils to the bite site, a process further enhanced by the *Ae. aegypti* salivary protein NeSt1. NeSt1 directly stimulates neutrophils and macrophages to produce neutrophil-attracting chemokines such as CXCL2, enhancing neutrophil infiltration at the bite site (Hastings et al., 2019).

Although neutrophils are not permissive to arbovirus infection, they release additional chemokines, including CCL2, which promote recruitment of myeloid cells such as macrophages and monocytes (Lefteri et al., 2022). These myeloid cells are highly permissive to arbovirus infection, acting as additional targets at the bite site and driving virus dissemination. Arboviruses such as ZIKV, DENV, and SFV replicate more efficiently in the presence of saliva-induced leukocyte recruitment than when introduced by needle inoculation alone (Lefteri et al., 2022; Pingen et al., 2016; Styer et al., 2011).

Salivary proteins can also directly interact with myeloid cells, conditioning them to become more permissive to arboviral infection by activating the autophagy pathway. *Ae. aegypti* venom allergen-1 (AaVA-1), binds to receptors on myeloid linage cells and is subsequently internalized into the cytoplasm, where it interacts with leucine-rich pentatricopeptide repeat–containing protein (LRPPRC), an inhibitor of autophagy (Sun et al., 2020). Under normal conditions, LRPPRC sequesters Beclin-1, preventing autophagy. Binding of AaVA-1 to LRPPRC releases Beclin-1, triggering the autophagic pathway and promoting viral uptake and replication (Sun et al., 2020). *In vivo*, mice infected with ZIKV via bites from AaVA-1-deficient mosquitoes showed lower viremia and increased survival compared to bites from wild-type mosquitoes (Sun et al., 2020).

5.4 Mosquito saliva suppresses early antiviral response

Mosquito saliva can modulate host antiviral signaling pathways (Fig. 2). Lymphotoxin-β receptor inhibitor (LTRIN), a ~15-kDa protein from *Ae. aegypti* saliva, binds the lymphotoxin-β receptor (LTβR) on macrophages and dendritic cells (Jin et al., 2018). LTβR dimerization activates NF-κB, triggering the production of pro-inflammatory cytokines and other antiviral genes. By preventing LTβR dimerization, LTRIN inhibits NF-κB signaling, reducing pro-inflammatory cytokine responses in experimental systems (Jin et al., 2018). These findings support a role for LTRIN in modulating antiviral and inflammatory pathways; however, the extent to which LTβR-dependent signaling contributes to the early bite-site responses versus later immune activation and antiviral defense likely depends on tissue and temporal context.

In addition to dampening innate immunity, mosquito saliva skews the adaptive immune response toward a Th2-dominated profile as seen in CHIKV-infected mice (Thangamani et al., 2010). Salivary proteins suppress Th1-associated cytokines and enhance production of IL-4 and

IL-10, creating an environment that favors humoral responses over cytotoxic or antiviral cellular immunity (Schneider et al., 2004; Zeidner et al., 1999). This Th2 bias limits effective antiviral T cell activity and, together with the early suppression of NF-κB and interferon signaling, establishes a bite site that is highly permissive to viral replication and dissemination.

5.5 Antiviral functions of mosquito salivary proteins

Unlike several salivary proteins that enhance viral infection, the D7 family of mosquito salivary proteins has been shown to exhibit antiviral effects. Beyond their anti-inflammatory functions, D7 proteins have also been shown to interact directly with arboviral particles and interfere with infection (Conway et al., 2016). Recombinant *Ae. aegypti* D7L1 protein binds to the envelope protein of DENV, preventing viral attachment to host cells and resulting in significantly reduced DENV titers in both in vitro and in vivo models (Conway et al., 2016). Similarly, other *Ae. aegypti* salivary proteins, including AaVA-1, serpin, and apyrase, have been reported to bind to ZIKV particles, but these interactions did not lead to measurable decreases in viral titers in cell culture, making *Aedes* D7 proteins unique in their antiviral activity (Valenzuela-Leon et al., 2022).

However, the role of D7 proteins in arbovirus infection is highly dependent on mosquito species. Immunization with recombinant D7 from *Cx. tarsalis* unexpectedly enhanced disease severity following mosquito-transmitted WNV infection. These contrasting effects may reflect functional differences in ligand-binding activity and immunomodulatory properties between *Aedes* and *Culex* D7 proteins, which have been shown to exhibit distinct binding specificities (Martin-Martin et al., 2020b). For example, *Ae. aegypti* D7 proteins primarily bind biogenic amines and inflammatory lipid mediators such as leukotrienes (Calvo et al., 2009; Martin-Martin et al., 2021), whereas *Culex* D7 proteins exhibit distinct ligand-binding activities including ADP and ATP scavenging (Martin-Martin et al., 2020a), suggesting genus-specific adaptations that may differentially influence host immune responses and arbovirus infection outcomes. Vaccinated mice showed increased mortality, which was recapitulated by passive transfer of immune sera, indicating an antibody-mediated effect (Reagan et al., 2012). This was associated with reduced Th1 responses and increased IL-10–producing $CD4^+$ T cells, suggesting that anti-D7 immunity can skew host responses to favor viral pathogenesis (Reagan et al., 2012).

Similarly, the salivary protein aegyptin illustrates the complexity of saliva–virus interactions. Co-inoculation of aegyptin with DENV reduced viral titers locally and systemically while increasing pro-inflammatory cytokines, including GM-CSF, IFN-γ, IL-5, and IL-6 (McCracken et al., 2014). These findings suggest that some salivary factors can limit early viral replication through immune activation, and that viruses may counteract this effect, for example by downregulating aegyptin expression in infected mosquitoes (McCracken et al., 2014).

5.6 Other pathogens

Mosquito saliva creates a transient microenvironment at the bite site that not only promotes arboviral infection but also facilitates the establishment and transmission of other mosquito-borne pathogens, including *Plasmodium* parasites. In the case of *Plasmodium* sporozoites, specific salivary proteins such as AgSAP, sporozoite-associated mosquito saliva protein 1 (SAMSP1), and *An. gambiae* TRIO (AgTRIO) interact directly with the parasite to enhance infection, each acting through distinct mechanisms. AgSAP binds to heparan sulfate on the surface of mammalian cells, inhibiting T-cell activation and reducing local pro-inflammatory signaling, thereby creating an environment conducive to sporozoite invasion (Arora et al., 2021). SAMSP1, in contrast, enhances sporozoite gliding motility and cell traversal while limiting neutrophil recruitment to the bite site, enabling more efficient migration toward the liver (Chuang et al., 2021). The precise function of AgTRIO remains undefined; however, antibodies targeting this protein significantly reduce *Plasmodium* infection in murine models, suggesting a crucial role in sporozoite transmission (Dragovic et al., 2018).

While most salivary proteins promote parasite establishment, some act in opposition. The salivary gland protein mosquito gamma interferon–inducible lysosomal thiol reductase (mosGILT) binds directly to sporozoites and reduces their gliding motility and cell traversal ability, thereby limiting sporozoite infectivity. Silencing *mosGILT* expression in mosquitoes results in increased liver-stage infection, confirming its role as an inhibitory factor in malaria transmission (Schleicher et al., 2018). Beyond their effects at the bite site, salivary proteins also influence parasite development within the mosquito midgut, as discussed in Section 6.

5.7 Saliva, a determinant of vector competence

Vector competence refers to the intrinsic ability of a vector, such as mosquitoes, to acquire, disseminate, and transmit a pathogen to a vertebrate host (Beerntsen et al., 2000). This capacity has been shaped by the long coevolution of blood-feeding insects, their vertebrate hosts, and the pathogens they transmit. Mosquito saliva plays a central role in this process. Over evolutionary time, mosquito-borne pathogens, including arboviruses and *Plasmodium* parasites, have adapted to exploit salivary functions to enhance transmission and survival. Thus, mosquito saliva represents a molecular interface where evolutionary pressures on both vector and pathogen converge, shaping vector competence.

Mosquitoes of the genera *Aedes* and *Culex* are the principal vectors of arboviruses. *Aedes* species are competent vectors for viruses such as DENV, SFV, and CHIKV, while *Culex* species efficiently transmit Japanese encephalitis virus (JEV; *Orthoflavivirus japonicum*) and WNV. These mosquitoes have evolved a salivary protein profile that promotes arboviral infection. Proteins such as adenosine deaminase modulate platelet activity and cytokine signaling, while LTRIN suppresses NF-κB-mediated inflammatory responses, creating a permissive microenvironment at the bite site for viral replication (Jin et al., 2018; Mu et al., 2023). Most notably, *Aedes* saliva contains sialokinin, which increases vascular permeability and has been shown to be essential for viral infection and dissemination (Lefteri et al., 2022). By contrast, *Anopheles* mosquitoes—the primary vectors of *Plasmodium* parasites—have followed a different evolutionary trajectory. Their saliva contains proteins such as apyrase, AAPP, and anophelin, which efficiently prevent platelet aggregation and inflammation but lack arbovirus-enhancing components such as sialokinin (Francischetti et al., 1999; Lefteri et al., 2022; Martin-Martin et al., 2022; Pala et al., 2024; Yoshida et al., 2008). Consequently, *Anopheles* saliva does not induce the same level of vascular leakage or augment viral infection as observed in *Aedes* mosquitoes, consistent with their poor competence for most arboviruses. However, their saliva and its regulatory roles favor malaria transmission.

An exception to this dichotomy is O'nyong-nyong virus (ONNV), an arbovirus transmitted by *Anopheles*, which appears uniquely adapted to this salivary milieu and is not enhanced by *Aedes* saliva (Lefteri et al., 2022). Such specificity underscores how distinct pathogen groups—arboviruses with *Aedes*, *Plasmodium* with *Anopheles*— have coevolved with the specific saliva of the mosquitoes they infect, making saliva a critical evolutionary driver of vector competence and pathogen adaptation.

Beyond these inter-species differences, vector competence can also vary at the intra-species level. Natural variation in salivary protein expression among mosquito populations represents an underappreciated source of complexity in saliva-mediated transmission. Laboratory-reared mosquitoes may differ from wild-caught individuals in salivary composition due to differences in diet, microbiome, and genetic background. Geographic strains of the same species can also exhibit differences in salivary protein profiles that may influence vector competence. For example, distinct *Ae. albopictus* populations display population-specific sequence diversity in the D7 salivary protein alongside differences in CHIKV replication, highlighting that salivary factors and viral infection outcomes can vary even across geographically proximate populations. Future studies examining salivary proteome variation across wild and laboratory mosquito populations will be essential for translating laboratory findings to natural transmission settings (Kaczmarek et al., 2020).

6. Role of mosquito saliva in the mosquito midgut

6.1 Saliva reflux into the midgut

Mosquito salivary proteins have been well characterized for their role in facilitating blood feeding at the bite site. However, it has been shown that mosquito saliva can influence similar processes in the midgut (Klug et al., 2023; Pala et al., 2024). The hemostatic and immune functions of the ingested blood have been shown to remain active for several hours post feeding (Khattab et al., 2015; Pala et al., 2024; Simon et al., 2013). Mosquitoes and other hematophagous arthropods ingest a significant amount of their saliva, which greatly increases during blood feeding (Alves e Silva et al., 2025; Pala et al., 2024; Soares et al., 2006). Saliva then mixes with the ingested blood (Pala et al., 2024; Soares et al., 2006) and although relatively poorly understood, it has been hypothesized that the biochemical properties of many of these salivary proteins are retained within the blood bolus. This can have a profound influence on pathogen transmission and the biology of the midgut by suppressing harmful blood defense mechanisms like clotting, platelet aggregation, and complement activation which could damage digestive tissues and impair blood digestion (Klug et al., 2023; Pala et al., 2024). For a depiction of the potential roles of mosquito saliva in regulating blood bolus biology and mosquito midgut function, please refer to Fig. 1 in the review by Alves e Silva et al. (Alves e Silva et al., 2025).

A mass spectrometric analysis of the early *Ae. aegypti* peritrophic matrix revealed that 10% of its protein content is salivary-derived, incorporating proteins like D7, *aegyptin*, D7 long1, C-type lectins, antigen-5, apyrases, and serpins (Whiten et al., 2018). This provides additional evidence that salivary proteins are ingested during blood feeding. Mosquito saliva is thus proposed to be a major regulator of the midgut environment, potentially influencing the outcomes of arbovirus transmission to the mosquito.

6.2 Influence on midgut physiology and pathogen transmission

Ingesting blood containing host innate defenses presents a significant challenge to mosquitoes, requiring a coordinated set of responses. This can be divided into early and late response phases, with early responses counteracting host hemostatic processes, coagulation, and immune activation, whereas late processes facilitating blood digestion while protecting the midgut epithelium (Alves e Silva et al., 2025).

The host complement system, one of the first lines of defense against invading pathogens, remains active in the mosquito midgut for up to three hours after blood feeding and can damage digestive tissues (Barros et al., 2009). Complement also impacts arboviral (Londono-Renteria et al., 2016) and *Plasmodium* infection as shown for *Plasmodiium berghei and Plasmodium yoelii* in rodent models (Margos et al., 2001; Tsuboi et al., 1995), and reduces midgut microbiome levels (Khattab et al., 2015). Consistent with this, *Ae. aegypti* mosquitoes fed on DENV mixed with non-heat-inactivated human serum showed significantly lower viremia than those fed with heat-inactivated serum (Londono-Renteria et al., 2016), indicating that human complement may limit midgut arbovirus infection.

To counteract these effects, some mosquitoes evolved mechanisms to evade complement activity, including co-opting the host complement regulator factor H (Khattab et al., 2015), and the expression of anti-complement salivary proteins. For example, *An. albimanus* salivary albicin inhibits complement activation through the inhibition of the C3bBb convertase (Andersen et al., 2024), a function conserved in its *An. freeborni* homolog SG7. AF (Strayer et al., 2021). Additionally, *Anopheles* salivary complement inhibitors suppress the generation of proinflammatory anaphylatoxins C3a and C5a, which may reduce membrane attack complex formation and protect the midgut epithelium from complement-mediated damage (Andersen et al., 2024; Strayer et al., 2021).

In contrast, saliva from *Ae. aegypti* does not appear to exhibit complement inhibitory activity (Cavalcante et al., 2003; Pereira-Filho et al., 2020). However, in *Cx. quinquefasciatus*, female SGE inhibit both the classical and lectin complement pathways (de Melo Lara et al., 2021). These findings suggest that complement inhibition may represent a more important salivary adaptation in *Cx. quinquefasciatus* than in *Ae. aegypti*. Thus, salivary protein-mediated inactivation of human complement may protect the midgut epithelium from complement-mediated damage while potentially enhancing arbovirus infection in *Cx. quinquefasciatus*.

Salivary AgApyrase promotes fibrinolysis and inhibits coagulation within the ingested blood meal (Pala et al., 2024)(Fig. 2). It has been shown to enhance fibrinolysis by directly interacting with and activating tissue plasminogen activator (tPA), facilitating the conversion of plasminogen into plasmin, the protease responsible for fibrin degradation. It inhibits coagulation through its nucleotidase activity, by hydrolyzing extracellular ATP and ADP into AMP and inorganic phosphate, depleting ADP, a key mediator of platelet activation and aggregation. This ultimately reduces blood bolus viscosity which facilitates malaria parasite transmission and could allow for efficient blood digestion (Pala et al., 2024). Additionally, some *Ae. aegypti* salivary proteins are hypothesized to aid in blood meal digestion based on their anti-hemostatic functions at the bite site (Alves e Silva et al., 2025), namely apyrase, D7 proteins, and aegyptin that prevents platelet aggregation by blocking collagen-receptor interactions (Calvo et al., 2007). These proteins are proposed to reduce clot formation, inflammation, and blood bolus viscosity, potentially facilitating blood meal digestion.

Ingested leukocytes present an additional challenge. Activated neutrophils can release extracellular traps (NETs) that damage the midgut epithelium (Pala et al., 2024). These NETs also promote blood coagulation, creating a positive feedback loop in which activated platelets and fibrin further enhance NET formation (Gould et al., 2014). *An. gambiae* salivary apyrase was found to modulate neutrophil activity by decreasing NET formation (Pala et al., 2024). These NETs can also impact malaria parasite transmission in the midgut. Given the role NET formation in controlling arbovirus infection (Garishah et al., 2021; Hiroki et al., 2020) salivary apyrase may similarly facilitate mosquito midgut arbovirus infection.

As previously discussed, several other *Aedes* and *Culex* salivary proteins facilitate host arbovirus infection by regulating physiological responses in the skin. These proteins could perform similar functions in the midgut, for example, AaVA-1 promotes DENV and ZIKV transmission by activating autophagy in monocyte-lineage cells at the bite site, facilitating early viral replication and dissemination (Sun et al., 2020). Additionally, *Aedes* sialokinin enhances arbovirus infection by directly increasing vascular permeability at the bite site, which promotes rapid influx of virus-permissive immune cells and facilitates viral replication and dissemination (Lefteri et al., 2022). Similarly, other salivary proteins like CLIPA3 serine protease (Conway et al., 2014), Factor Xa-inhibitors (Stark and James, 1995), and the D7 proteins (Martin-Martin et al., 2023) modulate physiological and immunological responses in the skin. Thus, based on their role at the bite site, these proteins are hypothesized to also influence arbovirus infection within the mosquito midgut. However, since arboviruses can infect the midgut epithelium within minutes of blood ingestion, they could potentially evade these early hemostatic and inflammatory reactions. Thus, further studies are required to elucidate the potential involvement of salivary proteins in midgut arbovirus infection.

This highlights the need for further research to characterize the role of additional salivary proteins in the mosquito midgut, as they may represent promising targets for vaccine development or transgenic mosquito strategies aimed at prophylactic and transmission-blocking interventions.

6.3 Murine systems to model natural arbovirus transmission

As discussed, the substantial impact of ingested salivary proteins on modulating the host hemostatic and immune responses in the midgut, and pathogen transmission underscores the need for further investigation. However, most viral transmission studies rely on artificial membrane feeder systems where mosquitoes feed on an infectious blood meal of virus containing cell culture supernatant supplemented with a phagostimulant like ATP and defibrinated or anticoagulant-treated (EDTA or heparin) blood. In this setting, many host hemostatic and immune system components are either absent or inactive, unlike during a natural infection, which thereby limits the biological processes that can be explored.

The use of murine systems could facilitate the study of arboviral transmission in this context, providing a model to elucidate how the salivary proteins ingested by the mosquito during blood feeding regulate these responses and transmission dynamics within the midgut. In particular, given the role of the type-I interferon response in limiting DENV infection establishment (Züst et al., 2014), a mouse strain deficient in type-I IFN receptors (IRF3/7$^{-/- \ -/-}$) and therefore a diminished type-I IFN response was evaluated to study transmission of a non-mouse adapted DENV strain to *Ae. aegpyti* mosquitoes (Christofferson et al., 2013). In this model, naïve mosquitoes acquired infection after feeding on viremic mice, highlighting the potential of IRF3/7$^{-/- \ -/-}$ mice for transmission studies. In another approach, artificial viremia was induced in immunocompetent C3H mice by injection of DENV infected K562 erythroleukemia cells (Runtuwene et al., 2014). These mice also supported *Ae. aegypti* DENV infection, further demonstrating their potential as a model system for studying host-to-vector transmission.

This ultimately underscores the importance of utilizing existing and developing novel murine models that more closely mimic natural transmission conditions. Such models may help uncover how ingested mosquito salivary proteins regulate the midgut environment and influence arbovirus transmission. However, murine models do not fully recapitulate human infection biology, as immune responses, skin architecture, cytokine signaling, and susceptibility to arboviruses can differ substantially between mice and humans (Zellweger and Shresta, 2014). In addition, many arbovirus studies rely on immunocompromised mouse strains, particularly those deficient in type-I interferon signaling, which may alter early antiviral and inflammatory responses (Trinchieri, 2010). Thus, complementary approaches including human skin explants like the one established for Zika and Mayaro viruses (Esterly et al., 2022), organotypic artificial skin models, and ex vivo feeding systems provide additional options with physiologically relevant platforms for studying saliva-mediated arbovirus transmission and midgut infection dynamics.

7. Epidemiological and biomedical significance

7.1 Biomarkers of mosquito exposure

Assessing human-mosquito contact is essential for monitoring mosquito populations, estimating the risk of arboviral transmission, and guiding vector control strategies. The injected saliva after a mosquito bite elicits antibody responses in humans (Brummer-Korvenkontio et al., 1997). These responses provide a measurable proxy for exposure and, importantly, for virus transmission risk. Initial studies used mosquito SGE or saliva to determine immunoreactivity (Remoue et al., 2007, 2006; Waitayakul et al., 2006), but variability, scalability, and cross reactivity (Doucoure et al., 2012; Moreira et al., 2001; Ndille et al., 2012) caused limitations. The identification of genus-specific salivary antigens in *Anopheles* and *Aedes* mosquitoes (Arcà et al., 2005; Calvo et al., 2004; Ribeiro et al., 2007) has enabled the development of immunoassays based on individual, genus-restricted salivary antigens.

There have been several studies showing the potential of individual salivary proteins as biomarkers for human exposure to *Aedes* mosquito bites. Among these, the *Ae. aegypti* 34 kDa salivary protein has been extensively used to evaluate human exposure to *Ae. aegypti* bites. In a study conducted across seven villages in southern Benin, IgG responses to a peptide derived from the N-terminal region of this protein (Nterm-34 kDa) varied substantially among individuals and between villages and were strongly associated with rainfall patterns, with higher antibody levels observed during the rainy season when mosquito exposure was higher (Ndille et al., 2012). These findings suggest that IgG responses to Nterm-34 kDa could serve as a reliable biomarker for detecting variation in human exposure to *Ae. aegypti* bites. Building on this approach, the same biomarker was used to evaluate the effectiveness of vector control interventions in an urban area of Saint-Denis at La Réunion Island. IgG levels decreased within two weeks after intervention and remained low for at least four weeks (Ndille et al., 2016), highlighting its potential as a short-term indicator of vector control efficacy.

Beyond temporal dynamics, antibody responses to Nterm-34 kDa have also been used to investigate spatial heterogeneity in mosquito-human contact. In a cross-sectional study conducted in Saint-Louis, Senegal, IgG responses varied significantly between neighborhoods, revealing heterogeneous exposure to *Aedes* mosquitoes within the urban environment. Spatial analysis further identified localized "hot spots" of exposure, highlighting the sensitivity of this biomarker for detecting fine-scale variation in vector exposure and identifying populations at increased risk of mosquito-borne viral diseases (Sagna et al., 2019). Consistent with this, a retrospective study conducted in Vientiane, Lao PDR, showed higher IgG responses in less urbanized peripheral areas compared to highly urbanized zones, mirroring the spatial distribution of dengue virus infections, thereby supporting the use of Nterm-34 kDa-specific responses as indicators of dengue transmission risk (Ndille et al., 2014).

In a murine model, IgG responses against the 34k2 salivary proteins, members of the same 34 kDa protein family, from *Ae. albopictus* (al34k2) and *Ae. aegypti* (ae34k2) reliably reflected exposure to mosquito bites (Buezo Montero et al., 2019). Consistent with this, human serological studies demonstrated that IgG responses to al34k2 could detect seasonal and geographic variation in *Ae. albopictus* exposure (Buezo Montero et al., 2020), with further validation in populations from Réunion Island and Bolivia supporting its use as a biomarker of exposure and for evaluating vector control interventions (Buezo Montero et al., 2022).

The *Ae. albopictus* salivary protein Ag5–3 has been proposed as a novel serological marker of exposure, with human IgG responses correlating with mosquito density and seasonal transmission patterns (Dipaola et al., 2025). Similarly, antibodies against the *Ae. aegypti* salivary proteins AeD7L1 and AeD7L2 have been identified as sensitive and specific markers of human exposure (Chea et al., 2024). In a longitudinal pediatric cohort in Cambodia, IgG responses to recombinant AeD7L1 and2 strongly correlated with whole SGE, showed clear seasonal variation, and minimal cross-reactivity with non-*Aedes* mosquitoes, supporting their use to measure exposure. Notably, baseline antibody levels were also associated with dengue clinical outcomes, suggesting a potential link between anti-saliva immunity and disease manifestation (Chea et al., 2024). These findings further support the potential of salivary protein antibody responses as sensitive indicators of human-mosquito contact and tools for monitoring the efficacy of vector control interventions.

7.2 Saliva-based vaccine development

Most arbovirus vaccine strategies have involved targeting viral antigens (Bano and Kumar, 2023; López-Camacho et al., 2020; Ramasamy et al., 2018; Sun et al., 2023). However, these approaches have proven difficult due to antigenic diversity, multiple co-circulating viruses, and rapid mutation rates, particularly for RNA viruses, which comprise most medically relevant mosquito-borne arboviruses. An alternative strategy targets mosquito salivary proteins, based on the recognition that mosquito saliva contains bioactive molecules that enhance viral transmission by altering host immune responses (Mu et al., 2023; Shrivastava et al., 2024; Wang et al., 2024). A key advantage of mosquito saliva-based vaccines is their potential to block transmission of several arboviruses (e.g., DENV, WNV, CHIKV, ZIKV) transmitted by the same vector. Additionally, unlike viral antigens, mosquito salivary proteins are relatively conserved, potentially reducing the risk of vaccine escape due to genetic variation.

Preclinical studies support this concept. Passive immunization in mice against specific mosquito salivary proteins can modify infection outcomes by neutralizing saliva-driven immune programming at the bite site. Antibodies targeting *Ae. aegypti* bacteria-responsive protein 1 (AgBR1) suppress early bite-site inflammation, particularly neutrophil infiltration, and partially protect mice from mosquito-transmitted ZIKV and WNV infection (Uraki et al., 2019b, 2019a). Similarly, targeting NeSt1 reduces pro-IL-1β/CXCL2 signaling, limits macrophage infiltration at the bite site, and protects mice from early ZIKV replication and disease (Hastings et al., 2019); with combined targeting of AgBR1 and NeSt1 further enhancing survival and reducing viremia (Marin-Lopez et al., 2021).

Additional salivary proteins have emerged as promising targets. For example, *Aa*NRP promotes neutrophil and myeloid recruitment via TLR1/TLR4-MyD88-NF-κB signaling, creating a local environment that facilitates both DENV and ZIKV transmission. Disruption of this saliva-driven signaling axis through pharmacologic inhibition reduced viral replication and transmission in murine models (Wang et al., 2024). Likewise, AaVA-1 promotes DENV and ZIKV transmission by facilitating early viral replication and dissemination. Its genetic disruption results in reduced viremia in infected mice following mosquito transmission (Sun et al., 2020).

Translational efforts have progressed into early human trials. The salivary peptide-based vaccine *An. gambiae* saliva vaccine (AGS-v), originally derived from four synthetic peptides (AGS-6, AGS-30, AGS-31, and AGS-35) from conserved regions of *An. gambiae* salivary proteins, was shown to be safe in a randomized Phase I clinical trial, inducing antigen-specific IgG and IFN-γ-producing T-cell responses (Manning et al., 2020). An updated formulation, AGS-v PLUS, included an additional peptide (AGS-20) derived from the conserved PRS-1 salivary protein found across multiple mosquito genera, including *Cx. quinquefasciatus*, *Ae. aegypti*, and *Ae. albopictus* (Friedman-Klabanoff et al., 2022). In a Phase I clinical trial, vaccinated participants were exposed to *Ae. aegypti* and *Ae. albopictus* bites. Transcriptomic analysis of skin biopsies collected post mosquito feeding revealed that vaccination enhanced Th1- and $CD8^+$ T-cell responses and upregulated antiviral pathways, including interferon signaling, alongside neutrophil-driven inflammation (Barningham et al., 2025). These findings suggest that vaccines targeting mosquito salivary components can reprogram the cutaneous immune environment at the bite site in ways that may limit arbovirus establishment and transmission.

Interestingly, immunization against AgApyrase reduces *P. berghei* transmission from infected mice to mosquitoes, disrupting further transmission cycles (Pala et al., 2024). Comparable transmission-blocking approaches using salivary antigens have not been shown for arboviruses, highlighting the need for further studies.

Despite these advances, challenges remain. Immunization against some salivary proteins can enhance viral transmission in murine models (Reagan et al., 2012), underscoring the need for careful antigen selection. In addition, saliva-based vaccines are unlikely to provide protection against non-vector routes of transmission, such as blood transfusion and sexual or vertical transmission (Uraki et al., 2019b).

Collectively, these studies illustrate that targeting immunization strategies against mosquito salivary proteins represents a promising vector-directed approach, with potential broad applicability across multiple mosquito-borne diseases.

7.3 Pharmacological potential of salivary compounds

The properties and functions of hematophagous arthropod salivary proteins offer an interesting avenue for therapeutic development. For example, this has been extensively shown for salivary ixolaris, from the tick *Ixodes scapularis*, which exhibits antithrombotic activity (Oliveira et al., 2012), reduces tumor growth and metastasis (de Oliveira et al., 2012), blocks tumor signaling pathways (Carneiro-Lobo et al., 2012), and inhibits glioblastoma growth (Carneiro-Lobo et al., 2009) in model systems.

Mosquito salivary proteins display similar broad pharmacological potential. *Ae. aegypti* cecropins (*Aeae*Cec1–5) suppress TNF-α, IL-1β, IL-6, and NO production, reduced lung inflammation and edema, and improved outcomes in mice infected with *Escherichia coli* or *Pseudomonas aeruginosa*. These benefits were linked to inhibition of MAPK and NF-κB signaling and partial neutralization of LPS, highlighting their therapeutic potential (Wei et al., 2018). Additionally, a truncated *Anopheles* cecropin, An-cecB-1, has been shown to exhibit antimalarial activity in vitro against drug-resistant *Plasmodium falciparum* strains, and in vivo in mice using the rodent malaria parasite *P. berghei* (Lou et al., 2024).

Cx. pipiens pallens saliva, including the peptide sialokinin, inhibit TRPV1 and TRPA1 ion channels in vitro in heterologous systems and native sensory neurons, and reduced pain responses in mouse models. These findings suggest that mosquito saliva contains bioactive components capable of suppressing nociceptive signaling, highlighting their potential as novel antinociceptive agents (Derouiche et al., 2022).

Ae. aegypti SGE have been shown to exhibit broad anti-inflammatory effects across multiple disease models. In dextran sulfate sodium (DSS)-induced colitis, a murine model for inflammatory bowel disease, SGE treatment improved disease outcome and postmortem scores, reduced peripheral lymphocytes, decreased pro-inflammatory cytokines (IFN-γ, TNF-α, IL-1β, and IL-5), and additionally diminished intestinal inflammation. Notably, early administration of SGE conferred protection against disease relapse upon DSS re-challenge (Sales-Campos et al., 2015).

In a murine cecum ligation and puncture (CLP) model for polymicrobial sepsis, SGE pretreatment reduced bacterial load and neutrophil infiltration, while increasing NO production, antioxidant defenses, and survival, highlighting its ability to modulate inflammation and oxidative stress. (De Souza Gomes et al., 2018). Similarly, *Ae. aegypti* saliva ameliorates acetaminophen-induced hepatotoxicity in mice, reducing liver damage, serum alanine aminotransferase levels, pro-inflammatory cytokines (TNF-α, IL-6, IL-1β, and IL-10), and inflammatory immune cell (NKT cells, macrophages, and dendritic cells) infiltration in the liver (Assis et al., 2021).

As previously discussed, recombinant *Ae. aegypti* D7 protein prevents DENV2 binding and/or infection in permissive cell types, as well as infection and dissemination in mice (Conway et al., 2016). This highlights the potential of D7-based anti-viral therapeutics.

Together, these findings emphasize the importance of further evaluating mosquito saliva as a potential source for novel therapeutic compounds.

8. Conclusion and future directions

Collectively, current evidence positions mosquito saliva as a central regulator of arbovirus transmission, shaping infection at the bite site through coordinated effects on vascular permeability, immune signaling, and cellular recruitment. Mosquito salivary proteins evolved to facilitate efficient blood feeding by counteracting host hemostatic and immune defenses, and arboviruses exploit these functions to enhance transmission. However, these interactions are not uniformly beneficial to the virus and instead reflect a context-dependent balance in which individual salivary components can either promote or limit infection.

Beyond the skin, the potential role of salivary proteins within the mosquito midgut remains poorly defined. The co-ingestion of saliva with the blood meal introduces these proteins into the blood bolus, where they may retain anti-hemostatic and immunomodulatory functions and influence early pathogen dynamics. However, whether these activities impact arbovirus infection in mosquitoes has not been clearly established, in part due to experimental approaches that do not fully capture the physiological complexity of natural blood feeding.

Future work should prioritize defining the role of salivary proteins within the mosquito midgut, particularly their influence on arbovirus infection dynamics following blood ingestion. Experimental systems that incorporate natural blood feeding and preserve physiological host responses will be essential to capture these interactions with accuracy. Studies should also examine how salivary proteins interact with the mosquito microbiome, which is increasingly recognized as a modulator of vector competence, to determine whether salivary-microbiome interactions influence midgut susceptibility to arbovirus infection. The growing appreciation that salivary proteins shape early infection in a vector-specific and context-dependent manner strengthens the rationale for targeting these molecules in pan-arboviral vaccine and therapeutic strategies, and for developing salivary protein antibodies as practical biomarkers of vector exposure in epidemiological surveillance.

References

Agarwal, A., Joshi, G., Nagar, D.P., Sharma, A.K., Sukumaran, D., Pant, S.C., et al., 2016. Mosquito saliva induced cutaneous events augment Chikungunya virus replication and disease progression. Infect. Genet. Evol. 40, 126–135.

Alvarenga, P.H., Andersen, J.F., 2022. An overview of D7 protein structure and physiological roles in blood-feeding Nematocera. Biology 12, 39.

Alvarenga, P.H., Francischetti, I.M.B., Calvo, E., Sá-Nunes, A., Ribeiro, J.M.C., Andersen, J.F., 2010. The function and three-dimensional structure of a thromboxane A2/cysteinyl leukotriene-binding protein from the saliva of a mosquito vector of the malaria parasite. PLoS Biol. 8, e1000547.

Alves e Silva, T.L., Joseph, R.E., Vega-Rodriguez, J., 2025. Beyond the bite: how mosquito salivary proteins modulate midgut biology and malaria parasite transmission. Curr. Opin. Insect Sci. 69, 101363.

Andersen, J.F., Lei, H., Strayer, E.C., Pham, V., Ribeiro, J.M.C., 2024. Mechanism of complement inhibition by a mosquito protein revealed through cryo-EM. Commun. Biol. 7, 649.

Arcà, B., Lombardo, F., Valenzuela, J.G., Francischetti, I.M.B., Marinotti, O., Coluzzi, M., et al., 2005. An updated catalogue of salivary gland transcripts in the adult female mosquito, Anopheles gambiae. J. Exp. Biol. 208, 3971–3986.

Arora, G., Sajid, A., Chuang, Y.-M., Dong, Y., Gupta, A., Gambardella, K., et al., 2021. Immunomodulation by mosquito salivary protein AgSAP contributes to early host infection by Plasmodium. mBio 12, e0309121.

Assis, J.B., Cogliati, B., Esteves, E., Capurro, M.L., Fonseca, D.M., Sá-Nunes, A., 2021. Aedes aegypti mosquito saliva ameliorates acetaminophen-induced liver injury in mice. PLoS One 16, e0245788.

Bano, N., Kumar, A., 2023. Immunoinformatics study to explore dengue (DENV-1) proteome to design multi-epitope vaccine construct by using CD4+ epitopes. J. Genet. Eng. Biotechnol. 21, 128.

Barningham, L., Carr, I.M., Jossi, S., Cole, M., Ponce, A., Short, M., et al., 2025. AGS-v PLUS, a mosquito salivary peptide vaccine, modulates the response to Aedes mosquito bites in humans. Vaccines 13, 1026.

Barros, V.C., Assumpção, J.G., Cadete, A.M., Santos, V.C., Cavalcante, R.R., Araújo, R.N., et al., 2009. The role of salivary and intestinal complement system inhibitors in the midgut protection of triatomines and mosquitoes. PLoS One 4, e6047.

Beerntsen, B.T., Champagne, D.E., Coleman, J.L., Campos, Y.A., James, A.A., 1999. Characterization of the Sialokinin I gene encoding the salivary vasodilator of the yellow fever mosquito, Aedes aegypti. Insect Mol. Biol. 8, 459–467.

Beerntsen, B.T., James, A.A., Christensen, B.M., 2000. Genetics of mosquito vector competence. Microbiol. Mol. Biol. Rev. 64, 115–137.

Bizzarro, B., Barros, M.S., Maciel, C., Gueroni, D.I., Lino, C.N., Campopiano, J., et al., 2013. Effects of Aedes aegypti salivary components on dendritic cell and lymphocyte biology. Parasit. Vectors 6, 329.

Boppana, V.D., Thangamani, S., Adler, A.J., Wikel, S.K., 2009. SAAG-4 is a novel mosquito salivary protein that programs host CD4+ T cells to express IL-4. Parasite Immunol. 31, 287–295.

Brummer-Korvenkontio, H., Palosuo, T., François, G., Reunala, T., 1997. Characterization of Aedes communis, Aedes aegypti and Anopheles stephensi mosquito saliva antigens by immunoblotting. Int. Arch. Allergy Immunol. 112, 169–174.

Buezo Montero, S., Gabrieli, P., Montarsi, F., Borean, A., Capelli, S., De Silvestro, G., et al., 2020. IgG antibody responses to the Aedes albopictus 34k2 salivary protein as novel candidate marker of human exposure to the tiger mosquito. Front. Cell. Infect. Microbiol. 10, 377.

Buezo Montero, S., Gabrieli, P., Poinsignon, A., Zamble, B.Z.H., Lombardo, F., Remoue, F., et al., 2022. Human IgG responses to the Aedes albopictus 34k2 salivary protein: analyses in Réunion Island and Bolivia confirm its suitability as marker of host exposure to the tiger mosquito. Parasit. Vectors 15, 260.

Buezo Montero, S., Gabrieli, P., Severini, F., Picci, L., Di Luca, M., Forneris, F., et al., 2019. Analysis in a murine model points to IgG responses against the 34k2 salivary proteins from Aedes albopictus and Aedes aegypti as novel promising candidate markers of host exposure to Aedes mosquitoes. PLoS Negl. Trop. Dis. 13, e0007806.

Calvo, E., Andersen, J., Francischetti, I.M., deL, Capurro, M., deBianchi, A.G., et al., 2004. The transcriptome of adult female Anopheles darlingi salivary glands. Insect Mol. Biol. 13, 73–88.

Calvo, E., Mans, B.J., Andersen, J.F., Ribeiro, J.M.C., 2006. Function and evolution of a mosquito salivary protein family. J. Biol. Chem 281, 1935–1942.

Calvo, E., Mans, B.J., Ribeiro, J.M.C., Andersen, J.F., 2009. Multifunctionality and mechanism of ligand binding in a mosquito antiinflammatory protein. Proc. Natl. Acad. Sci. U. S. A. 106, 3728–3733.

Calvo, E., Mizurini, D.M., Sá-Nunes, A., Ribeiro, J.M.C., Andersen, J.F., Mans, B.J., et al., 2011. Alboserpin, a factor Xa inhibitor from the mosquito vector of yellow fever, binds heparin and membrane phospholipids and exhibits antithrombotic activity. J. Biol. Chem. 286, 27998–28010.

Calvo, E., Tokumasu, F., Marinotti, O., Villeval, J.-L., Ribeiro, J.M.C., Francischetti, I.M.B., 2007. Aegyptin, a novel mosquito salivary gland protein specifically binds to collagen and prevents its interaction with glycoprotein VI, integrin $\alpha 2\beta 1$ and von Willebrand factor. J. Biol. Chem. 282, 26928–26938.

Carneiro-Lobo, T.C., Konig, S., Machado, D.E., Nasciutti, L.E., Forni, M.F., Francischetti, I.M.B., et al., 2009. Ixolaris, a tissue factor inhibitor, blocks primary tumor growth and angiogenesis in a glioblastoma model. J. Thromb. Haemost. 7, 1855–1864.

Carneiro-Lobo, T.C., Schaffner, F., Disse, J., Ostergaard, H., Francischetti, I.M.B., Monteiro, R.Q., et al., 2012. The tick-derived inhibitor Ixolaris prevents tissue factor signaling on tumor cells. J. Thromb. Haemost. 10, 1849–1858.

Cavalcante, R.R., Pereira, M.H., Gontijo, N.F., 2003. Anti-complement activity in the saliva of phlebotomine sand flies and other haematophagous insects. Parasitology 127, 87–93.

Chagas, A.C., Ramirez, J.L., Jasinskiene, N., James, A.A., Ribeiro, J.M.C., Marinotti, O., et al., 2014. Collagen-binding protein, Aegyptin, regulates probing time and blood feeding success in the dengue vector mosquito, Aedes aegypti. Proc. Natl. Acad. Sci. U. S. A. 111, 6946–6951.

Champagne, D.E., Ribeiro, J.M., 1994. Sialokinin I and II: vasodilatory tachykinins from the yellow fever mosquito Aedes aegypti. Proc. Natl. Acad. Sci. U. S. A. 91, 138–142.

Champagne, D.E., Smartt, C.T., Ribeiro, J.M., James, A.A., 1995. The salivary gland-specific apyrase of the mosquito Aedes aegypti is a member of the 5'-nucleotidase family. Proc. Natl. Acad. Sci. U. S. A. 92, 694–698.

Chea, S., Willen, L., Nhek, S., Ly, P., Tang, K., Oristian, J., et al., 2024. Antibodies to Aedes aegypti D7L salivary proteins as a new serological tool to estimate human exposure to Aedes mosquitoes. Front. Immunol. 15, 1368066.

Choumet, V., Attout, T., Chartier, L., Khun, H., Sautereau, J., Robbe-Vincent, A., et al., 2012. Visualizing non infectious and infectious Anopheles gambiae blood feedings in naive and saliva-immunized mice. PLoS One 7, e50464.

Christofferson, R.C., McCracken, M.K., Johnson, A.-M., Chisenhall, D.M., Mores, C.N., 2013. Development of a transmission model for dengue virus. Virol. J. 10, 127.

Chuang, Y.-M., Agunbiade, T.A., Tang, X.-D., Freudzon, M., Almeras, L., Fikrig, E., 2021. The effects of a mosquito salivary protein on sporozoite traversal of host cells. J. Infect. Dis. 224, 544–553.

Conway, M.J., Londono-Renteria, B., Troupin, A., Watson, A.M., Klimstra, W.B., Fikrig, E., et al., 2016. Aedes aegypti D7 saliva protein inhibits dengue virus infection. PLoS Negl. Trop. Dis. 10, e0004941.

Conway, M.J., Watson, A.M., Colpitts, T.M., Dragovic, S.M., Li, Z., Wang, P., et al., 2014. Mosquito saliva serine protease enhances dissemination of dengue virus into the mammalian host. J. Virol. 88, 164–175.

Cox, J., Mota, J., Sukupolvi-Petty, S., Diamond, M.S., Rico-Hesse, R., 2012. Mosquito bite delivery of dengue virus enhances immunogenicity and pathogenesis in humanized mice. J. Virol. 86, 7637–7649.

Cross, M.L., Cupp, E.W., Enriquez, F.J., 1994. Differential modulation of murine cellular immune responses by salivary gland extract of Aedes aegypti. Am. J. Trop. Med. Hyg. 51, 690–696.

Daniel, J.L., Dangelmaier, C., Jin, J., Ashby, B., Smith, J.B., Kunapuli, S.P., 1998. Molecular basis for ADP-induced platelet activation: I. Evidence for three distinct ADP receptors on human platelets. J. Biol. Chem. 273, 2024–2029.

de Melo Lara, L., Pereira-Filho, A.A., Mateus Pereira, R.H., Ferreira Malta, L.G., D'Ávila Pessoa, G.C., Koerich, L.B., et al., 2021. Adaptations to haematophagy: investigations on how male and female Culex quinquefasciatus (Diptera: Culicidae) deal with human complement activation after a blood meal. Insect Biochem. Mol. Biol. 139, 103650.

de Oliveira, A.D.S., Lima, L.G., Mariano-Oliveira, A., Machado, D.E., Nasciutti, L.E., Andersen, J.F., et al., 2012. Inhibition of tissue factor by ixolaris reduces primary tumor growth and experimental metastasis in a murine model of melanoma. Thromb. Res. 130, e163–e170.

De Souza Gomes, R., Navegantes-Lima, K.C., Monteiro, V.V.S., De Brito Oliveira, A.L., Rodrigues, D.V.S., Reis, J.F., et al., 2018. Salivary gland extract from Aedes aegypti improves survival in murine polymicrobial sepsis through oxidative mechanisms. Cells 7, 182.

Depinay, N., Hacini, F., Beghdadi, W., Peronet, R., Mécheri, S., 2006. Mast cell-dependent down-regulation of antigen-specific immune responses by mosquito bites. J. Immunol. 176, 4141–4146.

Derouiche, S., Li, T., Sakai, Y., Uta, D., Aoyagi, S., Tominaga, M., 2022. Inhibition of transient receptor potential vanilloid 1 and transient receptor potential ankyrin 1 by mosquito and mouse saliva. Pain 163, 299–307.

Dipaola, M.G., Perugini, E., Mancini, G., Gennari, N., Serini, P., Bevivino, G., et al., 2025. A novel biomarker of human exposure to Aedes albopictus based on the Ag5-3 salivary protein from the tiger mosquito. Parasit. Vectors 18, 470.

Doucoure, S., Mouchet, F., Cornelie, S., DeHecq, J.S., Rutee, A.H., Roca, Y., et al., 2012. Evaluation of the human IgG antibody response to Aedes albopictus saliva as a new specific biomarker of exposure to vector bites. PLoS Negl. Trop. Dis. 6, e1487.

Dragovic, S.M., Agunbiade, T.A., Freudzon, M., Yang, J., Hastings, A.K., Schleicher, T.R., et al., 2018. Immunization with AgTRIO, a protein in Anopheles saliva, contributes to protection against Plasmodium infection in mice. Cell Host Microbe 23, 523–535.

Esterly, A.T., Lloyd, M.G., Upadhyaya, P., Moffat, J.F., Thangamani, S., 2022. A human skin model for assessing arboviral infections. J. I. D. Innov. 2, 100128.

Figueiredo, A.C., de Sanctis, D., Gutiérrez-Gallego, R., Cereija, T.B., Macedo-Ribeiro, S., Fuentes-Prior, P., et al., 2012. Unique thrombin inhibition mechanism by anophelin, an anticoagulant from the malaria vector. Proc. Natl. Acad. Sci. U. S. A. 109, E3649–E3658.

Flayer, C.H., Kernin, I.J., Matatia, P.R., Zeng, X., Yarmolinsky, D.A., Han, C., et al., 2024. A γδ T cell–IL-3 axis controls allergic responses through sensory neurons. Nature 634, 440–446.

Fontaine, A., Pascual, A., Orlandi-Pradines, E., Diouf, I., Remoué, F., Pagès, F., et al., 2011. Relationship between exposure to vector bites and antibody responses to mosquito salivary gland extracts. PLoS One 6, e29107.

Francischetti, I.M.B., 2010. Platelet aggregation inhibitors from hematophagous animals. Toxicon. 56, 1130–1144.

Francischetti, I.M.B., Valenzuela, J.G., Ribeiro, J.M.C., 1999. Anophelin: kinetics and mechanism of thrombin inhibition. Biochemistry 38, 16678–16685.

Friedman-Klabanoff, D.J., Birkhold, M., Short, M.T., Wilson, T.R., Meneses, C.R., Lacsina, J.R., et al., 2022. Safety and immunogenicity of AGS-v PLUS, a mosquito saliva peptide vaccine against arboviral diseases: a randomized, double-blind, placebo-controlled phase 1 trial. eBioMedicine 86, 104375.

Fujimura, M., Miyake, Y., Uotani, K., Kanamori, K., Matsuda, T., 1988. Secondary release of thromboxane A2 in aerosol leukotriene C4-induced bronchoconstriction in guinea pigs. Prostaglandins 35, 427–435.

Garishah, F.M., Rother, N., Riswari, S.F., Alisjahbana, B., Overheul, G.J., van Rij, R.P., et al., 2021. Neutrophil extracellular traps in dengue are mainly generated NOX-independently. Front. Immunol. 12, 629167.

Gould, T.J., Vu, T.T., Swystun, L.L., Dwivedi, D.J., Mai, S.H.C., Weitz, J.I., et al., 2014. Neutrophil extracellular traps promote thrombin generation through platelet-dependent and platelet-independent mechanisms. Arterioscler. Thromb. Vasc. Biol. 34, 1977–1984.

Griffiths, R.B., Gordon, R.M., 1952. An apparatus which enables the process of feeding by mosquitoes to be observed in the tissues of a live rodent; together with an account of the ejection of saliva and its significance in malaria. Ann. Trop. Med. Parasitol. 46, 311–319.

Guerrero, D., Vo, H.T.M., Lon, C., Bohl, J.A., Nhik, S., Chea, S., et al., 2022. Evaluation of cutaneous immune response in a controlled human in vivo model of mosquito bites. Nat. Commun. 13, 7036.

Gurera, D., Bhushan, B., Kumar, N., 2018. Lessons from mosquitoes' painless piercing. J. Mech. Behav. Biomed. Mater. 84, 178–187.

Hastings, A.K., Uraki, R., Gaitsch, H., Dhaliwal, K., Stanley, S., Sproch, H., et al., 2019. Aedes aegypti NeSt1 protein enhances Zika virus pathogenesis by activating neutrophils. J. Virol. 93 e00395-19.

Henrique, M.O., Neto, L.S., Assis, J.B., Barros, M.S., Capurro, M.L., Lepique, A.P., et al., 2019. Evaluation of inflammatory skin infiltrate following Aedes aegypti bites in sensitized and non-sensitized mice reveals saliva-dependent and immune-dependent phenotypes. Immunology 158, 47–59.
Hiroki, C.H., Toller-Kawahisa, J.E., Fumagalli, M.J., Colon, D.F., Figueiredo, L.T.M., Fonseca, B.A.L.D., et al., 2020. Neutrophil extracellular traps effectively control acute Chikungunya virus infection. Front. Immunol. 10, 3108.
Isawa, H., Yuda, M., Orito, Y., Chinzei, Y., 2002. A mosquito salivary protein inhibits activation of the plasma contact system by binding to factor XII and high molecular weight kininogen. J. Biol. Chem. 277, 27651–27658.
Jin, L., Guo, X., Shen, C., Hao, X., Sun, P., Li, P., et al., 2018. Salivary factor LTRIN from Aedes aegypti facilitates the transmission of Zika virus by interfering with the lymphotoxin-β receptor. Nat. Immunol. 19, 342–353.
Kaczmarek, M.E., Herzog, N.L., Noval, M.G., Zuzworsky, J., Shah, Z., Bajwa, W.I., et al., 2020. Distinct New York City Aedes albopictus mosquito populations display differences in salivary gland protein D7 diversity and chikungunya virus replication. Viruses 12, 698.
Khattab, A., Barroso, M., Miettinen, T., Meri, S., 2015. Anopheles midgut epithelium evades human complement activity by capturing factor H from the blood meal. PLoS Negl. Trop. Dis. 9, e0003513.
Klug, D., Gautier, A., Calvo, E., Marois, E., Blandin, S.A., 2023. The salivary protein Saglin facilitates efficient midgut colonization of Anopheles mosquitoes by malaria parasites. PLoS Pathog. 19, e1010538.
Kuijpers, M.J.E., Schulte, V., Bergmeier, W., Lindhout, T., Brakebusch, C., Offermanns, S., et al., 2003. Complementary roles of glycoprotein VI and alpha2beta1 integrin in collagen-induced thrombus formation in flowing whole blood ex vivo. FASEB J 17, 685–687.
Lefteri, D.A., Bryden, S.R., Pingen, M., Terry, S., McCafferty, A., Beswick, E.F., et al., 2022. Mosquito saliva enhances virus infection through sialokinin-dependent vascular leakage. Proc. Natl. Acad. Sci. U. S. A. 119, e2114309119.
Li, Z., Ji, C., Cheng, J., Åbrink, M., Shen, T., Kuang, X., et al., 2022. Aedes albopictus salivary proteins adenosine deaminase and 34k2 interact with human mast cell specific proteases tryptase and chymase. Bioengineered 13, 13752–13766.
Londono-Renteria, B., Grippin, C., Cardenas, J.C., Troupin, A., Colpitts, T.M., 2016. Human C5a protein participates in the mosquito immune response against dengue virus. J. Med. Entomol. 53, 505–512.
López-Camacho, C., De Lorenzo, G., Slon-Campos, J.L., Dowall, S., Abbink, P., Larocca, R.A., et al., 2020. Immunogenicity and efficacy of Zika virus envelope domain III in DNA, protein, and ChAdOx1 adenoviral-vectored vaccines. Vaccines 8, 307.
Lou, J., Zhang, D., Wu, J., Zhu, G., Zhang, M., Tang, J., et al., 2024. Antimalarial activity of cecropin antimicrobial peptides derived from Anopheles mosquitoes. Antimicrob. Agents Chemother. 68, e00311-24.
Manning, J.E., Oliveira, F., Coutinho-Abreu, I.V., Herbert, S., Meneses, C., Kamhawi, S., et al., 2020. Safety and immunogenicity of a mosquito saliva peptide-based vaccine: a randomised, placebo-controlled, double-blind, phase 1 trial. Lancet 395, 1998–2007.
Mans, B.J., Calvo, E., Ribeiro, J.M.C., Andersen, J.F., 2007. The crystal structure of D7r4, a salivary biogenic amine-binding protein from the malaria mosquito Anopheles gambiae. J. Biol. Chem. 282, 36626–36633.
Margos, G., Navarette, S., Butcher, G., Davies, A., Willers, C., Sinden, R.E., et al., 2001. Interaction between host complement and mosquito-midgut-stage Plasmodium berghei. Infect. Immun. 69, 5064–5071.
Marin-Lopez, A., Wang, Y., Jiang, J., Ledizet, M., Fikrig, E., 2021. AgBR1 and NeSt1 antisera protect mice from Aedes aegypti-borne Zika infection. Vaccine 39, 1675–1679.

Martin-Martin, I., Kern, O., Brooks, S., Smith, L.B., Valenzuela-Leon, P.C., Bonilla, B., et al., 2021. Biochemical characterization of AeD7L2 and its physiological relevance in blood feeding in the dengue mosquito vector, Aedes aegypti. FEBS J 288, 2014–2029.

Martin-Martin, I., Kojin, B.B., Aryan, A., Williams, A.E., Molina-Cruz, A., Valenzuela-Leon, P.C., et al., 2023. Aedes aegypti D7 long salivary proteins modulate blood feeding and parasite infection. mBio 14, e0228923.

Martin-Martin, I., Leon, P.C.V., Amo, L., Shrivastava, G., Iniguez, E., Aryan, A., et al., 2022. Aedes aegypti sialokinin facilitates mosquito blood feeding and modulates host immunity and vascular biology. Cell Rep 39, 110648.

Martin-Martin, I., Paige, A., Valenzuela Leon, P.C., Gittis, A.G., Kern, O., Bonilla, B., et al., 2020a. ADP binding by the Culex quinquefasciatus mosquito D7 salivary protein enhances blood feeding on mammals. Nat. Commun. 11, 2911.

Martin-Martin, I., Smith, L.B., Chagas, A.C., Sá-Nunes, A., Shrivastava, G., Valenzuela-Leon, P.C., et al., 2020b. Aedes albopictus D7 salivary protein prevents host hemostasis and inflammation. Biomolecules 10, 1372.

McCracken, M.K., Christofferson, R.C., Grasperge, B.J., Calvo, E., Chisenhall, D.M., Mores, C.N., 2014. Aedes aegypti salivary protein 'aegyptin' co-inoculation modulates dengue virus infection in the vertebrate host. Virology 468–470, 133–139.

McCracken, M.K., Gromowski, G.D., Garver, L.S., Goupil, B.A., Walker, K.D., Friberg, H., et al., 2020. Route of inoculation and mosquito vector exposure modulate dengue virus replication kinetics and immune responses in rhesus macaques. PLoS Negl. Trop. Dis. 14, e0008191.

Mendes-Sousa, A.F., Queiroz, D.C., Vale, V.F., Ribeiro, J.M.C., Valenzuela, J.G., Gontijo, N.F., et al., 2016. An inhibitor of the alternative pathway of complement in saliva of New World anopheline mosquitoes. J. Immunol. 197, 599–610.

Mendes-Sousa, A.F., Vale, V.F., Queiroz, D.C., Pereira-Filho, A.A., da Silva, N.C.S., Koerich, L.B., et al., 2018. Inhibition of the complement system by saliva of Anopheles (Nyssorhynchus) aquasalis. Insect Biochem. Mol. Biol. 92, 12–20.

Moreira, C.K., Marrelli, M.T., Lima, S.P., Marinotti, O., 2001. Analysis of salivary gland proteins of the mosquito Anopheles darlingi (Diptera: Culicidae). J. Med. Entomol. 38, 763–767.

Mu, X., Lin, Z., Sun, Y., Chen, L., Lv, Q., Ji, C., et al., 2023. Aedes albopictus salivary adenosine deaminase is an immunomodulatory factor facilitating dengue virus replication. Sci. Rep. 13, 16660.

Narayanan, D., Sim, S.L., Kumari, S., 2025. Skin immunity and inflammation: cellular interactions and communication. Clin. Transl. Immunol. 14, e70053.

Ndille, E.E., Doucoure, S., Damien, G., Mouchet, F., Drame, P.M., Cornelie, S., et al., 2012. First attempt to validate human IgG antibody response to Nterm-34kDa salivary peptide as biomarker for evaluating exposure to Aedes aegypti bites. PLoS Negl. Trop. Dis. 6, e1905.

Ndille, E.E., Doucoure, S., Poinsignon, A., Mouchet, F., Cornelie, S., D'Ortenzio, E., et al., 2016. Human IgG antibody response to Aedes Nterm-34kDa salivary peptide, an epidemiological tool to assess vector control in chikungunya and dengue transmission area. PLoS Negl. Trop. Dis. 10, e0005109.

Ndille, E.E., Dubot-Pérès, A., Doucoure, S., Mouchet, F., Cornelie, S., Sidavong, B., et al., 2014. Human IgG antibody response to Aedes aegypti Nterm-34 kDa salivary peptide as an indicator to identify areas at high risk for dengue transmission: a retrospective study in urban settings of Vientiane city, Lao PDR. Trop. Med. Int. Health 19, 576–580.

Oka, K., 1989. Correlation of Aedes albopictus bite reaction with IgE antibody assay and lymphocyte transformation test to mosquito salivary antigens. J. Dermatol. 16, 341–347.

Oka, K., Ohtaki, N., 1989. Clinical observations of mosquito bite reactions in man: a survey of the relationship between age and bite reaction. J. Dermatol. 16, 212–219.

Oka, K., Ohtaki, N., Igawa, K., Yokozeki, H., 2018. Study on the correlation between age and changes in mosquito bite response. J. Dermatol. 45, 1471–1474.
Oka, K., Ohtaki, N., Yasuhara, T., Nakajima, T., 1989. A study of mosquito salivary gland components and their effects on man. J. Dermatol. 16, 469–474.
Oliveira, A.D.S. de, Lima, L.G., Mariano-Oliveira, A., Machado, D.E., Nasciutti, L.E., Andersen, J.F., et al., 2012. Inhibition of tissue factor by ixolaris reduces primary tumor growth and experimental metastasis in a murine model of melanoma. Thromb. Res. 130, e163–e170.
Opasawatchai, A., Yolwong, W., Thuncharoen, W., Inrueangsri, N., Itsaradisaikul, S., Sasisakulporn, C., et al., 2020. Novel salivary gland allergens from tropical mosquito species and IgE reactivity in allergic patients. World Allergy Organ. J. 13, 100099.
Pala, Z.R., Alves e Silva, T.L., Minai, M., Crews, B., Patino-Martinez, E., Carmona-Rivera, C., et al., 2024. Mosquito salivary apyrase regulates blood meal hemostasis and facilitates malaria parasite transmission. Nat. Commun. 15, 8194.
Peng, Z., Simons, F.E., 1998. A prospective study of naturally acquired sensitization and subsequent desensitization to mosquito bites and concurrent antibody responses. J. Allergy Clin. Immunol. 101, 284–286.
Peng, Z., Simons, F.E.R., 2004. Mosquito allergy: immune mechanisms and recombinant salivary allergens. Int. Arch. Allergy Immunol. 133, 198–209.
Peng, Z., Yang, M., Simons, F.E., 1996. Immunologic mechanisms in mosquito allergy: correlation of skin reactions with specific IgE and IgG antibodies and lymphocyte proliferation response to mosquito antigens. Ann. Allergy Asthma Immunol. 77, 238–244.
Pereira-Filho, A.A., Mateus Pereira, R.H., da Silva, N.C.S., Ferreira Malta, L.G., Serravite, A.M., Carvalho de Almeida, C.G., et al., 2020. The gut anti-complement activity of Aedes aegypti: investigating new ways to control the major human arboviruses vector in the Americas. Insect Biochem. Mol. Biol. 120, 103338.
Pingen, M., Bryden, S.R., Pondeville, E., Schnettler, E., Kohl, A., Merits, A., et al., 2016. Host inflammatory response to mosquito bites enhances the severity of arbovirus infection. Immunity 44, 1455–1469.
Ramasamy, V., Arora, U., Shukla, R., Poddar, A., Shanmugam, R.K., White, L.J., et al., 2018. A tetravalent virus-like particle vaccine designed to display domain III of dengue envelope proteins induces multi-serotype neutralizing antibodies in mice and macaques which confer protection against antibody dependent enhancement in AG129 mice. PLoS Negl. Trop. Dis. 12, e0006191.
Reagan, K.L., Machain-Williams, C., Wang, T., Blair, C.D., 2012. Immunization of mice with recombinant mosquito salivary protein D7 enhances mortality from subsequent West Nile virus infection via mosquito bite. PLoS Negl. Trop. Dis. 6, e1935.
Reiter, N., Reiter, M., Altrichter, S., Becker, S., Kristensen, T., Broesby-Olsen, S., et al., 2013. Anaphylaxis caused by mosquito allergy in systemic mastocytosis. Lancet 382, 1380.
Remoue, F., Alix, E., Cornelie, S., Sokhna, C., Cisse, B., Doucoure, S., et al., 2007. IgE and IgG4 antibody responses to Aedes saliva in African children. Acta Trop. 104, 108–115.
Remoue, F., Cisse, B., Ba, F., Sokhna, C., Herve, J.-P., Boulanger, D., et al., 2006. Evaluation of the antibody response to Anopheles salivary antigens as a potential marker of risk of malaria. Trans. R. Soc. Trop. Med. Hyg. 100, 363–370.
Reunala, T., Brummer-Korvenkontio, H., Palosuo, K., Miyanij, M., Ruiz-Maldonado, R., Löve, A., et al., 1994a. Frequent occurrence of IgE and IgG4 antibodies against saliva of Aedes communis and Aedes aegypti mosquitoes in children. Int. Arch. Allergy Immunol. 104, 366–371.
Reunala, T., Brummer-Korvenkontio, H., Palosuo, T., 1994b. Are we really allergic to mosquito bites? Ann. Med. 26, 301–306.

Ribeiro, J.M.C., Arcà, B., Lombardo, F., Calvo, E., Phan, V.M., Chandra, P.K., et al., 2007. An annotated catalogue of salivary gland transcripts in the adult female mosquito, Aedes aegypti. BMC Genomics 8, 6.
Ribeiro, J.M.C., Francischetti, I.M.B., 2003. Role of arthropod saliva in blood feeding: sialome and post-sialome perspectives. Annu. Rev. Entomol. 48, 73–88.
Ribeiro, J.M.C., Nussenzveig, R.H., 1993. The salivary catechol oxidase/peroxidase activities of the mosquito Anopheles albimanus. J. Exp. Biol. 179, 273–287.
Ribeiro, J.M.C., Nussenzveig, R.H., Tortorella, G., 1994. Salivary vasodilators of Aedes triseriatus and Anopheles gambiae (Diptera: Culicidae). J. Med. Entomol. 31, 747–753.
Ribeiro, J.M.C., Rossignol, P.A., Spielman, A., 1984. Role of mosquito saliva in blood vessel location. J. Exp. Biol. 108, 1–7.
Ribeiro, J.M.C., Valenzuela, J.G., 1999. Purification and cloning of the salivary peroxidase/catechol oxidase of the mosquito Anopheles albimanus. J. Exp. Biol. 202, 809–816.
Ribeiro, J.M.C., Valenzuela, J.G., 2003. The salivary purine nucleosidase of the mosquito, Aedes aegypti. Insect Biochem. Mol. Biol. 33, 13–22.
Ribeiro, J.M.C., Rossignol, P.A., Spielman, A., 1985. Salivary gland apyrase determines probing time in anopheline mosquitoes. J. Insect Physiol. 31, 689–692.
Rossignol, P.A., Ribeiro, J.M.C., Spielman, A., 1984. Increased intradermal probing time in sporozoite-infected mosquitoes. Am. J. Trop. Med. Hyg. 33, 17–20.
Runtuwene, L.R., Konishi, E., Yamanaka, A., Makino, Y., Suzuki, Y., Takasaki, T., et al., 2014. Dengue transmission model by means of viremic adult immuno-competent mouse. Parasit. Vectors 7, 143.
Sagna, A.B., Kassié, D., Couvray, A., Adja, A.M., Hermann, E., Riveau, G., et al., 2019. Spatial assessment of contact between humans and Anopheles and Aedes mosquitoes in a medium-sized African urban setting, using salivary antibody-based biomarkers. J. Infect. Dis. 220, 1199–1208.
Sales-Campos, H., De Souza, P.R., Basso, P.J., Ramos, A.D., Nardini, V., Chica, J.E.L., et al., 2015. Aedes aegypti salivary gland extract ameliorates experimental inflammatory bowel disease. Int. Immunopharmacol. 26, 13–22.
Schleicher, T.R., Yang, J., Freudzon, M., Rembisz, A., Craft, S., Hamilton, M., et al., 2018. A mosquito salivary gland protein partially inhibits Plasmodium sporozoite cell traversal and transmission. Nat. Commun. 9, 2908.
Schmid, M.A., Glasner, D.R., Shah, S., Michlmayr, D., Kramer, L.D., Harris, E., 2016. Mosquito saliva increases endothelial permeability in the skin, immune cell migration, and dengue pathogenesis during antibody-dependent enhancement. PLoS Pathog 12, e1005676.
Schneider, B.S., Higgs, S., 2008. The enhancement of arbovirus transmission and disease by mosquito saliva is associated with modulation of the host immune response. Trans. R. Soc. Trop. Med. Hyg. 102, 400–408.
Schneider, B.S., Soong, L., Zeidner, N.S., Higgs, S., 2004. Aedes aegypti salivary gland extracts modulate anti-viral and TH1/TH2 cytokine responses to Sindbis virus infection. Viral Immunol. 17, 565–573.
Scurtu, L.G., Simionescu, O., 2021. Soluble factors and receptors involved in skin innate immunity — what do we know so far? Biomedicines 9, 1795.
Shrivastava, G., Valenzuela-Leon, P.C., Botello, K., Calvo, E., 2024. Aedes aegypti saliva modulates inflammasome activation and facilitates flavivirus infection in vitro. iScience 27, 108620.
Shrivastava, G., Valenzuela-Leon, P.C., Chagas, A.C., Kern, O., Botello, K., Zhang, Y., et al., 2022. Alboserpin, the main salivary anticoagulant from the disease vector Aedes albopictus, displays anti-FXa-PAR signaling in vitro and in vivo. Immunohorizons 6, 373–383.
Simon, N., Lasonder, E., Scheuermayer, M., Kuehn, A., Tews, S., Fischer, R., et al., 2013. Malaria parasites co-opt human factor H to prevent complement-mediated lysis in the mosquito midgut. Cell Host Microbe 13, 29–41.

Smith, L.B., Duge, E., Valenzuela-León, P.C., Brooks, S., Martin-Martin, I., Ackerman, H., et al., 2022. Novel salivary antihemostatic activities of long-form D7 proteins from the malaria vector Anopheles gambiae facilitate hematophagy. J. Biol. Chem. 298, 101971.

Soares, A.C., Carvalho-Tavares, J., Gontijo, N.D.E.F., Dos Santos, V.C., Teixeira, M.M., Pereira, M.H., 2006. Salivation pattern of Rhodnius prolixus (Reduviidae; Triatominae) in mouse skin. J. Insect Physiol. 52, 468–472.

Stark, K.R., James, A.A., 1995. A factor Xa-directed anticoagulant from the salivary glands of the yellow fever mosquito Aedes aegypti. Exp. Parasitol. 81, 321–331.

Strayer, E.C., Lu, S., Ribeiro, J., Andersen, J.F., 2021. Salivary complement inhibitors from mosquitoes: structure and mechanism of action. J. Biol. Chem. 296, 100083.

Styer, L.M., Lim, P.-Y., Louie, K.L., Albright, R.G., Kramer, L.D., Bernard, K.A., 2011. Mosquito saliva causes enhancement of West Nile virus infection in mice. J. Virol. 85, 1517–1527.

Sun, H., Acharya, D., Paul, A.M., Lai, H., He, J., Bai, F., et al., 2023. Antibody-dependent enhancement activity of a plant-made vaccine against West Nile virus. Vaccines 11, 197.

Sun, P., Nie, K., Zhu, Y., Liu, Y., Wu, P., Liu, Z., et al., 2020. A mosquito salivary protein promotes flavivirus transmission by activation of autophagy. Nat. Commun. 11, 260.

Thangamani, S., Higgs, S., Ziegler, S., Vanlandingham, D., Tesh, R., Wikel, S., 2010. Host immune response to mosquito-transmitted chikungunya virus differs from that elicited by needle inoculated virus. PLoS One 5, e12137.

Trinchieri, G., 2010. Type I interferon: friend or foe? J. Exp. Med. 207, 2053–2063.

Tsuboi, T., Cao, Y.M., Torii, M., Hitsumoto, Y., Kanbara, H., 1995. Murine complement reduces infectivity of Plasmodium yoelii to mosquitoes. Infect. Immun. 63, 3702–3704.

Uraki, R., Hastings, A.K., Brackney, D.E., Armstrong, P.M., Fikrig, E., 2019a. AgBR1 antibodies delay lethal Aedes aegypti-borne West Nile virus infection in mice. NPJ Vaccines 4, 23.

Uraki, R., Hastings, A.K., Marin-Lopez, A., Sumida, T., Takahashi, T., Grover, J.R., et al., 2019b. Aedes aegypti AgBR1 antibodies modulate early Zika virus infection of mice. Nat. Microbiol. 4, 948–955.

Valenzuela-Leon, P.C., Shrivastava, G., Martin-Martin, I., Cardenas, J.C., Londono-Renteria, B., Calvo, E., 2022. Multiple salivary proteins from Aedes aegypti mosquito bind to the Zika virus envelope protein. Viruses 14, 221.

Visser, I., Koenraadt, C.J.M., Koopmans, M.P.G., Rockx, B., 2023. The significance of mosquito saliva in arbovirus transmission and pathogenesis in the vertebrate host. One Health 16, 100506.

Vogt, M.B., Lahon, A., Arya, R.P., Kneubehl, A.R., Spencer Clinton, J.L., Paust, S., et al., 2018. Mosquito saliva alone has profound effects on the human immune system. PLoS Negl. Trop. Dis. 12, e0006439.

Waitayakul, A., Somsri, S., Sattabongkot, J., Looareesuwan, S., Cui, L., Udomsangpetch, R., 2006. Natural human humoral response to salivary gland proteins of Anopheles mosquitoes in Thailand. Acta Trop. 98, 66–73.

Wanasen, N., Nussenzveig, R.H., Champagne, D.E., Soong, L., Higgs, S., 2004. Differential modulation of murine host immune response by salivary gland extracts from the mosquitoes Aedes aegypti and Culex quinquefasciatus. Med. Vet. Entomol. 18, 191–199.

Wang, Z., Nie, K., Liang, Y., Niu, J., Yu, X., Zhang, O., et al., 2024. A mosquito salivary protein-driven influx of myeloid cells facilitates flavivirus transmission. EMBO J 43, 1690–1721.

Wei, L., Yang, Y., Zhou, Y., Li, M., Yang, H., Mu, L., et al., 2018. Anti-inflammatory activities of Aedes aegypti cecropins and their protection against murine endotoxin shock. Parasit. Vectors 11, 470.

Whiten, S.R., Ray, W.K., Helm, R.F., Adelman, Z.N., 2018. Characterization of the adult Aedes aegypti early midgut peritrophic matrix proteome using LC-MS. PLoS One 13, e0194734.
Whittle, B.J.R., Kauffman, G.L., Moncada, S., 1981. Vasoconstriction with thromboxane A2 induces ulceration of the gastric mucosa. Nature 292, 472–474.
Yang, D., Han, Z., Oppenheim, J.J., 2017. Alarmins and immunity. Immunol. Rev. 280, 41–56.
Yonei, Y., Guth, P.H., 1990. Effect of a leukotriene receptor antagonist on LTC4 vasoconstriction in rat stomach. Am. J. 259, G147–G154.
Yoshida, S., Sudo, T., Niimi, M., Tao, L., Sun, B., Kambayashi, J., et al., 2008. Inhibition of collagen-induced platelet aggregation by anopheline antiplatelet protein, a saliva protein from a malaria vector mosquito. Blood 111, 2007–2014.
Zeidner, N.S., Higgs, S., Happ, C.M., Beaty, B.J., Miller, B.R., 1999. Mosquito feeding modulates Th1 and Th2 cytokines in flavivirus susceptible mice: an effect mimicked by injection of sialokinins, but not demonstrated in flavivirus resistant mice. Parasite Immunol. 21, 35–44.
Zellweger, R.M., Shresta, S., 2014. Mouse models to study dengue virus immunology and pathogenesis. Front. Immunol. 5, 151.
Züst, R., Toh, Y.-X., Valdés, I., Cerny, D., Heinrich, J., Hermida, L., et al., 2014. Type I interferon signals in macrophages and dendritic cells control dengue virus infection: implications for a new mouse model to test dengue vaccines. J. Virol. 88, 7276–7285.

Further reading

Fong, S.-W., Tan, J.J.L., Sridhar, V., Amrun, S.N., Neo, V.K.X., Wong, N., et al., 2025. Mosquito salivary sialokinin reduces monocyte activation and chikungunya virus-induced inflammation via neurokinin receptors. Nat. Commun. 16, 8644.
Kojin, B.B., Martin-Martin, I., Araújo, H.R.C., Bonilla, B., Molina-Cruz, A., Calvo, E., et al., 2021. Aedes aegypti SGS1 is critical for Plasmodium gallinaceum infection of both the mosquito midgut and salivary glands. Malar. J. 20, 11.
Marin-Lopez, A., Huck, J.D., Esterly, A.T., Azcutia, V., Rosen, C., Garcia-Milian, R., et al., 2024. The human CD47 checkpoint is targeted by an immunosuppressive Aedes aegypti salivary factor to enhance arboviral skin infectivity. Sci. Immunol. 9, eadk9872.

CHAPTER EIGHT

Mosquito-microbiota interactions affect viral diversity and evolution

Lucy de Guilhem de Lataillade[a,1], Hugo D. Perdomo[a,1], Laura E. Brettell[b], and Grant L. Hughes[a,*]

[a]Departments of Vector Biology and Tropical Disease Biology, Centre for Neglected Tropical Diseases, Liverpool School of Tropical Medicine, Liverpool, United Kingdom
[b]School of Science, Engineering and Environment, University of Salford, Manchester, United Kingdom
*Corresponding author. e-mail address: Grant.Hughes@lstmed.ac.uk

Contents

Abstract

Mosquitoes exist as holobionts in which resident and transient microbes shape development, immunity, and impact other symbionts, including co-localised viruses. Because most arboviruses must replicate in the mosquito midgut before disseminating to the haemocoel and salivary glands, the gut represents both a critical barrier to infection and a major bottleneck that can reshape viral genetic diversity. This chapter examines how mosquito–microbiota interactions intersect with arbovirus population dynamics and evolution across the transmission cycle. We first provide a background on arbovirus evolution and its drivers, and then examine the unique barriers these viruses face. We then draw on examples of microbiota-virus interactions to understand how they mesh into an arbovirus evolutionary context, either indirectly, for example via immune priming of the host, or by more direct pressures such as microbial secretion of metabolites that modulate infection or resource utilisation. Finally, we discuss emerging symbiotic control approaches and outline methodological priorities to better quantify, and ultimately understand, microbiota-driven selection to guide robust, evolution-informed vector control strategies.

[1] These authors contributed equally

Advances in Insect Physiology, Volume 70
ISSN 0065-2806, https://doi.org/10.1016/bs.aiip.2026.06.003

1. Introduction

Multicellular organisms and microorganisms are continuously evolving together. A collective of bacteria, viruses and fungi, termed the microbiota, not only co-exists in unison with a host but also contributes to critical biological functions. In return, the host provides an environmental niche and resources for microbiota, thereby establishing a symbiotic relationship. The definition of symbiosis, or of symbionts themselves, is not universal and can vary across fields. Here, we use the broadest definition of symbiosis: the interaction between two or more organisms of different species that persists for at least part of their lifetime (de Bary, 1879). Symbionts can have different relationships with their host, spanning from mutualistic to pathogenic. This broad definition allows us to encompass all the diversity of host-symbiont relationships. It should be noted that under this definition, in theory, arboviral pathogens could also be classified as symbionts. In mosquitoes, symbionts have been shown to affect development, feeding, reproduction, immunity, and vectorial capacity (Souza et al., 2019). Here, we mainly focus on the interactions occurring in the gut of mosquitoes, given that it is a critical barrier for arboviral infection and a bottleneck of viral diversity. The gut is an environment where microbiota have the opportunity to interact with both the host and the arbovirus. We will posit how those interactions and pressures are linked to viral evolution, but will draw upon knowledge from other systems, given that there are gaps in our current understanding in this field.

1.1 Virus transmission in a mosquito and its impact on diversity and evolution

For many arboviruses, transmission between vertebrate hosts depends on mosquito vectors. The transmission cycle is initiated when a mosquito acquires the virus during a blood meal from an infected vertebrate host. Following ingestion, the virus infects and replicates within the mosquito midgut, subsequently overcomes the midgut barrier to enter the haemolymph. From there, the virus disseminates to secondary tissues, where further replication occurs, and ultimately invades the salivary glands. Transmission is completed when the virus is inoculated into a new vertebrate host during a subsequent blood-feeding. Each stage of this transmission cycle imposes distinct selective pressures on viral populations, thereby driving viral evolution and potentially altering viral fitness. While increases in viral fitness benefit the virus, they can have diverse, sometimes fundamentally different consequences for hosts. These outcomes may include modulation of virulence or the emergence of infection in new host species (Amin et al., 2020; Bellone et al., 2020; Badgett et al., 2002; Noval et al., 2019). The impact of such changes can range from reduced mosquito survival following infection to the emergence of novel viral pandemics driven by host shifts into human populations (Amin et al., 2020; Ciota et al., 2013). These considerations emphasize the importance of viral evolution and highlight the need to better understand both the processes involved and the factors that influence them.

Point mutations in arboviruses occasionally lead to epidemics. Most mosquito-borne viruses possess RNA genomes and rely on RNA-dependent RNA polymerases for replication, which are inherently error-prone. This low replication fidelity generates substantial genetic diversity, a feature often associated with increased adaptive potential (Lauring and Andino, 2010; Vignuzzi et al., 2005). Due to the nature of the arboviral transmission cycle, alternating between vertebrate hosts and arthropod vectors, these viruses experience repeated population bottlenecks alongside shifting selective pressures. Bottlenecks involve stochastic reductions in population size that can restrict genetic diversity (Fitzmeyer et al., 2023; Forrester et al., 2012; Grubaugh et al., 2016). In mosquitoes, key bottlenecks occur during initial infection, midgut escape, and salivary gland invasion (Lequime et al., 2016). Despite these constraints, arboviruses often maintain higher genetic diversity in the mosquito vector than in vertebrate hosts (Jerzak et al., 2005). Alternating selective pressures further complicate viral evolution, as mutations that are advantageous in the insect vector may be deleterious in the vertebrate host, and vice versa. For example, dengue virus (DENV) passaged repeatedly in mosquito C6/36 cells showed increased fitness in the insect cells but reduced fitness in vertebrate cells (Vasilakis et al., 2009). Similarly, West Nile virus (WNV) adapted in mosquito vectors exhibits diminished fitness in avian hosts (Grubaugh et al., 2016). This intricate evolutionary landscape results in overall evolutionary rates that are an order of magnitude lower than those of single-host RNA viruses (Costa et al., 2012). Nevertheless, arboviruses continue to accumulate mutations over time, some of which can significantly influence transmission dynamics. Notably, the emergence of dominant variants with enhanced outbreak potential has been linked to mosquito-associated selection, as demonstrated by the A226V mutation of Chikungunya virus (CHIKV) E1 protein that led to increased transmission potential to *Aedes albopictus* in the Indian Ocean region (Tsetsarkin et al., 2011), and by genotype replacement events of Japanese encephalitis virus (JEV) (Schuh et al., 2014).

The mosquito host imposes selective pressures that shape viral evolution. When a mosquito, like any other organism, becomes infected with a pathogen, it can respond in two ways: (1) resistance, the ability to reduce viral load, and (2) tolerance, the ability to limit infection-associated damage without necessarily reducing viral load (Miller et al., 2005; Perdomo et al., 2026). These distinct strategies have different evolutionary consequences for viruses. Resistance exerts direct selective pressure by limiting viral replication, whereas tolerance may impose neutral or even permissive conditions that allow viral persistence (Finlay and McFadden, 2006). The evolutionary implications of tolerance have been explored in theoretical frameworks, but experimental validation remains limited. In contrast, there is growing empirical evidence that resistance mechanisms in mosquitoes influence viral evolution. For instance, mosquito cells exhibiting activation of the JAK–STAT and ERK pathways, mediated by insulin/insulin-like growth factor 1 signalling, have been associated with the emergence of specific viral variants, particularly in the envelope and NS5 proteins (Char et al., 2024). Similarly, WNV has been shown to accumulate point mutations in genomic regions heavily targeted by RNA interference (RNAi), suggesting direct immune-driven selection (Brackney et al., 2009).

The availability of extensive genetic tools in *Drosophila melanogaster* has enabled detailed investigation of host–virus interactions in this organism as a model. Experimental evolution studies with WNV in flies demonstrated that modulation of RNAi activity significantly alters viral diversity, with stronger antiviral responses promoting increased diversification (Lezcano et al., 2023). Interestingly, studies involving Drosophila C virus (DCV), a virus infecting *D. melanogaster* in nature, showed that disruption of key antiviral components of the RNAi pathway, such as Dicer-2, leads to increased viral diversity, opposite to what has been seen with WNV (Mongelli et al., 2022). Similar trends have been observed when other immune-related genes are knocked out, including those traditionally associated with antibacterial defence. Collectively, these findings indicate that viral evolution is shaped by a broad range of host immune effectors, including those not classically linked to antiviral responses, underscoring the complexity of the selective pressures imparted by the host.

2. Gut-associated symbionts

The midgut is a critical tissue for bacterial and viral symbiont colonisation and dissemination. There, the microbiota composition and abundance can be heavily influenced by diet or developmental stage (Muturi et al., 2018). Microbiota profoundly influences viral infection in mosquitoes by distinct modes of action (Katak et al., 2023). Primarily, activation of the host immune system by the microbiota influences viral outcomes (Li et al., 2024). In addition, intrinsic toxin production by bacteria can directly affect viral infection (Wu et al., 2019), or indirectly modulate infection through effects on the host (Pan et al., 2012). Furthermore, thickening of the peritrophic matrix during a blood meal, which can be enhanced by bacteria, is also known to affect infection dynamics (Song et al., 2025; Xi et al., 2008). These interactions indicate that modulating the microbiome can alter virus transmission (Santos et al., 2022). However, the focus of many studies is generally on a particular bacterial taxon, with the aim of identifying taxa that reduce viral infection, which could be suitable for vector control. Notably, several bacteria have been shown to influence virus, and even, parasite development. *Enterobacter, Chromobacterium, Rosenbergiella,* and *Beauvaria bassiania* inhibit DENV, Zika (ZIKV) and La Crosse (LACV) viruses by producing specific proteins or metabolites (Joyce et al., 2011; Saraiva et al., 2018; Sun et al., 2024; Yu et al., 2022; Zhang et al., 2024). *Acinetobacter* also induces carryover effects (Dickson et al., 2017), where bacterial infection of the gut in larvae can inhibit adult arboviral infection later in life through mechanisms not yet fully understood (Giraud et al., 2022). *Bacillus*, *Elizabethkingia*, and *Asaia* are all known to inhibit Plasmodium development through mechanisms that potentially could also affect viruses (Capone et al., 2013; Chen et al., 2015; Yu et al., 2020), with *Asaia* being investigated for paratransgenesis applications. However, some bacteria induce different phenotypes depending on the genetics and background of the pathogen and mosquito host. For instance, strains of *Serratia* increase susceptibility to DENV and CHIKV in *Aedes aegypti* by secreting a toxin that facilitates viral infection of the mosquito gut (Apte-Deshpande et al., 2014; Wu et al., 2019), while others enhance resistance to the *Plasmodium* parasites and ZIKV (Bai et al., 2019; do Nascimento et al., 2022). This duality has parallels to the concept of pathobionts (Chandra et al., 2021), where symbiont-host interactions are malleable and are shaped not only by their surrounding environment but also by the genetics of each party. Pathobionts are thus normally commensal symbionts which can emerge as pathogens to their host under specific conditions. Similar context-dependent phenotypes likely exist in the tripartite interactions among a host, microbiota, and the pathogen, whereby the outcome of pathogen infection and transmission is likely influenced not only by all three players, but also their surrounding environment.

Further complicating the dissection of these interactions is the challenge in characterising microbes. Amplicon sequencing approaches, from which much of our current understanding is derived, only discriminate to the family or genus level at best. Metagenomic approaches can reveal greater taxonomic and functional information, but at a higher cost and requiring more advanced bioinformatic analysis, meaning metagenomic studies are less common. Whole genome sequencing provides high-resolution genetic information of isolates, which is invaluable in the characterisation of taxa with control potential, as small genomic changes are known to affect pathogen infection outcomes (Bando et al., 2013). Thus, bacterial strain diversity is likely an underappreciated factor influencing infection outcomes (Nysten et al., 2024). Examining how each specific bacterial strain affects mosquito phenotypes should include studies on vectorial capacity, and virus diversity which in turn informs us in regard to evolution impacts. Additionally, understanding how robust these phenotypes are across diverse microbiomes and mosquito hosts will be essential for deploying microbial-based vector control strategies.

3. The impacts of microbiota and host on viral evolution are intertwined

Considering these interactions, microbiota have the potential to be a driving force influencing viral evolution through direct and indirect effects, including activation of the immune system, metabolite secretion, and resource competition. Viruses and symbionts share a similar ecological niche within the mosquito, whether intracellularly with *Wolbachia* or within the gut, and this co-localisation offers opportunities for interaction and for microbiota to exert evolutionary pressures on invading arboviruses. The arthropod immune system, which is thought to have evolved in response to microbiota and is essential for maintaining microbiota homeostasis in the host, also impacts DENV and other viruses (Crespo-Bellido and Duffy, 2023; Hosen et al., 2026). DENV is known to be impacted by the Toll pathway (Xi et al., 2008), which can be elicited by microbiota (Carvalho et al., 2012). Host secretion of antimicrobial peptides in response to *Pantoea, Proteus,* and *Paenibacillus* also inhibited DENV (Ramirez et al., 2012). Similarly, the Immunodeficiency pathway (IMD) is elicited by the microbiota, which in turn reduces Sindbis virus (SINV) titres (Barletta et al., 2017). Broadly, activation of the immune system, which reduces viral load and diversity, is intrinsically linked to microbial interactions that also elicit immune responses (Cai and Christophides, 2024). Specifically, the Toll or IMD pathway, are known to be elicited by bacteria and impact viral survival and dissemination within the mosquito (Guégan et al., 2018; Weaver et al., 2021). More direct interactions include secretions of effector molecules. Sphingosine, a known effector molecule in antimicrobial pathways in mammals, is produced by *Enterobacter hormaechei* in a population of *Ae. albopictus* and suppresses ZIKV infection in the gut (Sun et al., 2024). Similarly, in *Ae. aegypti*, *Chromobacterium* produces an aminopeptidase which degrades the E protein of DENV (Saraiva et al., 2018). Another strain of *Chromobacterium* was also found to produce a lipase with broad antiviral activity against WNV, JEV, DENV, and ZIKV (Yu et al., 2022). The influence of microbiota on viral infection and dissemination can strongly influence viral evolution (Coffey et al., 2008). Direct and indirect impacts could affect viral diversity, randomly causing a stochastic drift of the population, or, on the contrary, exerting a strong selection pressure to evade certain pathways or anti-microbial peptides (AMPs).

Virus-microbiome interactions and the underlying drivers of evolution are intertwined.

Viral infection in mosquitoes can impact microbiota but the changes that occur may be different for divergent hosts which likely possess distinct microbiomes. As such, some bacteria may be modulated by infection while others are not. A cohort of field-collected mosquitoes exposed to ZIKV exhibited a higher bacterial load, but also a decreased relative abundance of certain taxa compared to blood-fed mosquitoes unexposed to ZIKV, indicating potential for selection pressure (Cansado-Utrilla et al., 2025). Indeed, the coexistence of intracellular bacteria and viruses within cells could also induce selection pressures due to competition for resources, as suggested by the density-dependent effect of *Wolbachia*-mediated pathogen blocking (Lu et al., 2012). Conversely, in other hosts, it has been observed that viruses may utilise microbial-derived resources to enhance their replication. In mice, measles virus uses bacterial Lipo-polysaccharides (LPS) (Robinson et al., 2014), and enteroviruses utilise bacterial peptidoglycans or LPS to replicate and diversify (Kuss et al., 2011). This dependence on bacterial resources can lead to co-evolution. Despite that, in humans, the microbiome is generally considered antagonistic to virus diversification due to immune priming and the secretion of AMPs, with several examples highlighting bacteria-induced secretion of interferons to attenuate CHIKV and Influenza A virus (González and Elena, 2021). These phenomena are expected to drive evolution (Mazel-Sanchez et al., 2019; Stevens et al., 2021; Weiss and Aksoy, 2011).

4. Evolution of antiviral resistance to symbiotic control

The intracellular vertically transmitted bacterium *Wolbachia* induces phenotypes in the infected host that are desirable for vector control strategies. Their deployment into the field, either by releasing solely males or both males and females has revolutionized vector control. Population suppression approaches inundate native mosquito populations with *Wolbachia*-infected males, which induce sterility when mating with their native female counterparts that lack *Wolbachia* or possess an incompatible *Wolbachia* strain(s). Alternatively, population replacement approaches exploit both the ability of *Wolbachia* to spread through native populations when both sexes are released and the ability of *Wolbachia* to interfere with pathogen transmission in infected females. Importantly, both approaches have been demonstrated to be successful in the field and led to significant epidemiological outcomes (Beebe et al., 2021; Crawford et al., 2020; Lim et al., 2026; Nazni et al., 2019; Utarini et al., 2021). A significant reduction of 83 % in dengue haemorrhagic fever cases was seen when comparing fully *w*Mel-treated compared to untreated periods (Indriani et al., 2023). Given that the replacement strategy impairs virus transmission, there is, theoretically, an opportunity for the development of viral resistance to *Wolbachia,* which would reduce the efficacy of the intervention (Edenborough et al., 2021).

RNA viruses have rapid mutation rates, which could potentially lead to the emergence of resistance to *Wolbachia* blocking. However, it is important to consider that a *Wolbachia*-blocking resistant virus would need to be functional in both mosquitoes and vertebrates. This host-switching necessity of arboviruses creates a strong bottleneck, further hindering the emergence of variants capable of infection in dual hosts (Holmes, 2003). Using a serial passage system in mosquitoes that avoided the vertebrate host, a mutation in the envelope of DENV-1 was repeatedly observed in *Wolbachia*-infected mosquitoes (Thi Hue Kien et al., 2023). This mutation was lost when passages were performed in *Wolbachia*-uninfected *Ae. aegypti* or human cells, further emphasising the constraints of viral evolution imposed on arboviruses by two hosts. Despite this, the study demonstrated that *Wolbachia* has the potential to influence viral genotype. The antiviral effect is known to be less effective against DENV1 compared to the other DENV serotypes (Flores et al., 2020), and it is feasible that this incomplete DENV suppression may promote the emergence of resistant variants (Pawlotsky, 2016). Similarly, the heterogeneous nature of *Wolbachia* localization in the mosquito host may enable tissue-specific reservoirs whereby virus may escape blocking. The host also plays a role in the blocking phenotype and can be selected upon. Experimental evolution of *Ae. aegypti* infected with *Wolbachia,* selecting for high, low, or random blocking phenotypes, demonstrated that the fitness of *Ae. aegypti* is reduced when selection was imposed for a lower blocking phenotype (Ford et al., 2019). Mosquito survival is important for vectorial capacity, given the extrinsic incubation period of the virus.

Other systems may shed light on the influence of *Wolbachia* on viral populations and evolution. *Wolbachia*-infected flies infected with DCV had silent mutations, but these did not impact the infectivity of the virus (Martinez et al., 2019). *Culex pipiens* mosquitoes naturally infected with *w*Pip also are infected with *Culex pipiens* densovirus (CpDV), which is vertically transmitted to offspring at a prevalence of around 90%. Antibiotic curing of the symbionts reduced both CpDV transmission and density (Altinli et al., 2019). Given their ubiquity and the high vertical transmission rates of both the endosymbiont and the Insect Specific Viruses (ISVs), these systems might be opportune to examine *Wolbachia*-virus interactions from an evolutionary perspective.

To help predict if the viruses have the capacity to overcome *Wolbachia*'s antiviral effects, it is critical to understand the mechanism(s) by which the bacterium exerts this phenotype. There is evidence for several distinct mechanisms, including enhancement of basal immunity, sequestration of nutrients such as cholesterol and modulation of small RNAs (Terradas and McGraw, 2017). It is likely that mechanisms differ depending on the virus, the mosquito host, and the *Wolbachia* strain and each specific interaction needs to be studied in detail to understand what mechanism is influencing each blocking phenotype. Intriguingly, in an *Anopheles* – O'nyong 'nyong virus (ONNV) system, nutrient sequestration and immune signalling were functionally linked, whereby cholesterol interacts with the Toll immune pathway to modulate *Wolbachia* blocking (Pujhari et al., 2023). A greater understanding of the *Wolbachia*-mediated blocking phenotype will enable tailored and specific approaches to investigate the evolution of virus resistance in *Wolbachia*-based control strategies. This can then facilitate strategies to mitigate or overcome resistance if it were to emerge, such as the use of distinct *Wolbachia* strains that employ divergent blocking mechanism(s).

An alternative vector control approach exploits transgenic bacteria engineered to interfere with pathogen transmission in the mosquito host, termed paratransgenesis. While most paratransgenic approaches aim to reduce the burden of malaria, they can also be used to interfere with arboviruses (Altinli et al., 2025; French et al., 2025; Ratcliffe et al., 2022). One specific example of transgenic *Serratia* has shown the ability to block *Plasmodium* or arboviruses depending on if it is ingested by *Anopheles* or *Aedes* mosquitoes, respectively (Hu et al., 2025). Contrary to *Wolbachia*-based control, where the mode of action is not fully elucidated and likely differs between hosts, *Wolbachia* strains, and pathogens, the mechanism that mediates the anti-pathogen effect in paratransgenic control approaches is engineered into the microbe. Given the pathogen blocking mechanism is unknown, experiments can be designed, and outcomes modelled to investigate and predict the evolution of resistance to a particular effector molecule and/or intervention strategy. While the priority of research in this field is to develop effective anti-pathogen approaches using self-sustaining microbes, it is important to consider how arboviral pathogens could evolve resistance to these interventions. To that end, strategies using distinct effector molecules or RNAi that target both host genes and pathways antagonistic to virus infection, or that are tiled across the arboviral genomes in the case of small RNAs, would likely mitigate the evolutionary pressure against the approach. However, such strategies could be evaluated in vitro and validated in vivo to examine the evolutionary potential of the virus in response to the antiviral activity.

5. Other perspectives and future directions

The mosquito is inextricably associated with both its microbiome and the arboviral pathogens it transmits. To date, both direct and indirect effects of the microbiota on mosquito – arbovirus interactions have been identified, and new mechanisms continue to emerge. However, the consequences of these interactions for arboviral evolution remain poorly understood, as relatively few studies have directly addressed this question. The limited number of investigations examining the selective pressures imposed by specific symbionts on mosquito–virus interactions have yielded intriguing results, underscoring the need for greater attention to this field. A principal approach to uncovering the direct effects of microbiota on viral evolution is through experimental evolution studies; however, these are often time-consuming and costly. Consequently, much of the current evidence regarding the influence of the microbiota on viral evolution is indirect, frequently relying on correlations between microbiota-induced immune pressures and observed viral genetic changes. Encouragingly, recent technological advances offer new opportunities to address these limitations. Although 16S rRNA amplicon sequencing remains widely used in microbiome research, shotgun metagenomic sequencing is becoming more common place, facilitating a more comprehensive characterization of microbial communities under different experimental conditions. It is important to note that this chapter focuses primarily on the bacterial component of the microbiome and its influence on arboviral evolution. However, the microbiome also encompasses fungi and viruses, which likely play significant roles in shaping these interactions. The emphasis on bacterial communities largely reflects the historical focus of the field, but a broader application of metagenomic sequencing approaches is advancing our understanding of the contributions of other microbial groups, including fungi and viruses. Other methodological advances, such as the ability to manipulate and transplant mosquito microbiota to establish defined microbial communities (Coon et al., 2022; Hegde et al., 2024; Zhao et al., 2023), will provide a robust framework for future research. Moving forward, studies could draw on experimental evolution approaches to investigate how specific microbiome components and overall community diversity influence viral evolution.

References

Altinli, M., Leonard, S.P., Kohl, A., Hughes, G.L., 2025. Bacteria-mediated dsRNA delivery for mosquito-borne virus control. Trends Parasitol. 41 (10), 880–893. https://doi.org/10.1016/j.pt.2025.07.015.

Altinli, M., Soms, J., Ravallec, M., Justy, F., Bonneau, M., Weill, M., et al., 2019. Sharing cells with *Wolbachia*: the transovarian vertical transmission of *Culex pipiens* densovirus. Environ. Microbiol. 21 (9), 3284–3298. https://doi.org/10.1111/1462-2920.14511.

Amin, M., Amin, M., Amin, M., Sorour, M.K., Kasry, A., 2020. Comparing the binding interactions in the receptor binding domains of SARS-CoV-2 and SARS-CoV. J. Phys. Chem. Lett. 11 (12), 4897–4900. https://doi.org/10.1021/ACS.JPCLETT.0C01064.

Apte-Deshpande, A.D., Paingankar, M.S., Gokhale, M.D., Deobagkar, D.N., 2014. *Serratia odorifera* mediated enhancement in susceptibility of *Aedes aegypti* for chikungunya virus. Indian. J. Med. Res. 139 (5), 762. https://pmc.ncbi.nlm.nih.gov/articles/PMC4140042/.

Badgett, M.R., Auer, A., Carmichael, L.E., Parrish, C.R., Bull, J.J., 2002. Evolutionary dynamics of viral attenuation. J. Virol. 76 (20), 10524–10529. https://doi.org/10.1128/JVI.76.20.10524-10529.2002.

Bai, L., Wang, L., Vega-Rodríguez, J., Wang, G., Wang, S., 2019. A gut symbiotic bacterium *Serratia marcescens* renders mosquito resistance to Plasmodium infection through activation of mosquito immune responses. Front. Microbiol. 10, 1580. https://doi.org/10.3389/fmicb.2019.01580.

Bando, H., Okado, K., Guelbeogo, W.M., Badolo, A., Aonuma, H., Nelson, B., et al., 2013. Intra-specific diversity of *Serratia marcescens* in *Anopheles* mosquito midgut defines Plasmodium transmission capacity. Sci. Rep. 3 (1), 1641. https://doi.org/10.1038/srep01641.

Barletta, A.B.F., Nascimento-Silva, M.C.L., Talyuli, O.A.C., Oliveira, J.H.M., Pereira, L.O.R., Oliveira, P.L., et al., 2017. Microbiota activates IMD pathway and limits Sindbis infection in *Aedes aegypti*. Parasites & Vectors 10 (1), 103. https://doi.org/10.1186/s13071-017-2040-9.

Beebe, N.W., Pagendam, D., Trewin, B.J., Boomer, A., Bradford, M., Ford, A., et al., 2021. Releasing incompatible males drives strong suppression across populations of wild and *Wolbachia*-carrying *Aedes* aegypti in Australia. Proc. Natl Acad. Sci. U S A. 118 (41), e2106828118. https://doi.org/10.1073/PNAS.2106828118.

Bellone, R., Lequime, S., Jupille, H., Göertz, G.P., Aubry, F., Mousson, L., et al., 2020. Experimental adaptation of dengue virus 1 to *Aedes albopictus* mosquitoes by in vivo selection. Sci. Rep. 10 (1), 18404. https://doi.org/10.1038/s41598-020-75042-4.

Brackney, D.E., Beane, J.E., Ebel, G.D., 2009. RNAi targeting of west Nile virus in mosquito midguts promotes virus diversification. PLOS Pathog. 5 (7), e1000502. https://doi.org/10.1371/journal.ppat.1000502.

Cai, J.A., Christophides, G.K., 2024. Immune interactions between mosquitoes and microbes during midgut colonization. Curr. Opin. Insect Sci. 63 (11), 101195. https://doi.org/10.1016/j.cois.2024.101195.

Cansado-Utrilla, C., Saldaña, M.A., Golovko, G., Khanipov, K., Watson, R.K., Wild, A.L., et al., 2025. Mosquito host background impacts microbiome-Zika virus interactions in field- and laboratory-reared *Aedes aegypti*. Anim. Microbiome. 7 (1), 116. https://doi.org/10.1186/S42523-025-00482-0.

Capone, A., Ricci, I., Damiani, C., Mosca, M., Rossi, P., Scuppa, P., et al., 2013. Interactions between *Asaia*, Plasmodium and *Anopheles*: new insights into mosquito symbiosis and implications in Malaria symbiotic control. Parasites & Vectors 6 (1), 182. https://doi.org/10.1186/1756-3305-6-182.

Carvalho, F.A., Aitken, J.D., Vijay-Kumar, M., Gewirtz, A.T., 2012. Toll-like receptor-gut microbiota interactions: perturb at your own risk!. Annu. Rev. Physiol. 74 (74, 2012), 177–198. https://doi.org/10.1146/annurev-physiol-020911-153330.

Chandra, H., Sharma, K.K., Tuovinen, O.H., Sun, X., Shukla, P., 2021. Pathobionts: mechanisms of survival, expansion, and interaction with host with a focus on Clostridioides difficile. Gut Microbes 13 (1), 1979882. https://doi.org/10.1080/19490976.2021.1979882.

Char, A.B., Trammell, C.E., Fawcett, S., Chauhan, M., Debebe, Y., Céspedes, N., et al., 2024. Sustained antiviral insulin signaling during West Nile virus infection results in viral mutations. Front. Cell. Infect. Microbiol. 14, 1492403. https://doi.org/10.3389/fcimb.2024.1492403.

Chen, S., Bagdasarian, M., Walker, E.D., 2015. *Elizabethkingia anophelis*: molecular manipulation and interactions with mosquito hosts. Appl. Environ. Microbiol. 81 (6), 2233–2243. https://doi.org/10.1128/AEM.03733-14.

Ciota, A.T., Ehrbar, D.J., Matacchiero, A.C., Van Slyke, G.A., Kramer, L.D., 2013. The evolution of virulence of West Nile virus in a mosquito vector: implications for arbovirus adaptation and evolution. BMC Evolut. Biol. 13 (1), 71. https://doi.org/10.1186/1471-2148-13-71.

Coffey, L.L., Vasilakis, N., Brault, A.C., Powers, A.M., Tripet, F.R., Weaver, S.C., 2008. Arbovirus evolution in vivo is constrained by host alternation. Proc. Natl Acad. Sci. 105 (19), 6970–6975. https://doi.org/10.1073/pnas.0712130105.

Coon, K.L., Hegde, S., Hughes, G.L., 2022. Interspecies microbiome transplantation recapitulates microbial acquisition in mosquitoes. Microbiome. 10 (1), 58. https://doi.org/10.1186/S40168-022-01256-5.

Costa, R.L., Voloch, C.M., Schrago, C.G., 2012. Comparative evolutionary epidemiology of dengue virus serotypes. Infection, Genet. Evol. 12 (2), 309–314. https://doi.org/10.1016/j.meegid.2011.12.011.

Crawford, J.E., Clarke, D.W., Criswell, V., Desnoyer, M., Cornel, D., Deegan, B., et al., 2020. Efficient production of male *Wolbachia*-infected *Aedes aegypti* mosquitoes enables large-scale suppression of wild populations. Nat. Biotechnol. 38 (4), 482–492. https://doi.org/10.1038/s41587-020-0471-x.

Crespo-Bellido, A., Duffy, S., 2023. The how of counter-defense: viral evolution to combat host immunity. Curr. Opin. Microbiol. 74, 102320. https://doi.org/10.1016/j.mib.2023.102320.

de Bary, H.A., 1879. Die Erscheinung der Symbiose. Verlag Karl J. Trübner, Strassburg.

Dickson, L.B., Jiolle, D., Minard, G., Moltini-Conclois, I., Volant, S., Ghozlane, A., et al., 2017. Carryover effects of larval exposure to different environmental bacteria drive adult trait variation in a mosquito vector. Sci. Adv. 3 (8), e1700585. https://doi.org/10.1126/sciadv.1700585.

do Nascimento, R.M., Campolina, T.B., Chaves, B.A., Delgado, J.L.F., Godoy, R.S.M., Pimenta, P.F.P., et al., 2022. The influence of culture-dependent native microbiota in Zika virus infection in *Aedes aegypti*. Parasites & Vectors 15 (1), 57. https://doi.org/10.1186/S13071-022-05160-7.

Edenborough, K.M., Flores, H.A., Simmons, C.P., Fraser, J.E., 2021. Using *Wolbachia* to eliminate dengue: will the virus fight back? J. Virol. 95 (13). https://doi.org/10.1128/JVI.02203-20.

Finlay, B.B., McFadden, G., 2006. Anti-immunology: evasion of the host immune system by bacterial and viral pathogens. Cell 124 (4), 767–782. https://doi.org/10.1016/j.cell.2006.01.034.

Fitzmeyer, E.A., Gallichotte, E.N., Weger-Lucarelli, J., Kapuscinski, M.L., Abdo, Z., Pyron, K., et al., 2023. Loss of West Nile virus genetic diversity during mosquito infection due to species-dependent population bottlenecks. IScience 26 (10). https://doi.org/10.1016/j.isci.2023.107711.

Flores, H.A., de Bruyne, J.T., O'Donnell, T.B., Nhu, V.T., Giang, N.T., Trang, H.T.X., et al., 2020. Multiple Wolbachia strains provide comparative levels of protection against dengue virus infection in *Aedes aegypti*. PLOS Pathog. 16 (4), e1008433. https://doi.org/10.1371/JOURNAL.PPAT.1008433.

Ford, S.A., Allen, S.L., Ohm, J.R., Sigle, L.T., Sebastian, A., Albert, I., et al., 2019. Selection on *Aedes aegypti* alters *Wolbachia*-mediated dengue virus blocking and fitness. Nat. Microbiol. 4 (11), 1832–1839. https://doi.org/10.1038/s41564-019-0533-3.

Forrester, N.L., Guerbois, M., Seymour, R.L., Spratt, H., Weaver, S.C., 2012. Vector-borne transmission imposes a severe bottleneck on an RNA virus population. PLoS Pathog. 8 (9). https://doi.org/10.1371/journal.ppat.1002897.

French, S., Da Silva, R., Storm, J., Wastika, C.E., Cullen, I., Ten Have, M., et al., 2025. Exploiting venom toxins in paratransgenesis to prevent mosquito-borne disease. Parasites & Vectors. https://doi.org/10.1186/s13071-025-06663-9.

Giraud, É., Varet, H., Legendre, R., Sismeiro, O., Aubry, F., Dabo, S., et al., 2022. Mosquito-bacteria interactions during larval development trigger metabolic changes with carry-over effects on adult fitness. Mol. Ecol. 31 (5), 1444–1460. https://doi.org/10.1111/MEC.16327.

González, R., Elena, S.F., 2021. The interplay between the host microbiome and pathogenic viral infections. MBio. 12 (6). https://doi.org/10.1128/MBIO.02496-21;WGROUP:STRING:PUBLICATION.

Grubaugh, N.D., Ruckert, C., Armstrong, P.M., Bransfield, A., Anderson, J.F., Ebel, G.D., et al., 2016. Transmission bottlenecks and RNAi collectively influence tick-borne flavivirus evolution. Virus Evol. 2 (2). https://doi.org/10.1093/ve/vew033.

Grubaugh, N.D., Weger-Lucarelli, J., Murrieta, R.A., Fauver, J.R., Garcia-Luna, S.M., Prasad, A.N., et al., 2016. Genetic drift during systemic arbovirus infection of mosquito vectors leads to decreased relative fitness during host switching. Cell Host Microbe 19 (4), 481–492. https://doi.org/10.1016/j.chom.2016.03.002.

Guégan, M., Zouache, K., Démichel, C., Minard, G., Tran Van, V., Potier, P., Mavingui, P., Valiente Moro, C., 2018. The mosquito holobiont: fresh insight into mosquito-microbiota interactions. Microbiome. 6 (1), 49. https://doi.org/10.1186/s40168-018-0435-2.

Hegde, S., Brettell, L.E., Quek, S., Etebari, K., Saldaña, M.A., Asgari, S., et al., 2024. *Aedes aegypti* gut transcriptomes respond differently to microbiome transplants from field-caught or laboratory-reared mosquitoes. Environ. Microbiol. 26 (2), e16576. https://doi.org/10.1111/1462-2920.16576.

Holmes, E.C., 2003. Patterns of intra- and interhost nonsynonymous variation reveal strong purifying selection in dengue virus. J. Virol. 77 (20), 11296–11298. https://doi.org/10.1128/JVI.77.20.11296-11298.2003.

Hosen, M.E., Supti, S.J., Horwood, P.F., Sarker, S., 2026. Adaptive evolution of dengue virus: interplay between viral genetics and host immunity. Rev. Med. Virol. 36 (1). https://doi.org/10.1002/rmv.70082.

Hu, W., Gao, H., Cui, C., Wang, L., Wang, Y., Li, Y., et al., 2025. Harnessing engineered symbionts to combat concurrent malaria and arboviruses transmission. Nat. Commun. 16 (1), 2104. https://doi.org/10.1038/s41467-025-57343-2.

Indriani, C., Tanamas, S.K., Khasanah, U., Ansari, M.R., Rubangi, Tantowijoyo, W., et al., 2023. Impact of randomised w mel Wolbachia deployments on notified dengue cases and insecticide fogging for dengue control in Yogyakarta City. Glob. Health Action 16 (1). https://doi.org/10.1080/16549716.2023.2166650.

Jerzak, G., Bernard, K.A., Kramer, L.D., Ebel, G.D., 2005. Genetic variation in West Nile virus from naturally infected mosquitoes and birds suggests quasispecies structure and strong purifying selection. J. Gen. Virol. 86 (8), 2175–2183. https://doi.org/10.1099/vir.0.81015-0.

Joyce, J.D., Nogueira, J.R., Bales, A.A., Pittman, K.E., Anderson, J.R., 2011. Interactions between La crosse virus and bacteria isolated from the digestive tract of *Aedes albopictus* (Diptera: Culicidae). J. Med. Entomol. 48 (2), 389–394. https://doi.org/10.1603/ME09268.

Katak, R. de M., Cintra, A.M., Burini, B.C., Marinotti, O., Souza-Neto, J.A., Rocha, E.M., 2023. Biotechnological potential of microorganisms for mosquito population control and reduction in vector competence. Insects 14 (9), 718. https://doi.org/10.3390/INSECTS14090718.

Kuss, S.K., Best, G.T., Etheredge, C.A., Pruijssers, A.J., Frierson, J.M., Hooper, L.V., et al., 2011. Intestinal microbiota promote enteric virus replication and systemic pathogenesis. Science 334 (6053), 249–252. https://doi.org/10.1126/science.1211057.

Lauring, A.S., Andino, R., 2010. Quasispecies theory and the behavior of RNA viruses. PLOS Pathog. 6 (7), e1001005. https://doi.org/10.1371/journal.ppat.1001005.

Lequime, S., Fontaine, A., Ar Gouilh, M., Moltini-Conclois, I., Lambrechts, L., 2016. Genetic drift, purifying selection and vector genotype shape dengue virus intra-host genetic diversity in mosquitoes. PLOS Genet. 12 (6), e1006111. https://doi.org/10.1371/journal.pgen.1006111.

Lezcano, O.M., Fuhrmann, L., Ramakrishnan, G., Beerenwinkel, N., Huynen, M.A., Van Rij, R.P., 2023. Parallel evolution and enhanced virulence upon in vivo passage of an RNA virus in *Drosophila melanogaster*. Virus Evol. 9 (2). https://doi.org/10.1093/ve/vead074.

Li, M., Zhou, Y., Cheng, J., Wang, Y., Lan, C., Shen, Y., 2024. Response of the mosquito immune system and symbiotic bacteria to pathogen infection. Parasites & Vectors 17 (1), 69. https://doi.org/10.1186/S13071-024-06161-4.

Lim, J.T., Chong, C.-S., Chang, C.-C., Mailepessov, D., Dickens, B., Lai, Y.L., Project Wolbachia–Singapore Consortium, 2026. Dengue suppression by male *Wolbachia*-infected mosquitoes. N. Engl. J. Med. https://doi.org/10.1056/NEJMOA2503304.

Lu, P., Bian, G., Pan, X., Xi, Z., 2012. *Wolbachia* induces density-dependent inhibition to dengue virus in mosquito cells. PLoS Negl. Trop. Dis. 6 (7), e1754. https://doi.org/10.1371/journal.pntd.0001754.

Martinez, J., Bruner-Montero, G., Arunkumar, R., Smith, S.C.L., Day, J.P., Longdon, B., et al., 2019. Virus evolution in *Wolbachia*- infected *Drosophila*. Proc. Biol. Sci. 286 (1914). https://doi.org/10.1098/rspb.2019.2117.

Mazel-Sanchez, B., Yildiz, S., Schmolke, M., 2019. Ménage à trois: virus, host, and microbiota in experimental infection models. Trends Microbiol. 27 (5), 440–452. https://doi.org/10.1016/j.tim.2018.12.004.

Miller, M.R., White, A., Boots, M., 2005. The evolution of host resistance: tolerance and control as distinct strategies. J. Theor. Biol. 236 (2), 198–207. https://doi.org/10.1016/j.jtbi.2005.03.005.

Mongelli, V., Lequime, S., Kousathanas, A., Gausson, V., Blanc, H., Nigg, J., et al., 2022. Innate immune pathways act synergistically to constrain RNA virus evolution in *Drosophila melanogaster*. Nat. Ecol. & Evolution. 6 (5), 565–578. https://doi.org/10.1038/s41559-022-01697-z.

Muturi, E.J., Dunlap, C., Ramirez, J.L., Rooney, A.P., Kim, C.H., 2018. Host blood-meal source has a strong impact on gut microbiota of *Aedes aegypti*. FEMS Microbiol. Ecol. 95 (1), 213. https://doi.org/10.1093/femsec/fiy213.

Nazni, W.A., Hoffmann, A.A., NoorAfizah, A., Cheong, Y.L., Mancini, M.V., Golding, N., et al., 2019. Establishment of *Wolbachia* Strain *w*AlbB in Malaysian populations of *Aedes aegypti* for dengue control. Curr. Biol. 29 (24), 4241–4248.e5. https://doi.org/10.1016/j.cub.2019.11.007.

Noval, M.G., Rodriguez-Rodriguez, B.A., Rangel, M.V., Stapleford, K.A., 2019. Evolution-driven attenuation of alphaviruses highlights key glycoprotein determinants regulating viral infectivity and dissemination. Cell Rep. 28 (2), 460–471.e5. https://doi.org/10.1016/j.celrep.2019.06.022.

Nysten, J., Sofras, D., Van Dijck, P., 2024. One species, many faces: the underappreciated importance of strain diversity. PLOS Pathog. 20 (1), e1011931. https://doi.org/10.1371/journal.ppat.1011931.

Pan, X., Zhou, G., Wu, J., Bian, G., Lu, P., Raikhel, A.S., et al., 2012. *Wolbachia* induces reactive oxygen species (ROS)-dependent activation of the toll pathway to control dengue virus in the mosquito *Aedes aegypti*. Proc. Natl. Acad. Sci. U. S. A. 109 (1). https://doi.org/10.1073/pnas.1116932108.

Pawlotsky, J.M., 2016. Hepatitis C virus resistance to direct-acting antiviral drugs in interferon-free regimens. Gastroenterology 151 (1), 70–86. https://doi.org/10.1053/J.GASTRO.2016.04.003.

Perdomo, H.D., Khorramnejad, A., Chen, F., Lozada-Chávez, A.N., Bonizzoni, M., 2026. Tolerance to persistent cell-fusing agent virus infection in Aedes aegypti mosquitoes correlates with maintenance of lipid reserves. BMC Biol. 24 (1). https://doi.org/10.1186/S12915-026-02553-3.

Pujhari, S., Hughes, G.L., Pakpour, N., Suzuki, Y., Rasgon, J.L., 2023. *Wolbachia*-induced inhibition of O′nyong nyong virus in Anopheles mosquitoes is mediated by Toll signaling and modulated by cholesterol. BioRxiv, 2023.05.31.543096. https://doi.org/10.1101/2023.05.31.543096.

Ramirez, J.L., Souza-Neto, J., Cosme, R.T., Rovira, J., Ortiz, A., Pascale, J.M., et al., 2012. Reciprocal tripartite interactions between the *Aedes aegypti* midgut microbiota, innate immune system and dengue virus influences vector competence. PLoS Negl. Trop. Dis. 6 (3), 1–11. https://doi.org/10.1371/journal.pntd.0001561.

Ratcliffe, N.A., Furtado Pacheco, J.P., Dyson, P., Castro, H.C., Gonzalez, M.S., Azambuja, P., et al., 2022. Overview of paratransgenesis as a strategy to control pathogen transmission by insect vectors. Parasites & Vectors 15 (1), 112. https://doi.org/10.1186/s13071-021-05132-3.

Robinson, C.M., Jesudhasan, P.R., Pfeiffer, J.K., 2014. Bacterial lipopolysaccharide binding enhances virion stability and promotes environmental fitness of an enteric virus. Cell Host Microbe. 15 (1), 36–46. https://doi.org/10.1016/j.chom.2013.12.004.

Santos, N.A.C. dos, Magi, F.N., Andrade, A.O., Bastos, A. da S., Pereira, S., et al., 2022. Assessment of antibiotic treatment on *Anopheles darlingi* survival and susceptibility to plasmodium vivax. Front. Microbiol. 13, 971083. https://doi.org/10.3389/fmicb.2022.971083.

Saraiva, R.G., Fang, J., Kang, S., Angleró-Rodríguez, Y.I., Dong, Y., Dimopoulos, G., 2018. Aminopeptidase secreted by *Chromobacterium* sp. Panama inhibits dengue virus infection by degrading the E protein. PLOS Negl. Trop. Dis. 12 (4), e0006443. https://doi.org/10.1371/journal.pntd.0006443.

Schuh, A.J., Ward, M.J., Leigh Brown, A.J., Barrett, A.D.T., 2014. Dynamics of the emergence and establishment of a newly dominant genotype of Japanese encephalitis virus throughout Asia. J. Virol. 88 (8), 4522–4532. https://doi.org/10.1128/jvi.02686-13.

Song, X., Zhou, H., Wang, J., 2025. Cell wall components of gut commensal bacteria stimulate peritrophic matrix formation in malaria vector mosquitoes through activation of the IMD pathway. PLOS Biol. 23 (1), e3002967. https://doi.org/10.1371/JOURNAL.PBIO.3002967.

Souza, R.S., Virginio, F., Riback, T.I.S., Suesdek, L., Barufi, J.B., Genta, F.A., 2019. Microorganism-based larval diets affect mosquito development, size and nutritional reserves in the yellow fever mosquito *Aedes aegypti* (Diptera: Culicidae). Front. Physiol. 10, 424594. https://doi.org/10.3389/fphys.2019.00152.

Stevens, E.J., Bates, K.A., King, K.C., 2021. Host microbiota can facilitate pathogen infection. PLOS Pathog. 17 (5), e1009514. https://doi.org/10.1371/journal.ppat.1009514.

Sun, X., Wang, Y., Yuan, F., Zhang, Y., Kang, X., Sun, J., et al., 2024. Gut symbiont-derived sphingosine modulates vector competence in *Aedes* mosquitoes. Nat. Commun. 15 (1), 8221. https://doi.org/10.1038/s41467-024-52566-1.

Terradas, G., McGraw, E.A., 2017. *Wolbachia*-mediated virus blocking in the mosquito vector *Aedes aegypti*. Curr. Opin. Insect Sci. 22, 37–44. https://doi.org/10.1016/J.COIS.2017.05.005.

Thi Hue Kien, D., Edenborough, K., Da Silva Goncalves, D., Thuy Vi, T., Casagrande, E., Thi Le Duyen, H., et al., 2023. Genome evolution of dengue virus serotype 1 under selection by *Wolbachia pipientis* in *Aedes aegypti* mosquitoes. Virus Evol. 9 (1). https://doi.org/10.1093/VE/VEAD016.

Tsetsarkin, K.A., Chen, R., Leal, G., Forrester, N., Higgs, S., Huang, J., et al., 2011. Chikungunya virus emergence is constrained in Asia by lineage-specific adaptive landscapes. Proc. Natl. Acad. Sci. U S A. 108 (19), 7872–7877. https://doi.org/10.1073/pnas.1018344108.

Utarini, A., Indriani, C., Ahmad, R.A., Tantowijoyo, W., Arguni, E., Ansari, M.R., et al., 2021. Efficacy of *Wolbachia*-infected mosquito deployments for the control of dengue. N. Engl. J. Med. 384 (23), 2177–2186. https://doi.org/10.1056/NEJMOA2030243.

Vasilakis, N., Deardorff, E.R., Kenney, J.L., Rossi, S.L., Hanley, K.A., Weaver, S.C., 2009. Mosquitoes put the brake on arbovirus evolution: experimental evolution reveals slower mutation accumulation in mosquito than vertebrate cells. PLOS Pathog. 5 (6), e1000467. https://doi.org/10.1371/journal.ppat.1000467.

Vignuzzi, M., Stone, J.K., Arnold, J.J., Cameron, C.E., Andino, R., 2005. Quasispecies diversity determines pathogenesis through cooperative interactions in a viral population. Nature 439 (7074), 344–348. https://doi.org/10.1038/nature04388.

Weaver, S.C., Forrester, N.L., Liu, J., Vasilakis, N., 2021. Population bottlenecks and founder effects: implications for mosquito-borne arboviral emergence. Nat. Rev. Microbiol. 19 (3), 184–195. https://doi.org/10.1038/s41579-020-00482-8.

Weiss, B., Aksoy, S., 2011. Microbiome influences on insect host vector competence. Trends Parasitol. 27 (11), 514–522. https://doi.org/10.1016/j.pt.2011.05.001.

Wu, P., Sun, P., Nie, K., Zhu, Y., Shi, M., Xiao, C., et al., 2019. A gut commensal bacterium promotes mosquito permissiveness to arboviruses. Cell Host Microbe 25 (1), 101–112.e5. https://doi.org/10.1016/j.chom.2018.11.004.

Wu, P., Yu, X., Wang, P., Cheng, G., 2019. Arbovirus lifecycle in mosquito: acquisition, propagation and transmission. Expert. Rev. Mol. Med. 21, e1. https://doi.org/10.1017/erm.2018.6.

Xi, Z., Ramirez, J.L., Dimopoulos, G., 2008. The *Aedes aegypti* toll pathway controls dengue virus infection. PLoS Pathog. 4 (7), e1000098. https://doi.org/10.1371/journal.ppat.1000098.

Yu, X., Tong, L., Zhang, L., Yang, Y., Xiao, X., Zhu, Y., et al., 2022. Lipases secreted by a gut bacterium inhibit arbovirus transmission in mosquitoes. PLOS Pathog. 18 (6), e1010552. https://doi.org/10.1371/journal.ppat.1010552.

Yu, S., Wang, P., Qin, J., Zheng, H., Wang, J., Liu, T., et al., 2020. *Bacillus sphaericus* exposure reduced vector competence of *Anopheles dirus* to Plasmodium yoelii by upregulating the Imd signaling pathway. Parasites Vectors 13 (1), 446. https://doi.org/10.1186/S13071-020-04321-W.

Zhang, L., Wang, D., Shi, P., Li, J., Niu, J., Chen, J., et al., 2024. A naturally isolated symbiotic bacterium suppresses flavivirus transmission by Aedes mosquitoes. Science 384 (6693). https://doi.org/10.1126/science.adn9524.

Zhao, S.Y., Hughes, G.L., Coon, K.L., 2023. A cryopreservation method to recover laboratory- and field-derived bacterial communities from mosquito larval habitats. PLOS Negl. Trop. Dis. 17 (4), e0011234. https://doi.org/10.1371/JOURNAL.PNTD.0011234.

CHAPTER NINE

Evolutionary constraints and design principles in CRISPR-based mosquito biocontrol

Eddie Hill, Robyn Raban, and Omar S. Akbari*
Department of Cell and Developmental Biology, School of Biological Sciences, University of California, San Diego, La Jolla, CA, United States
*Corresponding author. e-mail address: oakbari@ucsd.edu

Contents

Abstract

CRISPR-based genetic biocontrol technologies have emerged as powerful tools that are poised to control mosquito populations and reduce the transmission of vector-borne diseases. Non-gene drive CRISPR systems offer self-limiting and controllable approaches for population suppression, while gene drives, which bias inheritance to spread engineered traits, provide the potential for scalable and sustained intervention following smaller releases. It is increasingly recognized that the performance of these systems is fundamentally shaped by evolutionary and ecological processes, and that their success depends on how effectively these constraints are incorporated into system design. Here, we synthesize emerging principles underlying the evolutionary dynamics of CRISPR-based biocontrol technologies, including resistance evolution, standing genetic variation, and spatial population structure. We further outline strategies to enhance evolutionary robustness, such as multiplexed targeting and constrained genomic site selection, as well as design approaches that improve controllability and confinement, including confinable or self-limiting architectures. We

Advances in Insect Physiology, Volume 70
ISSN 0065-2806, https://doi.org/10.1016/bs.aiip.2026.05.001

also discuss how advances in population genomics, quantitative modeling, and data-driven approaches are improving predictions of gene drive behavior under realistic ecological conditions relevant to mosquito vectors. Together, these insights underscore the importance of integrating evolutionary principles into the design of effective, controllable, and durable genetic biocontrol strategies for mosquito disease control.

1. Introduction

Since the earliest stages of human civilization, pest species—particularly disease-transmitting arthropods—have profoundly shaped human populations (Winegard, 2020). Among these, mosquitoes of the *Aedes* and *Anopheles* genera constitute some of the most significant global health threats, owing to their role in transmitting pathogens that cause malaria, dengue, and other arboviral diseases. Efforts to control mosquito populations have a long history, ranging from chemical interventions to ecological management strategies. Among these, genetic approaches have emerged as particularly promising, offering the potential for species-specific, sustainable, and scalable solutions for mosquito population control.

Early genetic biocontrol strategies were developed in the pretransgenic era, most notably the sterile insect technique (SIT), in which irradiated sterile males are released to mate with wild females, resulting in nonviable offspring (Knipling, 1955; Dyck et al., 2005). This method achieved remarkable success in certain contexts, such as the eradication of the New World screwworm from North America (Wyss, 2000). SIT has also been adapted for mosquito control, including pilot and operational programs targeting *Anopheles*, *Aedes*, and *Culex* species (Lutrat et al., 2019). However, its large-scale effectiveness has been limited in many species due to challenges in sex-sorting males and females, reduced mating competitiveness of sterilized insects resulting from irradiation and adult handling, and the need for continuous large-scale releases (Wyss, 2000; Robinson et al., 2009). These limitations motivated the search for alternative genetic control strategies, including both self-sustaining and self-limiting systems (Alphey, 2014).

The development of programmable genome editing technologies, particularly CRISPR–Cas systems (Jinek et al., 2012), has transformed the field of genetic biocontrol by enabling precise, efficient, and scalable manipulation of insect genomes (Raban et al., 2023; Champer et al., 2016; Esvelt et al., 2014). CRISPR-based approaches allow targeted DNA cleavage at user-defined loci, which can be repaired through endogenous pathways such as homology-directed repair (HDR), providing a versatile platform for engineering both conventional transgenes and inheritance-biasing systems.

Within this framework, the conceptual foundation for inheritance-biasing systems lies in naturally occurring "selfish genetic elements," collectively referred to as gene drive systems, which bias their own inheritance and enable propagation through populations even when they impose fitness costs on their hosts. Gene drives occur widely in nature and include diverse mechanisms such as meiotic drive elements, toxin–antidote systems, transposable elements, and homing endonucleases (Burt and Trivers, 2008). Their ability to propagate through populations despite selective disadvantages has inspired the development of synthetic gene drive systems to control or modify wild populations.

Prior to the advent of CRISPR, early efforts to engineer genetic biocontrol systems—often inspired by naturally occurring selfish genetic elements—gave rise to strategies such as underdominance-based systems (Akbari et al., 2013; Buchman et al., 2018) and toxin–antidote elements like Medea (Akbari et al., 2014; Buchman et al., 2018; Chen et al., 2007), which have thus far been developed and validated in *Drosophila* model systems, but may provide a foundation for future adaptation to mosquito species. The advent of CRISPR further enabled the development of engineered underdominance systems, including Synthetic Postzygotic Isolation Exploiting CRISPR-based Incompatibilities for Engineering Species (SPECIES) and Engineered Genetic Incompatibility (EGI), which use programmable genome editing to create reproductive barriers between modified and wild-type populations, also first demonstrated in *Drosophila* (Buchman et al., 2021; Maselko et al., 2020), and represent promising candidates for extension into mosquito vectors with further development. Building on its programmability, CRISPR-based homing-based gene drive systems (CHGDs) emerged as a particularly influential class of self-perpetuating technologies. CHGDs leverage targeted DNA cleavage and HDR to copy themselves into homologous chromosomes, thereby achieving super-Mendelian inheritance (Champer et al., 2016; Esvelt et al., 2014). This molecular simplicity, combined with high programmability, has accelerated the development of CHGD technologies for applications such as population suppression and population modification in disease vectors like mosquitoes.

In mosquitoes, CHGDs have demonstrated the capacity to either reduce population size by targeting genes essential for fertility or viability (Kyrou et al., 2018), or to modify populations by introducing anti-pathogen effector genes that block disease transmission (Raban et al., 2026; Carballar-Lejarazú et al., 2020). For example, engineered systems can impair female fertility or skew sex ratios, leading to population collapse under certain experimental conditions. Alternatively, gene drives could be used to spread traits that reduce vector competence, offering a strategy to control deadly diseases such as malaria, dengue, and chikungunya without eliminating the population. Comparable genetic control strategies in agricultural pests have focused more heavily on population suppression approaches, highlighting a key distinction in application goals between mosquito and non-mosquito systems.

Despite their promise, CHGDs operate within complex evolutionary and ecological landscapes. Their spread depends on a balance between drive efficiency and associated fitness costs, and they are subject to resistance evolution, ecological variability, and population structure (Noble et al., 2017). Error-prone DNA repair pathways can generate mutations at target sites, producing alleles that evade further cleavage and compromise drive effectiveness (Hammond et al., 2017). Additionally, environmental factors and standing genetic variation can influence drive dynamics, making predictions of field performance challenging (North et al., 2019; Beaghton et al., 2016). These challenges are particularly well characterized in mosquito systems, where extensive laboratory and modeling studies have provided insight into spatial spread, resistance dynamics, and ecological constraints. As CHGD technologies advance toward potential real-world applications, integrating molecular understanding with population and evolutionary dynamics will be essential for predicting performance and guiding design. Ultimately, these factors will shape the efficacy, safety, and appropriate role of gene drives in next-generation strategies for controlling vector-borne diseases and other global challenges.

2. Non-gene drive CRISPR technologies

Although CHGDs have advanced rapidly in laboratory settings, no transgene-based gene drive system has yet been deployed in the wild due to unresolved technical, ecological, political, social, and regulatory challenges (Committee on Gene Drive Research in Non-Human Organisms:

Recommendations for Responsible Conduct et al., 2016). In contrast, a range of self-limiting genetic biocontrol strategies—many of which predate CRISPR—have already demonstrated effectiveness in field applications. Classical approaches such as the SIT and the release of insects carrying dominant lethal alleles (RIDL) have achieved substantial reductions in mosquito vector populations (Alphey et al., 2010; Bouyer et al., 2020). Variants such as female-specific RIDL (fsRIDL), which generate nonviable or flightless females through sex-specific regulatory elements, coupled with tetracycline-repressible expression systems to enable conditional control during mass rearing, have further improved efficiency by selectively targeting the reproductive capacity of mosquito populations (Spinner et al., 2022; Carvalho et al., 2015; Fu et al., 2010).

Extending these foundational approaches, CRISPR-based technologies have enabled a new generation of self-limiting systems that leverage precise, programmable DNA cleavage rather than classical transgenic or homology-based mechanisms. These approaches exploit the precision and programmability of CRISPR to induce targeted genetic disruption in a controlled, non-propagating manner, offering scalable alternatives to CHGDs for population suppression. One prominent example is the precision-guided sterile insect technique (pgSIT), which leverages CRISPR-mediated gene disruption to generate sterile males (Kandul et al., 2019b; Li et al., 2021; Apte et al., 2024; Li et al., 2024; Kandul et al., 2022) (Fig. 1A).

pgSIT has been developed and validated in multiple insect species, including *Aedes* and *Anopheles* mosquitoes demonstrating its portability across species (Kandul et al., 2019b; Li et al., 2021; Apte et al., 2024; Li et al., 2024; Kandul et al., 2022). The system relies on crossing two homozygous engineered strains—one expressing Cas9 and the other guide RNAs (gRNAs)—to induce genome editing early in development. This results in widespread biallelic disruption of genes essential for female viability and male fertility, effectively producing functionally dominant phenotypes in a single generation (Kandul et al., 2019b; Li et al., 2021; Apte et al., 2024; Li et al., 2024; Kandul et al., 2022). This mechanism, termed "lethal mosaicism," arises from a combination of maternal deposition and zygotic expression of CRISPR components, promoting embryonic cleavage and target gene disruption. Depending on target gene selection, female lethality can occur at embryonic or early larval stages, reducing rearing costs and improving operational efficiency. Because the CRISPR components are maintained in separate parental lines, Cas9 and gRNAs may be expressed individually, but genome editing activity is only induced

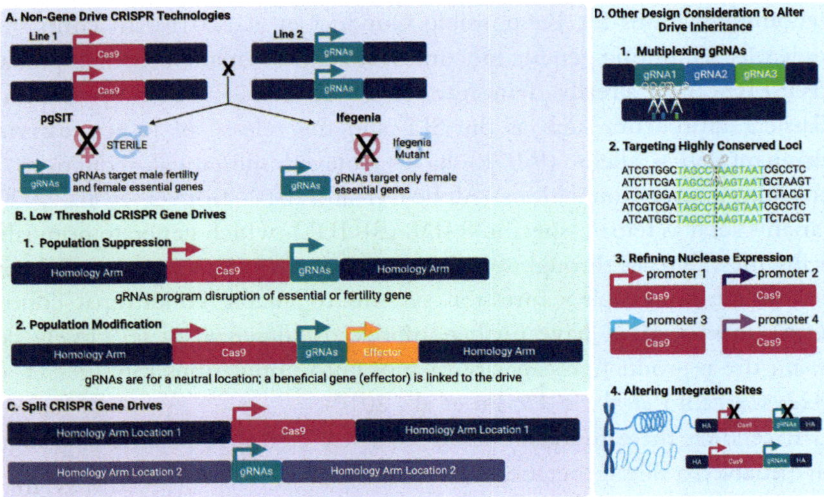

Fig. 1 Overview of key architectures of CRISPR-based genetic control technologies for mosquito vectors. (A) Non–gene drive CRISPR systems, such as precision-guided sterile insect technique (pgSIT) and Ifegenia, are binary designs comprising one line expressing Cas9 and another expressing gRNAs targeting genes essential for male fertility and female viability (pgSIT) or female-specific survival (Ifegenia). Upon crossing, these systems generate sterile male mosquitoes (pgSIT) or males carrying female-lethal mutations (Ifegenia). Temperature-inducible variants of pgSIT have also been developed using heat-shock promoters (Kandul et al., 2021) or temperature-sensitive nucleases such as Cas12a (Nguyen et al., 2025), enabling consolidation into a single strain in insect model systems; however, these temperature-inducible configurations have not yet been demonstrated in mosquitoes. (B) Low-threshold (self-propagating) CRISPR-based homing gene drives (CHGDs) encode tightly linked Cas9 and gRNA components, ensuring co-inheritance and enabling spread through mosquito populations over time and space. (B1) Population suppression drives target genes essential for fertility or viability, often in a sex-specific manner, to reduce or eliminate populations. (B2) Population modification drives introduce effector genes that reduce vector competence, such as molecules that block pathogen transmission. (C) Split CHGDs separate Cas9 and gRNA components onto different genetic elements, preventing guaranteed co-inheritance. Drive activity occurs only when both components are present, resulting in more spatially and temporally confinable dynamics compared to low-threshold drives. (D) Across both low-threshold and split architectures in mosquito systems, multiple design refinements have been developed to enhance evolutionary stability. These include multiplexed gRNAs targeting multiple sites, selection of highly conserved and functionally constrained loci, optimization of Cas9 expression, and evaluation of alternative genomic insertion sites. Together, these strategies aim to improve homing efficiency and reduce the formation and spread of resistance alleles arising from error-prone DNA repair in mosquito populations. *Created in BioRender.*

when these components are combined in the same individual following crossing, rendering each parental line effectively genetically inactive ("off") with respect to editing. This functional separation minimizes unintended editing during strain maintenance and enhances long term strain stability during mass rearing. In contrast to gene drives, in which CRISPR editing is continuously active and can impose strong selective pressure for resistance alleles, pgSIT operates transiently in a single generation, thereby reducing the likelihood of resistance evolution.

However, this binary design introduces operational constraints, as large-scale implementation requires the parallel maintenance of separate strains and strict prevention of cross-contamination (Kandul et al., 2019a). These requirements can increase infrastructure complexity, labor demands, and cost, representing a trade-off relative to single-strain systems. Maintenance of homozygous Cas9- and gRNA-expressing parental lines may impose fitness costs arising from transgene expression, genomic insertion effects, off-target activity, or unintended somatic expression, which could reduce fecundity or colony stability over time. While off-target cleavage rates are generally low in well-optimized systems, they are unlikely to be zero and may contribute to subtle fitness or developmental effects. Accordingly, careful promoter selection, genomic insertion site optimization, and colony management strategies are critical for minimizing these effects and maintaining high productivity in mass-rearing settings.

Despite these logistical and biological considerations, the binary architecture also confers potential advantages. Released individuals are F1 hybrids, which may reduce some inbreeding-associated costs and potentially benefit from increased heterosis (hybrid vigor) relative to true-breeding parental strains; however, the extent to which this translates into improved field performance or mating competitiveness remains species- and construct-dependent and must be evaluated empirically. Additionally, pgSIT eliminates the need for irradiation, thereby avoiding radiation-induced fitness costs, although this represents a shift in the source of fitness effects rather than their complete elimination. An additional unresolved issue is how sterile-male designs targeting genes such as β2-tubulin or *zpg* influence post-mating dynamics. Because these targets can impair fertility through defective spermatogenesis or sperm function, females that mate with pgSIT males may retain the capacity to remate, and sperm competition under non-zero multiple-mating rates could reduce suppression efficiency. Over longer timescales, variation in female remating propensity could itself become a substrate for selection, representing a potential behavioral route of resistance.

However, in *An. gambiae*, pgSIT males have been shown to successfully transfer mating plugs to females, which are known to reduce female remating, suggesting that post-mating refractoriness may be at least partially maintained in this system (Apte et al., 2024). Nevertheless, the extent to which these dynamics vary across species, ecological contexts, and genetic targets remains an important area for further study.

Importantly, as a self-limiting (i.e., non-heritable) system, pgSIT males are reproductive "dead ends" that do not persist in the environment. Because its effects are confined to a single generation, released sterile males compete with wild males for mates, reducing viable offspring and driving progressive population suppression over time and space, although ecological and behavioral adaptations may still influence long-term outcomes. The modular design of pgSIT enables flexible targeting of genes and phenotypes, facilitating adaptation across species and ecological contexts. Multiplexed gRNA strategies can further enhance penetrance and reduce incomplete disruption, and experimental studies across multiple species demonstrate the robustness and adaptability of this approach (Kandul et al., 2019b; Li et al., 2021; Apte et al., 2024; Li et al., 2024; Kandul et al., 2022).

Beyond its molecular design, pgSIT also offers practical advantages for large-scale implementation. Similar to fsRIDL, pgSIT can be deployed through the release of eggs rather than adults, reducing logistical handling requirements, lowering costs, and facilitating scalable distribution. This approach is most readily applicable to *Aedes* species, where egg collection, storage, and distribution are well established, but may be more challenging to implement in *Anopheles* mosquitoes due to differences in egg biology and handling constraints. This feature can simplify large-scale release logistics in urban and peri-urban environments where egg-based deployment is feasible. In contrast to traditional SIT, where each released sterile male must be individually produced and sorted (approximately a 1:1 production-to-release ratio), pgSIT introduces an amplification step that scales with the fecundity of the parental female, as each sorted female can generate hundreds of sterile male progeny through a single cross. It should be noted, however, that pgSIT does not inherently increase production efficiency relative to other true-breeding engineered systems, as these approaches rely on mass rearing; rather, its advantages lie in system stability, deployment flexibility and the ability to generate sterile males without irradiation or release-stage sex sorting. In this context, pgSIT represents one of the only scalable genetic approaches capable of producing sterile males without relying on radiation, which can induce widespread chromosomal damage and associated fitness costs, offering a potentially more consistent and tunable alternative to conventional sterilization methods.

Importantly, pgSIT could be integrated into existing SIT infrastructure, leveraging established mass-rearing and release frameworks to support deployment. Nevertheless, implementation requires efficient sex-sorting during the production phase to establish parental crosses. While this shifts complexity upstream, several scalable sex-sorting approaches have been developed (Papathanos et al., 2009), including robotic pupal sorters (Mamai et al., 2024), Complex Object Parametric Analyzer and Sorter (COPAS) platforms (Marois, 2024), and AI-assisted vision-based systems (Crawford et al., 2020). Fluorescence-based systems such as SEPARATOR (Weng et al., 2023, 2024) have also been demonstrated in mosquitoes. However, integration of these approaches with pgSIT is not necessarily straightforward, as compatibility depends on genomic insertion sites, marker configurations, and construct interactions. For example, SEPARATOR-based strategies may require the development of new constructs to avoid marker overlap and subsequent re-validation of pgSIT performance in the modified genetic background. In addition, achieving the throughput required for large-scale deployment may necessitate the development of faster, high-capacity sorting platforms optimized for fluorescence-based systems such as SEPARATOR. Similarly, cisgenic genetic sexing strain (CGSS) approaches (Davydova et al., 2026), while conceptually promising, have not yet been developed in mosquito species and would require substantial adaptation and testing prior to integration with pgSIT systems.

As such, upstream sex sorting remains an important engineering and validation challenge rather than a fully resolved component of current pgSIT implementations. In mosquitoes, where sexual dimorphism is often limited at early developmental stages, this constraint further emphasizes the need for robust, scalable sorting technologies. Additional considerations include species-specific optimization of target genes and achieving sufficient release ratios to ensure effective population suppression. Together, these factors highlight how system design shapes both operational performance and evolutionary robustness in self-limiting genetic control strategies.

To further improve scalability and reduce the need to sex-sort the parental cross, temperature-inducible pgSIT (TI-pgSIT) has been developed as an extension of the pgSIT framework, in which Cas endonuclease expression is placed under heat-inducible control within a single genetic background (Kandul et al., 2021). In this system, gRNAs are constitutively expressed, while nuclease activity is triggered by temperature elevation. At permissive temperatures, the strain remains viable and fertile, facilitating colony maintenance, whereas induction activates targeted disruption of genes required for female viability and male fertility. This design conceptually reduces production complexity by eliminating the need to maintain and cross separate parental lines, thereby simplifying colony management. Importantly, TI-pgSIT has thus far been demonstrated only in *Drosophila melanogaster*, where it serves as a proof-of-concept for consolidating the binary pgSIT system into a single inducible strain, and has not yet been validated in mosquito vectors. A Cas12a-based TI-pgSIT variant has also been reported in *D. melanogaster* (Nguyen et al., 2025), highlighting the potential flexibility in nuclease choice for further system optimization. Future work will be required to translate TI-pgSIT into mosquito systems, where its ability to simplify production workflows could have significant implications for large-scale vector control.

In addition to sterile male strategies, CRISPR has enabled the development of non-driving female elimination systems, such as Inherited Female Elimination by Genetically Encoded Nucleases Interrupting Alleles (Ifegenia) (Smidler et al., 2023; Smidler and Akbari, 2025) (Fig. 1A). This self-limiting binary system employs separate Cas9- and gRNA-expressing lines that, when combined, induce targeted disruption of genes essential for female development, resulting in female-specific lethality. Males remain viable and can transmit the genetic components to subsequent generations, enabling sustained suppression through repeated releases without relying on gene drive mechanisms. As currently implemented, however, Ifegenia does not eliminate the need for upstream sex sorting or controlled parental crosses, because the system is maintained in separate lines rather than as a single true-breeding strain. In this respect, its production logic is similar to that of binary pgSIT systems, and any reduction in sorting requirements would depend on future consolidation into a stable true-breeding configuration. Collectively, these non–gene drive CRISPR-based technologies represent a versatile and increasingly mature class of genetic biocontrol tools. While they require repeated releases and do not spread autonomously like gene drives, they offer important advantages in terms of stability, controllability, reversibility, and potentially simpler regulatory pathways. As such, they are likely to play a critical role both as standalone interventions and as complementary strategies in integrated mosquito control programs. In contrast, gene drives are designed to spread through populations and therefore face distinct evolutionary dynamics and challenges that must be carefully considered.

3. Mechanisms and architectures of CHGDs

CHGDs are engineered genetic systems that bias inheritance by exploiting the cellular DNA repair machinery (Esvelt et al., 2014; Champer et al., 2016). At their core, these systems encode a programmable endonuclease (typically Cas9) and one or more gRNAs that target a specific genomic locus. Following Cas9-mediated double-strand break formation at the homologous wild-type allele, the cell can repair the lesion through HDR, using the drive-containing chromosome as a template. This process results in the copying—or "homing"—of the CHGD cassette into the homologous chromosome, effectively converting heterozygous cells into homozygotes and enabling super-Mendelian inheritance across generations.

The efficiency of this process depends on several factors, including the timing and tissue specificity of Cas9 expression, the balance between HDR and error-prone repair pathways such as end joining (EJ), and the genomic context of the target site (Fig. 2). Germline-restricted expression of Cas9 is typically favored, as it maximizes HDR rates while minimizing somatic mutation and associated fitness costs (Hammond et al., 2016; Kandul et al., 2020; Champer et al., 2018). Conversely, cleavage events occurring outside of the germline often lead to EJ-mediated repair, generating mutations that can impede drive propagation. In a recent meta-analysis of CHGDs across multiple taxa, Cas9 promoter choice had surprisingly limited predictive power for drive performance, despite being a major focus in the field (Verkuijl et al., 2026). This result was attributed to confounding effects of species differences, laboratory practices, and other design variables, as well as the fact that overall performance is shaped more by the combined influence of multiple design elements (e.g., insertion site and construct architecture) than by expression timing alone. Similar trends were observed when analyses were restricted to mosquito systems.

Recent work has diversified the architectures of CHGDs for mosquito control, reflecting efforts to improve efficiency, limit resistance, and tailor population-level outcomes. Classical CHGDs targeting fertility or viability genes disrupt essential loci—often those required for female fertility or development—thereby suppressing mosquito populations (Kyrou et al., 2018; Hammond et al., 2016; Burt, 2003) (Fig. 1B). By targeting haplosufficient genes such that only individuals lacking functional copies are affected, these systems can impose substantial reproductive load. However, they remain vulnerable to the emergence of resistance alleles that restore gene function while evading cleavage (Burt and Crisanti, 2018; Hammond et al., 2017).

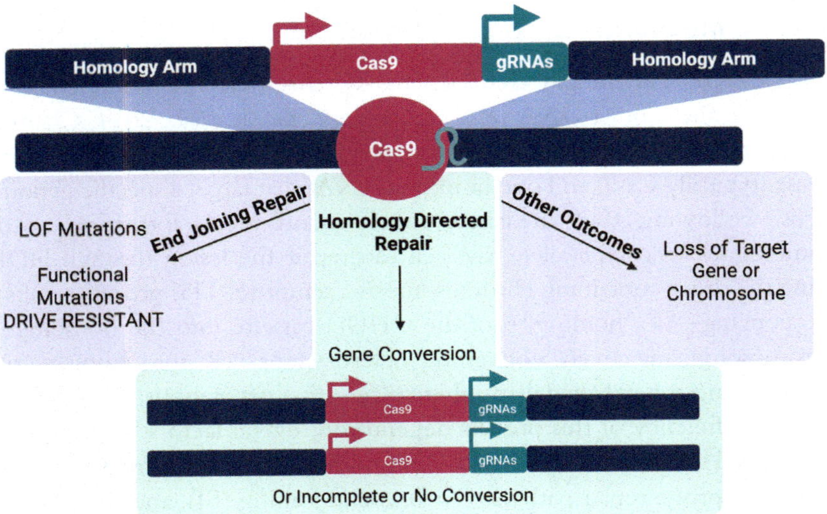

Fig. 2 Mechanisms of CHGD resistance arising from DNA repair pathways in mosquito populations. Homology directed repair (HDR) enables CRISPR-based homing gene drive (CHGD) propagation by copying the drive cassette into the homologous wild-type chromosome following Cas9 (fuchsia) and gRNA (teal)-mediated DNA cleavage, thereby converting heterozygotes into homozygotes. However, this process is not always efficient in mosquito germlines: HDR can result in incomplete copying of the drive (partial conversion) or fail entirely, leaving the allele unmodified. In some cases, unrepaired DNA breaks may lead to loss of the targeted chromosome, which is often lethal and can contribute to biased inheritance of the drive. In contrast, end-joining (EJ) repair pathways, including non-homologous end joining, frequently generate insertions or deletions (indels) at the target site. These loss of function (LOF) mutations can disrupt gene function and may be deleterious if the targeted locus is essential in mosquitoes, leading to their removal from the population. Critically, some indels preserve gene function while altering the target sequence, producing alleles that are resistant to further cleavage. The accumulation and spread of such resistance alleles represent a major evolutionary barrier to gene drive persistence and can ultimately lead to drive failure in mosquito populations. *Created by BioRender.*

In contrast, population modification drives aim to spread effector genes that reduce vector competence, such as anti-pathogen molecules that interfere with malaria or dengue transmission (Gantz et al., 2015; Carballar-Lejarazú et al., 2020; Raban et al., 2026; Hoermann et al., 2022) (Fig. 1B). These systems typically impose lower fitness costs, which can enhance their persistence in populations but may reduce the strength of selection driving spread. Their success depends critically on the efficacy, durability, and evolutionary robustness of the effector, as incomplete or unstable activity can compromise long-term disease control (Marshall et al., 2019).

Table 1 Overview of CRISPR based mosquito biocontrol systems and their key design features. This table summarizes major classes of CRISPR-based genetic biocontrol strategies, including self-limiting and self-propagating systems, and compares their mechanisms, inheritance patterns, design strategies, strengths, and limitations. Systems shown include precision-guided sterile insect technique (pgSIT) and temperature-inducible pgSIT (TI-pgSIT), which rely on non-heritable CRISPR-mediated gene disruption; Ifegenia and RIDL, which employ self-limiting lethal strategies; and homing-based gene drives (CHGD/IGD), split gene drives, cleave-and-rescue (ClvR), and homing-and-rescue (HomeR) systems, which vary in their inheritance dynamics and mechanisms of spread. The "Built-in mosquitoes?" column indicates whether each system has been experimentally demonstrated in mosquito species. Collectively, these approaches illustrate the diversity of CRISPR-based designs for mosquito control, highlighting trade-offs between scalability, containment, stability, and susceptibility to resistance.

CRISPR Based -Biocontrol Systems

System Type	Mechanism	Inheritance	Design Strategy	Strengths	Limitations	Built In Mosquitoes?
pgSIT	CRISPR disruption	Non-heritable	Binary design	Containment	Repeated releases	Yes
TI-pgSIT	Inducible CRISPR	Non-heritable	Inducible control	Scalable	Repeated releases	No
Ifegenia	Female lethality	Self-limiting	Target essential genes	Males make males	Persistence	Yes
CHGD/IGD	Homing HDR	Super-Mendelian	Homing based	Self-sustaining	Rapid Resistance	Yes
Split GD	Separated components	Self-limiting	Component split/ Homing based	Control	Lower efficiency/ rapid resistance	Yes
ClvR	Toxin-antidote	Self-propagating	Rescue allele/ cleavage based	Stable	Complex to build	No
HomeR	Homing+rescue	Super-Mendelian	Recoded rescue genes/Homing based	Stable	Complex to build	No

Additional design innovations—including confineable split-drive systems that separate Cas9 and gRNA components to restrict spread (Nash et al., 2019; Li et al., 2020), multiplexed gRNA strategies (Anderson et al., 2024; Godall et al., 2025), optimization of germline Cas9 expression (Green et al., 2023), and targeting of the transgene integration site (Reid et al., 2022)—have further expanded the CHGD toolkit. More recent work has also explored alternative architectural strategies, including integral gene drives (IGD) that leverage endogenous regulatory elements to improve expression stability and reduce genomic perturbation, as well as hybrid suppression–modification designs (Yen et al., 2025). Moreover, a homing gene drive targeting the ultra-conserved microRNA gene mir-184 in *An. gambiae* has been shown to combine population suppression with population modification by imposing fitness costs on homozygous females while simultaneously enabling the spread of antimalarial effector genes (Verkuijl et al., 2025). Collectively, these approaches expand the CHGD toolkit and highlight trade-offs in efficiency, resistance, and confinement across mosquito systems (Table 1).

4. Emerging gene drive architectures

Recent advances in CRISPR-based gene drives have expanded strategies for mosquito population modification, with increasing emphasis on designs that improve controllability and limit resistance. Homing-and-rescue systems, such as HomeR (Kandul et al., 2021), couple Cas9-mediated homing with a recoded rescue sequence that restores the function of an essential target gene following insertion. By targeting essential genes and providing functional rescue, these systems constrain the formation of loss-of-function resistance alleles, as mutations that disrupt the target site are often deleterious. However, incomplete or imprecise homing events can still generate functional resistance alleles, representing a limitation for long-term stability. Although HomeR systems have been developed and validated in model species, they have not yet been implemented in mosquito vectors and remain promising candidates for future adaptation.

Similarly, non-homing cleave-and-rescue (ClvR) systems decouple cleavage from homing by using Cas9 to disrupt an essential gene while simultaneously supplying a resistant rescue allele (Oberhofer et al., 2024, 2021b; Oberhofer et al., 2019; Oberhofer et al., 2021a; Champer et al., 2020). Unlike homing-based gene drives, ClvR systems do not rely on HDR for propagation, but instead spread through a toxin–antidote mechanism in which only individuals inheriting the rescue allele remain viable. This design allows them to function in contexts where HDR is inefficient and contributes to their relative genetic stability by limiting the spread of loss-of-function resistance alleles. However, ClvR systems remain susceptible to gene conversion or recombination events that can separate the rescue element from the drive or generate resistance alleles, representing a potential failure mode. In addition, mutations at gRNA target sites that prevent cleavage can give rise to resistant alleles that evade disruption, posing a further constraint on long-term system efficacy. Like HomeR, ClvR systems have been demonstrated in model species and remain promising candidates for future adaptation to mosquito vectors. They can also be engineered to exhibit threshold-dependent dynamics, offering greater control over spatial spread and confinement.

More broadly, these approaches, along with other self-limiting designs such as daisy-chain gene drives (Noble et al., 2019; Guo et al., 2025; Verkuijl et al., 2022), represent important extensions of CRISPR-based population control strategies aimed at improving containment and reversibility. However, these systems remain largely untested in mosquito vectors, and incomplete target disruption or unintended DNA repair outcomes may still generate resistant alleles or partial-rescue phenotypes under certain genetic and ecological conditions.

In parallel, distinct CRISPR-based strategies have emerged that operate through transmission distortion rather than inheritance bias. CRISPR-based sex distortion systems have been demonstrated in mosquitoes, including X-shredding and X-poisoning strategies that bias sex ratios toward males by selectively damaging the X chromosome during spermatogenesis (Galizi et al., 2014; Haber et al., 2024). These approaches provide an alternative suppression modality that does not rely on population-wide inheritance, instead acting through transmission ratio distortion, and further illustrate the expanding diversity of CRISPR-based genetic control strategies.

Despite these advances, all gene drive systems operate within an evolutionary framework shaped by mutation, selection, and genetic drift. The same mechanisms that enable biased inheritance also impose strong selective pressures that can promote resistance, alter fitness landscapes, and influence population structure. As such, understanding the mechanistic basis of gene drive function is inseparable from anticipating and managing their evolutionary consequences. Collectively, these designs reflect a shift from purely inheritance-driven systems toward strategies that explicitly account for and incorporate evolutionary constraints.

4.1 Evolutionary constraints on gene drive systems

A central challenge for CHGDs is their interaction with evolutionary processes that shape their spread, stability, and long-term efficacy. Chief among these is the emergence of resistance alleles, which arise primarily through error-prone DNA repair pathways such as end joining (EJ), generating mutations at target sites that prevent further cleavage (Champer et al., 2016; Raban et al., 2023) (Fig. 2). While HDR can convert drive heterozygotes into homozygotes, it can also fail or produce incomplete conversions, resulting in partial drive elements. These processes generate resistance and reshape fitness landscapes, fundamentally influencing gene drive dynamics.

Standing genetic variation within mosquito populations can further reduce targeting efficiency if polymorphisms exist at gRNA binding sites. Sequence variation in flanking regions may also influence cleavage and repair outcomes by altering chromatin context or microhomology patterns that bias repair toward resistance-generating alleles. Once generated, resistance alleles—particularly those that restore gene function while evading cleavage—can be strongly favored by selection and spread rapidly, as demonstrated in *Anopheles* laboratory populations (Hammond et al., 2017). Together, these findings indicate that resistance is an expected evolutionary outcome rather than an exception.

In contrast, self-limiting systems such as pgSIT are not subject to inheritance-driven resistance dynamics, as they do not rely on sustained propagation through populations. Because their effects are confined to a single generation, mutations at CRISPR target sites in wild populations are unlikely to compromise system performance. As a result, these approaches are expected to impose more limited sustained selective pressure for resistance evolution, highlighting a fundamental distinction in how evolutionary processes shape the long-term efficacy of self-limiting versus self-propagating genetic control strategies.

Beyond resistance, gene drive dynamics are shaped by population genetic processes. Although gene drives can spread despite fitness costs due to biased inheritance (Unckless et al., 2017), mosquito populations are spatially structured and connected by migration. Gene flow can facilitate spread across landscapes but may also introduce resistance alleles from untreated populations, complicating outcomes (North et al., 2019). Gene drives can also impose fitness costs through disruption of essential genes, metabolic burdens from transgene expression, or off-target effects (Marshall et al., 2017). More broadly, such fitness costs and resulting selective pressures are not unique to gene drives, but are shared across many population suppression strategies—including self-limiting genetic approaches, as well as conventional interventions like insecticides or environmental modification—which may similarly drive compensatory changes in life history traits. These pressures may select for compensatory changes in life history traits, such as altered mating behavior or increased reproductive output, potentially reducing suppression efficacy over time.

At broader scales, gene drive deployment may generate ecological and evolutionary feedback. Population suppression could alter ecosystem dynamics, including predator–prey relationships and interspecific competition (Committee on Gene Drive Research in Non-Human Organisms: Recommendations for Responsible Conduct et al., 2016). At the same time, gene drives may initiate evolutionary "arms races," in which resistance evolves in response to control efforts, paralleling patterns observed with insecticides.

Finally, gene drives can influence genetic diversity and adaptive potential. While homing drives may spread across multiple genetic backgrounds—resulting in soft sweep–like dynamics—they can still reduce diversity near the drive locus under strong selection. In contrast, toxin–antidote systems such as ClvR may drive more classical selective sweeps through rapid haplotype replacement. Given the substantial standing genetic variation present in mosquito populations (Anopheles gambiae, 1000 Genomes Consortium et al., 2017), these processes highlight the importance of considering population genetics and ecology when predicting gene drive outcomes.

5. Toward evolutionarily robust gene drives

A fundamental challenge is that resistance is, to some extent, an inevitable outcome of CHGDs, as continuous or repeated nuclease activity creates persistent selective pressure for mutations that evade cleavage. One key strategy to mitigate this is multiplexing, in which multiple gRNAs target multiple sites within a gene or across functionally linked loci (Fig. 1D) (Champer et al., 2020; Marshall et al., 2017; Godall et al., 2025). This reduces the probability that single mutational events generate functional resistance alleles, thereby delaying—though not eliminating—the emergence of resistance and improving drive durability (Marshall et al. 2017). Another important design principle is the selection of target sites within constrained genomic regions (Fig. 1D). By focusing on highly conserved sequences or functionally essential domains, the range of viable resistance mutations can be limited, as most alterations at these sites are likely to be deleterious. This approach restricts the evolutionary pathways available for escape and can enhance long-term system stability (Champer et al., 2016).

Advances in predictive modeling have further strengthened the ability to anticipate evolutionary trajectories. Population genetic and eco-evolutionary models are increasingly used to simulate gene drive spread under realistic conditions, incorporating spatial structure, migration, and stochasticity (Noble et al., 2017; Marshall and Akbari, 2018; Sánchez C. et al., 2020). Integration of machine learning approaches with population genomic data may further enable data-driven optimization of target site selection, prediction of resistance pathways, and refinement of system parameters prior to deployment (Schrider and Kern, 2018; Korfmann et al., 2023). Recent cross-species analyses suggest that gene drive performance is strongly influenced by species-specific biological and genetic context (Verkuijl et al., 2026). Together, these findings emphasize that effective gene drive design must be tailored to the target system, balancing efficiency, stability, and resistance within realistic evolutionary and ecological constraints.

6. Implications for vector control

CRISPR-based gene drives represent a promising addition to the existing toolkit for controlling mosquito-borne diseases, offering the potential for scalable and species-specific interventions. Unlike conventional methods such as insecticides or habitat reduction, gene drives can propagate autonomously through populations, enabling sustained effects following limited releases. In principle, this capacity for self-propagation could help overcome key limitations of current strategies, including the widespread emergence of insecticide resistance and the logistical challenges associated with repeated interventions.

However, the effectiveness of gene drives in real-world settings will depend critically on their interaction with evolutionary dynamics. Resistance evolution remains the most significant threat to long-term efficacy, as mutations that evade cleavage can rapidly spread under strong selective pressure. In addition, spatial heterogeneity in mosquito populations—including variation in population density, migration patterns, and environmental conditions—can lead to uneven spread and potentially limit the impact of gene drive systems. Together, these factors underscore the importance of incorporating local population structure and ecological context into both the design and deployment of gene drive strategies.

Given these challenges, gene drives are unlikely to function optimally as standalone solutions. Instead, integrated vector control approaches that combine gene drives with existing interventions—such as insecticide-treated nets, environmental management, and biological control—may provide more robust and sustainable outcomes. In addition, incorporating scalable self-limiting genetic technologies, alongside gene drives may further enhance flexibility and controllability, allowing for complementary modes of population suppression that operate on different temporal and evolutionary scales. For example, self-limiting approaches could be deployed initially to substantially reduce population size, followed by gene drive release into the remaining smaller population, potentially improving efficiency and reducing the likelihood of resistance emergence. Such strategies could reduce selection pressure on any single intervention, delay the emergence of resistance, and enhance overall effectiveness. Moving forward, integrating gene drives into broader vector control frameworks will require coordinated efforts across disciplines, including evolutionary biology, public health, and policy, to ensure both efficacy and responsible implementation. Together, these considerations indicate that successful deployment will depend not only on technological innovation, but also on embedding gene drives within adaptive, evolution-aware vector control strategies.

7. Conclusions and outlook

CRISPR-based genetic biocontrol technologies have emerged as promising tools for vector control, enabling the deliberate manipulation of inheritance and population dynamics in mosquito populations. Over the past several years, rapid advances in both non–gene drive and gene drive CRISPR systems have been accompanied by a deeper understanding of the evolutionary and ecological processes that govern their performance. Together, these advances underscore that the success of these technologies cannot be evaluated solely in terms of molecular efficiency, but must be considered within the broader context of evolutionary dynamics, population structure, and ecological complexity.

A central theme emerging from recent work is that resistance evolution, genetic diversity, and spatial heterogeneity are not peripheral challenges, but fundamental features of the ecological systems targeted by mosquito biocontrol strategies. These factors can profoundly influence the spread, stability, and long-term efficacy of engineered constructs, often in ways that are difficult to predict without integrating population genetics and ecology. As such, future progress will depend on moving beyond static design paradigms toward frameworks that explicitly anticipate and incorporate evolutionary responses, requiring evolution-aware design principles that account for uncertainty, variability, and ongoing evolutionary change to achieve durable biocontrol outcomes.

Encouragingly, the field is already shifting in this direction. Advances in targeting strategies, constrained genomic site selection, and binary, self-limiting or confinable architectures reflect a growing emphasis on engineering evolutionary robustness. At the same time, improvements in population genomics and population modeling are enabling more accurate and context-specific predictions of gene drive behavior across diverse species and environments. These tools will be essential for identifying optimal target sites, forecasting resistance pathways, and designing systems that are robust and resilient under realistic ecological and evolutionary conditions.

However, it is important to recognize that CHGDs remain under active development and have not yet been deployed in the field. Their eventual implementation will require not only further technical validation but also careful navigation of regulatory frameworks, public engagement, and broader societal considerations (Taitingfong et al., 2022; Schairer et al., 2022; Long et al., 2020; Schairer et al., 2019). Self-limiting CRISPR-based approaches such as pgSIT are at a similar pre-deployment stage and will require comparable regulatory scrutiny and stakeholder engagement. However, because pgSIT is non-heritable and does not persist in the environment, it may present a more tractable pathway for phased testing and potential field implementation, relative to self-propagating gene drive systems. Accordingly, these approaches may offer opportunities for earlier translational evaluation, while gene drive technologies continue to mature.

In this context, self-limiting CRISPR-based approaches such as pgSIT may represent a particularly tractable path forward. Because these systems do not rely on sustained inheritance or long-term genetic persistence in wild populations, they are inherently less vulnerable to resistance dynamics that can compromise the efficacy of self-propagating gene drives over time. While homing gene drives offer the potential for autonomous spread, their long-term success remains tightly coupled to evolutionary constraints, including the emergence of resistance alleles and population heterogeneity. By contrast, self-limiting systems operate through repeated, controllable releases, potentially providing a more robust and predictable framework for near-term implementation. In addition, their non-heritable and transient nature may present fewer regulatory and societal barriers, as they offer greater controllability, reversibility, and confinement relative to self-propagating gene drive systems. Importantly, recent modeling and economic analyses suggest that pgSIT-based interventions can be cost-effective relative to existing malaria control strategies, with estimated costs of approximately \$0.36–3.03 per person per year and \$11–94 per disability-adjusted life year (DALY) averted. These conservative estimates fall within—and in some scenarios up to ~50% lower than—the cost range of widely used interventions such as insecticide-treated nets and indoor residual spraying, depending on transmission setting and implementation scale (Gendron, Raban, Mondal, M Sánchez C, et al., 2025; Gendron, Raban, Mondal, Sánchez C, et al., 2025). These models also predict substantial public health benefits, including prevention of thousands of infections and hundreds of deaths annually in endemic regions, while enabling scalable, locally deployable production frameworks.

Looking forward, the most effective applications of CRISPR-based biocontrol will likely arise from coordinated strategies that combine gene drives with complementary approaches, including self-limiting systems such as pgSIT and established vector control interventions. Such integrated frameworks can reduce selective pressures on any single system, enhance overall effectiveness, and provide flexibility in responding to evolving ecological and evolutionary challenges.

Ultimately, realizing the full potential of these technologies will require sustained interdisciplinary collaboration across molecular biology, evolutionary genetics, ecology, public health, and policy, coupled with adaptive management strategies that incorporate monitoring, feedback, and iterative redesign. By aligning technological innovation with a rigorous understanding of evolutionary processes—and by strategically deploying tools that are ready today while advancing those still in development—it may be possible to build vector control systems that are not only effective, but also predictable and evolutionarily resilient over both the short and long term. In the end, the success of CRISPR-based mosquito control will depend not only on what we can engineer, but on how effectively we anticipate evolutionary responses while deploying solutions that are scalable, cost-effective, and societally acceptable.

Acknowledgments

This work was supported in part by 1R01AI190001 and 1R21AI196522 awarded to O.S.A. E.H. also received support from AI007384 (HOPE T32 Training Grant).

Ethical conduct of research

None to declare.

Author contributions

O.S.A. conceptualized the study. All authors reviewed, edited, and approved the manuscript.

Competing interests

O.S.A. is a founder of Agragene, Inc., Synvect, Inc., and Cloak Biotechnologies with equity interest. The terms of this arrangement have been reviewed and approved by the University of California at San Diego, in accordance with its conflict of interest policies. All other authors declare no competing interests.

References

Akbari, O.S., Chen, C.-H., Marshall, J.M., Huang, H., Antoshechkin, I., Hay, B.A., 2014. Novel synthetic medea selfish genetic elements drive population replacement in Drosophila; a theoretical exploration of medea-dependent population suppression. ACS Synth. Biol. 3 (12), 915–928.

Akbari, O.S., Matzen, K.D., Marshall, J.M., Huang, H., Ward, C.M., Hay, B.A., 2013. A synthetic gene drive system for local, reversible modification and suppression of insect populations. Curr. Biology: CB 23 (8), 671–677.

Alphey, L., 2014. Genetic control of mosquitoes. Annu. Rev. Entomol. 59 (1), 205–224.

Alphey, L., Benedict, M., Bellini, R., et al., 2010. Sterile-insect methods for control of mosquito-borne diseases: an analysis. Vector Borne Zoonotic Dis. (Larchmont, N. Y.) 10 (3), 295–311.

Anderson, M.A.E., Gonzalez, E., Edgington, M.P., et al., 2024. A multiplexed, confinable CRISPR/Cas9 gene drive can propagate in caged aedes aegypti populations. Nat. Commun. 15 (1), 729.

Anopheles gambiae 1000 Genomes Consortium, Data analysis group, Partner working group, et al., 2017. Genetic diversity of the african malaria vector anopheles gambiae. Nature 552 (7683), 96–100.

Apte, R.A., Smidler, A.L., Pai, J.J., et al., 2024. Eliminating Malaria vectors with precision-guided sterile males. Proc. Natl. Acad. Sci. U S Am. 121 (27) e2312456121.

Beaghton, A., Beaghton, P.J., Burt, A., 2016. Gene drive through a landscape: reaction-diffusion models of population suppression and elimination by a sex ratio distorter. Theor. Popul. Biol. 108 (April), 51–69.

Bouyer, J., Culbert, N.J., Dicko, A.H., et al., 2020. Field performance of sterile male mosquitoes released from an uncrewed aerial vehicle. Sci. Robot. 5 (43). https://doi.org/10.1126/scirobotics.aba6251.

Buchman, A.B., Ivy, T., Marshall, J.M., Akbari, O.S., Hay, B.A., 2018. Engineered reciprocal chromosome translocations drive high threshold, reversible population replacement in drosophila. ACS Synth. Biol. 7 (5), 1359–1370.

Buchman, A., Marshall, J.M., Ostrovski, D., Yang, T., Akbari, O.S., 2018. Synthetically engineered gene drive system in the worldwide crop pest. Proc. Natl Acad. Sci. U S Am. 115 (18), 4725–4730.

Buchman, A., Shriner, I., Yang, T., et al., 2021. Engineered reproductively isolated species drive reversible population replacement. Nat. Commun 12 (1), 3281.

Burt, A., 2003. Site-specific selfish genes as tools for the control and genetic engineering of natural populations. Proc. Biol. Sci. 270 (1518), 921–928.

Burt, A., Crisanti, A., 2018. Gene drive: evolved and synthetic. ACS Chem. Biol. 13 (2), 343–346.

Burt, A., Trivers, R., 2008. Genes in Conflict: The Biology of Selfish Genetic Elements. Harvard University Press.

Carballar-Lejarazú, R., Ogaugwu, C., Tushar, T., et al., 2020. Next-generation gene drive for population modification of the malaria vector mosquito. Proc. Natl Acad. Sci. U S Am. 117 (37), 22805–22814.

Carvalho, D.O., McKemey, A.R., Garziera, L., et al., 2015. Suppression of a field population of aedes aegypti in Brazil by sustained release of transgenic male mosquitoes. PLoS Neglected Tropical Dis. 9 (7), e0003864.

Champer, J., Buchman, A., Akbari, O.S., 2016. Cheating evolution: engineering gene drives to manipulate the fate of wild populations. Nat. Reviews. Genet. 17 (3), 146–159.

Champer, J., Kim, I.K., Champer, S.E., Clark, A.G., Messer, P.W., 2020. Performance analysis of novel toxin-antidote crispr gene drive systems. BMC Biol. 18 (1), 27.

Champer, J., Liu, J., Oh, S.Y., et al., 2018. Reducing resistance allele formation in CRISPR gene drive. Proc. Natl Acad. Sci. U S Am. 115 (21), 5522–5527.

Champer, S.E., Oh, S.Y., Liu, C., et al., 2020. Computational and experimental performance of CRISPR homing gene drive strategies with multiplexed gRNAs. Sci. Adv. 6 (10), eaaz0525.

Chen, C.-H., Huang, H., Catherine, M., Ward, et al., 2007. A synthetic maternal-effect selfish genetic element drives population replacement in drosophila. Sci. (N. York, N. Y.) 316 (5824), 597–600.

Committee on Gene Drive Research in Non-Human Organisms: Recommendations for Responsible Conduct, 2016. Board on life sciences, division on earth and life studies, and national academies of sciences, engineering, and medicine. Gene Drives on the Horizon: Advancing Science, Navigating Uncertainty, and Aligning Research with Public Values. National Academies Press, US.

Crawford, J.E., Clarke, D.W., Criswell, V., et al., 2020. Efficient production of male wolbachia-infected aedes aegypti mosquitoes enables large-scale suppression of wild populations. Nat. Biotechnol. 38 (4), 482–492.

Davydova, S., Liu, J., Kandul, N.P., et al., 2026. Generating cisgenic sexing strains in insect pests. Commun. Biol. 9 (1). https://doi.org/10.1038/s42003-026-09624-9.

Dyck, V.A., Hendrichs, J., Robinson, A.S. (Eds.), 2005. Sterile Insect Technique: Principles and Practice in Area-Wide Integrated Pest Management, 2005th ed. Springer PDF.

Esvelt, K.M., Smidler, A.L., Catteruccia, F., Church, G.M., 2014. Concerning RNA-guided gene drives for the alteration of wild populations. eLife 3 (July). https://doi.org/10.7554/eLife.03401.

Fu, G., Lees, R.S., Nimmo, D., et al., 2010. Female-specific flightless phenotype for mosquito control. Proc. Natl Acad. Sci. U S Am. 107 (10), 4550–4554.

Galizi, R., Doyle, L.A., Menichelli, M., et al., 2014. A synthetic sex ratio distortion system for the control of the human malaria mosquito. Nat. Commun. 5 (1), 3977.

Gantz, V.M., Jasinskiene, N., Tatarenkova, O., et al., 2015. Highly efficient Cas9-mediated gene drive for population modification of the malaria vector mosquito anopheles stephensi. Proc. Natl. Acad. Sci. U S Am. 112 (49), E6736–E6743.

Gendron, W.A.C., Raban, R., Mondal, A., Sánchez C, H.M., et al., 2025. Estimated cost and operational structure of pgSIT malaria vector control programs in selected West African countries. Sci. Afr. 29 (September) (None).

Gendron, W.A.C., Raban, R., Mondal, A., M Sánchez C, H., et al., 2025. Evaluating the cost of malaria elimination by anopheles gambiae precision guided SIT in the upper river region, the gambia. PLOS Glob. Public. Health 5 (7), e0004903.

Godall, M.L., Lewis Shackleford, M.P., Edgington, et al., 2025. Integrating multiplexing into confineable gene drives effectively overrides resistance in *Anopheles Stephensi*. bioRxiv. August 15. https://doi.org/10.1101/2025.08.15.670499.

Green, E.I., Jaouen, E., Klug, D., et al., 2023. A population modification gene drive targeting both saglin and lipophorin impairs plasmodium transmission in anopheles mosquitoes. eLife 12, e93142. https://doi.org/10.7554/elife.93142.

Guo, J., Chen, W., Champer, J., 2025. Experimental demonstration of daisy chain gene drive and modelling of daisy suppression systems. bioRxiv. September 24. https://doi.org/10.1101/2025.09.20.677490.

Haber, An, D., Arien, Y., Lamdan, L.B., et al., 2024. Targeting mosquito X-chromosomes reveals complex transmission dynamics of sex ratio distorting gene drives. Nat. Commun. 15 (1), 4983.

Hammond, A., Galizi, R., Kyrou, K., et al., 2016. A CRISPR-Cas9 gene drive system targeting female reproduction in the malaria mosquito vector anopheles gambiae. Nat. Biotechnol 34 (1), 78–83.

Hammond, A.M., Kyrou, K., Bruttini, M., et al., 2017. The creation and selection of mutations resistant to a gene drive over multiple generations in the malaria mosquito. PLoS Genet. 13 (10), e1007039.

Hoermann, A., Habtewold, T., Selvaraj, P., et al., 2022. Gene drive mosquitoes can aid malaria elimination by retarding plasmodium sporogonic development. Sci. Adv. 8 (38), eabo1733.

Jinek, M., Krzysztof Chylinski, I.F., Michael Hauer, J.A., Doudna, Emmanuelle Charpentier, 2012. A programmable Dual-RNA-guided DNA endonuclease in adaptive bacterial immunity. Sci. (N. York, N. Y.) 337 (6096), 816–821.

Kandul, N.P., Liu, J., Akbari, O.S., 2021. Temperature-inducible precision-guided sterile insect technique. CRISPR J. 4 (6), 822–835.

Kandul, N.P., Liu, J., Buchman, A., et al., 2022. Precision guided sterile males suppress populations of an invasive crop pest. GEN. Biotechnol. 1 (4), 372–385.

Kandul, N.P., Liu, J., Buchman, A., Gantz, V.M., Bier, E., Akbari, O.S., 2020. Assessment of a split homing based gene drive for efficient knockout of multiple genes. G3 Genes|Genomes|Genetics. 10 (2), 827–837.

Kandul, N.P., Liu, J., Sanchez C, H.M., Wu, S.L., Marshall, J.M., Akbari, O.S., 2019a. Reply to 'Concerns about the feasibility of using "precision Guided Sterile Males" to control insects'. Nat. Commun. 10 (1), 3955.

Kandul, N.P., Liu, J., Sanchez C, H.M., Wu, S.L., Marshall, J.M., Akbari, O.S., 2019b. Transforming insect population control with precision guided sterile males with demonstration in flies. Nat. Commun. 10 (1), 84.

Knipling, E.F., 1955. Possibilities of insect control or eradication through the use of sexually sterile males1. J. Economic Entomol. 48 (4), 459–462.

Korfmann, K., Gaggiotti, O.E., Fumagalli, M., 2023. Deep learning in population genetics. Genome Biol. Evolution 15 (2). https://doi.org/10.1093/gbe/evad008.

Kyrou, K., Hammond, A.M., Galizi, R., et al., 2018. A CRISPR-Cas9 gene drive targeting doublesex causes complete population suppression in caged anopheles gambiae mosquitoes. Nat. Biotechnol. 36 (11), 1062–1066.

Li, M., Kandul, N.P., Sun, R., et al., 2024. Targeting sex determination to suppress mosquito populations. eLife 12 (January). https://doi.org/10.7554/eLife.90199.

Li, M., Yang, T., Bui, M., et al., 2021. Suppressing mosquito populations with precision guided sterile males. Nat. Commun. 12 (1), 5374.

Li, M., Yang, T., Kandul, N.P., et al., 2020. Development of a confinable gene drive system in the human disease vector. eLife 9 (January). https://doi.org/10.7554/eLife.51701.

Long, K.C., Alphey, L., Annas, G.J., et al., 2020. Core commitments for field trials of gene drive organisms. Sci. (N. York, N. Y.) 370 (6523), 1417–1419.

Lutrat, C., Giesbrecht, D., Marois, E., Whyard, S., Baldet, T., Bouyer, J., 2019. Sex sorting for pest control: it's raining men. Trends Parasitol. 35 (8), 649–662.

Mamai, W., Bueno-Masso, O., Wallner, T., et al., 2024. Efficiency assessment of a novel automatic mosquito pupae sex separation system in support of area-wide male-based release strategies. Sci. Rep. 14 (1), 9170.

Marois, E., 2024. Sorting Mosquito Larvae with a COPAS machine. Cold Spring Harb. Protoc. 2024 (6) db.prot108307.

Marshall, J.M., Akbari, O.S., 2018. Can CRISPR-based gene drive be confined in the wild? A question for molecular and population biology. ACS Chem. Biol. 13 (2), 424–430.

Marshall, J.M., Buchman, A., Sánchez C, H.M., Akbari, O.S., 2017. Overcoming evolved resistance to population-suppressing homing-based gene drives. Sci. Rep. 7 (1), 3776.

Marshall, J.M., Raban, R.R., Kandul, N.P., Edula, J.R., León, T.M., Akbari, O.S., 2019. Winning the tug-of-war between effector gene design and pathogen evolution in vector population replacement strategies. Front. Genet. 10 (October), 1072.

Maselko, M., Feltman, N., Upadhyay, A., et al., 2020. Engineering multiple species-like genetic incompatibilities in insects. Nat. Commun. 11 (1), 4468.

Nash, A., Urdaneta, G.M., Beaghton, A.K., et al., 2019. Integral gene drives for population replacement. Biol. Open. 8 (1). https://doi.org/10.1242/bio.037762.

Nguyen, C., Omotayo, A.I., Sanz Juste, S., Feng, X., López Del Amo, V., 2025. A temperature-sensitive CRISPR-Cas12a system for sterile insect technique. Nat. Commun. 16 (1), 9677.

Noble, C., Min, J., Olejarz, J., et al., 2019. Daisy-chain gene drives for the alteration of local populations. Proc. Natl Acad. Sci. U S Am. 116 (17), 8275–8282.

Noble, C., Olejarz, J., Esvelt, K.M., Church, G.M., Nowak, M.A., 2017. Evolutionary dynamics of CRISPR gene drives. Sci. Adv. 3 (4), e1601964.

North, A.R., Burt, A., Godfray, H.C.J., 2019. Modelling the potential of genetic control of malaria mosquitoes at national scale. BMC Biol. 17 (1), 26.

Oberhofer, G., Ivy, T., Hay, B.A., 2019. Cleave and rescue, a novel selfish genetic element and general strategy for gene drive. Proc. Natl Acad. Sci. U S Am. 116 (13), 6250–6259.

Oberhofer, G., Ivy, T., Hay, B.A., 2021a. Gene drive that results in addiction to a temperature-sensitive version of an essential gene triggers population collapse in Drosophila. Proc. Natl Acad. Sci. U S Am. 118 (49). https://doi.org/10.1073/pnas.2107413118.

Oberhofer, G., Ivy, T., Hay, B.A., 2021b. Split versions of cleave and rescue selfish genetic elements for measured self limiting gene drive. PLoS Genet. 17 (2), e1009385.

Oberhofer, G., Johnson, M.L., Ivy, T., Antoshechkin, I., Hay, B.A., 2024. Cleave and rescue gamete killers create conditions for gene drive in plants. Nat. Plants 10 (6), 936–953.

Papathanos, P.A., Bossin, H.C., Benedict, M.Q., et al., 2009. Sex separation strategies: past experience and new approaches. Malar. J. 8 (2), S5.

Raban, R., James, A.A., Akbari, O.S., 2026. Advances in CRISPR gene drives for mosquito population control. Curr. Opin. Microbiol. 90 (April), 102712.

Raban, R., Marshall, J.M., Hay, B.A., Akbari, O.S., 2023. Manipulating the destiny of wild populations using CRISPR. Annu. Rev. Genet. 57 (November), 361–390.

Reid, W., Williams, A.E., Sanchez-Vargas, I., et al., 2022. Assessing single-locus CRISPR/Cas9-based gene drive variants in the mosquito *Aedes aegypti* via single-generation crosses and modeling. G3 (Bethesda, Md.) 12 (12). https://doi.org/10.1093/g3journal/jkac280.

Robinson, A.S., Vreysen, M.J.B., Hendrichs, J., Feldmann, U., 2009. Enabling technologies to improve area-wide integrated pest management programmes for the control of screwworms. Med. Veterinary Entomol. 23 (1), 1–7 (June).

Sánchez C, H.M., Bennett, J.B., Wu, S.L., Rašić, G., Akbari, O.S., Marshall, J.M., 2020. Modeling confinement and reversibility of threshold-dependent gene drive systems in spatially-explicit aedes aegypti populations. BMC Biol. 18 (1), 50.

Schairer, C.E., Taitingfong, R., Akbari, O.S., Bloss, C.S., 2019. A typology of community and stakeholder engagement based on documented examples in the field of novel vector control. PLoS Neglected Tropical Dis. 13 (11), e0007863.

Schairer, C.E., Triplett, C., Akbari, O.S., Bloss, C.S., 2022. California residents' perceptions of gene drive systems to control mosquito-borne disease. Front. Bioeng. Biotechnol. 10 (March), 848707.

Schrider, D.R., Kern, A.D., 2018. Supervised machine learning for population genetics: a new paradigm. Trends Genetics: TIG 34 (4), 301–312.

Smidler, A.L., Akbari, O.S., 2025. CRISPR technologies for the control and study of malaria-transmitting anopheline mosquitoes. Parasites & Vectors 18 (1), 252.

Smidler, A.L., James, J.P., Apte, R.A., et al., 2023. A confinable female-lethal population suppression system in the malaria vector. Sci. Adv. 9 (27) eade8903.

Spinner, S.A.M., Barnes, Z.H., Puinean, A.M., et al., 2022. New self-sexing strain eliminates barriers to scalable and sustainable vector control for governments and communities in dengue-prone environments. Front. Bioeng. Biotechnol. 10 (October), 975786.

Taitingfong, R.I., Triplett, C., Vásquez, V.N., et al., 2022. Exploring the value of a global gene drive project registry. Nat. Biotechnol. 41 (1), 9–13.

Unckless, R.L., Clark, A.G., Messer, P.W., 2017. Evolution of resistance against CRISPR/Cas9 gene drive. Genetics 205 (2), 827–841.

Verkuijl, S.A.N., Anderson, M.A.E., Alphey, L., Bonsall, M.B., 2022. Daisy-chain gene drives: the role of low cut-rate, resistance mutations, and maternal deposition. PLoS Genet. 18 (9), e1010370.

Verkuijl, S.A.N., Del Corsano, G., Capriotti, P., et al., 2025. A suppression-modification gene drive for malaria control targeting the ultra-conserved RNA gene Mir-184. Nat. Commun. 16 (1), 3923.

Verkuijl, S.A.N., Edward, R.I.-C., Liu, B., Bonsall, M.B., Leftwich, P.T., Windbichler, N., 2026. Cross-species meta-analysis reveals determinants of homing gene drive performance. BioRxiv. March 3. https://doi.org/10.64898/2026.02.28.708699.

Weng, S.-C., Antoshechkin, I., Marois, E., Akbari, O.S., 2023. Efficient sex separation by exploiting differential alternative splicing of a dominant marker in aedes aegypti. PLoS Genet. 19 (11), e1011065.

Weng, S.-C., Chen, F., Li, M., et al., 2024. Establishing a dominant early larval sex-selection strain in the asian malaria vector anopheles stephensi. Infect. Dis. Poverty 13 (1), 83.

Winegard, T.C., 2020. The Mosquito: A Human History of Our Deadliest Predator. Penguin.

Wyss, J.H., 2000. Screwworm eradication in the Americas. Ann. N. Y. Acad. Sci. 916, 186–193.

Yen, P.-S., Verkuijl, S.A.N.R., Capriotti, P., et al., 2025. The nanosd integral gene drive enables population modification of the malaria vector anopheles gambiae. G3 (Bethesda, Md.). ahead of print, October 29. https://doi.org/10.1093/g3journal/jkaf246.

CHAPTER TEN

microRNAs at the vector-virus interface: Molecular mediators in mosquito-pathogen dynamics

Sassan Asgari*
School of the Environment and School of Chemistry and Molecular Biosciences, The University of Queensland, Brisbane, QLD, Australia
*Corresponding author. e-mail address: s.asgari@uq.edu.au

Contents

Abstract

MicroRNAs (miRNAs) are small non-coding RNAs that play vital roles in post-transcriptional gene regulation by controlling target gene expression. In mosquitoes, miRNAs have been shown to regulate genes involved in development, blood digestion, immunity, and host-pathogen interactions. Although many conserved miRNAs are found in mosquitoes and other insects, there are also mosquito-specific miRNAs, which may be adaptations to their blood-feeding lifestyle and their role as vectors of pathogens. Research indicates that viral infections in mosquitoes can alter their miRNA profiles, influencing various biological pathways that may either promote or hinder viral replication. Additionally, miRNAs have been found to mediate interactions between mosquitoes and the endosymbiont *Wolbachia*. Some miRNAs affected by *Wolbachia* might contribute to its virus-blocking mechanism by regulating antiviral genes. Although the use of miRNAs in mosquito or disease control has been limited, there have been proof-of-concept attempts to utilise them to reduce vector competence or aid vaccine production. This chapter specifically focuses on the role of mosquito miRNAs in virus-mosquito interactions and explores their potential to reduce the transmission of mosquito-borne viruses.

Advances in Insect Physiology, Volume 70
ISSN 0065-2806, https://doi.org/10.1016/bs.aiip.2026.04.001

1. Introduction

microRNAs (miRNAs) are small non-coding RNAs ranging from 18 to 24 nucleotides (nt) in length and are known to regulate gene expression by interacting with target genes through sequence complementarity. miRNAs are produced from primary miRNA (pri-miRNA) genes located throughout different regions of the genome (Bartel, 2009). They may reside in non-coding regions or within introns of the genome. Pri-miRNA contains one or more stem-loop structures that are cleaved by Drosha in association with Pasha at the stem, resulting in the formation of precursor miRNA (pre-miRNA), approximately 70 nt long (Fig. 1). The pre-miRNA is subsequently transported into the cytoplasm through the nuclear membrane, facilitated by Exportin-5 in association with Ran. The stem-loop head is then cleaved by the ribonuclease Dicer-1, producing a miRNA duplex. This short dsRNA structure triggers the assembly of the RNA-induced silencing complex (RISC), in which Argonaute 1 (AGO1) protein plays a key role. Although AGO2 mainly mediates short-interfering RNA (siRNA) induced silencing in insects, miRNAs can also be loaded onto the AGO2 protein associated with a different RISC from that of the AGO1-associated RISC (Ghildiyal et al., 2010; Yang et al., 2014). Following loading onto RISC, one strand of the duplex is degraded, guiding the complex to target mRNAs. In some cases, both arms of the miRNA, known as the 5p and 3p arms, can be functional. The outcome of the miRNA-RISC interaction with target mRNA can differ, ranging from mRNA degradation to translational repression or, in some cases, an increase in the target gene's transcript levels (Bruno et al., 2011; Orom et al., 2008; Vasudevan et al., 2007). Therefore, the effect of miRNA on the target is not always repression. Based on that assumption, most studies have searched for negative correlations between the miRNA of interest and the expression levels of its potential targets, overlooking potential positive interactions. The target sequences may be located in the 5'UTR, exon, or 3'UTR, although initially it was thought that target sequences exclusively reside in the 3'UTR region of mRNAs (Hafner et al., 2010; Helwak et al., 2013; Reczko et al., 2012).

Along with the canonical miRNA pathway described above, there are some non-canonical pathways for miRNA production. For example, miRNA could be generated by bypassing Dicer-1 (e.g. cleavage by AGO2) or avoiding the cleavage by Drosha (e.g. mirtrons generated through intron splicing) (Okamura et al., 2007; Yang and Lai, 2011).

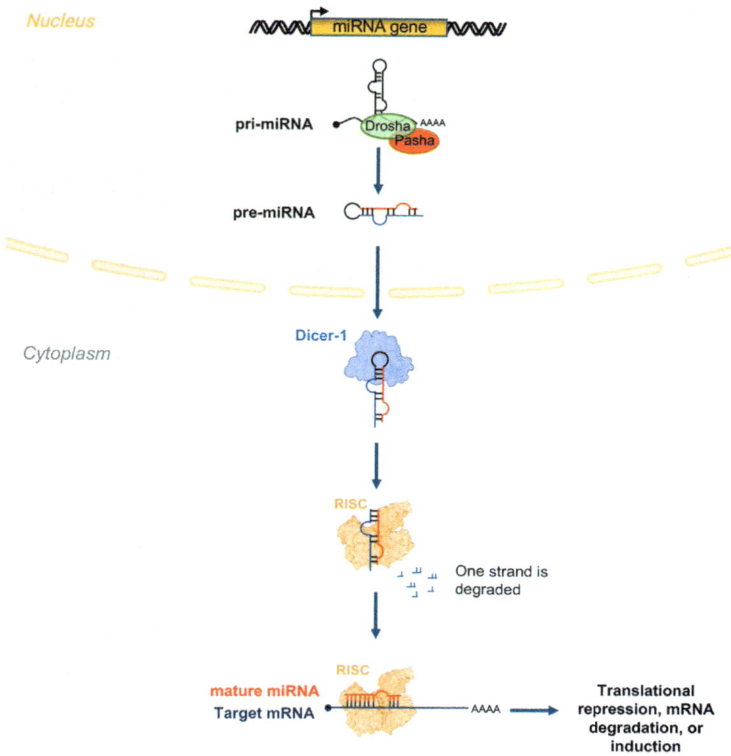

Fig. 1 miRNA biogenesis pathway. miRNA genes are expressed by RNA polymerase gene with a 5′-cap and a polyA tail (primary miRNA, pre-miRNA). They may contain one or several stem loops that are recognised by the ribonuclease Drosha in association with Pasha and cleaved at the base. The generated precursor miRNA (pre-miRNA) is then transported into the cytoplasm, where its hairpin head is cleaved by Dicer-1. The resulting short RNA duplex triggers the formation of the RNA-induced silencing complex (RISC), which includes the argonaute protein. One of the strands is degraded and the mature miRNA guides the RISC complex to the target mRNA by sequence complementarity, which is usually not perfect. The outcome is variable, ranging from mRNA degradation and translational repression to induction in some instances. A combination of Biorender and PowerPoint tools were used to generate the figure.

miRNAs are primarily recognised as fine-tuners of gene expression, adding an extra layer to the various stages of gene regulation. They are involved in numerous biological processes in insects, including development and differentiation, metabolism, adaptation, insecticide resistance, behaviour, caste determination, immunity, and host-pathogen interactions [reviewed in (Asgari, 2013, 2015, 2018; Lucas and Raikhel, 2013; Lucas et al., 2015a)]. It is important to note that each miRNA may target one or multiple genes. Conversely, a target gene can be regulated by multiple miRNAs, increasing its functional complexity and pleiotropic effects. Further, while some miRNAs are highly conserved across insect species, others may be group- or species-specific, resulting in either conserved or unique functions within insects. This review mainly aims to provide insights into the role of miRNAs in how mosquitoes interact with the viruses they transmit.

2. Conservation and divergence of miRNAs and the biogenesis components in mosquitoes

The core molecular machinery involved in miRNA biogenesis is highly conserved across insects, including mosquitoes. Key components include Drosha, Pasha (DGCR8 ortholog), Dicer-1, Loquacious (LOQS), and AGO1, all of which are present and functionally conserved in mosquitoes. The high level of conservation underscores the essential role of miRNA regulation in fundamental cellular and developmental processes of mosquitoes. However, evolutionary divergence and species-specific adaptations in mosquitoes have introduced differences compared to other insects, such as *Drosophila*. For instance, LOQS-PD is specific to *Drosophila* and is not found in any mosquitoes, whereas LOQS2 is specific to *Aedes* mosquito species, likely arising from two independent duplication events of the first double-stranded RNA-binding domain of LOQS (Estevez-Castro et al., 2024; Haac et al., 2015). LOQS2 is likely diversifying as part of the mosquitoes' antiviral response and, unlike LOQS, may not play a significant role in the miRNA pathway. Nonetheless, LOQS2 interacts with LOQS and R2D2, regulating systemic replication of dengue virus (DENV) and Zika virus (ZIKV) in *Aedes aegypti* mosquitoes (Estevez-Castro et al., 2024). Notably, LOQS plays a role in both siRNA and miRNA pathways by interacting with both AGO1 and AGO2, independent of its interaction with (Haac et al., 2015) Dicer.

Although many miRNAs are conserved between mosquitoes and other insects, some have been identified as being specific to mosquitoes based on available miRNA sequences. Li et al. (2009) identified 13 mosquito-specific miRNAs; however, as more data have been added to miRBase, a database for annotated miRNAs, homologs of some of these miRNAs have been identified in other insects and included in the database, such as miR-1175 and miR-1889. Despite this, several miRNAs still appear to be exclusive to mosquitoes. These include miR-2940-46, miR-1174, miR-1890, miR-1891, and some novel, uncharacterized miRNAs (Hu et al., 2015). A search in miRBase at the time of writing this chapter indicated that miR-2940 is specific to *Ae. aegypti* and has not been found in other mosquitoes or insects (Kozomara and Griffiths-Jones, 2014). The function of miR-2940 has been demonstrated in mosquito-pathogen interactions, which will be discussed below.

Among the mosquito-specific miRNAs mentioned above, the functions of miR-1174 and miR-1890 have been determined, although not specifically related to mosquito-virus interactions. miR-1174 is essential for blood digestion and development. A reduction in the abundance of miR-1174 through silencing resulted in defects in sugar absorption and blood digestion by fine-tuning the transcript levels of its target gene, serine hydroxymethyltransferase (Liu et al., 2014). miR-1890 targets the 3'UTR of the juvenile hormone (JH)-controlled chymotrypsin-like SP, JHA15. Depleting this miRNA led to problems with blood digestion, ovary development, and egg deposition (Lucas et al., 2015b). The role of miR-1891 in reproduction was also investigated in *Aedes albopictus* mosquitoes, in which inhibiting this miRNA reduced lifespan and fecundity compared with controls (Puthiyakunnon et al., 2013). Additionally, this miRNA was found to be upregulated in biting *Culex pipiens* mosquitoes before they took a blood meal (Marzec et al., 2023). Overall, these mosquito-specific miRNAs likely represent genomic innovations that enable mosquitoes to respond to environmental challenges, such as host-seeking behaviours, blood meal digestion, vector competence, and pathogen infection. In particular, viral infection may likely create evolutionary pressure for miRNA diversification to optimise the host response or enhance viral replication [reviewed in (Girardi et al., 2018)]. Novel miRNAs, in particular, are less conserved in sequence, expressed at lower levels, and evolve rapidly under less restrictive conditions or positive selection (Liu et al., 2008). Viral selective pressures might influence the emergence of novel miRNAs as part of immune or antiviral evolutionary adaptations, particularly in tissues involved in dissemination and transmission, such as the midgut and salivary glands. Although speculative, this presents an intriguing hypothesis worth testing in the ongoing evolutionary arms race between the virus and the host. The birth and rapid demise of miRNA families have been observed in mammals and insects, and these events may be influenced by selective pressures (Lu et al., 2008; Meunier et al., 2013).

The miRNA biogenesis machinery is highly conserved in mosquitoes and across insects in general, but lineage-specific features such as LOQS2 and a subset of mosquito-restricted miRNAs indicate functional divergence related to mosquito biology. Several of these miRNAs influence traits relevant to vector competence, yet there is currently no direct evidence that viral infection triggers the emergence or diversification of miRNA genes in mosquitoes. Therefore, although long-term virus–host interactions are a plausible selective force shaping mosquito miRNA repertoires, this remains an open hypothesis that requires explicit evolutionary testing.

3. Impact of virus infection on mosquito miRNA repertoires

Viral infections often result in changes in miRNA abundance in mosquitoes. These alterations could be part of the mosquito's antiviral responses to infection, or a strategy used by the virus to evade immune responses and aid its replication by regulating proviral and antiviral genes [reviewed in (Asgari, 2014)]. These effects are mainly studied by examining how viral infections alter the host's miRNA profile using high-throughput methods, such as microarrays (now less common) and small RNA-Seq (see below). Several studies have documented the impacts of arboviruses and insect-specific viruses in mosquitoes at whole-mosquito, tissue-specific, or cell line levels (summarised in Table 1). However, it is somewhat challenging to compare these studies directly, given differences in mosquito species, post-infection sampling times, virus types, and whether whole mosquitoes or tissues were analysed.

In DENV serotype 2-infected *Ae. aegypti* mosquitoes, 5, 3, and 23 miRNAs were differentially abundant at 2, 4, and 9 days post-infection (dpi), respectively (Campbell et al., 2014). Among those, miR-308–3p and miR-305–5p (9 dpi) overlap with those in ZIKV-infected *Ae. aegypti* mosquitoes at 7 and 14 dpi (Saldaña et al., 2017); in both mosquito-virus interactions, the two miRNAs showed downregulation. In *Ae. albopictus* mosquitoes infected with DENV2, overlapping differentially abundant miRNAs with ZIKV-infected mosquitoes are miR-2940–3p (depleted in DENV, but enriched in ZIKV), miR-263a-5p (depleted in both), miR-308–5p (enriched in both), miR-989 (depleted in DENV, but enriched in ZIKV), and miR-2941 (depleted in both) (Liu et al., 2015). In another study, comparing uninfected and DENV2-infected *Ae. albopictus*

Table 1 Studies that examined the effect of virus infection on mosquito miRNA profiles using high-throughput approaches.

Mosquito species	Virus	Tissue (times post-infection)	References
Aedes aegypti	dengue virus	Whole mosquito (2, 4, 9 dpi) C6/36 cell line	Campbell et al. (2014) Avila-Bonilla et al. (2017)
	chikungunya virus	Saliva and salivary glands (14 dpi)	Fiorillo et al. (2022)
	Zika virus	Whole mosquito (2, 7, 14 dpi) Midgut and salivary glands (2, 4, 6 dpi)	Saldaña et al. (2017), Zhu et al. (2021)
	Palm Creek virus	Whole mosquito (2, 6, 12 dpi)	Lee et al. (2017)
	Ross River virus	Whole mosquito (2, 6, 12 dpi)	Sinclair and Asgari, (2020)
Aedes albopictus	chikungunya virus	Singh's cell line (24 hpi)	Shrinet et al. (2014)
	dengue virus	Midgut (1 dpi) pools of 100 midguts and no replicates Whole mosquitoes (7 dpi) pool of 14 mosquitoes no replicates Saliva (10 dpi), pool of 50 mosquitoes and no replicates, virus injected to mosquitoes C6/36 cells (3 dpi)	Liu et al. (2016), Su et al. (2017), Su et al. (2019) Liu et al. (2015), Maharaj et al., 2015, Azlan et al. (2022)
	bluetongue virus	Singh's cell line (12 hpi)	Xing et al. (2016)
Anopheles stephensi	Mayaro virus	Whole mosquitoes (2, 7, 14 dpi)	Henderson et al. (2022)
Anopheles coluzzii	o'nyong nyong virus	Whole mosquitoes (3 dpi)	Carissimo et al. (2018)

midguts, three miRNAs (miR-2941, miR-989, miR-2943) were differentially abundant (Liu et al., 2016); the first two also showed changes in abundance in ZIKV-infected *Ae. aegypti* mosquitoes (Saldaña et al., 2017). In *Culex quinquefasciatus* mosquitoes infected with West Nile virus (WNV), miR-989 was depleted by 2.8-fold (Skalsky et al., 2010); however, this miRNA was enriched by 1.8-fold at 2 and 14 dpi in *Ae. aegypti* mosquitoes infected with ZIKV (Saldaña et al., 2017).

Based on available studies, differentially expressed miRNAs tend to overlap in mosquitoes infected with viruses from the same family, especially within the same host species. For instance, in *Ae. albopictus* mosquitoes infected with chikungunya virus (CHIKV), miR-100, miR-283, miR-305-3p, and miR-927 were among the highly differentially abundant miRNAs (Shrinet et al., 2014). However, none of these was differentially regulated in *Ae. aegypti* mosquitoes infected with DENV2 or ZIKV (Campbell et al., 2014; Saldaña et al., 2017). There was a small overlap of differentially expressed *Ae. aegypti* miRNAs when infected with Ross River virus (RRV), an alphavirus, as well as those infected with DENV2 (miR-8–5p and miR-2940–3p) and ZIKV (aae-miR-989-3p) (Sinclair and Asgari, 2020). This overlap might be because the same host mosquito was used across the three studies; although the RRV study analysed miRNAs at the tissue level (midgut and fat body), while the other two used whole mosquitoes.

In *Ae. aegypti* mosquitoes infected with the insect-specific virus Palm Creek virus, there were minimal changes in the miRNA profile, with only aae-miR-2940–5p showing a significant increase in abundance (Lee et al., 2017). This might be because insect-specific viruses are often non-pathogenic to mosquitoes and cause minimal alterations in host physiology. Interestingly, in *Anopheles coluzzii* mosquitoes infected with o'nyong virus and analysed at 3 dpi, only one miRNA (aga-chr2L_23370) showed differential abundance by enrichment between infected and uninfected mosquitoes (Carissimo et al., 2018). Possibly, including more time points might have revealed additional differentially abundant miRNAs. However, in a study carried out on *An. stephensi* mosquitoes infected with Mayaro virus at three time points (2, 7, and 14 dpi), there were eight differentially abundant miRNAs between infected and uninfected groups across the three time points (Henderson et al., 2022). Notably, three of the eight miRNAs, miR-286b, miR-2944b, and miR-309, overlapped with miRNAs that were differentially abundant in *Ae. aegypti* mosquitoes infected with ZIKV (Saldaña et al., 2017).

Aside from the impact of viral infections on the abundance of mosquito miRNAs, DENV2 infection was found to change post-transcriptional modifications of miRNAs in Aag2 cells by altering isomiR populations, particularly through arm-switching and alterations at the 3p end of miRNAs (Etebari et al., 2015). IsomiRs are sequence variants of canonical miRNAs that originate from the same miRNA precursor but differ slightly in length or sequence due to variations in processing by enzymes such as Dicer or Drosha, or through post-transcriptional nucleotide additions or trimming (Neilsen et al., 2012). These variants can exhibit different stabilities, target specificities, or regulatory functions compared with the canonical miRNA (Burroughs et al., 2010; Tan et al., 2014). miRNA arm switching refers to a change in the relative abundance of the 5p and 3p arms of a miRNA duplex during different developmental stages, infection, tissues, or conditions, leading to alterations in the target gene set regulated by that miRNA (Heydarzadeh et al., 2021). Both phenomena (alterations in isomiR populations and arm switching) may potentially alter the target gene repertoire and, consequently, virus replication.

Overall, viral infections often alter mosquito miRNA profiles, impacting both miRNA levels and processing, including isomiR populations. These changes have been seen across various mosquito-virus systems, tissues, and experimental settings. Some similarities in differentially abundant miRNAs are observed among related viruses or within the same mosquito species, but broad comparisons are limited due to differences in host species, virus families, sampling times, and analytical methods. Additionally, environmental factors, mosquito age, and the genetic background of both the virus and mosquitoes may influence host competence. Moreover, many of these studies are primarily descriptive. While such descriptive research has been valuable for cataloguing infection-associated changes in mosquito miRNA profiles, their largely correlative nature, limited functional validation, and reliance on experimental context limit mechanistic and evolutionary insights. The biological significance of most infection-related miRNA and isomiR alterations remains uncertain, and it is unclear which responses are adaptive host mechanisms, viral strategies, or incidental effects. This underscores a major gap in understanding how viral infection impacts mosquito miRNA landscapes.

4. Regulation of viral infection by mosquito miRNAs

Multiple studies have examined the overall impact of interrupting the miRNA pathway on viral infection, with varied results. For example, in the case of pathogenic alphaviruses Semliki Forest virus, o'nyong virus, and CHIKV, silencing *Ago1* did not significantly affect their replication, whether assessed in vitro, in vivo, or both (Keene et al., 2004; McFarlane et al., 2014; Schnettler et al., 2013). Conversely, in *Ae. aegypti* AF5 cells, silencing *Ago1* led to a notable increase in the replication of the insect-specific agua salud alphavirus, although not to the same extent as when *Ago2* (which is involved in both the siRNA and miRNA pathways) was silenced (Altinli et al., 2022). It is understood that miRNAs can be loaded onto AGO2 (Ghildiyal et al., 2010; Hussain et al., 2013a; Rubio et al., 2013; Yang et al., 2014); thus, when *Ago1* is silenced, miRNAs may be redirected to AGO2, compensating for the absence of AGO1, which may explain why silencing *Ago1* often results in minimal effects on virus replication. Regarding the orthoflavivirus WNV (Kunjin strain), silencing *Dicer-1* in *Ae. aegypti* Aag2 cells led to reduced replication of the virus (Hussain et al., 2012). However, it should be noted that, conversely, siRNAs cannot be loaded into AGO1-RISC, and thus siRNAs cannot be redirected to the complex, compensating for the absence of AGO2 in AGO2-depletion studies.

Apart from examining the overall impact of the miRNA pathway on virus infection, several studies have documented the role of specific miRNAs in mosquito-virus interactions following the identification of differentially abundant miRNAs in response to viral infection. miR-2940 is a mosquito-specific miRNA that produces functional 5p and 3p arms. In *Ae. aegypti* Aag2 cells and *Ae. albopictus* C6/36 cells infection with WNV and CHIKV, respectively, led to decreases in miR-2940–5p levels (Shrinet et al., 2014; Slonchak et al., 2014). This decrease was found to negatively regulate the metalloprotease m41 ftsh (FTSH) in *Ae. aegypti* cells (Slonchak et al., 2014). FTSH is a proviral protein and a target of the miRNA that it positively regulates (Slonchak et al., 2014; Zhang et al., 2013); therefore, its suppression after virus infection, as part of the antiviral response, may create a less conducive environment for virus replication. The *DNA methyltransferase Dnmt2* gene, which was shown to facilitate DENV2 replication in Aag2 cells, was identified as another target of aae-miR-2940–5p, negatively regulating the gene (Zhang et al., 2013).

Atlastin (AaATL) was shown to be induced in DENV2-infected *Ae. aegypti* mosquito cells and functions as a proviral protein (Hussain et al., 2022). The gene encoding Atlastin was identified as a target of aae-miR-989-3p. Transfection with the miRNA mimic led to reduced DENV2 replication, suggesting that aae-miR-989-3p may negatively regulate AaATL expression (Hussain et al., 2022). Considering the induction of this miRNA has been observed in ZIKV-infected *Ae. aegypti* and DENV-infected C6/36 cells, an increase in the miRNA levels following virus infection could regulate the expression of AaATL as a proviral protein (Campbell et al., 2014; Su et al., 2019).

In *Ae. albopictus* C6/36 cells, miR-927 was found to have proviral effects during DENV2 infection, as increases and decreases in miR-927 abundance led to increased and decreased DENV2 replication, respectively (Avila-Bonilla et al., 2020). *Ae. albopictus filamin* was determined as a direct target of miR-927. Considering flamin is an actin-binding protein involved in the Toll pathway, the authors demonstrated that a reduction in flamin results in the downregulation of immune effector molecules controlled by the Toll pathway, such as Cecropin A, Cecropin G, and Defensin D (Avila-Bonilla et al., 2020; Edwards et al., 1997). As a consequence, DENV replicates more efficiently.

In Aag2 cells infected with CHIKV, ubiquitin-related modifier (URM) was identified as a target of an induced miRNA, aal-miR-2b (Dubey et al., 2017). The target sequences are located in the 3'UTR of the gene, and the interaction results in negative regulation. Inhibition of the miRNA led to increased CHIKV replication. Consistent with the induction of aal-miR-2b following virus infection, URM transcript levels decreased, confirming the negative interaction. The authors proposed that by increasing the abundance of the miRNA upon CHIKV infection, *Ae. aegypti* reduces the transcript levels of URM, which is involved in tRNA thiolation, thereby resulting in less efficient CHIKV replication (Dubey et al., 2017).

Mosquito miRNAs may also directly target viral genomes. Yen et al. predicted several *Ae. aegypti* miRNAs that could target the genomes of CHIKV, DENV, and ZIKV (Yen et al., 2019). Another study carried out a similar investigation using *Ae. aegypti* and *Culex quinquefasciatus* miRNAs targeting the 3'UTR of DENV1–4, YFV, ZIKV, JEV, MVEV, USUV, WNV1, and WNV2, respectively (Avila-Bonilla and Salas-Benito, 2024). However, the interactions were not experimentally validated in either study. Of note, orthoflaviviruses exhibit complex secondary structures in their

3'UTRs, which may impede miRNA interactions with target sequences in this region (Slonchak et al., 2014). Dubey et al., on the other hand, experimentally demonstrated that aae-miR-2944b-5p from *Ae. aegypti* interacts with the 3'UTR of CHIKV (Dubey et al., 2019). When an aae-miR-2944b-5p inhibitor (antagomir) was applied both in vitro and in vivo, CHIKV replication was significantly increased, suggesting the antiviral role of aae-miR-2944b-5p. Another study showed that *Ae. albopictus* aal-miR-252 suppresses DENV2 replication in C6/36 cells by targeting sequences that encode the virus envelope protein (Yan et al., 2014). In contrast to these two studies, Zhou et al. revealed that the induced *Ae. albopictus* aal-miR-281 in the midgut following DENV2 infection interacts with the 3'UTR of DENV2 and facilitates the virus's replication (Zhou et al., 2014). This could be an example of the utilisation of a mosquito miRNA to benefit the virus. Similar examples exist, such as human hsa-miR-122, which binds to the 5'UTR of hepatitis C virus in the liver and enhances its replication by facilitating the translation of its genome (Jangra et al., 2010; Shimakami et al., 2012).

5. miRNAs in mosquito immune responses

Blood feeding is a key event in mosquito biology that triggers oogenesis and egg production (Nouzova et al., 2019; Roy et al., 2017). It also significantly impacts the mosquito transcriptome and miRNA profile, many of which may regulate immune-related genes (Lucas and Raikhel, 2013; Roy et al., 2015). For example, aae-miR-375 was found to be notably induced in blood-fed *Ae. aegypti* mosquitoes compared to those not fed on blood (Hussain et al., 2013b). Several potential targets of this miRNA were identified, with their expression influenced by it, including immune genes *cactus* and *REL1*. The induction of this miRNA increased *cactus*, which inhibits NF-kB activation, which usually induces antimicrobial peptide expression, and decreased *REL1*, an NF-kB factor involved in the Toll pathway (Hussain et al., 2013b). Over-expressing *REL1* in *Ae. aegypti* mosquitoes led to a marked rise in the expression of various immune genes, which significantly overlapped with the pattern seen in *cactus*-depleted mosquitoes via RNAi (Zou et al., 2011). aae-miR-375 was also shown to promote DENV2 replication in Aag2 cells, likely because it suppresses immune effectors by altering the expression of *cactus* and *REL1* genes (Hussain et al., 2013b). The implications of miRNA regulation of NF-kB in mosquitoes remain unknown.

Viruses can also suppress mosquitoes' immunity by downregulating immune-related genes. For example, DENV infection results in suppression of *cecropin* and *defensin* levels in Aag2 cells, and Semliki Forest virus, and Sindbis virus inhibit the Toll pathway (Fragkoudis et al., 2008; Sanders et al., 2005; Sim and Dimopoulos, 2010). Again, it is likely that changes in miRNA abundances due to blood feeding or viral infection play a role in regulating these immune-related genes.

Based on the studies above, blood feeding and viral infection significantly alter mosquito miRNA profiles, with several miRNAs involved in regulating innate immune pathways. Specifically, blood meal-induced miRNAs, such as aae-miR-375, influence key parts of the Toll pathway by regulating genes like *cactus* and *REL1*, leading to altered immune gene expression and, in some cases, increased viral replication. Alongside evidence that arboviruses suppress immune effector genes, these findings suggest that miRNA-mediated regulation helps fine-tune mosquito immune responses during physiological and infection-related challenges. However, the extent to which infection-associated miRNA changes reflect virus-driven immune suppression or indirect consequences of blood feeding and infection remains largely unresolved.

6. miRNAs as mediators of *Wolbachia*-mosquito-virus interactions

Wolbachia pipientis is an endosymbiotic bacterium found in arthropods and nematodes (Hilgenboecker et al., 2008). The interaction ranges from obligatory (predominantly in nematodes) to facultative (in many insects). It is estimated that over 40–60% of insect species are naturally infected with various strains of *Wolbachia* (Weinert et al., 2015; Zug and Hammerstein, 2012). Its rapid spread among insect species is attributed to vertical transmission and reproductive manipulations of the host, particularly cytoplasmic incompatibility (CI), in which mating between *Wolbachia*-infected males and uninfected females results in no progeny. Consequently, *Wolbachia*-infected females have an advantage, as they produce viable progeny when mating with either *Wolbachia*-infected or uninfected males (Hochstrasser, 2023). Additionally, in 2008, it was discovered that *Wolbachia* protects *Drosophila* flies against viral infections (Hedges et al., 2008; Teixeira et al., 2008). Given the absence of a natural *Wolbachia* infection in *Ae. aegypti* mosquitoes, two strains of *Wolbachia*, *w*MelPop and *w*Mel, were transinfected from *D. melanogaster* into the mosquitoes (McMeniman et al., 2009). These mosquitoes were found to be resistant to the replication of several RNA viruses, including DENV, CHIKV, and ZIKV (Aliota et al., 2016; Moreira et al., 2009). Consequently, the World Mosquito Program (https://www.worldmosquitoprogram.org/) initiated the release of these *Wolbachia*-infected *Ae. aegypti* mosquitoes to alleviate the burden of dengue in various countries, achieving remarkable outcomes. The mechanism of inhibition appears to be dependent on the host, virus, and *Wolbachia* strain, potentially involving multiple processes such as induction of the immune system, competition for resources and space, alterations to host genes and miRNAs, degradation of viral RNA by host (XRN1), and *Wolbachia* factors (RNase H) (Asgari, 2014; Geoghegan et al., 2017; Hussain et al., 2023; Pan et al., 2012; Thomas et al., 2018; White et al., 2017).

Alterations in miRNAs have been observed in *Ae. aegypti* mosquitoes transinfected with *Wolbachia* (Bishop et al., 2022; Hussain et al., 2011; Mayoral et al., 2014). These alterations could affect virus replication if the abundances of pro- or antiviral miRNAs are affected. aae-miR-2940 is one of the highly differentially expressed miRNAs in *Wolbachia*-infected mosquitoes compared to uninfected mosquitoes (Hussain et al., 2011). Three targets of this miRNA have been identified, which play roles in *Wolbachia* maintenance and density, as well as in the replication of DENV2 in mosquito cells. For instance, aae-miR-2940 negatively regulates the expression of *DNA methyltransferase 2* (*AaDnmt2*), which is a proviral gene (Zhang et al., 2013). Significant induction of aae-miR-2940 in the presence of *Wolbachia*, and hence a reduction in AeDNMT2, may contribute to less efficient DENV replication. Another target of aae-miR-2940 is *pelo*, which was also shown to promote DENV2 replication in mosquito cells (Asad et al., 2018). Downregulation of *pelo* by an increase in the abundance of aae-miR-2940 in the presence of *Wolbachia* may also create a less favourable environment for DENV replication.

aae-miR-34–3p was found to be significantly induced in *w*AlbB-infected *Ae. aegypti* mosquitoes (She et al., 2024). This miRNA increased the expression of *defensin* and *cecropin* antiviral effector genes via the Toll pathway regulator *MyD88*. The results suggested a crosstalk between the classical immunity Toll pathway and the miRNA pathway as part of the virus blocking mechanism induced by *Wolbachia* in mosquitoes. This crosstalk is mediated by a long non-coding RNA (lncRNA), aae-lnc-2268, which functions as an aae-miR-34–3p sponge. *Wolbachia* suppresses the abundance of this lncRNA, leading to increased abundance of aae-miR-34-3p and, in turn, increased expression of the *MyD88* gene, which regulates the Toll pathway and the expression of antiviral effectors (She et al., 2024).

Wolbachia infection provides strong antiviral protection in mosquitoes and is associated with changes in host miRNA profiles, suggesting that miRNA-mediated regulation may play a role in blocking viruses. Specific miRNAs, including aae-miR-2940-5p and aae-miR-34-3p, influence proviral host factors and immune signalling pathways, connecting *Wolbachia* infection with decreased viral replication through effects on host gene regulation. However, the relative significance of miRNA-mediated mechanisms compared to other *Wolbachia*-induced antiviral processes, and their applicability across different *Wolbachia* strains, mosquito species, viruses, and field conditions, remains only partially understood.

7. Application of miRNAs for vector or disease control

While there have been several studies towards the application and translation of miRNA functions for virus inhibition, these have been limited. One such approach is to genetically modify mosquitoes to produce small inhibitory RNAs targeting the viral genome based on insect miRNA backbones. Tng et al. (2022) generated *Ae. aegypti* (Aag2, AF05, and AF319 *Dicer-2* KO from AF05) and *Ae. albopictus* (U4.4 and C6/36) cell lines producing miRNA-like small RNAs (with two mismatches) and short hairpin RNAs (shRNA, perfect base-pairing) with complementary sequences to the CHIKV genome (nsP2 viral protease) using a *Drosophila melanogaster* pre-miRNA-1 hairpin (Tng et al., 2022). These were accommodated in the intron of the polyubiquitin promoter sequence in the mosquito genome. The results showed reduced CHIKV replication in the cell lines, although shRNA designs were more effective at inhibiting virus replication than miRNA-like designs. Liu et al. (2021) generated transgenic *Ae. aegypti* mosquitoes to express virus-specific miRNA hairpins to target the DENV genome of all four serotypes using the blood-meal-inducible carboxypeptidase A promoter or the polyubiquitin promoter (Liu et al., 2021). DENVs replicated with lower efficiency in the modified mosquitoes, resulting in lower infection rates and viral titers across most serotypes. In addition, there was a significant reduction in virus transmission. Ramyasoma et al. (2020) produced *Ae. aegypti* mosquitoes encoding multiple miRNA-based shRNAs to target the four serotypes of DENV, also using the blood-meal-inducible carboxypeptidase A promoter for their induction. The modified mosquitoes had lower rates of infection, dissemination, and transmission of DENV2 and DENV4 than control wild-type mosquitoes (Ramyasoma et al., 2020).

Another application of mosquito miRNAs attempted has been in the field of orthoflavivirus vaccines. In an attempt to mitigate reversion of attenuated orthoflavivirus vaccines based on live attenuated viruses, sequences complementary to mosquito-specific miR-184 and miR-275 miRNAs were inserted into the 3'UTR or ORF region of DENV4, leading to inhibition of virus replication in mosquito cell lines and adult *Aedes* mosquitoes (Tsetsarkin et al., 2015). The same group also produced a modified Langat virus (a tick-borne orthoflavivirus) with target sequences inserted into its genome targeting miR-1, miR-275, and miR-279, restricting virus replication in tick-derived cells but not in simian Vero cells (Tsetsarkin et al., 2016). The authors suggested that this approach could be utilised to provide additional biosafety levels for vaccines.

8. Concluding remarks

miRNAs, as post-transcriptional regulators in mosquitoes, influence a wide range of biological processes, from development and immunity to metabolism and vector competence. Over the past decade, increasing evidence has highlighted the dynamic and multifaceted roles of mosquito miRNAs in shaping interactions with arboviruses such as DENV, ZIKV, CHIKV, and others. The evolutionary path of mosquito miRNAs reveals a balance between deep conservation and functional innovation. Comparative genomics not only uncovers key aspects of mosquito biology but also facilitates the discovery of lineage-specific miRNAs as potential targets for species-specific vector control strategies. Variations in miRNA profiles and their regulatory networks may explain differences in vectorial capacity, virus host range, and immune responses across species. Comparative studies could offer opportunities to identify crucial regulatory miRNAs that contribute to vector adaptation and pathogen transmission.

Some knowledge gaps concerning the role of miRNAs in mosquito-virus interactions include a lack of functional characterisation of most mosquito miRNAs, a limited understanding of miRNA-virus interactions across diverse viruses, since most studies have focused on a small number of viruses, and poorly understood context-dependent and tissue-specific regulation following virus infection. Further, most miRNA studies have been performed on whole mosquitoes or tissues, lacking granularity. Recently developed techniques could be utilised to explore the profiles of miRNAs, isomiRs, and their target RNAs in infected versus uninfected cells within a single tissue at single-cell resolution by integrating single-cell small RNA-Seq with single-cell RNA-Seq for mRNA transcripts (Hagemann-Jensen et al., 2018; Maji et al., 2025). Additionally, the crosstalk between miRNAs and other regulatory pathways, especially immune and antiviral pathways, remains understudied. Importantly, there has been limited translation into applied vector or disease control strategies. This translational gap can be addressed by developing tissue- or stage-specific miRNA-based gene drives, utilising paratransgenesis systems, or implementing RNAi-enhancement technologies to disrupt vector competence.

Acknowledgments

Work in Asgari's laboratory has been supported by the Australian Research Council (DP150101782, DP210101791) and the National Health and Medical Research Council (APP1062983, APP1027110). The author wishes to thank two anonymous reviewers for their helpful comments.

References

Aliota, M.T., Peinado, S.A., Velez, I.D., Osorio, J.E., 2016. The *w*Mel strain of *Wolbachia* reduces transmission of Zika virus by *Aedes aegypti*. Sci. Rep. 6, 28792.

Altinli, M., Leggewie, M., Badusche, M., Gyanwali, R., Scherer, C., Schulze, J., et al., 2022. Antiviral RNAi response against the Insect-specific Agua Salud Alphavirus. mSphere 7, e0100321.

Asad, S., Hussain, M., Hugo, L., Osei-Amo, S., Zhang, G., Watterson, D., et al., 2018. Suppression of the pelo protein by *Wolbachia* and its effect on dengue virus in *Aedes aegypti*. PLoS Neg. Trop. Dis. 12, e0006405.

Asgari, S., 2018. microRNAs as regulators of insect host-pathogen interactions and immunity. Adv. Insect Physiol. 55, 19–45.

Asgari, S., 2015. Regulatory role of cellular and viral microRNAs in insect-virus interactions. Curr. Opin. Insect Sci. 8, 104–110.

Asgari, S., 2014. Role of microRNAs in arbovirus/vector interactions. Viruses 6, 3514–3534.

Asgari, S., 2013. MicroRNA functions in insects. Insect Biochem Mol Biol 43, 388–397.

Avila-Bonilla, R.G., Salas-Benito, J.S., 2024. Computational screening to predict microRNA targets in the flavivirus 3' UTR genome: an approach for antiviral development. Int. J. Mol. Sci. 25, 10135.

Avila-Bonilla, R.G., Yocupicio-Monroy, M., Marchat, L.A., De Nova-Ocampo, M.A., Del Ángel, R.M., Salas-Benito, J.S., 2017. Analysis of the miRNA profile in C6/36 cells persistently infected with dengue virus type 2. Virus Res. 232, 139–151.

Avila-Bonilla, R.G., Yocupicio-Monroy, M., Marchat, L.A., Pérez-Ishiwara, D.G., Cerecedo-Mercado, D.A., Del Ángel, R.M., et al., 2020. miR-927 has pro-viral effects during acute and persistent infection with dengue virus type 2 in C6/36 mosquito cells. J. Gen. Virol. 101, 825–839.

Azlan, A., Yunus, M.A., Abdul Halim, M., Azzam, G., 2022. Revised annotation and characterization of novel *Aedes albopictus* miRNAs and their potential functions in dengue virus infection. Biology 11, 1536.

Bartel, D.P., 2009. MicroRNAs: target recognition and regulatory functions. Cell 23, 215–233.

Bishop, C., Hussain, M., Hugo, L.E., Asgari, S., 2022. Analysis of *Aedes aegypti* microRNAs in response to *Wolbachia w*AlbB infection and their potential role in mosquito longevity. Sci. Rep. 12, 15245.

Bruno, I.G., Karam, R., Huang, L., Bhardwaj, A., Lou, C.H., Shum, E.Y., et al., 2011. Identification of a microRNA that activates gene expression by repressing nonsense-mediated RNA decay. Mol. Cell 42, 500–510.

Burroughs, A.M., Ando, Y., de Hoon, M.J.L., Tomaru, Y., Nishibu, T., Ukekawa, R., et al., 2010. A comprehensive survey of 3' animal miRNA modification events and a possible role for 3' adenylation in modulating miRNA targeting effectiveness. Genome Res. 20, 1398–1410.

Campbell, C.L., Harrison, T., Hess, A.M., Ebel, G.D., 2014. MicroRNA levels are modulated in *Aedes aegypti* after exposure to Dengue-2. Insect Mol. Biol. 23, 132–139.

Carissimo, G., Pain, A., Belda, E., Vernick, K.D., 2018. Highly focused transcriptional response of *Anopheles coluzzii* to O'nyong nyong arbovirus during the primary midgut infection. BMC Genomics 19, 526.

Dubey, S.K., Shrinet, J., Jain, J., Ali, S., Sunil, S., 2017. *Aedes aegypti* microRNA miR-2b regulates ubiquitin-related modifier to control chikungunya virus replication. Sci. Rep. 7, 17666.

Dubey, S.K., Shrinet, J., Sunil, S., 2019. *Aedes aegypti* microRNA, miR-2944b-5p interacts with 3'UTR of chikungunya virus and cellular target vps-13 to regulate viral replication. PLoS Negl. Trop. Dis. 13, e0007429.

Edwards, D.N., Towb, P., Wasserman, S.A., 1997. An activity-dependent network of interactions links the Rel protein Dorsal with its cytoplasmic regulators. Development 124, 3855–3864.

Estevez-Castro, C.F., Rodrigues, M.F., Babarit, A., Ferreira, F.V., de Andrade, E.G., Marois, E., et al., 2024. Neofunctionalization driven by positive selection led to the retention of the loqs2 gene encoding an *Aedes* specific dsRNA binding protein. BMC Biol. 22, 14.

Etebari, K., Osei-Amo, S., Blomberg, S., Asgari, S., 2015. Dengue virus infection alters post-transcriptional modification of microRNAs in the mosquito vector *Aedes aegypti*. Sci. Rep. 5, 15968.

Fiorillo, C., Yen, P.-S., Colantoni, A., Mariconti, M., Azevedo, N., Lombardo, F., et al., 2022. MicroRNAs and other small RNAs in *Aedes aegypti* saliva and salivary glands following chikungunya virus infection. Sci. Rep. 12, 9536.

Fragkoudis, R., Chi, Y., Siu, R., Barry, G., Attarzadeh-Yazdi, G., Merits, A., et al., 2008. Semliki Forest virus strongly reduces mosquito host defence signaling. Insect Mol. Biol. 17, 647–656.

Geoghegan, V., Stainton, K., Rainey, S.M., Ant, T.H., Dowle, A.A., Larson, T., et al., 2017. Perturbed cholesterol and vesicular trafficking associated with dengue blocking in *Wolbachia*-infected *Aedes aegypti* cells. Nat. Commun. 8, 526.

Ghildiyal, M., Xu, J., Seitz, H., Weng, Z., Zamore, P.D., 2010. Sorting of *Drosophila* small silencing RNAs partitions microRNA* strands into the RNA interference pathway. RNA 16, 43–56.

Girardi, E., López, P., Pfeffer, S., 2018. On the importance of host microRNAs during viral infection. Front. Genet. 9, 439.

Haac, M., Anderson, M., Eggleston, H., Myles, K., Adelman, Z., 2015. The hub protein loquacious connects the microRNA and short interfering RNA pathways in mosquitoes. Nucleic Acids Res. 43, 3688–3700.

Hafner, M., Landthaler, M., Burger, L., Khorshid, M., Hausser, J., Berninger, P., et al., 2010. Transcriptome-wide identification of RNA-binding protein and microRNA target sites by PAR-CLIP. Cell 141, 129–141.

Hagemann-Jensen, M., Abdullayev, I., Sandberg, R., Faridani, O.R., 2018. Small-seq for single-cell small-RNA sequencing. Nat. Protoc. 13, 2407–2424.

Hedges, L.M., Brownlie, J.C., O'Neill, S.L., Johnson, K.N., 2008. *Wolbachia* and virus protection in insects. Science 322, 702.

Helwak, A., Kudla, G., Dudnakova, T., Tollervey, D., 2013. Mapping the human miRNA interactome by CLASH reveals frequent noncanonical binding. Cell 153, 654–665.

Henderson, C., Brustolin, M., Hegde, S., Dayama, G., Lau, N., Hughes, G.L., et al., 2022. Transcriptomic and small RNA response to Mayaro virus infection in *Anopheles stephensi* mosquitoes. PLoS Negl. Trop. Dis. 16, e0010507.

Heydarzadeh, S., Ranjbar, M., Karimi, F., Seif, F., Alivand, M.R., 2021. Overview of host miRNA properties and their association with epigenetics, long non-coding RNAs, and Xeno-infectious factors. Cell Biosci. 11, 43.

Hilgenboecker, K., Hammerstein, P., Schlattmann, P., Telschow, A., Werren, J.H., 2008. How many species are infected with *Wolbachia*?—A statistical analysis of current data. FEMS Microbiol. Lett. 218 215–220.

Hochstrasser, M., 2023. Molecular biology of cytoplasmic incompatibility caused by *Wolbachia* endosymbionts. Annu. Rev. Microbiol. 77, 299–316.

Hu, W., Criscione, F., Liang, S., Tu, Z., 2015. MicroRNAs of two medically important mosquito species: *Aedes aegypti* and *Anopheles stephensi*. Insect Mol. Biol. 24, 240–252.

Hussain, M., Bradshaw, T., Lee, M., Asgari, S., 2022. The involvement of Atlastin in dengue virus and *Wolbachia* infection in *Aedes aegypti* and its regulation by aae-miR-989. Microbiol. Spectr. 10, e0225822.

Hussain, M., Frentiu, F.D., Moreira, L.A., O'Neill, S.L., Asgari, S., 2011. Wolbachia utilizes host microRNAs to manipulate host gene expression and facilitate colonization of the dengue vector Aedes aegypti. Proc. Natl. Acad. Sci., USA 108, 9250–9255.

Hussain, M., O'Neill, S.L., Asgari, S., 2013a. Wolbachia interferes with the intracellular distribution of Argonaute 1 in the dengue vector *Aedes aegypti* by manipulating the host microRNAs. RNA Biol. 10, 1868–1875.

Hussain, M., Torres, S., Schnettler, E., Funk, A., Grundhoff, A., Pijlman, G.P., et al., 2012. West Nile virus encodes a microRNA-like small RNA in the 3' untranslated region which upregulates GATA4 mRNA and facilitates virus replication in mosquito cells. Nucleic Acids Res. 40, 2210–2223.

Hussain, M., Walker, T., O'Neill, S., Asgari, S., 2013b. Blood meal induced microRNA regulates development and immune associated genes in the Dengue mosquito vector, *Aedes aegypti*. Insect Biochem. Mol. Biol. 43, 146–152.

Hussain, M., Zhang, G., Leitner, M., Hedges, L.M., Asgari, S., 2023. *Wolbachia* RNase HI contributes to virus blocking in the mosquito *Aedes aegypti*. iScience 26, 105836.

Jangra, R.K., Yi, M., Lemon, S.M., 2010. Regulation of hepatitis C virus translation and infectious virus production by the microRNA miR-122. J. Virol. 84, 6615–6625.

Keene, K., Foy, B., Sanchez-Vargas, I., Beaty, B., Blair, C., Olson, K., 2004. RNA interference acts as a natural antiviral response to O'nyong-nyong virus (*Alphavirus*; *Togaviridae*) infection of *Anopheles gambiae*. Proc. Natl. Acad. Sci. USA 101, 17240–17245.

Kozomara, A., Griffiths-Jones, S., 2014. miRBase: annotating high confidence microRNAs using deep sequencing data. Nucleic Acids Res. 42, D68–D73.
Lee, M., Etebari, K., Hall-Mendelin, S., van den Hurk, A.F., Hobson-Peters, J., Vatipally, S., et al., 2017. Understanding the role of microRNAs in the interaction of *Aedes aegypti* mosquitoes with an insect-specific flavivirus. J. Gen. Virol. 98, 892–1903.
Li, S., Mead, E.A., Liang, L., Tu, Z., 2009. Direct sequencing and expression analysis of a large number of miRNAs in *Aedes aegypti* and a multi-species survey of novel mosquito miRNAs. BMC Genomics 10, 581.
Liu, N., Okamura, K., Tyler, D.M., Phillips, M.D., Chung, W.-J., Lai, E.C., 2008. The evolution and functional diversification of animal microRNA genes. Cell Res. 18, 985–996.
Liu, S., Lucas, K., Roy, S., Ha, J., Raikhel, A., 2014. Mosquito-specific microRNA-1174 targets serine hydroxymethyltransferase to control key functions in the gut. Proc. Natl. Acad. Sci. USA 111, 14460–14465.
Liu, W.-L., Hsu, C.-W., Chan, S.-P., Yen, P.-S., Su, M.P., Li, J.-C., et al., 2021. Transgenic refractory *Aedes aegypti* lines are resistant to multiple serotypes of dengue virus. Sci. Rep. 11, 23865.
Liu, Y., Zhou, Y., Wu, J., Zheng, P., Li, Y., Zheng, X., et al., 2015. The expression profile of *Aedes albopictus* miRNAs is altered by dengue virus serotype-2 infection. Cell Biosci. 5, 16.
Liu, Y.X., Li, F.X., Liu, Z.Z., Jia, Z.R., Zhou, Y.H., Zhang, H., et al., 2016. Integrated analysis of miRNAs and transcriptomes in *Aedes albopictus* midgut reveals the differential expression profiles of immune-related genes during dengue virus serotype-2 infection. Insect Sci. 23, 377–385.
Lu, J., Shen, Y., Wu, Q., Kumar, S., He, B., Shi, S., et al., 2008. The birth and death of microRNA genes in *Drosophila*. Nat. Genet. 40, 351–355.
Lucas, K., Raikhel, A.S., 2013. Insect microRNAs: biogenesis, expression profiling and biological functions. Insect Biochem. Mol. Biol. 43, 24–38.
Lucas, K.J., Zhao, B., Liu, S., Raikhel, A.S., 2015a. Regulation of physiological processes by microRNAs in insects. Curr. Opin. Insect Sci. 11, 1–7.
Lucas, K.J., Zhao, B., Roy, S., Gervaise, A.L., Raikhel, A.S., 2015b. Mosquito-specific microRNA-1890 targets the juvenile hormone-regulated serine protease JHA15 in the female mosquito gut. RNA Biol. 12, 1383–1390.
Maharaj, P., Widen, S., Huang, J., Wood, T., Thangamani, S., 2015. Discovery of mosquito saliva microRNAs during CHIKV infection. PLoS Negl. Trop. Dis. 9, e0003386.
Maji, R.K., Leisegang, M.S., Boon, R.A., Schulz, M.H., 2025. Revealing microRNA regulation in single cells. Trends Genet. 41, 522–536.
Marzec, S., Siperstein, A., Zhou, A., Holzapfel, C.M., Bradshaw, W.E., Meuti, M.E., et al., 2023. MicroRNA expression prior to biting in a vector mosquito anticipates physiological processes related to energy utilization, reproduction and immunity. Insects 14, 700.
Mayoral, J., Etebari, K., Hussain, M., Khromykh, A., Asgari, S., 2014. *Wolbachia* infection modifies the profile, shuttling and structure of microRNAs in a mosquito cell line. PLoS One 9, e96107.
McFarlane, M., Arias-Goeta, C., Martin, E., O'Hara, Z., Lulla, A., Mousson, L., et al., 2014. Characterization of *Aedes aegypti* innate-immune pathways that limit Chikungunya virus replication. PLoS Negl. Trop. Dis. 8, e2994.
McMeniman, C.J., Lane, R.V., Cass, B.N., Fong, A.W.C., Sidhu, M., Wang, Y.-F., et al., 2009. Stable introduction of a life-shortening *Wolbachia* infection into the mosquito *Aedes aegypti*. Science 323, 141–144.

Meunier, J., Lemoine, F., Soumillon, M., Liechti, A., Weier, M., Guschanski, K., et al., 2013. Birth and expression evolution of mammalian microRNA genes. Genome Res. 23, 34–45.
Moreira, L.A., Iturbe-Ormaetxe, I., Jeffery, J.A., Lu, G.J., Pyke, A.T., Hedges, L.M., et al., 2009. A *Wolbachia* symbiont in *Aedes aegypti* limits infection with Dengue, Chikungunya, and *Plasmodium*. Cell 139, 1268–1278.
Neilsen, C.T., Goodall, G.J., Bracken, C.P., 2012. IsomiRs–the overlooked repertoire in the dynamic microRNAome. Trends Genet. 28, 544–549.
Nouzova, M., Clifton, M.E., Noriega, F.G., 2019. Mosquito adaptations to hematophagia impact pathogen transmission. Curr. Opin. Insect Sci. 34, 21–26.
Okamura, K., Hagen, J.W., Duan, H., Tyler, D.M., Lai, E.C., 2007. The mirtron pathway generates microRNA-class regulatory RNAs in *Drosophila*. Cell 130, 89–100.
Orom, U.A., Nielsen, F.C., Lund, A.H., 2008. MicroRNA-10a binds the 5′UTR of ribosomal protein mRNAs and enhances their translation. Mol. Cell 30, 460–471.
Pan, X., Zhou, G., Wu, J., Bian, G., Lu, P., Raikhel, A.S., et al., 2012. *Wolbachia* induces reactive oxygen species (ROS)-dependent activation of the Toll pathway to control dengue virus in the mosquito *Aedes aegypti*. Proc. Natl. Acad. Sci. USA 109, E23–E31.
Puthiyakunnon, S., Yao, Y., Li, Y., Gu, J., Peng, H., Chen, X., 2013. Functional characterization of three microRNAs of the Asian tiger mosquito, *Aedes albopictus*. Parasit. Vectors 6, 230.
Ramyasoma, H.P.B.K.D., Dassanayake, R.S., Hapugoda, M., Capurro, M.L., Silva Gunawardene, Y.I.N., 2020. Multiple dengue virus serotypes resistant transgenic *Aedes aegypti* fitness evaluated under laboratory conditions. RNA Biol. 17, 918–929.
Reczko, M., Maragkakis, M., Alexiou, P., Grosse, I., Hatzigeorgiou, A.G., 2012. Functional microRNA targets in protein coding sequences. Bioinformatics 28, 771–776.
Roy, S., Saha, T.T., Johnson, L., Zhao, B., Ha, J., White, K.P., et al., 2015. Regulation of gene expression patterns in mosquito reproduction. PLoS Genet. 11, e1005450.
Roy, S., Smykal, V., Johnson, L., Saha, T., Zou, Z., Raikhel, A.S., 2017. Regulation of reproductive processes in female mosquitoes. Adv. Insect Physiol. 51, 115–144.
Rubio, M., Montañez, R., Perez, L., Milan, M., Belles, X., 2013. Regulation of atrophin by both strands of the miR-8 precursor. Insect Biochem. Mol. Biol. 43, 1009–1014.
Saldaña, M.A., Etebari, K., Hart, C.E., Widen, S.G., Wood, T.G., Thangamani, S., et al., 2017. Zika virus alters the microRNA expression profile and elicits an RNAi response in *Aedes aegypti* mosquitoes. PLoS Negl. Trop. Dis. 11, e0005760.
Sanders, H., Foy, B., Evans, A., Ross, L., Beaty, B., Olson, K., et al., 2005. Sindbis virus induces transport processes and alters expression of innate immunity pathway genes in the midgut of the disease vector, *Aedes aegypti*. Insect Biochem. Mol. Biol. 35, 1293–1307.
Schnettler, E., Donald, C.L., Human, S., Watson, M., Siu, R.W., McFarlane, M., et al., 2013. Knockdown of piRNA pathway proteins results in enhanced Semliki Forest virus production in mosquito cells. J. Gen. Virol. 94, 1680–1689.
She, L., Shi, M., Cao, T., Yuan, H., Wang, R., Wang, W., et al., 2024. *Wolbachia* mediates crosstalk between miRNA and Toll pathways to enhance resistance to dengue virus in *Aedes aegypti*. PLoS Pathog. 20, e1012296.
Shimakami, T., Yamane, D., Jangra, R.K., Kempf, B.J., Spaniel, C., Barton, D.J., et al., 2012. Stabilization of hepatitis C virus RNA by an Ago2-miR-122 complex. Proc. Natl. Acad. Sci. USA 109, 941–946.
Shrinet, J., Jain, S., Jain, J., Bhatnagar, R., Sunil, S., 2014. Next generation sequencing reveals regulation of distinct *Aedes* microRNAs during chikungunya virus development. PLoS Negl. Trop. Dis. 8, e2616.
Sim, S., Dimopoulos, G., 2010. Dengue virus inhibits immune responses in *Aedes aegypti* cells. PLoS One 5, e10678.

Sinclair, J.B., Asgari, S., 2020. Ross River virus provokes differentially expressed microRNA and RNA interference responses in *Aedes aegypti* mosquitoes. Viruses 12, 695.

Skalsky, R., Vanlandingham, D.L., Scholle, F., Higgs, S., Cullen, B.R., 2010. Identification of microRNAs expressed in two mosquito vectors, *Aedes albopictus* and *Culex quinquefasciatus*. BMC Genomics 11, 119.

Slonchak, A., Hussain, M., Torres Morales, S., Asgari, S., Khromykh, A.A., 2014. Expression of mosquito microRNA aae-miR-2940-5p is down-regulated in response to West Nile virus infection to restrict viral replication. J. Virol. 88, 8457–8467.

Su, J., Li, C., Zhang, Y., Yan, T., Zhu, X., Zhao, M., et al., 2017. Identification of microRNAs expressed in the midgut of *Aedes albopictus* during dengue infection. Parasit. Vectors 10, 63.

Su, J., Wang, G., Li, C., Xing, D., Yan, T., Zhu, X., et al., 2019. Screening for differentially expressed miRNAs in *Aedes albopictus* (Diptera: Culicidae) exposed to DENV-2 and their effect on replication of DENV-2 in C6/36 cells. Parasit. Vectors 12, 44.

Tan, G.C., Chan, E., Molnar, A., Sarkar, R., Alexieva, D., Isa, I.M., et al., 2014. 5' isomiR variation is of functional and evolutionary importance. Nucleic Acids Res. 42, 9424–9435.

Teixeira, L., Ferreira, A., Ashburner, M., 2008. The bacterial symbiont *Wolbachia* induces resistance to RNA viral infections in *Drosophila melanogaster*. PLoS Biol. 6, 2753–2763.

Thomas, S., Verma, J., Woolfit, M., O'Neill, S.L., 2018. *Wolbachia*-mediated virus blocking in mosquito cells is dependent on XRN1-mediated viral RNA degradation and influenced by viral replication rate. PLoS Pathog. 14, e1006879.

Tng, P.Y.L., Carabajal Paladino, L.Z., Anderson, M.A.E., Adelman, Z.N., Fragkoudis, R., Noad, R., et al., 2022. Intron-derived small RNAs for silencing viral RNAs in mosquito cells. PLoS Negl. Trop. Dis. 16, e0010548.

Tsetsarkin, K.A., Liu, G., Kenney, H., Bustos-Arriaga, J., Hanson, C.T., Whitehead, S.S., et al., 2015. Dual miRNA Targeting restricts host range and attenuates neurovirulence of flaviviruses. PLoS Pathog. 11, e1004852.

Tsetsarkin, K.A., Liu, G., Kenney, H., Hermance, M., Thangamani, S., Pletnev, A.G., 2016. Concurrent micro-RNA mediated silencing of tick-borne flavivirus replication in tick vector and in the brain of vertebrate host. Sci. Rep. 6, 33088.

Vasudevan, S., Tong, Y., Steitz, J.A., 2007. Switching from repression to activation: microRNAs can up-regulate translation. Science 318, 1931–1934.

Weinert, L.A., Araujo, E.V., Ahmed, M.Z., Welch, J.J., 2015. The incidence of bacterial endosymbionts in terrestrial arthropods. Proc. Biol. Sci. 282, 20150249.

White, P.M., Serbus, L.R., Debec, A., Codina, A., Bray, W., Guichet, A., et al., 2017. Reliance of *Wolbachia* on high rates of host proteolysis revealed by a genome-wide RNAi screen of *Drosophila* cells. Genetics 205, 1473–1488.

Xing, S., Du, J., Gao, S., Tian, Z., Zheng, Y., Liu, G., et al., 2016. Analysis of the miRNA expression profile in an *Aedes albopictus* cell line in response to bluetongue virus infection. Infect. Genet. Evol. 39, 74–84.

Yan, H., Zhou, Y., Liu, Y., Deng, Y., Chen, X., 2014. miR-252 of the Asian tiger mosquito *Aedes albopictus* regulates dengue virus replication by suppressing the expression of the dengue virus envelope protein. J. Med. Virol. 86, 1428–1436.

Yang, J.-S., Lai, E.C., 2011. Alternative miRNA biogenesis pathways and the interpretation of core miRNA pathway mutants. Mol. Cell 43, 892–903.

Yang, J.S., Smibert, P., Westholm, J.O., Jee, D., Maurin, T., Lai, E.C., 2014. Intertwined pathways for Argonaute-mediated microRNA biogenesis in *Drosophila*. Nucleic Acids Res. 42, 1987–2002.

Yen, P.-S., Chen, C.-H., Sreenu, V., Kohl, A., Failloux, A.-B., 2019. Assessing the potential interactions between cellular miRNA and arboviral genomic RNA in the Yellow fever mosquito, *Aedes aegypti*. Viruses 11, 540.

Zhang, G., Hussain, M., O'Neill, S.L., Asgari, S., 2013. *Wolbachia* uses a host microRNA to regulate transcripts of a methyltransferase, contributing to dengue virus inhibition in *Aedes aegypti*. Proc. Natl. Acad. Sci. U S A 110, 10276–10281.

Zhou, Y., Liu, Y., Yan, H., Li, Y., Zhang, H., Xu, J., et al., 2014. miR-281, an abundant midgut-specific miRNA of the vector mosquito *Aedes albopictus* enhances dengue virus replication. Parasit. Vectors 7, 488.

Zhu, C., Jiang, Y., Zhang, Q., Gao, J., Li, C., Li, C., et al., 2021. Vector competence of *Aedes aegypti* and screening for differentially expressed microRNAs exposed to Zika virus. Parasit. Vectors 14, 504.

Zou, Z., Souza-Neto, J.A., Xi, Z., Kokoza, V., Shin, S.W., Dimopoulos, G., et al., 2011. Transcriptome analysis of *Aedes aegypti* transgenic mosquitoes with altered immunity. PLoS Pathog. 7, e1002394.

Zug, R., Hammerstein, P., 2012. Still a host of hosts for *Wolbachia*: analysis of recent data suggests that 40% of terrestrial arthropod species are infected. PLoS One 7, e38544.

Printed and bound by CPI Group (UK) Ltd, Croydon, CR0 4YY
24/09/2026
02222378-0005